Creatine and Creatine Kinase in Health and Disease

Subcellular Biochemistry
Volume 46

SUBCELLULAR BIOCHEMISTRY

SERIES EDITOR

J. ROBIN HARRIS, University of Mainz, Mainz, Germany

ASSISTANT EDITORS

B.B. BISWAS, University of Calcutta, Calcutta, India

P. QUINN, King's College London, London, U.K.

Recent Volumes in this Series

Creatine and Creatine Kinase in Health and Disease

Subcellular Biochemistry Volume 46

Edited by

Gajja S. Salomons
VU University Medical Center, Amsterdam, The Netherlands

and

Markus Wyss
DSM Nutritional Products Ltd., Basel, Switzerland

Springer

A C.I.P. Catalogue record for this book is available from the Library of Congress.

ISBN 978-1-4020-6485-2 (HB)
ISBN 978-1-4020-6486-9 (e-book)

Published by Springer,
P.O. Box 17, 3300 AA Dordrecht, The Netherlands.

www.springer.com

Printed on acid-free paper

Dedicated to all mothers and fathers of this world

CONTENTS

FOREWORD – CREATINE AND CREATINE KINASE IN HEALTH AND DISEASE – A WORTHY PASSION

GAJJA S. SALOMONS[1] AND MARKUS WYSS[2]

[1] *VU University Medical Center, Department of Clinical Chemistry, Metabolic Unit, De Boelelaan 1117, 1081 HV Amsterdam, The Netherlands*

[2] *DSM Nutritional Products Ltd., Biotechnology R&D, Bldg. 203/17B, P.O. Box 3255, 4002 Basel, Switzerland*

To make a project a success, you typically need to have three things: a good idea, an enthusiastic team, and a committed 'customer' (in this case, the Springer publishing company). When J. Robin Harris, the Series Editor of *Subcellular Biochemistry*, first approached us with the idea of a book on "Creatine and Creatine Kinase in Health and Disease", we were both surprised and delighted. Next, we were impressed by the dedicated participation of many of the key players in creatine and creatine kinase research from all over the world. In fact, all authors we asked to participate accepted the invitation, and have submitted high-quality manuscripts in a timely manner. Consequently, the book is up to date, and both editors and authors are confident that this book will become an important contribution to expand the knowledge and funding of creatine and creatine kinase research. Finally, the continuous support of the Springer publishing company and, in particular, of Mike van den Bosch, Marie Johnson, Marlies Vlot, Max Haring and Priyaa Menon was important for the realization of this project. We are greatly indebted to all those who have contributed to this book for their enthusiasm, dedication, and efforts, and for their passion for creatine and creatine kinase research. Similarly, we greatly acknowledge our colleagues for their support and patience in situations where our attention was partially redirected to this book.

It is worthwhile mentioning that recently, two other notable books covering creatine and creatine kinase research have been published (Vial, 2006; Saks, 2007). Rather than being redundant, these books focus their attention on different aspects of creatine and creatine kinase research. Thus, the three books are complementary and, together, provide an even broader basis for comprehensive understanding of the roles of creatine and creatine kinase in health and disease.

REFERENCES

Saks, V.A., ed., 2007, Molecular System Bioenergetics – Energy for Life. Wiley-VCH, Weinheim, Germany.

Vial, C., ed., 2006, Creatine Kinase. NovaScience Publishers, New York, USA.

CONTRIBUTORS

Mayis Aliev
Cardiology Research Center, Institute of Experimental Cardiology, Laboratory of Cardiac Pathology, 121552 Moscow, Russia

Tiia Anmann
Laboratory of Bioenergetics, National Institute of Chemical Physics and Biophysics, 12618 Tallinn, Estonia

Claude Bachmann
Clinical Chemistry Laboratory, Centre Hospitalier Universitaire Vaudois and University of Lausanne, CH-1011 Lausanne, Switzerland

Roberta Battini
Department of Developmental Neuroscience, IRCCS Stella Maris, Calambrone, Pisa, Italy

Olivier Braissant
Clinical Chemistry Laboratory, Centre Hospitalier Universitaire Vaudois and University of Lausanne, CH-1011 Lausanne, Switzerland

David L. Christie
Molecular, Cell and Developmental Biology Section, School of Biological Sciences, University of Auckland, Auckland, New Zealand

Wim Derave
Department of Movement and Sport Sciences, Faculty of Medicine and Health Sciences, Ghent University, Ghent, Belgium

W. Ross Ellington
Institute of Molecular Biophysics and Department of Biological Science, Florida State University, Tallahassee, Florida 32306-4370, USA

Robert J. Ferrante

Geriatric Research Education and Clinical Center, Bedford VA Medical Center, Bedford, MA 01730, USA; and Neurology, Laboratory Medicine and Pathology, and Psychiatry Departments, Boston University School of Medicine, Boston, MA 02180, USA

Thomas Gastner

Degussa AG, D-83308 Trostberg, Germany

Rita Guzun

Laboratory of Fundamental and Applied Bioenergetics, INSERM U 884, Joseph Fourier University, BP53X – 38041, Grenoble Cedex 9, France

Arend Heerschap

Department of Radiology, Radboud University Nijmegen Medical Centre, 6500 HB Nijmegen, the Netherlands

Hugues Henry

Clinical Chemistry Laboratory, Centre Hospitalier Universitaire Vaudois and University of Lausanne, CH-1011 Lausanne, Switzerland

Peter Hespel

Research Center for Exercise and Health, Faculty of Kinesiology and Rehabilitation Sciences, K.U. Leuven, Leuven, Belgium

Ken-ichi Hosoya

Graduate School of Medicine and Pharmaceutical Sciences, University of Toyama, Toyama 930-0194, Japan

Dirk Isbrandt

Centre for Molecular Neurobiology Hamburg (ZMNH), University of Hamburg, D-20246, Hamburg, Germany

Tuuli Kaambre

Laboratory of Bioenergetics, National Institute of Chemical Physics and Biophysics, 12618 Tallinn, Estonia

Hermien E. Kan

Department of Radiology, Radboud University Nijmegen Medical Centre, 6500 HB Nijmegen, the Netherlands

Autumn M. Klein
Neurology Department, Brigham and Womens Hospital, Boston, MA 02115, USA; and Harvard Medical School, Boston, MA 02115, USA

Wesley McCall
School of Pharmacy, University of North Carolina at Chapel Hill, Chapel Hill, NC 27599, USA

Christine I.H.C. Nabuurs
Department of Radiology, Radboud University Nijmegen Medical Centre, 6500 HB Nijmegen, the Netherlands

Sumio Ohtsuki
Graduate School of Pharmaceutical Sciences, Tohoku University, Sendai 980-8578, Japan; and SORST, Japan Science and Technology Agency, Japan

Adam M. Persky
Division of Pharmacotherapy and Experimental Therapeutics, School of Pharmacy, University of North Carolina at Chapel Hill, Chapel Hill, NC 27599-7360, USA

Ivo Pischel
Finzelberg GmbH & Co. KG, D-56626 Andernach, Germany

Eric S. Rawson
Department of Exercise Science and Athletics, Bloomsburg University, Bloomsburg, PA 17815, USA

W. KlaasJan Renema
Department of Radiology, Radboud University Nijmegen Medical Centre, 6500 HB Nijmegen, the Netherlands

Valdur Saks
Laboratory of Fundamental and Applied Bioenergetics, INSERM U 884, Joseph Fourier University, 2280, BP53X – 38041, Grenoble Cedex 9, France; and Laboratory of Bioenergetics, National Institute of Chemical Physics and Biophysics, 12618 Tallinn, Estonia

Gajja S. Salomons
Department of Clinical Chemistry, Metabolic Unit, VU University Medical Center, 1081 HV Amsterdam, The Netherlands

Uwe Schlattner
Laboratory of Fundamental and Applied Bioenergetics, INSERM U 884, Joseph Fourier University, 2280, BP53X – 38041, Grenoble Cedex 9, France; and Institute of Cell Biology, ETH-Zurich, Hönggerberg HPM D24, CH-8093 Zurich, Switzerland

Andreas Schulze
University of Toronto, Department of Paediatrics, Division of Clinical and Metabolic Genetics, and Research Institute, The Hospital for Sick Children, Toronto, ON. M5G 1X8, Canada

Peter W. Schutz
Departments of Pediatrics, Pathology and Laboratory Medicine, University of British Columbia, Division of Biochemical Diseases, British Columbia Children's Hospital, Vancouver, B.C., V6H 3V4, Canada

Peeter Sikk
Laboratory of Bioenergetics, National Institute of Chemical Physics and Biophysics, 12618 Tallinn, Estonia

Sylvia Stockler
Department of Pediatrics, University of British Columbia, Division of Biochemical Diseases, British Columbia Children's Hospital, Vancouver, B.C., V6H 3V4, Canada

Tomohiko Suzuki
Laboratory of Biochemistry, Faculty of Science, Kochi University, Kochi 780-8520, Japan

Masanori Tachikawa
Graduate School of Medicine and Pharmaceutical Sciences, University of Toyama, Toyama 930-0194, Japan

Mark A. Tarnopolsky
Department of Pediatrics and Medicine (Neurology and Rehabilitation), Neuromuscular and Neurometabolic Clinic, Rm 2H26, McMaster University Medical Center, Hamilton, Ontario, Canada, L8N 3Z5

Tetsuya Terasaki
Graduate School of Pharmaceutical Sciences, Tohoku University, Japan; and SORST, Japan Science and Technology Agency, Japan

Marko Vendelin
Department of Mechanics and Applied Mathematics, Institute of Cybernetics, Tallinn Technical University, 12618 Tallinn, Estonia

Theo Wallimann
Institute of Cell Biology, ETH Zurich, Hönggerberg, CH-8093 Zurich, Switzerland

Bé Wieringa
Department of Cell Biology, Nijmegen Centre for Molecular Life Science, 6500 HB Radboud University Nijmegen Medical Centre, Nijmegen, the Netherlands

Markus Wyss
DSM Nutritional Products Ltd., Biotechnology R&D, CH-4002 Basel, Switzerland

CHAPTER 1

INTRODUCTION – CREATINE: CHEAP ERGOGENIC SUPPLEMENT WITH GREAT POTENTIAL FOR HEALTH AND DISEASE

THEO WALLIMANN

Institute of Cell Biology, ETH Zurich, Hönggerberg HPM-D24.1, Schafmattstrasse 18, CH-8093 Zurich, Switzerland

1. THE BOOK

The appearance in print of the present volume of the "Subcellular Biochemistry" Series entitled "Creatine and Creatine Kinase in Health and Disease", edited by Gajja S. Salomons and Markus Wyss, seems entirely timely. The importance and physiological significance of creatine kinase (CK) as well as the pleiotropic effects of creatine (Cr) and phosphocreatine (PCr) in health and disease have been largely underappreciated historically. Based on new discoveries in recent years, however, they are currently attracting much interest and even experience center stage attention, for instance with the recently announced large clinical Cr study with Parkinson's patients in the USA (Couzin, 2007). The comprehensive earlier review on "Creatine and Creatinine Metabolism" by Markus Wyss and Kaddurah-Daouk (2000), as well as the most recent review on Cr by John and Margaret Brosnan (2007), give a pre-taste of this new and exciting era of CK- and Cr-related research to come.

The volume presented here, as well as a new book on "Molecular Systems Bioenergetics: Energy for Life, Basic Principles, Organization and Dynamics of Cellular Energetics", edited by Valdur A. Saks (2007), provide a comprehensive overview of the field, with emphasis on complementary facets of this broad topic. The Saks book focuses on new data and theories derived mainly from basic science, taking a holistic systems biology approach to explain CK and related phosphotransfer systems, including mathematical modeling of metabolic and signaling networks related to bioenergetics. Conversely, the present volume links the basics of the CK

G.S. Salomons and M. Wyss (eds.), Creatine and Creatine Kinase in Health and Disease, 1–16.

system and of Cr action to their practical relevance for health, sports, rehabilitation, neuromuscular and neurodegenerative diseases, as well as for patients suffering from Cr deficiency syndromes. The discovery of the latter syndromes, the clinical manifestation of which is described here in detail, is living proof that CK and Cr are indeed essential for normal body function, especially for brain development and mental performance. The bridge from basic science – unravelling the structure and physiological function(s) of the CK isoenzymes and their substrate Cr in cells with high and fluctuating energy turnover (chapters 1–3; this chapter; Ellington and Suzuki, 2007; Saks *et al.*, 2007) – to clinical applications – including the power of Cr supplementation to positively influence human health (chapters 10–12; Tarnopolsky, 2007; Klein and Ferrante, 2007; Hespel and Derave, 2007) and actually prevent clinical symptoms in patients with certain forms of inherited Cr deficiency syndromes if treated early in life (chapters 8 and 9; Stockler *et al.*, 2007; Schulze and Battini, 2007) – is extremely rewarding, not to speak of the potential socioeconomic and health benefits of such cheap intervention which, apparently, lacks any serious side effects (chapter 14; Persky and Rawson, 2007).

In the present book, after a short excursion into the basics of CK evolution (chapter 2; Ellington and Suzuki, 2007) and the thermodynamics and modeling aspects of the CK phosphotransfer network (chapter 3; Saks *et al.*, 2007), topics on Cr synthesis, trafficking and metabolism in brain and across the blood-brain and blood-retina barriers are discussed in detail in chapters 4 and 5 (Braissant *et al.*, 2007; Tachikawa *et al.*, 2007), followed by new insights into the function of the Cr transporter (CRT) that is responsible for specific uptake of Cr into target cells (chapter 6; Christie, 2007). Then, the physiological consequences of CK and guanidinoacetate methyltransferase (GAMT) gene knock-outs in transgenic mouse models (chapter 7; Heerschap *et al.*, 2007), the pathophysiology and treatment of human Cr deficiency syndromes (chapters 8 and 9; Stockler *et al.*, 2007; Schulze and Battini, 2007), as well as the clinical use of Cr in neuromuscular and neurometabolic disorders (chapter 10; Tarnopolsky, 2007) are discussed, followed by a general presentation of the neuroprotective role of Cr (chapter 11; Klein and Ferrante, 2007). After these chapters, more applied topics, such as the ergogenic effects of Cr in sports and rehabilitation (chapter 12; Hespel and Derave, 2007), the pharmacokinetics of Cr (chapter 13; McCall and Persky, 2007), the safety of Cr supplementation in general (chapter 14; Persky and Rawson, 2007), as well as aspects concerned with chemical synthesis, purity and regulatory status of Cr as a nutritional supplement (chapter 15; Pischel and Gastner, 2007) are presented. The book concludes with an outlook on some promising future avenues of Cr-related research (chapter 16; Wyss *et al.*, 2007).

This volume truly reflects the present state of the art on CK and Cr, written by international experts with experience and recognition in the field. The present set of chapters represents a balanced mix from basic science and applied research to sport, rehabilitation and clinical subjects and thus makes good reading for CK and Cr specialists, but also for educated laymen and persons with a strong interest in Cr-related topics. This volume will serve as a solid basis for a new and exciting wave of innovative Cr research, which among others will encompass systems biology approaches, including proteomics and phospho-proteomics, to elucidate the pleiotropic effects of Cr supplementation on the cellular and whole-body level, with a great potential for unexpected novel findings concerning the multiple facets of the CK system, as well as of Cr supplementation. The fact that many decades of basic research on the functions of CK and Cr have finally led to practical applications for human health and disease, as well as to clinical treatments of patients, is rewarding both for basic scientists and clinicians. Thus, the present book makes good reading also from the perspective of science history and idea development.

2. HISTORICAL ASPECTS OF CREATINE KINASE AND CREATINE

2.1. Functions of Creatine Kinase and Phosphocreatine in Muscle Contraction

Historically, over the last century, the CK and Cr field went through alternating periods of excitement and depression, concerning new ideas and concepts of CK/PCr/Cr function, particularly in the area of muscle biochemistry. New discoveries related to the "true" function of PCr, the "energy-rich" version of Cr that had been discovered in 1927 simultaneously by Eggleton and Eggleton and by Fiske and Subbarow, met with great enthusiasm. For instance, the idea that PCr would represent the long sought-for "immediate" source of energy for muscle contraction, as suggested by the findings of Lundsgaard in 1930, led to the abolition of Otto Meyerhof's "lactate theory of muscle contraction". However, this excitement started to slowly wane after Lohmann and Lehmann discovered the CK reaction in the mid 1930ies, indicating that PCr could only be used as energy source for muscle contraction in the presence of adenine nucleotides, and in particular ADP. This view was supported in 1939 by the finding of Engelhardt and Lyubimova that myosin possesses ATPase activity, and in 1941 by the proof of the Nobel laureate Albert Szent-Györgyi that muscle contraction *in vitro* was inevitably associated with ATP utilization. However, this latter fact could never be confirmed in contracting intact muscle *in vivo*, but rather, as in the Lundsgaard experiment, contraction was always paralleled by a depletion in PCr. Thus, a long-lasting controversy was stirring up the muscle research community for many years to come, until Cain and Davies, who in 1962 used 2,4-dinitro-1-fluorobenzene (DNFB) to specifically inhibit the CK reaction in muscle, finally managed to prove that under these conditions, it was indeed MgATP, and not PCr, that was used as the immediate source of energy for muscle contraction *in vivo*. This data proved that PCr, the substrate of the CK reaction in conjunction with ADP, is used to regenerate the ATP hydrolyzed by the Ca^{2+}-regulated Mg^{2+}-dependent acto-myosin ATPase. This then concluded the "phosphocreatine theory" of muscle contraction, and the CK system got a rather dismal stamp for representing a trivial, simple, homogeneously distributed cellular ATP-buffering system. Accordingly, the muscle research community placed its focus fully on questions on how exactly ATP was used for muscle contraction and cell motility, taking for granted its replenishment for these processes (for reviews on these historical events, including a list of original publications, see (Lipmann, 1977, 1979; Mommaerts, 1969; Rapoport, 1977).

2.2. Development of the PCr-shuttle or CK/PCr-circuit Concept

By contrast, the PCr-shuttle or CK/PCr circuit concept that was developed in the 1970ies (see below) initially faced rather vicious scepticism that persisted for a lengthy period of time (Wallimann, 1996; Wiseman and Kushmerick, 1995). Over the years, however, this opposition seemed to slowly fade away (Chance *et al.*, 2006). The physiological concept of the PCr-shuttle or CK/PCr-circuit (for reviews, see Bessman and Carpenter, 1985; Bessman and Geiger, 1981; Saks *et al.*, 1978; Wallimann and Hemmer, 1994; Wallimann *et al.*, 1992, 2007) has followed the general destiny of many discoveries in science: first, it must not be true and therefore cannot be true; second, it may be true, but it is not important; and third, it is generally accepted but mostly trivial knowledge already around for years. It seems, however, that time has come that this concept reaches the "accepted" if not "generally accepted" status, since nowadays, it is discussed and visualized as back-up support figures without the need for citation, as for example in a recent review on "The failing heart – an engine out of fuel" (Neubauer, 2007).

As already mentioned, CK was considered a strictly soluble metabolic enzyme just for trivial buffering of cellular ATP levels according to its equilibrium constant (Meyer *et al.*, 1984). The facts that (i) CK existed as several tissue-specific and developmentally regulated cytosolic isoforms (Eppenberger *et al.*, 1964, 1967), that (ii) a CK isoform was identified that is located within mitochondria (Jacobs *et al.*, 1964; Jacobus and Lehninger, 1973), and that (iii) a small but significant fraction of soluble muscle-type MM-CK was shown to bind specifically to the sarcomeric M-band of skeletal muscle (Turner *et al.*, 1973; Wallimann, 1975), to the sarcoplasmic reticulum (SR) (Rossi *et al.*, 1990), as well as to the plasma membrane (Saks and Kupriyanov, 1982; for a recent review, see Wallimann *et al.*, 2007), were met with some resistance by the muscle research community. It was said that a soluble enzyme cannot be compartmentalized in a cell, that neither mitochondria nor myofibrils would depend functionally on CK bound at these locations, and – above all – that myofibrillar CK would certainly not play a structural role in the M-line architecture as proposed by us (Wallimann, 1983). As a matter of fact, the first notion of the existence of a PCr-shuttle in muscle, put forward by Martin Klingenberg and colleagues in 1964 (Jacobs *et al.*, 1964), was based on the identification of an unique mitochondrial CK (mtCK) isoform, while the evidence for myofibrillar-bound CK was only incidental. These rather intuitive ideas concerning a PCr-shuttle were further developed by Naegle *et al.* (1964), Bessman and Fonyo (1966), Scholte (1973) and Jacobus and Lehninger (1973) into an "acceptor control concept". According to this concept, extra-mitochondrial creatine is able to stimulate mitochondrial respiration, with the stimulation being mediated by mtCK. In other words, mitochondria are able to produce PCr as high-energy phosphate output by a process called "Cr-stimulated respiration" (Dolder *et al.*, 2003). With the unambiguous demonstration that a fraction of "cytosolic" muscle-type MM-CK is bound to the sarcomeric M-band, a more complete picture of the PCr-shuttle concept slowly emerged (Bessman and Geiger, 1981; Saks *et al.*, 1978; Schlattner *et al.*, 2006; Wallimann 1975; Wallimann *et al.*, 1977, 1992, 2007; Wegmann *et al.*, 1992; Wyss *et al.*, 1992; see also chapter 3; Saks *et al.*, 2007). Now, more than thirty years after these historical findings, many excellent publications from independent laboratories have corroborated (i) the concept of microcompartmentation of the CK isoenzymes, (ii) directional flux and transport of high-energy phosphates within a cell, (iii) functional coupling of mtCK to the adenine nucleotide translocase (ANT) of the inner and the voltage-gated anion carrier (VDAC = porin) of the outer mitochondrial membrane, thereby forming a mitochondrial energy channeling unit (Schlattner *et al.*, 2006; Wallimann *et al.*, 1998; see also chapter 3; Saks *et al.*, 2007), and (iv) the physiological function of the CK/PCr system in alleviating the diffusional limitations of adenine nucleotides, especially in polar cells such as spermatozoa (Kaldis *et al.*, 1997; Tombes and Shapiro, 1985), photoreceptor cells of the retina (Hemmer *et al.*, 1993) or inner ear sensory hair bundle cells (Shin *et al.*, 2007). The PCr-shuttle concept was also helped by integrating structural and functional aspects of the aesthetically rewarding three-dimensional structure of mitochondrial mtCK (Fritz-Wolf *et al.*, 1996; Schlattner *et al.*, 1998) and by experiments showing that muscle-type MM-CK was specifically anchored to the M-band by two lysine-charge clamps that are symmetrically exposed on the MM-CK dimer to make contact with the M-band proteins, myomesin and M-protein (Hornemann *et al.*, 2003; Wallimann *et al.*, 2007). It is important to note that the octameric structure of mtCK has been "invented" very early in evolution; that is, mtCK octamers are found already in certain species of sponges (Hoffman and Ellington, 2005; Hoffman *et al.*, 2006; see also chapter 2; Ellington and Suzuki, 2007), thus reinforcing the eminent importance of this octameric enzyme for cell and organ function.

2.3. Phenotypes of Creatine Kinase Knockout Mice

If CK, Cr and the PCr-shuttle were so important, how should we explain that the various CK isoenzyme knock-out mice are not lethal and show, at first glance, only relatively mild phenotypes (Steeghs *et al.*, 1995; see also chapter 7; Heerschap *et al.*, 2007)? As a matter of fact, these CK isoenzyme knock-out mice are around now for more than 10 years, and each year, by virtue of more in-depth studies and by using more sophisticated methods, some new severe and rather interesting impairments of normal cell and organ function are being discovered. The latest addition to this list is the finding that cytosolic brain-type BB-CK and ubiquitous mtCK double knock-out animals show significant impairment in inner ear physiology, exhibiting a great reduction in hearing threshold and abnormal tympanic function for body balance (Shin *et al.*, 2007). This is an interesting lesson for mouse geneticists, who initially thought that the severity of phenotypes, after knocking out a given gene, would be proportional to the "importance" of this gene. Fact is, however, that the principle successfully used in the aviation industry, namely that the most important functionalities have to be backed-up by several independent alternative mechanisms, also seems to apply to biological systems. In other words, it became obvious that tampering with the CK system leads to a host of compensatory events that can be uncovered when analyzing these transgenic animals with appropriate diligence. For example, an amazing proliferation of mitochondria is seen in glycolytic fast-twitch fibers of CK knock-out animals that makes them look like insect flight muscle fibers (Novotova *et al.*, 2006; van Deursen *et al.*, 1993; Ventura-Clapier *et al.*, 2004). This compensation mechanism represents an effective means of decreasing diffusion distances for ATP from mitochondria to myofibrils, since it is no longer PCr and Cr, but ATP and ADP, that have to be shuttled along this line in these muscle fibers. Such compensatory measures of cells and organs are teaching a great lesson, in particular on the plasticity of biological organisms with their remarkable ability to adapt to challenging situations. In addition, they can provide direct hints at the deleted gene's functions. Despite the above-mentioned back-up systems and compensation mechanisms, the phenotypes of some CK knock-out mice are severe enough to make their survival in the wild seem unlikely, due to impaired voluntary running capacity (Momken *et al.*, 2005), reduced cardiac performance at high work-load (Crozatier *et al.*, 2002), as well as a number of neuro-behavioral deficits (Jost *et al.*, 2002; Streijger *et al.*, 2004, 2005). In addition, double knock-out mice with a deletion in both the cytosolic BB-CK and the ubiquitous mtCK isoenzymes displayed severe problems to hold body temperature and to breed (own unpublished observation), which can be deadly in a harsh natural environment.

2.4. Discovery of Creatine and First Trials with Creatine Supplementation

The French scientist, Michel Eugène Chevreul, discovered Cr in 1832 as a new organic constituent that could be extracted from meat ("kreas" in Greek). In 1847, the German scientist, Justus von Liebig, chemically identified Cr as methyl-guanidino-acetic acid, a relatively simple guanidino compound. Today, we know that Cr is found in fresh meat and fish in concentrations ranging from 3 to 10 grams per kg wet weight. Justus von Liebig supported his laboratory largely by producing and selling meat broth, the famous Liebig's meat extract or Fleischbrühe in German, which contained about 8% Cr (Sulser, 1968). This was obviously the first attempt to bring Cr supplementation into the public domain with a small spin-off company, as one would call it today, with more such to come in the 1990ies for selling Cr to athletes. In the 1880ies, creatinine (Crn) was discovered and it was realized that this compound was likely the natural breakdown product of Cr. In 1926, Chanutin surmised, based on what was probably one of the first Cr supplementation trials in the history of mankind, that creatine is absorbed by the intestine and, thus, can be taken up rather quantitatively from alimentary sources such as fresh fish and meat (Chanutin, 1926). In 1927, about a century from the discovery of Cr, PCr was discovered (Eggleton and Eggleton, Fiske and Subbarow). For a review on the historic aspects of Cr, see Conway and Clark (1996).

Unfortunately, during the next decades, Cr supplementation was followed-up only with low-key research at most. Reminiscent of this time is that some body builders and weight-lifters resorted, as it was told, to "sweated beef", a method to extract Cr from meat by hot steam, resulting in a highly Cr-enriched meat juice that was anecdotally said among the members of the "scene" to be beneficial for muscle growth and performance. Not to forget is the so-called "Jewish medicine", a concentrated chicken soup from a fresh chicken boiled to perfection, used as traditionally inherited panacea within the Jewish community that served to "cure" almost everything. Finally, worth mentioning in the same context is the ritual consumption by the new mother of the after-birth placenta, which is also rich in Cr, a tradition followed by many ancestral civilizations. Could one of the most active ingredients in these concoctions have been Cr? Nobody knows, but in hindsight and in light of the present knowledge on the pleiotropic effects of Cr supplementation, it could have been the decisive active principle.

Serious, double-blinded and placebo-controlled research with Cr supplementation was started in the early 1990ies only, and the seminal papers on "Elevation of creatine in resting and exercised muscle of normal subjects by creatine supplementation" (Harris *et al.*, 1992) and on the "Influence of oral creatine supplementation on muscle torque during repeated bouts of maximal voluntary exercise in man" (Greenhaff *et al.*, 1993) initiated a big boost in Cr supplementation studies, mostly with athletes (chapter 12; Hespel and Derave, 2007). Interestingly, Roger Harris published his study in the very same year as Linford Christie (100 m dash) and Sally Gunnell (400 m hurdles) celebrated their Olympic victories, thereafter mentioning that they both ingested Cr. It seems that, once again, top athletes with the help of sports doctors did out-run the basic scientists.

Many of the several hundreds of publications from leading sports physiology laboratories around the world are clearly proving that Cr, in contrast to many other sports nutrition supplements, is a true ergogenic aid (chapter 12; Hespel and Derave, 2007). Nowadays, hundreds of thousands, if not millions of athletes of all proveniences are consuming Cr worldwide to boost physical performance, without any serious side effects, it seems (chapter 14; Persky and Rawson, 2007).

2.5. Human Patients with Cr Deficiency Syndromes

The ultimate proof that the CK system and Cr as such are essential for normal brain development and brain function came actually from the discovery of Cr deficiency patients (see chapters 8 and 9; Stockler *et al.*, 2007; Schulze and Battini, 2007). The findings in man were subsequently supported by a transgenic mouse model, where one of the Cr biosynthesis enzymes, guanidinoacetate methyltransferase (GAMT), had been ablated (Renema *et al.*, 2003; Schmidt *et al.*, 2004; see also chapter 7; Heerschap *et al.*, 2007). Patients which lack Cr in the brain, either due to a defect in Cr biosynthesis (Schulze, 2003; Stöckler-Ipsiroglu, 1997; Verhoeven *et al.*, 2005) or in the Cr transporter (deGrauw *et al.*, 2003; Kleefstra *et al.*, 2005; Rosenberg *et al.*, 2004), present with mental retardation, speech- and language delay, epilepsy and autistic-like behaviour (for details, see the relevant chapters 8 and 9 in this book; Stockler *et al.*, 2007; Schulze and Battini, 2007). Patients with a defect in Cr biosynthesis can be treated with Cr supplementation in combination with dietary restriction and/or additional supplements. These treatments, if started early, may prevent the development of clinical symptoms (Bianchi *et al.*, 2007; Schulze, 2003; Verbruggen *et al.*, 2007). These new and clinically relevant discoveries, supporting (i) the importance of the CK system and of Cr for normal physiological function of the human body, most importantly the brain, and (ii) justifying CK and Cr-related research in the realm of rare human diseases, gave a new wave of recognition to the CK/Cr system.

3. PLEIOTROPIC EFFECTS OF CREATINE

Common denominators for many muscular, neuromuscular and neurodegenerative diseases are (i) lowered energy status, that is decreased cellular energy reserves (PCr and ATP), (ii) accompanying chronic calcium overload, due to a misbalance in the energetics of calcium homeostasis, and (iii) concomitant formation of reactive oxygen species (ROS) (Rodriguez *et al.*, 2007; Tarnopolsky and Beal, 2001). Incidentally, the most obvious phenotype of knockout mice lacking CK in muscle were difficulties with calcium sequestration and muscle relaxation (Steeghs *et al.*, 1997). Accordingly, Cr supplementation was shown to improve calcium homeostasis in the *mdx* muscular dystrophy mouse model (Pulido *et al.*, 1998) and muscle relaxation in humans (Hespel *et al.*, 2002). These findings are easily explainable by the fact that the calcium pump ATPase of the SR is energetically very demanding and only works efficiently if a high local ATP/ADP ratio is maintained by the action of CK functionally coupled to this very SR calcium pump (Wallimann and Hemmer, 1994; Wallimann *et al.*, 2007). Cr supplementation in general increases the PCr pool of cells, provided that ATP synthesis is not impaired and that sufficient free ATP is available to reload the PCr buffer (Vandenberghe *et al.*, 1997). Thus indirectly, through an improved energy state, Cr would stabilize cellular calcium homeostasis and thereby prevent prolonged calcium overload which is thought to be responsible for the initiation of apoptosis, necrosis (Kroemer *et al.*, 2007) or cell destruction in general (Passaquin *et al.*, 2002), due to the initiation of mitochondrial permeability transition pore opening or due to the action of calcium-activated proteases, respectively.

As evidence for the above, *in vitro* treatment of cells as well as dietary supplementation of animals with Cr are highly beneficial with regard to protection from injury and enhancement of survival following noxious treatment (Brewer and Wallimann, 2000; Brustovetsky *et al.*, 2001; Sullivan *et al.*, 2000). Since Cr supplementation leads to an increase in intracellular PCr levels, a higher PCr/ATP ratio (Vandenberghe *et al.*, 1997) and thereby to a higher buffering capacity for intracellular ATP and ADP concentrations, one could explain at least part of the Cr-dependent protection of cells from various stressors, especially the neuroprotective effects of Cr (see chapters 8–11; Stockler *et al.*, 2007; Schulze and Battini, 2007; Tarnopolsky, 2007; Klein and Ferrante, 2007), by the classical beneficial effects of Cr in improving the cellular energy state. Cr supplementation mediates remarkable neuroprotection in experimental animal models of amyotrophic lateral sclerosis (ALS), Huntington's disease (HD), Parkinson's disease, and traumatic brain and spinal cord injury (see chapters 10 and 11 herein; Tarnopolsky, 2007; Klein and Ferrante, 2007). Prophylactic Cr administration also mediates neuroprotection in cerebral ischemia in mice (Adcock *et al.*, 2002; Prass *et al.*, 2007; Zhu *et al.*, 2004). Transgenic mice expressing high levels of BB-CK in liver cells, which normally express only very low levels of this enzyme, show a high degree of resistance to tumour necrosis factor-induced and hypoxia-induced apoptosis of liver cells when supplemented with dietary creatine (Hatano *et al.*, 2004). Furthermore, Cr exhibits a protective effect against ROS in cultured mammalian cells (Sestili *et al.*, 2006) and against UV-induced oxidative stress in intact human skin (Berneburg *et al.*, 2005; Lenz *et al.*, 2005).

It is thus believed that the presence of the CK system improves the stress resistance of cells, making them less susceptible to injury. Congruent with the above discussion, three non-exclusive mechanisms of action underlying the protective effect of creatine are presently discussed: (1) The energy buffer and storage function of the CK/PCr system is thought to keep the ATP/ADP ratio high during situations of stress, which otherwise might lead to cellular dysfunction and, eventually, cell death. A high cellular energy state and high local ATP/ADP ratios are particularly critical for efficient calcium pump function (Minajeva *et al.*, 1996; Wallimann and Hemmer, 1994). Thus, compromised cellular energetics lead to a unbalance in calcium homeostasis and, in the long run, to chronic calcium overload. A convincing proof for the utmost importance of the CK system for cellular calcium homeostasis is the phenotype of mice with ablated CK genes (de Groof *et al.*, 2002) that show problems with muscle relaxation and calcium re-uptake into the SR. (2) Cr is thought to act in concert with mtCK in inhibiting mitochondrial permeability transition pore opening, an early trigger of apoptosis (Dolder *et al.*, 2003; O'Gorman *et al.*, 1997). MtCK, a cubical-shaped octamer with a central channel, was shown to form a multienzyme complex with porin of the outer mitochondrial membrane and adenine nucleotide translocase (ANT) of the inner membrane (Brdiczka *et al.*, 1998; Schlattner *et al.*, 2006). This complex crosslinks the two mitochondrial membranes and forms a functionally coupled microcompartment for vectorial export of PCr into the cytosol. Disruption of the octameric structure of mtCK leads to an impairment of energy homeostasis and facilitated transition pore opening (O'Gorman *et al.*, 1997), which eventually results in activation of caspase cell-death pathways (Kroemer *et al.*, 2007). (3) A coordinated action of mtCK activity with oxidative phosphorylation is facilitated by tight functional coupling of mtCK with ANT (Dolder *et al.*, 2003), leading to the well-known phenomenon termed "creatine-stimulated respiration" that can also be demonstrated *in vivo*

(Kay *et al.*, 2000). Under physiological conditions, when mitochondrial respiration is directly stimulated by Cr, endogenous intramitochondrial adenine nucleotides are recycled inside mitochondria, and PCr is leaving the mitochondria as the energy-rich end product of respiration (Dolder *et al.*, 2003). The addition of Cr to respiring mitochondria also induces an optimal coupling of the respiratory chain to ATP generation by the F1-ATPase. This leads to a significantly reduced production by the respiratory chain of ROS in mitochondria, compared to mitochondria without Cr. Thus, the accompanying deleterious consequences of mitochondrial ROS for cell damage are minimized by Cr, as has recently been demonstrated convincingly with brain mitochondria (Meyer *et al.*, 2006).

In addition to the above, Cr is released from cells after hypo-osmotic swelling (Bothwell *et al.*, 2001) and, conversely, serves as an osmolyte that is taken up by cells under hypertonic stress (Alfieri *et al.*, 2006). Recent evidence indicates that Cr is not only synthesized in the brain where it is trafficking back and forth between various neuronal cell types and being taken up by neurons (chapters 4 and 5; Braissant *et al.*, 2007; Tachikawa *et al.*, 2007), but that it can also be released from neurons in an action-potential-dependent way, e.g. upon excitation of neurons (Almeida *et al.*, 2006). This exocytotic release of Cr, together with its action as a partial $GABA_A$ receptor agonist (De Deyn and Macdonald, 1990), indicates that Cr may act as a neuromodulator. The actions of Cr as an osmolyte and a neuromodulator, as well as its direct or indirect effects as an anti-oxidant (Berneburg *et al.*, 2005; Lenz *et al.*, 2005; Sestili *et al.*, 2006), represent non-energy related functions of Cr that may contribute to the pleiotropic effects observed with Cr.

Systems biology approaches involving the study of global gene expression of Cr-substituted versus Cr-depleted cells and organs or whole animals are likely to reveal novel and unexpected effects of Cr. A particularly interesting topic for study will be the connection of PCr and Cr to cellular signaling networks and, in particular, the AMPK signaling cascade. Recent breakthroughs in phosphoproteomics (Bodenmiller *et al.*, 2007) should facilitate investigation of whether Cr supplementation stimulates phosphorylation of AMPK and, thereby, triggers the AMPK signaling cascade.

4. CONCLUSIONS

After a continuous up-hill battle for most who have been active in the CK- and Cr-related research field, to obtain grants and funding over all those years, I would never have dreamed that, one day, some of the ideas and concepts elaborated by basic laboratory scientists would come to fruition with respect to human health and disease and that they could, potentially, have significant health and socio-economic benefits. This is certainly rewarding, and this spirit, together with a positive outlook for an interesting future for CK- and Cr-related research ahead, prevails throughout the book, including the concluding chapter (chapter 16; Wyss *et al.*, 2007). Again, this book is important, for it represents a timely synopsis of past achievements and new developments in the field. Significant advances have been made in deciphering some important aspects of the molecular structure and function of CK, the involvement of this enzyme in cellular energy distribution networks, and in revealing new and unexpected facts on the pleiotropic effects of Cr. These effects are important for normal cell function and can be exploited for practical application in health and disease. The reader can expect to learn much about the multifaceted role of CK and Cr in the context of cellular bioenergetics and will appreciate the broad spectrum of multidisciplinary approaches from molecular to cellular and finally to clinical Cr-related research.

Again, overall, this book makes a highly recommended reading for everybody interested in bioenergetics, energy homeostasis, as well as in theoretical and applied aspects of CK and Cr research, including the potential benefits of Cr supplementation for human health and disease. I can endorse the outlook given in the final chapter (chapter 16; Wyss *et al.*, 2007), where some interesting new aspects and possibilities concerning the future directions of Cr-related research are given, more appropriate Cr supplementation dosages are recommended, a list of clinical applications of Cr is provided, and potentially valuable new applications of Cr are presented for healthy people, both young and old. Thus, it is obvious that Cr has evolved from a simple "energy precursor metabolite" to a "true" ergogenic dietary supplement and, finally, to a promising cheap therapeutic agent with a potentially huge health and socio-economic impact. An exciting new era for the Cr field is conveyed not only for the Cr research community but also for the general reader of this book.

REFERENCES

Adcock, K.H., Nedelcu, J., Loenneker, T., Martin, E., Wallimann, T., and Wagner, B.P., 2002, Neuroprotection of creatine supplementation in neonatal rats with transient cerebral hypoxia-ischemia. *Dev. Neurosci.* **24:** 382–388.

Alfieri, R.R., Bonelli, M.A., Cavazzoni, A., Brigotti, M., Fumarola, C., Sestili, P., Mozzoni, P., De Palma, G., Mutti, A., Carnicelli, D., *et al.*, 2006, Creatine as a compatible osmolyte in muscle cells exposed to hypertonic stress. *J. Physiol.* **576:** 391–401.

Almeida, L.S., Salomons, G.S., Hogenboom, F., Jakobs, C., and Schoffelmeer, A.N., 2006, Exocytotic release of creatine in rat brain. *Synapse* **60:** 118–123.

Berneburg, M., Gremmel, T., Kurten, V., Schroeder, P., Hertel, I., von Mikecz, A., Wild, S., Chen, M., Declercq, L., Matsui, M., *et al.*, 2005, Creatine supplementation normalizes mutagenesis of mitochondrial DNA as well as functional consequences. *J. Invest. Dermatol.* **125:** 213–220.

Bessman, S.P., and Carpenter, C.L., 1985, The creatine-creatine phosphate energy shuttle. *Annu. Rev. Biochem.* **54:** 831–862.

Bessman, S.P., and Fonyo, A., 1966, The possible role of the mitochondrial bound creatine kinase in regulation of mitochondrial respiration. *Biochem. Biophys. Res. Commun.* **22:** 597–602.

Bessman, S.P., and Geiger, P.J., 1981, Transport of energy in muscle: the phosphorylcreatine shuttle. *Science* **211:** 448–452.

Bianchi, M.C., Tosetti, M., Battini, R., Leuzzi, V., Alessandri, M.G., Carducci, C., Antonozzi, I., and Cioni, G., 2007, Treatment monitoring of brain creatine deficiency syndromes: a 1H- and 31P-MR spectroscopy study. *Am. J. Neuroradiol.* **28:** 548–554.

Bodenmiller, B., Mueller, L.N., Mueller, M., Domon, B., and Aebersold, R., 2007, Reproducible isolation of distinct, overlapping segments of the phosphoproteome. *Nat. Meth.* **4:** 231–237.

Bothwell, J.H., Rae, C., Dixon, R.M., Styles, P., and Bhakoo, K.K., 2001, Hypo-osmotic swelling-activated release of organic osmolytes in brain slices: implications for brain oedema in vivo. *J. Neurochem.* **77:** 1632–1640.

Braissant, O., Bachmann, C., and Henry, H., 2007, Expression and function of AGAT, GAMT and CT1 in the mammalian brain. *Subcell. Biochem.* **46:** 67–81.

Brdiczka, D., Beutner, G., Ruck, A., Dolder, M., and Wallimann, T., 1998, The molecular structure of mitochondrial contact sites. Their role in regulation of energy metabolism and permeability transition. *Biofactors* **8:** 235–242.

Brewer, G.J., and Wallimann, T.W., 2000, Protective effect of the energy precursor creatine against toxicity of glutamate and beta-amyloid in rat hippocampal neurons. *J. Neurochem.* **74:** 1968–1978.

Brosnan, J.T., and Brosnan, M.E., 2007, Creatine: endogenous metabolite, dietary, and therapeutic supplement. *Annu. Rev. Nutr.* **27:** 241–261.

Brustovetsky, N., Brustovetsky, T., and Dubinsky, J.M., 2001, On the mechanisms of neuroprotection by creatine and phosphocreatine. *J. Neurochem.* **76:** 425–434.

Chance, B., Im, J., Nioka, S., and Kushmerick, M., 2006, Skeletal muscle energetics with PNMR: personal views and historic perspectives. *NMR Biomed.* **19:** 904–926.

Chanutin, A., 1926, The fate of creatine when administred to man. *J. Biol. Chem.* **67:** 29–41.

Christie, D.L., 2007, Functional insights into the creatine transporter. *Subcell. Biochem.* **46:** 99–118.

Conway, M., and Clark, J., 1996, Creatine and Creatine Phosphate: Scientific and Clinical Perspectives. Academic Press, San Diego, CA, USA.

Couzin, J., 2007, Clinical research: testing a novel strategy against Parkinson's disease. *Science* **315:** 1778.

Crozatier, B., Badoual, T., Boehm, E., Ennezat, P.V., Guenoun, T., Su, J., Veksler, V., Hittinger, L., and Ventura-Clapier, R., 2002, Role of creatine kinase in cardiac excitation-contraction coupling: studies in creatine kinase-deficient mice. *FASEB J.* **16:** 653–660.

De Deyn, P.P., and Macdonald, R.L., 1990, Guanidino compounds that are increased in cerebrospinal fluid and brain of uremic patients inhibit GABA and glycine responses on mouse neurons in cell culture. *Ann. Neurol.* **28:** 627–633.

de Groof, A.J., Fransen, J.A., Errington, R.J., Willems, P.H., Wieringa, B., and Koopman, W.J., 2002, The creatine kinase system is essential for optimal refill of the sarcoplasmic reticulum Ca^{2+} store in skeletal muscle. *J. Biol. Chem.* **277:** 5275–5284.

deGrauw, T.J., Cecil, K.M., Byars, A.W., Salomons, G.S., Ball, W.S., and Jakobs, C., 2003, The clinical syndrome of creatine transporter deficiency. *Mol. Cell. Biochem.* **244:** 45–48.

Dolder, M., Walzel, B., Speer, O., Schlattner, U., and Wallimann, T., 2003, Inhibition of the mitochondrial permeability transition by creatine kinase substrates. Requirement for microcompartmentation. *J. Biol. Chem.* **278:** 17760–17766.

Ellington, W.R., and Suzuki, T., 2007, Early evolution of the creatine kinase gene family and the capacity for creatine biosynthesis and membrane transport. *Subcell. Biochem.* **46:** 17–26.

Eppenberger, H.M., Dawson, D.M., and Kaplan, N.O., 1967, The comparative enzymology of creatine kinases. I. Isolation and characterization from chicken and rabbit tissues. *J. Biol. Chem.* **242:** 204–209.

Eppenberger, H.M., Eppenberger, M., Richterich, R., and Aebi, H., 1964, The ontogeny of creatine kinase isozymes. *Dev. Biol.* **10:** 1–16.

Fritz-Wolf, K., Schnyder, T., Wallimann, T., and Kabsch, W., 1996, Structure of mitochondrial creatine kinase. *Nature* **381:** 341–345.

Greenhaff, P.L., Casey, A., Short, A.H., Harris, R., Soderlund, K., and Hultman, E., 1993, Influence of oral creatine supplementation on muscle torque during repeated bouts of maximal voluntary exercise in man. *Clin. Sci. (London)* **84:** 565–571.

Harris, R.C., Soderlund, K., and Hultman, E., 1992, Elevation of creatine in resting and exercised muscle of normal subjects by creatine supplementation. *Clin. Sci. (London)* **83:** 367–374.

Hatano, E., Tanaka, A., Kanazawa, A., Tsuyuki, S., Tsunekawa, S., Iwata, S., Takahashi, R., Chance, B., and Yamaoka, Y., 2004, Inhibition of tumor necrosis factor-induced apoptosis in transgenic mouse liver expressing creatine kinase. *Liver Int.* **24:** 384–393.

Heerschap, A., Kan, H.E., Nabuurs, C.I.H.C., Renema, W.K., Isbrandt, D., and Wieringa, B., 2007, *In vivo* magnetic resonance spectroscopy of transgenic mice with altered expression of guanidinoacetate methyltransferase and creatine kinase isoenzymes. *Subcell. Biochem.* **46:** 119–148.

Hemmer, W., Riesinger, I., Wallimann, T., Eppenberger, H.M., and Quest, A.F., 1993, Brain-type creatine kinase in photoreceptor cell outer segments: role of a phosphocreatine circuit in outer segment energy metabolism and phototransduction. *J. Cell Sci.* **106:** 671–683.

Hespel, P., Op't Eijnde, B., and Van Leemputte, M., 2002, Opposite actions of caffeine and creatine on muscle relaxation time in humans. *J. Appl. Physiol.* **92:** 513–518.

Hespel, P., and Derave, W., 2007, Ergogenic effects of creatine in sports and rehabilitation. *Subcell. Biochem.* **46:** 245–259.

Hoffman, G.G., and Ellington, W.R., 2005, Over-expression, purification and characterization of the oligomerization dynamics of an invertebrate mitochondrial creatine kinase. *Biochim. Biophys. Acta* **1751:** 184–193.

Hoffman, G.G., Sona, S., Bertin, M., and Ellington, W.R., 2006, The role of an absolutely conserved tryptophan residue in octamer formation and stability in mitochondrial creatine kinases. *Biochim. Biophys. Acta* **1764:** 1512–1517.

Hornemann, T., Kempa, S., Himmel, M., Hayess, K., Furst, D.O., and Wallimann, T., 2003, Muscle-type creatine kinase interacts with central domains of the M-band proteins myomesin and M-protein. *J. Mol. Biol.* **332:** 877–887.

Jacobs, H., Heldt, H.W., and Klingenberg, M., 1964, High activity of creatine kinase in mitochondria from muscle and brain and evidence for a separate mitochondrial isoenzyme of creatine kinase. *Biochem. Biophys. Res. Commun.* **16:** 516–521.

Jacobus, W.E., and Lehninger, A.L., 1973, Creatine kinase of rat heart mitochondria. Coupling of creatine phosphorylation to electron transport. *J. Biol. Chem.* **248:** 4803–4810.

Jost, C.R., Van Der Zee, C.E., In 't Zandt, H.J., Oerlemans, F., Verheij, M., Streijger, F., Fransen, J., Heerschap, A., Cools, A.R., and Wieringa, B., 2002, Creatine kinase B-driven energy transfer in the brain is important for habituation and spatial learning behaviour, mossy fibre field size and determination of seizure susceptibility. *Eur. J. Neurosci.* **15:** 1692–1706.

Kaldis, P., Kamp, G., Piendl, T., and Wallimann, T., 1997, Functions of creatine kinase isoenzymes in spermatozoa. *Adv. Dev. Biochem.* **5:** 275–312.

Kay, L., Nicolay, K., Wieringa, B., Saks, V., and Wallimann, T., 2000, Direct evidence for the control of mitochondrial respiration by mitochondrial creatine kinase in oxidative muscle cells in situ. *J. Biol. Chem.* **275:** 6937–6944.

Kleefstra, T., Rosenberg, E.H., Salomons, G.S., Stroink, H., van Bokhoven, H., Hamel, B.C., and de Vries, B.B., 2005, Progressive intestinal, neurological and psychiatric problems in two adult males with cerebral creatine deficiency caused by an SLC6A8 mutation. *Clin. Genet.* **68:** 379–381.

Klein, A.M., and Ferrante, R.J., 2007, The neuroprotective role of creatine. *Subcell. Biochem.* **46:** 205–243.

Kroemer, G., Galluzzi, L., and Brenner, C., 2007, Mitochondrial membrane permeabilization in cell death. *Physiol. Rev.* **87:** 99–163.

Lenz, H., Schmidt, M., Welge, V., Schlattner, U., Wallimann, T., Elsasser, H.P., Wittern, K.P., Wenck, H., Stab, F., and Blatt, T., 2005, The creatine kinase system in human skin: protective effects of creatine against oxidative and UV damage in vitro and in vivo. *J. Invest. Dermatol.* **124:** 443–452.

Lipmann, F., 1977, Discovery of creatine phosphate in muscle. *Trends Biochem. Sci.* **2:** 21–22.

Lipmann, F., 1979, Discovery of the adenylic acid system in animal tissues. *Trends Biochem. Sci.* **4:** 22–24.

McCall, W., and Persky, A.M., 2007, Pharmacokinetics of creatine. *Subcell. Biochem.* **46:** 261–273.

Meyer, L.E., Machado, L.B., Santiago, A.P., da-Silva, W.S., De Felice, F.G., Holub, O., Oliveira, M.F., and Galina, A., 2006, Mitochondrial creatine kinase activity prevents reactive oxygen species generation: antioxidant role of mitochondrial kinase-dependent ADP re-cycling activity. *J. Biol. Chem.* **281:** 37361–37371.

Meyer, R.A., Sweeney, H.L., and Kushmerick, M.J., 1984, A simple analysis of the "phosphocreatine shuttle". *Am. J. Physiol.* **246:** C365–C377.

Minajeva, A., Ventura-Clapier, R., and Veksler, V., 1996, Ca^{2+} uptake by cardiac sarcoplasmic reticulum ATPase in situ strongly depends on bound creatine kinase. *Pflugers Arch.* **432:** 904–912.

Momken, I., Lechene, P., Koulmann, N., Fortin, D., Mateo, P., Doan, B.T., Hoerter, J., Bigard, X., Veksler, V., and Ventura-Clapier, R., 2005, Impaired voluntary running capacity of creatine kinase-deficient mice. *J. Physiol.* **565:** 951–964.

Mommaerts, W.F., 1969, Energetics of muscular contraction. *Physiol. Rev.* **49:** 427–508.

Naegle, S., Hockerts, T., and Boegelmann, G., 1964, The modification of the creatine phosphokinase equilibrium by myocardial pH decline under ischemic conditions. *Klin. Wochenschr.* **42:** 780–784.

Neubauer, S., 2007, The failing heart - an engine out of fuel. *N. Engl. J. Med.* **356:** 1140–1151.

Novotova, M., Pavlovicova, M., Veksler, V.I., Ventura-Clapier, R., and Zahradnik, I., 2006, Ultrastructural remodeling of fast skeletal muscle fibers induced by invalidation of creatine kinase. *Am. J. Physiol. Cell Physiol.* **291:** C1279–C1285.

O'Gorman, E., Beutner, G., Dolder, M., Koretsky, A.P., Brdiczka, D., and Wallimann, T., 1997, The role of creatine kinase in inhibition of mitochondrial permeability transition. *FEBS Lett.* **414:** 253–257.

Passaquin, A.C., Renard, M., Kay, L., Challet, C., Mokhtarian, A., Wallimann, T., and Ruegg, U.T., 2002, Creatine supplementation reduces skeletal muscle degeneration and enhances mitochondrial function in *mdx* mice. *Neuromusc. Disord.* **12:** 174–182.

Persky, A.M., and Rawson, E.S., 2007, Safety of creatine supplementation. *Subcell. Biochem.* **46:** 275–289.

Pischel, I., and Gastner, T., 2007, Creatine – its chemical synthesis, chemistry, and legal status. *Subcell. Biochem.* **46:** 291–307.

Prass, K., Royl, G., Lindauer, U., Freyer, D., Megow, D., Dirnagl, U., Stockler-Ipsiroglu, G., Wallimann, T., and Priller, J., 2007, Improved reperfusion and neuroprotection by creatine in a mouse model of stroke. *J. Cerebr. Blood Flow Metab.* **27:** 452–459.

Pulido, S.M., Passaquin, A.C., Leijendekker, W.J., Challet, C., Wallimann, T., and Ruegg, U.T., 1998, Creatine supplementation improves intracellular Ca^{2+} handling and survival in *mdx* skeletal muscle cells. *FEBS Lett.* **439:** 357–362.

Rapoport, S., 1977, Lohmann and ATP. *Trends Biochem. Sci.* **2:** 163–164.

Renema, W.K., Schmidt, A., van Asten, J.J., Oerlemans, F., Ullrich, K., Wieringa, B., Isbrandt, D., and Heerschap, A., 2003, MR spectroscopy of muscle and brain in guanidinoacetate methyltransferase (GAMT)-deficient mice: validation of an animal model to study creatine deficiency. *Magn. Reson. Med.* **50:** 936–943.

Rodriguez, M.C., MacDonald, J.R., Mahoney, D.J., Parise, G., Beal, M.F., and Tarnopolsky, M.A., 2007, Beneficial effects of creatine, CoQ10, and lipoic acid in mitochondrial disorders. *Muscle Nerve* **35:** 235–242.

Rosenberg, E.H., Almeida, L.S., Kleefstra, T., deGrauw, R.S., Yntema, H.G., Bahi, N., Moraine, C., Ropers, H.H., Fryns, J.P., deGrauw, T.J., *et al.*, 2004, High prevalence of SLC6A8 deficiency in X-linked mental retardation. *Am. J. Hum. Genet.* **75:** 97–105.

Rossi, A.M., Eppenberger, H.M., Volpe, P., Cotrufo, R., and Wallimann, T., 1990, Muscle-type MM creatine kinase is specifically bound to sarcoplasmic reticulum and can support Ca^{2+} uptake and regulate local ATP/ADP ratios. *J. Biol. Chem.* **265:** 5258–5266.

Saks, V.A., and Kupriyanov, V.V., 1982, Intracellular energy transport and control of cardiac contraction. *Adv. Myocardiol.* **3:** 475–497.

Saks, V.A., Rosenshtraukh, L.V., Smirnov, V.N., and Chazov, E.I., 1978, Role of creatine phosphokinase in cellular function and metabolism. *Can. J. Physiol. Pharmacol.* **56:** 691–706.

Saks, V.A., ed., 2007, Molecular System Bioenergetics – Energy for Life. Wiley-VCH, Weinheim, Germany.

Saks, V., Kaambre, T., Guzun, R., Anmann, T., Sikk, P., Schlattner, U., Wallimann, T., Aliev, M., and Vendelin, M., 2007, The creatine kinase phosphotransfer network: thermodynamic and kinetic considerations, the impact of the mitochondrial outer membrane and modelling approaches. *Subcell. Biochem.* **46:** 27–65.

Schlattner, U., Forstner, M., Eder, M., Stachowiak, O., Fritz-Wolf, K., and Wallimann, T., 1998, Functional aspects of the X-ray structure of mitochondrial creatine kinase: a molecular physiology approach. *Mol. Cell. Biochem.* **184:** 125–140.

Schlattner, U., Tokarska-Schlattner, M., and Wallimann, T., 2006, Mitochondrial creatine kinase in human health and disease. *Biochim. Biophys. Acta* **1762:** 164–180.

Schmidt, A., Marescau, B., Boehm, E.A., Renema, W.K., Peco, R., Das, A., Steinfeld, R., Chan, S., Wallis, J., Davidoff, M., *et al.*, 2004, Severely altered guanidino compound levels, disturbed body weight homeostasis and impaired fertility in a mouse model of guanidinoacetate N-methyltransferase (GAMT) deficiency. *Hum. Mol. Genet.* **13:** 905–921.

Scholte, H.R., 1973, On the triple localization of creatine kinase in heart and skeletal muscle cells of the rat: evidence for the existence of myofibrillar and mitochondrial isoenzymes. *Biochim. Biophys. Acta* **305:** 413–427.

Schulze, A., 2003, Creatine deficiency syndromes. *Mol. Cell. Biochem.* **244:** 143–150.

Schulze, A., and Battini, R., 2007, Pre-symptomatic treatment of creatine biosynthesis defects. *Subcell. Biochem.* **46:** 167–181.

Sestili, P., Martinelli, C., Bravi, G., Piccoli, G., Curci, R., Battistelli, M., Falcieri, E., Agostini, D., Gioacchini, A.M., and Stocchi, V., 2006, Creatine supplementation affords cytoprotection in oxidatively injured cultured mammalian cells via direct antioxidant activity. *Free Radic. Biol. Med.* **40:** 837–849.

Shin, J.B., Streijger, F., Beynon, A., Peters, T., Gadzala, L., McMillen, D., Bystrom, C., Van der Zee, C.E., Wallimann, T., and Gillespie, P.G., 2007, Hair bundles are specialized for ATP delivery via creatine kinase. *Neuron* **53:** 371–386.

Steeghs, K., Benders, A., Oerlemans, F., de Haan, A., Heerschap, A., Ruitenbeek, W., Jost, C., van Deursen, J., Perryman, B., Pette, D., *et al.*, 1997, Altered Ca^{2+} responses in muscles with combined mitochondrial and cytosolic creatine kinase deficiencies. *Cell* **89:** 93–103.

Steeghs, K., Oerlemans, F., and Wieringa, B., 1995, Mice deficient in ubiquitous mitochondrial creatine kinase are viable and fertile. *Biochim. Biophys. Acta* **1230:** 130–138.

Stöckler-Ipsiroglu, S., 1997, Creatine deficiency syndromes: a new perspective on metabolic disorders and a diagnostic challenge. *J. Pediatr.* **131:** 510–511.

Stockler, S., Schutz, P.W., and Salomons, G.S., 2007, Cerebral creatine deficiency syndromes: clinical aspects, treatment and pathophysiology. *Subcell. Biochem.* **46:** 149–166.

Streijger, F., Jost, C.R., Oerlemans, F., Ellenbroek, B.A., Cools, A.R., Wieringa, B., and Van der Zee, C.E., 2004, Mice lacking the UbCKmit isoform of creatine kinase reveal slower spatial learning acquisition, diminished exploration and habituation, and reduced acoustic startle reflex responses. *Mol. Cell. Biochem.* **256–257:** 305–318.

Streijger, F., Oerlemans, F., Ellenbroek, B.A., Jost, C.R., Wieringa, B., and Van der Zee, C.E., 2005, Structural and behavioural consequences of double deficiency for creatine kinases BCK and UbCKmit. *Behav. Brain Res.* **157:** 219–234.

Sullivan, P.G., Geiger, J.D., Mattson, M.P., and Scheff, S.W., 2000, Dietary supplement creatine protects against traumatic brain injury. *Ann. Neurol.* **48:** 723–729.

Sulser, H., 1968, Die Extraktstoffe des Fleisches. *Wissenschafts-Verlag Stuttgart.* **Handbuch II/2:** 1267–1304.

Tachikawa, M., Hosoya, K.-i., Ohtsuki, S., and Terasaki, T., 2007, A novel relationship between creatine transport at the blood-brain and blood-retinal barriers, creatine biosynthesis, and its use for brain and retinal energy homeostasis. *Subcell. Biochem.* **46:** 83–98.

Tarnopolsky, M.A., and Beal, M.F., 2001, Potential for creatine and other therapies targeting cellular energy dysfunction in neurological disorders. *Ann. Neurol.* **49:** 561–574.

Tarnopolsky, M.A., 2007, Clinical use of creatine in neuromuscular and neurometabolic disorders. *Subcell. Biochem.* **46:** 183–204.

Tombes, R.M., and Shapiro, B.M., 1985, Metabolite channeling: a phosphorylcreatine shuttle to mediate high energy phosphate transport between sperm mitochondrion and tail. *Cell* **41:** 325–334.

Turner, D.C., Wallimann, T., and Eppenberger, H.M., 1973, A protein that binds specifically to the M-line of skeletal muscle is identified as the muscle form of creatine kinase. *Proc. Natl. Acad. Sci. USA* **70:** 702–705.

van Deursen, J., Heerschap, A., Oerlemans, F., Ruitenbeek, W., Jap, P., ter Laak, H., and Wieringa, B., 1993, Skeletal muscles of mice deficient in muscle creatine kinase lack burst activity. *Cell* **74:** 621–631.

Vandenberghe, K., Goris, M., Van Hecke, P., Van Leemputte, M., Vangerven, L., and Hespel, P., 1997, Long-term creatine intake is beneficial to muscle performance during resistance training. *J. Appl. Physiol.* **83:** 2055–2063.

Ventura-Clapier, R., Kaasik, A., and Veksler, V., 2004, Structural and functional adaptations of striated muscles to CK deficiency. *Mol. Cell. Biochem.* **256–257:** 29–41.

Verbruggen, K.T., Sijens, P.E., Schulze, A., Lunsing, R.J., Jakobs, C., Salomons, G.S., and van Spronsen, F.J., 2007, Successful treatment of a guanidinoacetate methyltransferase deficient patient: Findings with relevance to treatment strategy and pathophysiology. *Mol. Genet. Metab.* **91:** 294–296.

Verhoeven, N.M., Salomons, G.S., and Jakobs, C., 2005, Laboratory diagnosis of defects of creatine biosynthesis and transport. *Clin. Chim. Acta* **361:** 1–9.

Wallimann, T., 1975, Creatine Kinase Isoenzymes and Myofibrillar Structure. PhD Thesis No. 5437, ETH Zurich, Switzerland.

Wallimann, T., 1983, Localization and Function of M-line-bound Creatine Kinase: Structural M-band Model and Creatine-Phosphate Shuttle. Habilitation, ETH Zurich, Switzerland.

Wallimann, T., 1996, 31P-NMR-measured creatine kinase reaction flux in muscle: a caveat! *J. Muscle Res. Cell Motil.* **17:** 177–181.

Wallimann, T., Dolder, M., Schlattner, U., Eder, M., Hornemann, T., O'Gorman, E., Ruck, A., and Brdiczka, D., 1998, Some new aspects of creatine kinase (CK): compartmentation, structure, function and regulation for cellular and mitochondrial bioenergetics and physiology. *Biofactors* **8:** 229–234.

Wallimann, T., and Hemmer, W., 1994, Creatine kinase in non-muscle tissues and cells. *Mol. Cell. Biochem.* **133–134:** 193–220.

Wallimann, T., Tokarska-Schlattner, M., Neumann, D., Epand, R.M., Epand, R.F., Andres, R.H., Widmer, H.R., Hornemann, T., Saks, V.A., Agarkova, I., and Schlattner, U., 2007, The phosphocreatine circuit: molecular and cellular physiology of creatine kinases, sensitivity to free radicals and enhancement by creatine supplementation. In: Molecular Systems Bioenergetics: Energy for Life, Basic Principles, Organization and Dynamics of Cellular Energetics, Saks, V.A., ed., Wiley-VCH, Weinheim, Germany.

Wallimann, T., Turner, D.C., and Eppenberger, H.M., 1977, Localization of creatine kinase isoenzymes in myofibrils. I. Chicken skeletal muscle. *J. Cell Biol.* **75:** 297–317.

Wallimann, T., Wyss, M., Brdiczka, D., Nicolay, K., and Eppenberger, H.M., 1992, Intracellular compartmentation, structure and function of creatine kinase isoenzymes in tissues with high and fluctuating energy demands: the 'phosphocreatine circuit' for cellular energy homeostasis. *Biochem. J.* **281:** 21–40.

Wegmann, G., Zanolla, E., Eppenberger, H.M., and Wallimann, T., 1992, In situ compartmentation of creatine kinase in intact sarcomeric muscle: the acto-myosin overlap zone as a molecular sieve. *J. Muscle Res. Cell Motil.* **13:** 420–435.

Wiseman, R.W., and Kushmerick, M.J., 1995, Creatine kinase equilibration follows solution thermodynamics in skeletal muscle. 31P NMR studies using creatine analogs. *J. Biol. Chem.* **270:** 12428–12438.

Wyss, M., and Kaddurah-Daouk, R., 2000, Creatine and creatinine metabolism. *Physiol. Rev.* **80:** 1107–1213.

Wyss, M., Smeitink, J., Wevers, R.A., and Wallimann, T., 1992, Mitochondrial creatine kinase: a key enzyme of aerobic energy metabolism. *Biochim. Biophys. Acta* **1102:** 119–166.

Wyss, M., Braissant, O., Pischel, I., Salomons, G.S., Schulze, A., Stockler, S., and Wallimann, T., 2007, Creatine and creatine kinase in health and disease – a bright future ahead? *Subcell. Biochem.* **46:** 309–334.

Zhu, S., Li, M., Figueroa, B.E., Liu, A., Stavrovskaya, I.G., Pasinelli, P., Beal, M.F., Brown, R.H., Jr., Kristal, B.S., Ferrante, R.J., and Friedlander, R.M., 2004, Prophylactic creatine administration mediates neuroprotection in cerebral ischemia in mice. *J. Neurosci.* **24:** 5909–5912.

CHAPTER 2

EARLY EVOLUTION OF THE CREATINE KINASE GENE FAMILY AND THE CAPACITY FOR CREATINE BIOSYNTHESIS AND MEMBRANE TRANSPORT

W. ROSS ELLINGTON[1] AND TOMOHIKO SUZUKI[2]

[1] *Institute of Molecular Biophysics and Department of Biological Science, Florida State University, Tallahassee, Florida 32306-4370, USA*

[2] *Laboratory of Biochemistry, Faculty of Science, Kochi University, Kochi 780-8520, Japan*

Abstract: The creatine kinase (CK)/phosphocreatine (PCr) energy buffering system is widespread in animal groups. Recent genomic sequencing and experimental results support the view that the capacity for creatine biosynthesis and membrane transport may have evolved quite early, perhaps coincident with CK. Conventional wisdom would suggest that CK evolved from an ancestral protein most similar to the CK homologue, arginine kinase. This early CK gene subsequently diverged into the cytoplasmic, mitochondrial and flagellar CK gene families. It is now clear that both the mitochondrial and cytoplasmic-flagellar genes were present prior to the divergence of sponges from the multi-cellular animal (metazoan) lineage, possibly as long as a billion years ago. Sponges constitute the most ancient, extant metazoan group. It is likely that the primary function of the CK-PCr system in these primitive animals was to mitigate reaction-diffusion constraints in highly polarized cells such as spermatozoa and choanocytes, the water current generating cells in sponges

1. CELLULAR REQUISITES FOR CK-BASED ATP, INORGANIC PHOSPHATE AND PROTON BUFFERING

Creatine kinase (CK) is a member of a highly conserved family of phosphoryl transfer enzymes called phosphagen (guanidino) kinases. Phosphagen kinases catalyze the reversible transfer of phosphate from a phosphorylated guanidine compound (phosphagen) to ADP yielding ATP and the guanidino substrate. In addition to CK, a broad range of other phosphagen kinases have been discovered and characterized including arginine, glycocyamine, hypotaurocyamine, lombricine, opheline, taurocyamine and thalassemine kinases (reviewed by Ellington, 2001).

G.S. Salomons and M. Wyss (eds.), Creatine and Creatine Kinase in Health and Disease, 17–26.

The CK reaction (phosphocreatine [PCr] + MgADP + H^+ ↔ creatine [Cr] + MgATP) fulfills a variety of physiological roles in cells displaying high and variable rates of ATP turnover including (a) temporal and spatial ATP buffering, (b) regulation of glycogenolysis through the sequestration and release of inorganic phosphate and (c) intracellular pH regulation via the liberation of inorganic phosphate, an effective proton buffer, upon net PCr hydrolysis (Ellington, 2001). Much of our understanding of these physiological roles as well as the nature of CK is based on studies of higher animals, mostly mammals and birds. However, the CK/PCr system is widespread in higher and lower invertebrate groups.

Creatine is isolated from the mainstream of amino acid metabolism and as such is a dead-end compound; its sole known metabolic fate is to be phosphorylated and dephosphorylated by the CK reaction (Walker, 1979). Creatine and phosphocreatine are subject to a spontaneous, non-enzymatic conversion to creatinine and phosphocreatinine which amounts to a significant drain on the cellular Cr/PCr pool (Walker, 1979). Clearly, a major requisite for a functional CK/PCr system is a source of Cr either from *de novo* biosynthesis and/or dietary uptake (Wyss and Kaddurah-Daouk, 2000). This former process is mediated by a biosynthetic pathway involving arginine:glycine amidinotransferase (AGAT) and guanidinoacetate methyltransferase (GAMT) (Wyss and Kaddurah-Daouk, 2000). If Cr is acquired from the diet, or its biosynthesis is compartmentalized within a given organism, for phosphorylation to take place Cr must be transported across plasma membranes into cells containing CK. A Na^+-and Cl^--dependent Cr transporter (CreaT) appears to facilitate this process (Wyss and Kaddurah-Daouk, 2000; Snow and Murphy, 2001). The last, essential requisite is the presence of CK. In higher organisms, the original CK gene has diverged into three CK isoform families, cytoplasmic, mitochondrial and flagellar (Cyt-, Mt- and FlgCKs), each targeted to different cellular micro-compartments (Wallimann *et al.*, 1992, Ellington, 2001).

This chapter will demonstrate that the bulk of the critical requisites for the functioning of the CK/PCr system – Cr biosynthetic capacity, CreaT and CK isoform diversity – were likely present early in the radiation of metazoans. Furthermore, we will argue that the driving force for the evolution of this system was related to problems of spatial mismatch of ATP supply and demand in highly polarized cells.

2. Cr/PCr AND Cr BIOSYNTHETIC CAPACITY ARE WIDESPREAD IN LOWER GROUPS

The presence of Cr and PCr in protochordates and invertebrates has long been recognized (Roche *et al.*, 1957; Robin, 1964; Van Pilsum *et al.*, 1972; Beis and Newsholme, 1975; reviewed in Ellington, 2001). While the capacity for Cr biosynthesis in vertebrates is well established (Walker, 1979; Wyss and Kaddurah-Daouk, 2000), until recently considerable doubt has existed as to whether lower groups such as protochordates and invertebrates possessed such a capacity. In an extensive survey of Cr levels and potential activities of AGAT and GAMT in higher and lower animals, Van Pilsum *et al.* (1972) were unable to detect significant

Cr biosynthetic enzyme activities in selected protochordates and invertebrates, even in groups that exhibited substantial tissue levels of Cr. These authors argued that these organisms likely maintained Cr/PCr pools from dietary sources and/or direct uptake from seawater in the case of marine species (Van Pilsum *et al.*, 1972). In support of the latter possibility, they subsequently demonstrated that the marine worm *Glycera dibranchiata*, a species that maintains high muscle Cr/PCr levels (Van Pilsum *et al.*, 1972), is capable of taking up Cr directly from seawater in a process that demonstrated transporter-like saturation kinetics (Van Pilsum *et al.*, 1975).

The potential for Cr biosynthesis in lower groups was recently re-examined by DeLigio and Ellington (2006). These authors argued that dietary uptake as a generic source of Cr is unlikely due to spectacular diversity of feeding strategies in these groups – strict herbivores, omnivores, raptors, detritus/deposit feeders and filter feeders (DeLigio and Ellington, 2006). Furthermore, they also discounted the potential for direct uptake from seawater given the low and variable natural levels of Cr and the fact that Cr would be subjected to the same non-enzymatic conversion to creatinine as takes place in living systems.

DeLigio and Ellington (2006) found that the protochordate tunicate (sea squirt) *Ciona intestinalis* expresses GAMT in four different tissues. When the cDNA for this GAMT was expressed in *E. coli*, the resulting recombinant GAMT displayed similar specific activities and apparent K_m for guanidinoacetate as typical mammalian GAMTs. Furthermore, quantitative PCR demonstrated that this GAMT is expressed primarily in the stomach and gonad, close to the site of packaging of Cr into the primitive-type spermatozoa of this organism (DeLigio and Ellington, 2006). In addition to the empirical studies, DeLigio and Ellington (2006) also conducted extensive mining of genomic and EST sequences available in GenBank and other on-line data repositories. GAMT sequences were found in the protochordate lancelet *Branchiostoma*, the sea urchin *Strongylocentrotus* and in the primitive colonial hydroid *Hydractinia*. Furthermore, AGAT sequences were found in *Ciona*, *Branchiostoma* and *Strongylocentrotus* (DeLigio and Ellington, 2006).

The available data support the view that the capacity for *de novo* Cr biosynthesis is likely widespread in the higher and lower invertebrate groups that maintain significant Cr/PCr pools and express CK. Validation of this assertion will come from the numerous genome and EST sequencing projects that are currently underway.

3. CreaT IS PRESENT IN PROTOCHORDATES AND LIKELY IN INVERTEBRATE GROUPS

CreaT couples the movement of Na^+ and Cl^- down their concentration gradients with the movement of Cr across the plasma membrane. CreaT is a member of a family of solute transporters (see chapter 6; Christie, 2007; Wyss and Kaddurah-Daouk, 2000). The annotated genomic database for the tunicate *Ciona* shows that this protochordate does have the gene that codes for CreaT, as shown in Figure 1. Although the solute specificity of the *Ciona* transporter has not been experimentally

```
            1                                                50
ci01001489  MNPQPSTGTR-CPSCCASNPCGDLPCNCNMVSAS--GQRETWDKKLDFIL
   Torpedo     MPSRAVRRCPGHLCKEMRAPRRAQPPDVPAGEPGSRVTWSRQMDFIM
            51                                              100
ci01001489  SCVGFAVGLGNVWRFPYLCYKNGGGAFLIPYLLFIVTSGVPVFFLETSLG
   Torpedo  SCVGFAVGLGNVWRFPYLCYKNGGGVFLIPYLLVAVFGGIPIFFLEISLG
            101                                             150
ci01001489  QFMKQGGIGIWNICPLMKGIGFASTVIVFFCNCYYILVLTWAIYYLYRSF
   Torpedo  QFMKAGGINAWNIAPLFKGLGYASMVIVFFCNTYYILVLTWSSFYLVQSF
            151                                             200
ci01001489  TAVLPWATCGNAWNTNGSTNTSLLNSTSSFNFTS-LSSPVIEFWEREVLK
   Torpedo  SSPLPWASCNNTWNTAACYEAGANASTEIYPPTAPAQSSIVQFWERRVLR
            201                                             250
ci01001489  ITSDITETGSIRWELAICLFIAWVLCYVCICKGVRATGKIVYFTALFPYV
   Torpedo  LSSGLGDVGEIGWELTLCLTATWMLVYFCIWKGVKTSGKVVYVTATFPYI
            251                                             300
ci01001489  VLLCLLIRGSMLPGALKGVEFYIKPNWTKLQEAQVWIDAGTQVFFSYAIG
   Torpedo  ILVILLVRGVTLHGAVQGIVYYLQPDWGKLGEAQVWIDAGTQIFFSYAIG
            301                                             350
ci01001489  LGALSALGSYNLYNNDCFKDCLILATVNSATSFFAGFVIFTFLGFMAHEQ
   Torpedo  LGTLTALGSYNQLHNDCYKDAFILSLVNSATSFFAGLVVFSILGFMAVEE
            351                                             400
ci01001489  RVPIERVAESGPGLAFIAYPKGVTLMPLSPLWSCLFFLMILMLGLDSQFV
   Torpedo  GVDISVVAESGPGLAFIAYPKAVTLMPFPQVWAVLFFIMLLCLGLGSQFV
            401                                             450
ci01001489  GVEGFLTAIIDMYPRVL--RPKRAIFAAVTCSVCFLVALSMITEGGMYVF
   Torpedo  GVEGFVTAILDLWPSKFSFRYLREVVVAMVICLSFLIDLSMITEGGMYIF
            451                                             500
ci01001489  QIFDYYSASGMTLLWMSLWECVTIAWIYGAERYYQDLKDMIGYRPPSLFK
   Torpedo  QIFDYYSASGTTLLWTAFWECVAVAWVYGGDRYLDDLAWMLGYRPWALVK
            501                                             550
ci01001489  YCWLFVTPAVALGILIFVLVSHKPLTYNRTYEYPWWGLLIGWVLALSSML
   Torpedo  WCWSVITPLVCMGIFTFHLVNYKPLTYNKTYTYPWWGEAIGWCLALASML
            551                                             600
ci01001489  CVPIVAAYRLAKAKGSTFLQRFSNAIRPQFKSHH-SEKYWDGKLMQNIFI
   Torpedo  CVPTTVLYSLSRGRGS-LKERWRKLTTPVWASHHLAYKMAGAKINQPCEG
            601           615
ci01001489  TE
   Torpedo  VVSCEEKVVIFESVL
```

Figure 1. Multalin (http://prodes.toulouse.inra.fr/multalin/) sequence alignment of a deduced amino acid sequence from a genomic sequence contig (ci01001489; http://genome.jgi-psf.org/ciona4/ciona4.home.html) of a CreaT from *Ciona* with the CreaT from the electric ray *Torpedo* (Guimbal and Kilimann, 1994).

validated, our BLAST searches consistently have shown that this protein is most similar to vertebrate CreaTs (Ellington, unpublished observations).

tBLASTn searches of the sea urchin *Strongylocentrotus* genomic database (http://www.ncbi.nlm.nih.gov/genome/guide/sea_urchin/) using the *Ciona* CreaT deduced amino acid sequence as a query have revealed a number of promising CreaT gene candidates (Ellington, unpublished observations). The previously mentioned observation that the marine worm *Glycera* has carrier-mediated capacity for Cr uptake (Van Pilsum *et al.*, 1975) supports the view that CreaT is also present in this protostome invertebrate. It should be noted that the apparent K_m for Cr uptake in *Glycera* was 37 μM (Van Pilsum *et al.*, 1975), which falls in the middle of the range of corresponding K_m values for vertebrate CreaTs (Wyss and Kaddurah-Daouk, 2000). The available evidence is highly suggestive that CreaTs are widely distributed in lower animal groups.

4. CK ISOFORM DIVERGENCE OCCURRED AT THE DAWN OF THE RADIATION OF THE METAZOA

Creatine kinases actually constitute a superfamily of three isoform gene families each targeted to different intracellular compartments – cytoplasmic (CytCK), mitochondrial (MtCK) and flagellar (FlgCK) CKs. The former two gene families code for proteins that exist as oligomers – CytCKs are dimers while typical MtCKs exist as both dimers and octamers, with the latter predominating under physiological conditions (Wallimann *et al.*, 1992; Wyss *et al.*, 1992). FlgCKs are "contiguous trimers" consisting of three fused, complete CK domains in a single polypeptide chain and most likely resulted from a gene duplication/fusion event followed by unequal crossing over (Wothe *et al.*, 1990). CytCKs, octameric MtCKs and FlgCKs evolved before the divergence of the protostome (most invertebrates such as arthropods, mollusks, worms) and deuterostome (echinoderms, protochordates, vertebrates) lineages, at least 650 million years ago (Ellington *et al.*, 1998; Suzuki *et al.*, 2004).

Sponges (phylum *Porifera*) are the most ancient, extant group of multi-cellular animals, having diverged from the metazoan lineage as much as a billion years ago (West and Powers, 1993). Early (Roche *et al.*, 1957; Robin and Guillou, 1980) and more recent (Ellington, 2000) studies showed that these organisms contain Cr and CK activity. Sona *et al.* (2004) cloned and expressed in *E. coli* the cDNAs for three CKs from the sponge *Tethya aurantia*. One of these CKs, denoted CK2, corresponded to a true MtCK and contained a mitochondrial targeting sequence. Interestingly, when CK2 was expressed, it was found to be an obligate dimer (Sona *et al.*, 2004). The other two CKs, denoted CK1 and CK3, displayed great sequence similarities to the domains of the FlgCKs from the marine worm *Chaetopterus*, the sea urchin *Strongylocentrotus* and the tunicate *Ciona*; these CKs were found to be dimeric in quaternary structure (Sona *et al.*, 2004).

A phylogenetic analysis of these *Tethya* CK sequences with other CKs is shown in Figure 2. Three distinct clusters are evident – mitochondrial, flagellar and cytoplasmic CKs. *Tethya* MtCK (CK2), not surprisingly, is at the base of the MtCK lineage, which consists of octameric MtCKs from protostomes and deuterostomes (Figure 2). Given its dimeric quaternary structure and basal position in the MtCK lineage, it is clear that the sponge MtCK is an immediate ancestor of MtCKs in higher groups. The CytCKs form a distinct lineage with the protostome CytCKs forming a distinct sub-cluster (Figure 2). The protochordate tunicate *Ciona* is at the base of the sub-cluster containing the muscle-type (M) and brain-type (B) CytCKs found in the vertebrates (Figure 2).

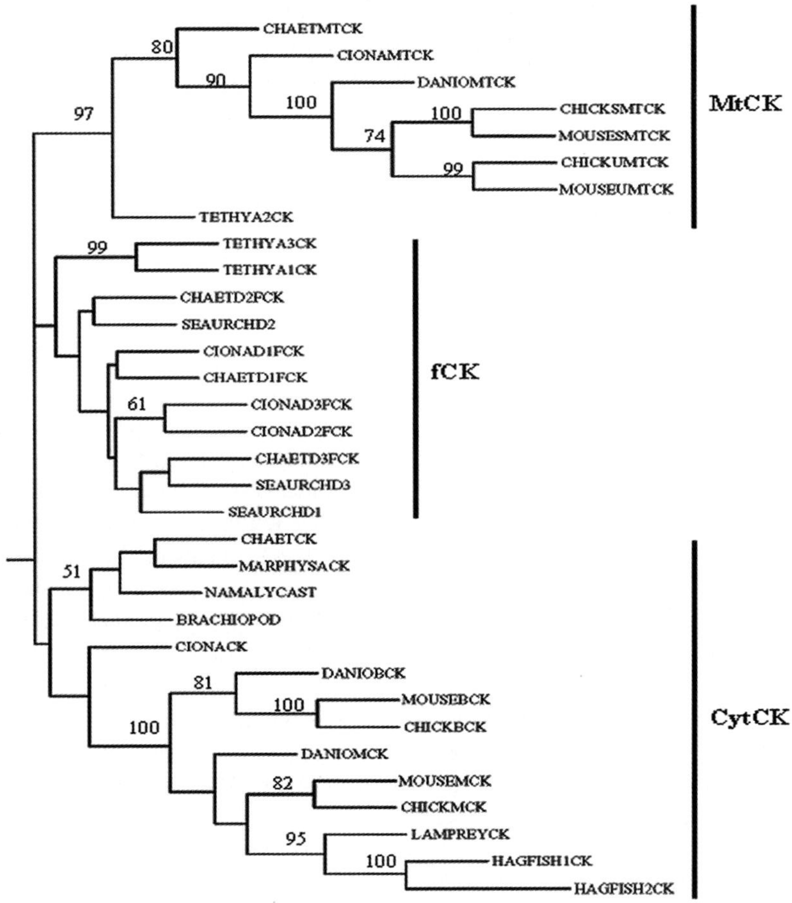

Figure 2. Neighbor Joining (NJ) tree showing the position of the sponge *Tethya* CKs in relation to the Mt-, Flg- and CytCK clusters [vertical lines] (from Sona *et al.*, 2004). FlgCKs are denoted fCK.

The two dimeric CK1 and CK3 from *Tethya* do not cluster with the CytCKs but rather are weakly associated with the FlgCK cluster (Figure 2). Previous phylogenetic analyses have shown that the FlgCKs are much more closely related to the CytCKs than MtCKs (Suzuki *et al.*, 2004). Given the sequence similarities between *Tethya* CK1 and CK3 and the domains of the FlgCKs as well as the phylogenetic results (Figure 2), these CKs have been referred to as protoflagellar CKs (protoflgCKs; Sona *et al.*, 2004). It is possible that these protoflgCKs in sponges constitute the immediate ancestors of the FlgCKs and are basal to the CytCKs, which evolved later.

CKs are thought to have evolved from an ancestral protein most similar to arginine kinase (AK) (Watts, 1971). AKs are considered basal because these proteins are typically functional monomers, utilize an unmodified amino acid substrate, arginine, and are widely distributed in the invertebrates and protozoa (Ellington, 2001; Uda *et al.*, 2006). CKs likely emerged as a result of an AK gene duplication event followed by divergence leading ultimately to the acquisition of creatine specificity and dimeric quaternary structure. Subsequent gene duplication events led to the evolution of the CK isoform system.

5. THE EARLY PHYSIOLOGICAL ROLE OF THE CK/PCr SYSTEM

The MtCK and protoflg/FlgCK lineages were present prior to the divergence of the most ancient, living group of multi-cellular animals – the sponges. CK isoform genes are clearly ancient. This observation sets the stage for a number of fundamental questions: (a) What were the driving forces that led to CK isoform diversity? And (b) what is (are) the physiological role(s) of the CK/PCr system in sponges and their immediate ancestors?

To answer the above questions, it is useful to point out an interesting set of puzzling patterns with respect to the distribution of phosphagens and their corresponding phosphagen kinases. Many protochordates and invertebrates display what has been termed the "pluriphosphagen" phenomenon (Robin, 1964; Ellington, 1989). That is, within a given organism, more than one phosphagen and its corresponding phosphagen kinase may be present. This is particularly true of marine worms where as many as three different phosphagen systems may be present in a single organism. However, a very consistent pattern of cellular distribution with respect to CK/PCr is seen. In marine worms, echinoderms and protochordates, regardless of which phosphagen system is present in somatic cells and eggs, the spermatozoa in these groups, without exception, express CK and accumulate significant pools of PCr and Cr (Ellington, 1989, 2001).

The CK/PCr system plays a critical role in sperm motility due to the extreme polarization of ATP source (mitochondria in the sperm mid-piece) and ATP sink (dynein ATPases in the flagellum), creating reaction-diffusion constraints. This role in spatial ATP buffering in spermatozoa has been validated experimentally (Tombes *et al.* , 1985; Van Dorsten *et al.*, 1995). Primitive-type sperm contain true Mt- and FlgCKs (Tombes and Shapiro, 1989); FlgCKs are localized in the flagellar membrane via a myristate anchor (Quest *et al.*, 1992). Thus, CK activities are localized at ATP source and sink providing the elements for a spatial ATP buffering and energy transport pathway as is shown in the right portion of Figure 3.

Sponges and other basal metazoans typically shed gametes into the water column for external fertilization. The primitive type sperm of these organisms are highly motile. Certainly, one likely site where Mt- and protoflgCKs could be expressed in sponges is in sperm. While the sponge protoflgCKs do not appear to undergo myristoylation, these proteins are most likely cytoplasmic and would be distributed throughout the cell, including the flagellum. Thus, these cells would have the CK elements necessary for a fully functional spatial ATP buffering system.

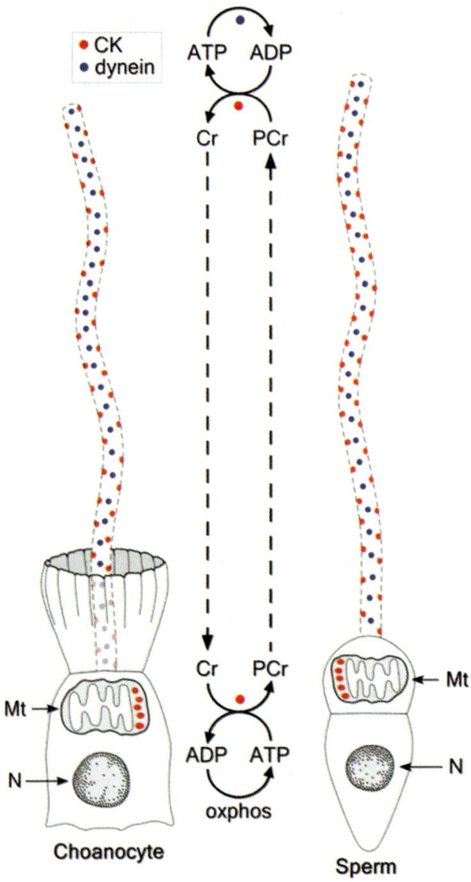

Figure 3. Spatial ATP buffering by the CK/PCr system in spermatozoa and sponge choanocytes. Abbreviations: Mt (mitochondrion); N (nucleus); oxphos (oxidative phosphorylation).

A second cell type in sponges may also be a likely site for CK isoform expression. Sponges have unique and highly specialized, flagellated cells called choanocytes (Figure 3), which generate the water currents in these organisms that provide nutrients and oxygen and remove wastes. Choanocytes form water-filled chambers within sponges and generate flow velocities as high as 22 cm/sec (Reiswig, 1975) and replace their entire body volume in water in 5 sec (Vogel, 1977). Flow generation is continuous. Mt- and protoflgCKs could potentially mediate spatial ATP buffering (see Figure 3). Choanocytes and sperm are developmentally related as it has been shown that sponge spermatozoa develop from choanocytes during gametogenesis (Kay and Reiswig, 1991).

The presence of these CK isoforms in sponge spermatozoa and choanocytes remains to be experimentally validated. However, it seems highly likely that CK isoform systems evolved in flagellated cells of early metazoans or their immediate protozoan ancestors as a means of mitigating reaction-diffusion constraints with respect to energy homeostasis. We have argued elsewhere that spatial ATP buffering was the early, primitive functional role of the CK reaction (Ellington, 2001). An extension of this argument is that the temporal ATP buffering role was acquired later with the advent of high power output neuromuscular systems.

6. SUMMARY

In this chapter we have shown that the CK/PCr system likely evolved in the earliest animals to facilitate energy transport in flagellated cells. The system of CK isoforms evolved very early with the Mt- and protoflg/FlgCKs being present in the most ancient, extant group of multi-cellular animals, the sponges. The selective pressure for the origin and divergence of these isoforms is related to the spatial ATP buffering role of the reaction in these cells. Available evidence would suggest that the enzymes of Cr biosynthesis, AGAT and GAMT, evolved very early, most likely coincident with the origin of CK. Furthermore, the capacity for membrane transport of Cr, as mediated by CreaT, may be present in invertebrate groups that maintain significant pools of Cr/PCr. Thus, the acquisition of CK, the CK isoform system and the capacity for Cr biosynthesis and transport can be viewed as a unique series of co-evolutionary events. The ultimate output of these events is manifested in the CK/PCr system and its critical physiological roles in higher organisms including man.

ACKNOWLEDGEMENTS

Some of the work described in this essay was supported by National Science Foundation grants IOB-0130024 and IOB-0542236 to WRE and Grants-in-Aid for Scientific Research in Japan 17570062 to TS. Figure 2 was reprinted from Biochem. Biophys. Res. Comm., Vol. 317, S. Sona, T. Suzuki and W.R. Ellington, Cloning and expression of mitochondrial and protoflagellar creatine kinases from a marine sponge: implications for the origin of intracellular energy transport systems, pp. 1207–1214, 2004, with permission from Elsevier.

REFERENCES

Beis, I., and Newsholme, E.A., 1975, The contents of adenine nucleotides, phosphagens and some glycolytic intermediates in resting muscles from vertebrates and invertebrates. *Biochem. J.* **152:** 23–32.

Christie, D.L., 2007, Functional insights into the creatine transporter. *Subcell. Biochem.* **46:** 99–118.

DeLigio, T., and Ellington, W.R, 2006, The capacity for the de novo biosynthesis of creatine is present in the tunicate *Ciona intestinalis* and is likely widespread in other protochordate and invertebrate groups. *Comp. Biochem. Physiol. D* **1:** 167–178.

Ellington, W.R., 1989, Phosphocreatine represents a thermodynamic and functional improvement over other muscle phosphagens. *J. Exp. Biol.* **143:** 177–194.

Ellington, W.R., 2000, A dimeric creatine kinase from a sponge: implications in terms of phosphagen kinase evolution. *Comp. Biochem. Physiol. B* **126:** 1–7.

Ellington, W.R., 2001, Evolution and physiological roles of phosphagen systems. *Annu. Rev. Physiol.* **63:** 289–325.

Ellington, W.R., Roux, K., and Pineda, A.O., 1998, Origin of octameric creatine kinases. *FEBS Lett.* **425:** 75–78.

Guimbal, C., and Kilimann, M.W., 1994, A creatine transporter cDNA from *Torpedo* illustrates structure-function relationships in the GABA/noradenaline transporter family. *J. Mol. Biol.* **241:** 317–324.

Kaye, H.R., and Reiswig, H.M., 1991, Sexual reproduction in four Caribbean commercial sponges. I. Reproductive cycles and spermatogenesis. *Invert. Reprod. Develop.* **19:** 1–12.

Quest, A.F.G., Chadwick, J.K., Wothe, D.D., McIlhinney, R.A.J., and Shapiro, B.M., 1992, Myristoylation of flagellar creatine kinase in the sperm phosphocreatine shuttle is linked to its membrane association properties. *J. Biol. Chem.* **267:** 15080–15085.

Reiswig, H.M., 1975, The aquiferous system of three demospongiae. *J. Morph.* **145:** 493–502.

Robin, Y., 1964, Biological distribution of guanidines and phosphagens in marine annelida and related phyla from California, with a note on pluriphosphagens. *Comp. Biochem. Physiol.* **12:** 347–367.

Robin, Y., and Guillou, Y., 1980, Quelque aspects du metabolisme energetique des eponges.*Comptes Rendus* **148:** 121–126.

Roche, J., van Thoai, N., and Robin, Y., 1957, Sur la presence de creatine chez les Invertebres et sa signification biologique. *Biochim. Biophys. Acta* **24:** 514–519.

Snow, R.J., and Murphy, R.M., 2001, Creatine and the creatine transporter. *Mol. Cell. Biochem.* **224:** 169–181.

Sona, S., Suzuki, T., and Ellington, W.R., 2004, Cloning and expression of mitochondrial and protoflagellar creatine kinases from a marine sponge: implications for the origin of intracellular energy transport systems. *Biochem. Biophys. Res. Comm.* **317:** 1207–1214.

Suzuki, T., Mizuta, C., Uda, K., Ishida, K., Mizuta, K., Sona, S., Compaan, D.M., and Ellington, W.R., 2004, Evolution and divergence of the genes for cytoplasmic, mitochondrial and flagellar creatine kinases. *J. Mol. Evol.* **59:** 218–226.

Tombes, R.M., Brokow, C.J., and Shapiro, B.M., 1985, Creatine kinase dependent energy transport in sea urchin spermatozoa. *Biophys. J.* **52:** 75–86.

Tombes, R.M., and Shapiro, B.M., 1989, Energy transport and cell polarity: relationship of phosphagen kinase activity to sperm function. *J. Exp. Zool.* **251:** 82–90.

Uda, K., Fujimoto, N., Akiyama, Y., Mizuta, K., Tanaka, K., Ellington, W.R., and Suzuki, T., 2006, Evolution of the arginine kinase gene family. *Comp. Biochem. Physiol. D* **1:** 209–218.

Van Dorsten, F.A., Wyss, M., Wallimann, T., and Nicolay, K., 1997, Activation of sea urchin motility is accompanied by an increase in creatine kinase exchange flux. *Biochem. J.* **325:** 411–416.

Van Pilsum, J.F., Stephens, G.C., and Taylor, D., 1972, Distribution of creatine, guanidinoacetate and the enzymes for their biosynthesis in the animal kingdom. *Biochem. J.* **126:** 325–345.

Van Pilsum, J.F., Taylor, DF., and Bans, L., 1975, Studies of the uptake of creatine from sea water by the marine annelid, *Glycera dibranchiata. Comp. Biochem. Physiol.* **51:** 611–617.

Vogel, S., 1977, Current-induced flow through living sponges in nature. *Proc. Natl. Acad. Sci. USA* **74:** 2069–2071.

Walker, J.B., 1979, Creatine biosynthesis, regulation and function. *Adv. Enzymol.* **50:** 177–242.

Wallimann, T., Wyss, M., Brdiczka, D., Nicolay, K., and Eppenberger, H.M., 1992, Intracellular compartmentation, structure and function of creatine kinase isoenzymes: the "phosphocreatine circuit" for cellular energy homeostasis. *Biochem. J.* **281:** 21–40.

Watts, D.C., 1971, Evolution of phosphagen kinases. In: E. Schoffeniels (ed.), Biochemical Evolution and the Origin of Life, North Holland, Amsterdam, pp. 150–173.

West, L., and Powers, D., 1993, Molecular phylogenetic position of the hexactinellid sponges in relation to the *Protista* and *Demospongiae. Mol. Mar. Biol. Biotechnol.* **2:** 71–75.

Wothe, D.D., Charbonneau, H., and Shapiro, B.M., 1990, The phosphorcreatine shuttle of sea urchin sperm: flagellar creatine kinase resulted from a gene triplication. *Proc. Natl. Acad. Sci. USA* **87:** 5203–5207.

Wyss, M., Smeitink, J., Wevers, R.A., and Wallimann, T., 1992, Mitochondrial creatine kinase: a key enzyme of aerobic energy metabolism. *Biochim. Biophys. Acta* **1102:** 119–166.

Wyss, M., and Kaddurah-Daouk, R., 2000, Creatine and creatinine metabolism. *Physiol. Rev.* **80:** 1107–1213.

CHAPTER 3

THE CREATINE KINASE PHOSPHOTRANSFER NETWORK: THERMODYNAMIC AND KINETIC CONSIDERATIONS, THE IMPACT OF THE MITOCHONDRIAL OUTER MEMBRANE AND MODELLING APPROACHES

VALDUR SAKS[1,2], TUULI KAAMBRE[2], RITA GUZUN[1], TIIA ANMANN[2], PEETER SIKK[2], UWE SCHLATTNER[1,3], THEO WALLIMANN[3], MAYIS ALIEV[4] AND MARKO VENDELIN[5]

[1]*Laboratory of Fundamental and Applied Bioenergetics, INSERM U 884, Joseph Fourier University, 2280, Rue de la Piscine, BP53X – 38041, Grenoble Cedex 9, France*

[2]*Laboratory of Bioenergetics, National Institute of Chemical Physics and Biophysics, Akadeemia tee 23, 12618 Tallinn, Estonia*

[3]*Institute of Cell Biology, ETH-Zurich, Hönggerberg HPM D24, Schafmattstrasse 18, CH-8093 Zurich, Switzerland*

[4]*Cardiology Research Center, Institute of Experimental Cardiology, Laboratory of Cardiac Pathology, 121552 Moscow, Russia*

[5]*Department of Mechanics and Applied Mathematics, Institute of Cybernetics, Tallinn Technical University, Akadeemia tee 21, 12618 Tallinn, Estonia*

Abstract: In this review, we summarize the main structural and functional data on the role of the phosphocreatine (PCr) – creatine kinase (CK) pathway for compartmentalized energy transfer in cardiac cells. Mitochondrial creatine kinase, MtCK, fixed by cardiolipin molecules in the vicinity of the adenine nucleotide translocator, is a key enzyme in this pathway. Direct transfer of ATP and ADP between these proteins has been revealed both in experimental studies on the kinetics of the regulation of mitochondrial respiration and by mathematical modelling as a main mechanism of functional coupling of PCr production to oxidative phosphorylation. In cells *in vivo* or in permeabilized cells *in situ*, this coupling is reinforced by limited permeability of the outer membrane of the mitochondria for adenine nucleotides due to the contacts with cytoskeletal proteins. Due to these mechanisms, at least 80% of total energy is exported from mitochondria by PCr molecules. Mathematical modelling of intracellular diffusion and energy transfer shows that the main function of the PCr CK pathway is to connect different pools (compartments) of ATP and, by this way, to overcome the local restrictions and diffusion limitation of adenine nucleotides due to the high degree of structural organization of cardiac cells

G.S. Salomons and M. Wyss (eds.), Creatine and Creatine Kinase in Health and Disease, 27–65.

1. GENERAL DESCRIPTION OF THE CREATINE KINASE SYSTEM IN MUSCLE CELLS

High work requirements and the unique contractile function of the heart have made cardiac muscle the most intensively studied object in bioenergetics. Among other items, the role of the creatine kinase (CK) system has been thoroughly studied in cardiac cells during several decades in many laboratories. In this review, we summarize the data regarding the role of the CK system in heart, with the aim to emphasize the most important evidences of functioning of this system. Its role is equally important both in skeletal and smooth muscle, in brain, and in many other types of cells, as described elsewhere (Wallimann and Hemmer, 1994; Ames, 2000).

Energy metabolism of cardiac cells is based on aerobic oxidation of fatty acids and carbohydrate substrates, coupled to ATP production by mitochondrial oxidative phosphorylation (Neely and Morgan, 1974; Williamson *et al.*, 1979; Opie, 1998; Stanley *et al.*, 2005; Taegtmeyer *et al.*, 2005). Then, the energy is supplied to the sites of its utilization via structurally and functionally organized phosphotransfer networks (Wallimann *et al.*, 1992; Dzeja and Terzic, 2003; Saks *et al.*, 1994, 2006a). The cardiac cell is structurally and functionally highly ordered, with very regular crystal-like arrangement of mitochondria and a unitary (modular) nature of energy metabolism (Saks *et al.*, 2001; Seppet *et al.*, 2001; Vendelin *et al.*, 2005; Weiss *et al.*, 2006), organized into intracellular energetic units, ICEUs (Figure 1). All CK isoenzymes are compartmentalized in the cells, within these ICEUs. Mitochondrial isoforms of CK were discovered in Klingenberg's laboratory in 1964 (Klingenberg, 1964), shortly after the discovery of the cytosolic MM- and BB-CK isoforms (see Eppenberger *et al.*, 1967). In muscle cells, significant fractions of the MM isozyme are connected structurally to the myofibrils, to the membrane of sarcoplasmic reticulum (SR) and to the sarcolemma (Wallimann *et al.*, 1992, 1998; Dzeja *et al.*, 1998; Wyss and Kaddurah-Daouk, 2000; Dzeja and Terzic, 2003; Bessman and Geiger, 1981; Scholte, 1973; Saks *et al.*, 1974, 1977, 2006b; Sharov *et al.*, 1977; Wallimann and Hemmer, 1994; Schlegel *et al.*, 1988a; Haas and Strauss, 1990; Fritz-Wolf *et al.*, 1996; Qin *et al.*, 1998; Schlattner *et al.*, 2000; Burklen *et al.*, 2006). This intracellular compartmentation of the CK isozymes forms the structural basis for an energy transfer network – the so-called phosphocreatine (PCr) circuit. Below, we first describe the structural evidence for the CK pathway and then the mechanisms of functioning for each of the CK isoenzymes.

2. STRUCTURAL STUDIES

The molecular structure and structure-function relationships of creatine kinase isoenzymes have been studied in much detail by some of us and other groups (Fritz-Wolf *et al.*, 1996; McLeish and Kenyon, 2005; Schlattner *et al.*, 1998; Schlegel *et al.*, 1988b; Schnyder *et al.*, 1994). In particular, the MtCK isoenzymes and their proteolipid complexes have been analyzed by various approaches. This

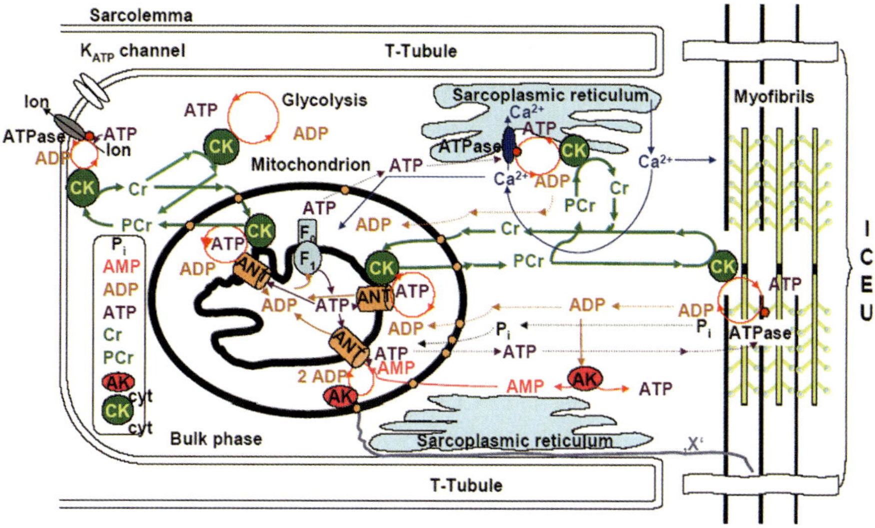

Figure 1. The coupled CK and adenylate kinase (AK) reactions within the intracellular energetic units (ICEUs) in muscle cells. By interaction with cytoskeletal elements, the mitochondria and sarcoplasmic reticulum are precisely fixed with respect to the structure of a sarcomere of myofibrils, i.e. between two Z-lines and, correspondingly, between two T-tubules. Calcium is released from the sarcoplasmic reticulum into the ICEU space in the vicinity of mitochondria and sarcomeres to activate contraction and mitochondrial dehydrogenases. Adenine nucleotides within the ICEU do not equilibrate rapidly with adenine nucleotides in the bulk water phase. ANT is mitochondrial adenine nucleotide translocase, and F_0F_1 is the ATP synthase complex. The mitochondria, sarcoplasmic reticulum, MgATPase of myofibrils as well as ATP-sensitive systems in the sarcolemma are interconnected by metabolic channeling of reaction intermediates and by high-energy phosphate transfer within the ICEU by the CK-PCr system. Protein factors (still unknown and marked as "X"), most probably connected to the cytoskeleton, fix the position of mitochondria and probably also control the permeability of the voltage-dependent anion carrier (VDAC) channels for ADP and ATP. Adenine nucleotides within the ICEU and the bulk water phase may be connected by some more rapidly diffusing metabolites such as Cr and PCr. Synchronization of the functioning of ICEUs within the cell may occur by the same metabolites (for example, P_i or PCr) and/or by synchronized release of calcium during the excitation-contraction coupling process. This scheme is an artwork of Christian Linke, a student of the Erasmus programme.

work has been summarized in several recent reviews (McLeish and Kenyon, 2005; Schlattner *et al.*, 1998, 2006; Schlattner and Wallimann, 2004, 2006; Wallimann *et al.*, 1992). Only a brief summary is given here, and the reader is referred to the cited reviews for a more detailed description and original references.

The solution of atomic structures of all four CK isoenzymes during the last decade, most of them with participation of some of us (Eder *et al.*, 1999, 2000; Fritz-Wolf *et al.*, 1996; Rao *et al.*, 1998), as well as of the transition state structure of a related arginine kinase (Zhou *et al.*, 1998) were an enormous contribution to the field. It allowed for the first time to put earlier data in a structure-function framework and to design new experiments that specifically target CK functions, for example, the specific association of CK isoenzymes with partner proteins involved in ATP-delivering or -consuming processes.

Mitochondrial creatine kinase differs from cytosolic CK not only by its organellar localization, but also by two important other properties (Figure 2). First, while the cytosolic CK isozymes are strictly dimeric, MtCK forms not only dimers but is able to reversibly associate into highly ordered, cuboidal octamers, the predominant form *in vitro* as well as *in vivo*. Second, octameric MtCK behaves as a peripheral membrane protein, interacting with different protein and lipid components of the mitochondrial membranes.

The MtCK octamer is stabilized *in vivo* by high protein concentrations in the mitochondrial intermembrane space and its binding to mitochondrial membranes. Kinetic and thermodynamic properties of octamer/dimer transitions were suggested as regulatory parameters for energy transduction in mitochondrial intermembrane contact sites. However, octamer/dimer transitions are too slow for fast metabolic adaptations, although they might play a role in long-term adaptations or modulation of energy metabolism. In contrast, octamer/dimer transitions may be crucial in pathological situations (Stachowiak *et al.*, 1998; Schlattner *et al.*, 2006).

Octameric MtCK is characterized by specific membrane-binding properties that are absent or less developed in dimeric MtCK. Also cytosolic CK isoenzymes are mainly soluble, and only a fraction is bound to particular subcellular structures like myofibrils, sarcoplasmic reticulum or plasma membrane (Wallimann *et al.*, 1992). By contrast, octameric MtCK behaves as a typical peripheral membrane protein (Schlattner *et al.*, 2006). It occurs in cristae and the peripheral intermembrane space of mitochondria (Figure 3) and shows a strong affinity for acidic phospholipids, in particular cardiolipin (diphosphatidyl glycerol) in the outer leaflet of the inner membrane, and to mitochondrial porin (VDAC, voltage dependent anion channel) in the outer membrane. Thus, by its size and its binding properties, MtCK can bridge the intermembrane space. There is also good evidence for a functional interaction of MtCK with the transmembrane protein adenine nucleotide translocator (ANT) in the inner mitochondrial membrane. However, the association between these two proteins seems to be indirect and mediated by cardiolipin. Both MtCK and ANT show a high affinity for cardiolipin ($K_D < 100\,nM$), which may lead to close co-localization in cardiolipin membrane patches. It is well known that cardiolipin-protein interactions are important for subunit assembly and complex formation of mitochondrial inner membrane proteins. Two recent studies are in full support of the MtCK-cardiolipin-ANT model. In collaboration with Epand's group, we could show that octameric MtCK can indeed induce cardiolipin clusters in model membranes (Epand *et al.*, 2007), and that the resulting proteolipid complexes can even play a role in transmembrane lipid transfer (Epand *et al.*, 2006).

The proteolipid complexes containing octameric MtCK, situated in the peripheral intermembrane space and the cristae (Figure 3), are perfectly suited to sustain the channeling of substrates and products of the reaction, thus avoiding dissipation

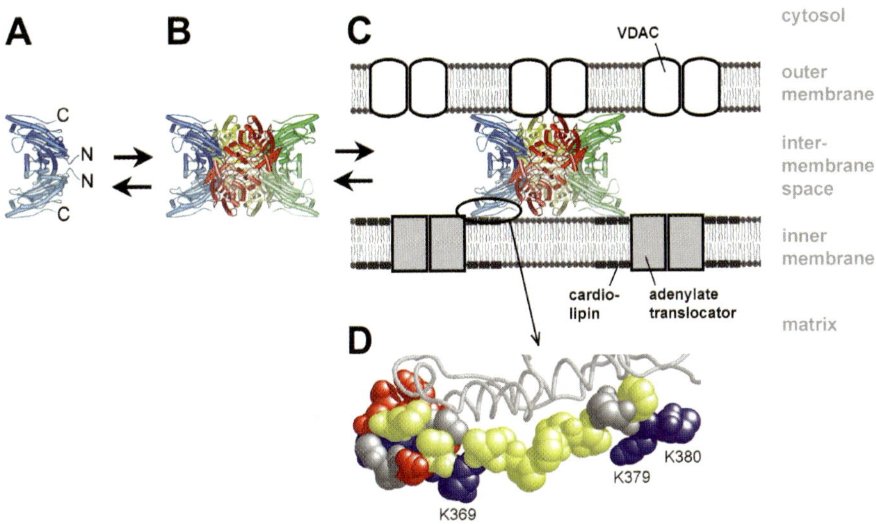

Figure 2. Proteolipid-complex formation of MtCK in the mitochondrial intermembrane space. (A) Nascent MtCK is imported into the mitochondrial intermembrane space, the mitochondrial targeting sequence is cleaved off, and folding is accompanied by assembly into dimers (blue banana-shaped molecule). (B) MtCK dimers rapidly associate into cube-like octamers with four-fold symmetry; this is a reversible association, but octamers are strongly favored at the pH values and MtCK concentrations present at this location. (C) Octameric MtCK binds to both mitochondrial membranes and assembles into proteolipid complexes containing the voltage-dependent anion carrier (VDAC = porin) and the adenine nucleotide transporter (ANT). Principal high-affinity receptors for MtCK are cardiolipin in the inner membrane (black rectangles in the vicinity of ANT) and VDAC together with other anionic phospholipids in the outer membrane. (D) Enlarged C-terminal membrane- (cardiolipin-) binding domain of MtCK, showing the involved basic lysine residues (indicated as K369, K379 and K380 of the C-terminus of MtCK) that bind to acidic phospholipids (cardiolipin). Interaction of MtCK with ANT is most likely indirect and involves common cardiolipin patches (see dark rectangles in the vicinity of ANT in C). Due to the symmetrical nature of the MtCK octamer, MtCK mediates contacts between the inner and outer membrane through four C-terminal proteolipid complexes with the outer membrane and four with the inner membrane, one of them depicted by a circle in C. The membrane-bound state of octameric MtCK is strongly favored by the large membrane interaction surface and the high affinity of MtCK for cardiolipin ($K_D = 80\,nM$) and VDAC ($K_D < 100\,nM$). Binding of dimers to phospholipid membranes occurs with much lower affinity and probably involves association into octamers while bound to the membrane. (A)-(C) show the X-ray structure of chicken sarcomeric MtCK in cartoon representation; each dimer is represented in a different color; (D) shows human sMtCK in space-filling representation with the following amino acid color code: red – acidic, blue – basic, yellow – hydrophobic. Figure modified from Schlattner *et al.*, 2004, 2006; Schlattner and Wallimann, 2006 with kind permission from J. Biol. Chem. and Nova Publishers.

of ATP into the bulk solution and driving the reaction towards phosphocreatine generation (for details see Figure 3). In contrast to octamers, dimeric MtCK has only low affinity to membranes and is much less efficient for cardiolipin clustering and channeling of energy-rich metabolites in mitochondrial contact sites (Epand *et al.*, 2007; Khuchua *et al.*, 1998).

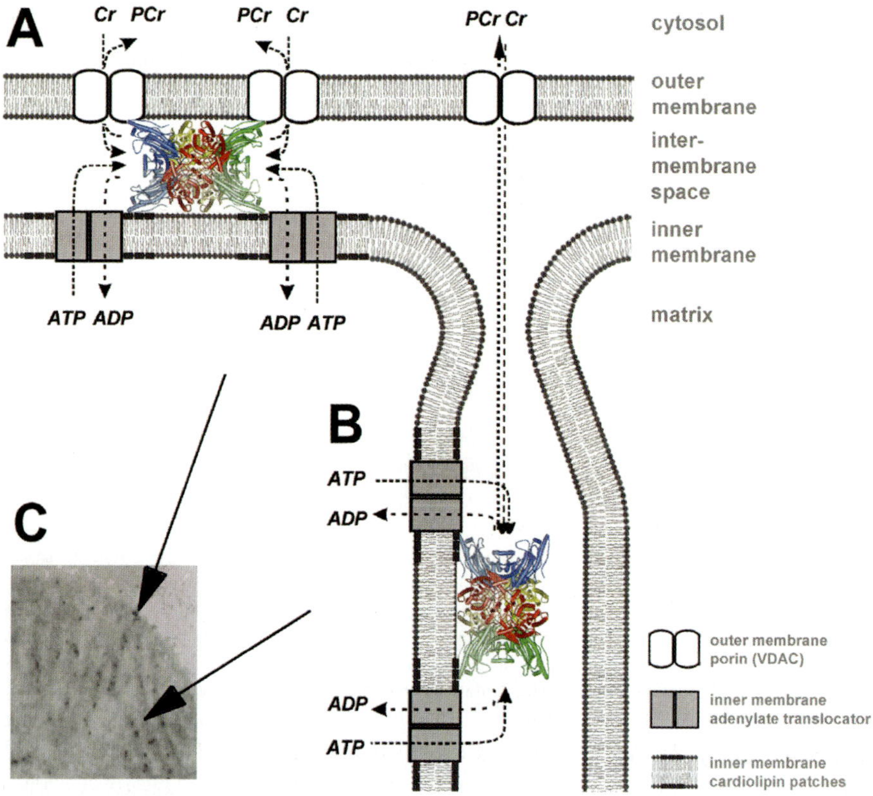

Figure 3. MtCK microcompartments in mitochondria: dual localization of MtCK and putative metabolite routes. MtCK is found at two mitochondrial locations: (A) the peripheral intermembrane space, mainly in so-called mitochondrial contact sites in association with adenine nucleotide transporter (ANT) and voltage-dependent anion carrier (VDAC), as well as (B) along the cristae space in association with ANT only. (C) Visualization by post-embedding immuno-gold labeling of MtCK in a mitochondrion from a photoreceptor cell of chicken retina. The proteolipid complexes assembled by MtCK (see Figure 2) create microcompartments that allow for a direct exchange (called metabolite channeling or functional coupling) of MtCK substrates and products, which is depicted by arrows: (A) In contact sites, octameric MtCK binds simultaneously to inner (IM) and outer mitochondrial membranes (OM), due to the identical top and bottom faces of the octamer. Binding partner in the inner membrane is the two-fold negatively charged cardiolipin, which allows a functional interaction with the ANT that is situated in cardiolipin membrane patches. In the outer membrane, MtCK interacts with other acidic phospholipids and, in a calcium-dependent manner, directly with the pore-forming VDAC. In contact sites, metabolite channeling allows for a constant supply of substrates to and removal of products from the active sites of MtCK. By preventing the dissipation of mitochondrially generated ATP into the cytosol, its energy content is directly transferred by MtCK to creatine to yield phosphocreatine, a metabolically inert compound. Phosphocreatine is then diffusing via VDAC into the cytosol, where it is available to CKs at various locations for *in situ* regeneration of ATP. *Vice versa*, intramitochondrial regeneration of ADP stimulates oxidative phosphorylation. (B) In cristae, ATP/ADP exchange occurs as in contact sites, while creatine and phosphocreatine have to diffuse along the cristae space to reach VDAC. Figure modified from Schlattner and Wallimann, 2004, 2006; Schlattner *et al.*, 2006 with kind permission from J. Biol. Chem. and Nova Publishers.

3. FUNCTIONAL STUDIES

3.1. Mitochondrial Creatine Kinases

In 1939, Belitzer and Tsybakova showed that in muscle homogenates, oxygen consumption was stimulated by creatine (Cr) and resulted in PCr production with a ratio of PCr/O of approximately 3 (Belitzer and Tsybakova, 1939). This result was the earliest indication for a functional coupling between MtCK and oxidative phosphorylation (see below). It is interesting to note that in their experiments on well washed skeletal muscle homogenates, Belitzer and Tsybakova observed strong stimulation of respiration by Cr without addition of exogenous adenine nucleotides. Much later, the same effect was described by Kim and Lee (1987) for isolated pig heart mitochondria and by Dolder *et al.* (2003) for liver mitochondria from transgenic mice expressing active MtCK. The results of these latter studies could be explained by very efficient use of endogenous adenine nucleotides by the coupled MtCK reaction. Bessman and Fonyo (1966) as well as Vial *et al.* (1972) showed in studies on isolated heart muscle mitochondria that in the presence of ATP, addition of Cr significantly increased the state 4 respiration rate. In 1973, Jacobus and Lehninger studied the kinetics of the stimulatory effect of Cr on the state 4 respiration rate and found that at its physiological concentration of 10–15 mM, Cr stimulated the respiration maximally to the state 3 level (Jacobus and Lehninger, 1973). The details of this effective mechanism of PCr production coupled to oxidative phosphorylation in isolated rat heart mitochondria were studied by applying kinetic analysis with some elements of mathematical modelling (Saks *et al.*, 1975; Jacobus and Saks, 1982). The results showed that oxidative phosphorylation controls PCr production in heart mitochondria. When uncoupled from oxidative phosphorylation (if the latter is not activated, for example), the MtCK reaction does not differ kinetically and thermodynamically from other CK isoenzymes: the reaction always favours ATP production and – according to the Haldane relationship – ADP and PCr binding is more effective due to higher affinities than that of ATP and Cr, respectively (Saks *et al.*, 1974, 1975). When the calculated predicted rates of the reaction were compared with the experimental ones, good agreement was seen in the absence of oxidative phosphorylation but not when the latter was activated: under conditions of oxidative phosphorylation, the MtCK reaction was strongly shifted towards PCr synthesis (Saks *et al.*, 1975). This suggested that ATP produced by mitochondrial oxidative phosphorylation is a much more effective substrate for MtCK than exogenous MgATP in the medium, and it was proposed that this is due to direct ANT-mediated transfer of ATP from the matrix space to MtCK, which should be located in close proximity to ANT to make such direct channeling possible (Saks *et al.*, 1975; Jacobus and Saks, 1982).

3.1.1. Kinetic evidence for the functional coupling between MtCK and ANT by direct transfer of ATP

A simple kinetic protocol was developed for demonstration of the mechanism of functional coupling between MtCK and ANT, which is shown in Figure 4 for isolated mitochondria. In these experiments, isolated mitochondria were incubated in the presence of 10 mM Cr, and the rate of PCr production in the MtCK reaction was measured by either one of two approaches: (1) spectrophotometrically by a coupled enzyme assay (involving phosphoenolpyruvate, pyruvate kinase and lactate dehydrogenase; PEP-PK-LDH) while oxidative phosphorylation was inhibited by rotenone (10 μM) and oligomycin (1 μM); or (ii) by measuring the rate of Cr-dependent oxygen consumption, followed by calculation of the rates of PCr production by using experimentally determined P/O ratios (Figure 4A). Figure 4B shows that in the presence of oxidative phosphorylation, the rate of the MtCK reaction in direction of PCr and ADP synthesis is significantly higher than in its absence, due to significantly decreased apparent K_m values for ATP in the CK reaction (Figure 4C) (Saks *et al.*, 1975; Jacobus and Saks, 1982).

To describe better the mechanism of this phenomenon, a complete kinetic analysis of the CK reaction in isolated rat heart mitochondria under both conditions – i.e. with and without oxidative phosphorylation (Jacobus and Saks, 1982) – was found to be very useful and informative. Figure 5 shows the principles and value of this approach. Figure 5A shows the complete kinetic scheme of the CK reaction, assuming a Bi-Bi quasi-equilibrium random-type reaction mechanism, according to Cleland's classification (Cleland, 1963). The rate equation of this reaction is:

$$V = \frac{Vm.[Cr].[MgATP]}{Kia.Kb + Ka.[Cr] + Kb.[MgATP] + [Cr].[MgATP]}$$

To determine the values of the dissociation constants for all substrates, the initial rates in both directions (in the absence of products) were determined for six different concentrations of both substrates (Figure 5B and D). The data were analyzed by performing, successively, primary (Figure 5C) and secondary analyses (Figure 5E)

Figure 4. Comparison of normalised CK kinetics in isolated cardiac mitochondria with and without oxidative phosphorylation (± ox.phos.), in the presence of 10 mM Cr. MIM, MOM and ANT are mitochondrial inner membrane, mitochondrial outer membrane, and adenine nucleotide transporter, respectively. (A) Scheme of experiments to measure the kinetics of the mitochondrial creatine kinase (MtCK) reaction in isolated cardiac mitochondria. On the left: studies of MtCK kinetics with oxidative phosphorylation; reaction rates were calculated from oxygen consumption rates and P/O ratios. On the right: studies of MtCK kinetics in the absence of oxidative phosphorylation; reaction rates were measured with a coupled enzyme assay (PEP-PK-LDH) after inhibition of the respiratory chain (RC) and ATP synthase by rotenone (10 μM) and oligomycin (1 μM), respectively. (B) MtCK kinetics with oxidative phosphorylation (+ox.phos.) were measured by respirometry in mitomed solution (final concentrations: 110 mM sucrose, 60 mM K-lactobionate, 0.5 mM dithiothreitol, 0.5 mM EGTA, 3 mM $MgCl_2$, 20 mM taurine, 3 mM KH_2PO_4, 20 mM K-HEPES, 5 mM glutamate, 2 mM malate, 2 mg/ml essentially fatty acid free bovine serum albumin and pH = 7.1), followed by normalization of the data (for $V_m = 304 \pm 12$ nmol O_2·mg protein^{-1}·min^{-1}). MtCK kinetics in the absence of oxidative phosphorylation (-ox.phos.) were measured in the same medium by spectrophotometry with the system described on the right-hand side of Figure 4A. (C) Normalised MtCK kinetics in double-reciprocal plots for illustrating the decrease in apparent K_m for ATP with oxidative phosphorylation.

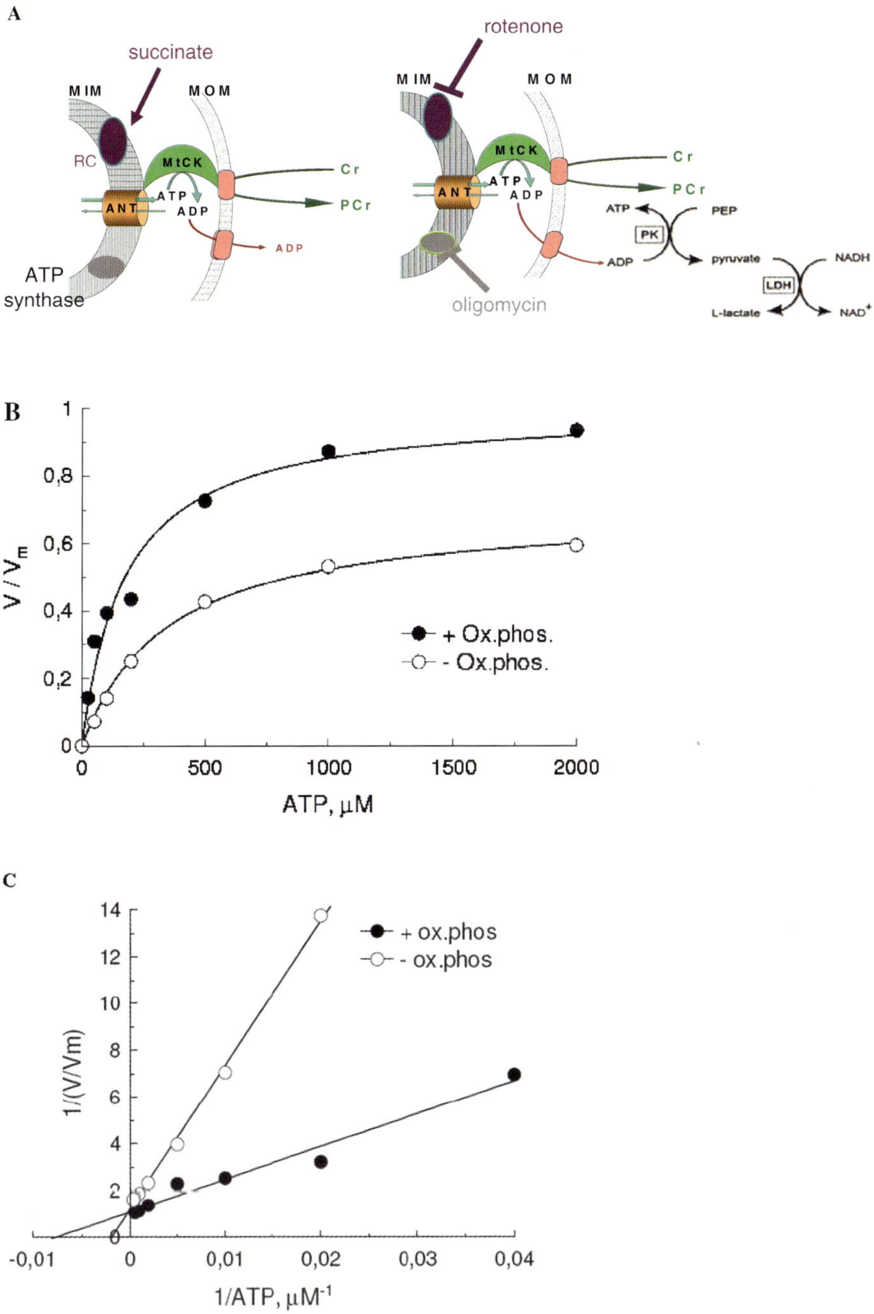
A
succinate
MIM
MOM
RC
MtCK
ANT
ATP
ADP
Cr
PCr
ADP
ATP
synthase
rotenone
MIM
MOM
MtCK
ANT
ATP
ADP
Cr
PCr
ATP
PEP
PK
ADP
pyruvate
NADH
LDH
L-lactate
NAD+
oligomycin
B
V / Vm
ATP, µM
+ Ox.phos.
- Ox.phos.
C
1/(V/Vm)
1/ATP, µM-1
+ ox.phos
- ox.phos

by a simple linearization method in double reciprocal plots, according to the following equations.

$$1/\mathrm{v} = \frac{1}{[\mathrm{Cr}]}\left(\frac{\mathrm{Kb}}{\mathrm{Vm}}\left(\frac{\mathrm{Kia}}{[\mathrm{MgATP}]}+1\right)\right)+\frac{1}{\mathrm{Vm}}\left(\frac{\mathrm{Ka}}{[\mathrm{MgATP}]}+1\right)$$

This family of straight lines could be used for a secondary analysis of ordinate intercepts, i_o:

$$i_o = \frac{1}{\mathrm{Vm}}\left(1+\frac{\mathrm{Ka}}{[\mathrm{MgATP}]}\right)$$

or slopes:

$$\mathrm{slope} = \frac{\mathrm{Kb}}{\mathrm{Vm}}\left(\frac{\mathrm{Kia}}{[\mathrm{MgATP}]}+1\right)$$

to find directly the values of dissociation constants K_a and K_{ia} for MgATP, respectively, as shown in Figure 5F. Similar analysis of 1/v versus 1/[MgATP] and secondary analysis allowed to determine the dissociation constants for Cr, K_{ib} and K_b (Figure 5G). This linearization method is very illustrative, since secondary analysis directly shows the reciprocal values of the dissociation constants for both substrates, ATP (Figure 5F) and Cr (Figure 5G). While the kinetic constants for guanidino substrates – Cr and PCr – were the same in both conditions, oxidative phosphorylation had a specific effect on the kinetic parameters for adenine nucleotides; in particular, the K_a value was decreased by an order of magnitude. The Haldane relationship for the CK reaction was no longer valid, showing the involvement of some other processes – oxidative phosphorylation and ANT (Jacobus and Saks, 1982). The most likely explanation is the direct transfer of ATP from ANT to MtCK due to their spatial proximity which results also in increased uptake of ADP from MtCK (reversed direct transfer); as a result, the turnover of adenine nucleotides is increased manifold at low external concentrations of MgATP, thereby maintaining high rates of oxidative phosphorylation and coupled PCr production in the presence of high Cr concentrations.

The conclusions of the privileged access of mitochondrial ATP to MtCK and increased mitochondrial turnover of adenine nucleotides in the presence of Cr were directly confirmed by Barbour *et al.* (1984) with the use of an isotopic method, and with a thermodynamic approach by De Furia *et al.* (1980), Saks *et al.* (1985) and Soboll *et al.* (1994). Finally, an effective competitive enzyme method for studying the functional coupling phenomenon, namely the pathway of ADP movement from MtCK back to mitochondria, was developed by Gellerich *et al.* (1982, 1987, 1994, 1998, 2002). These authors used an external PEP-PK system to trap ADP and thus to compete with ANT for this substrate. This competitive enzyme system was never able to suppress more than 50% of Cr-stimulated respiration in isolated heart mitochondria, thereby showing the rather

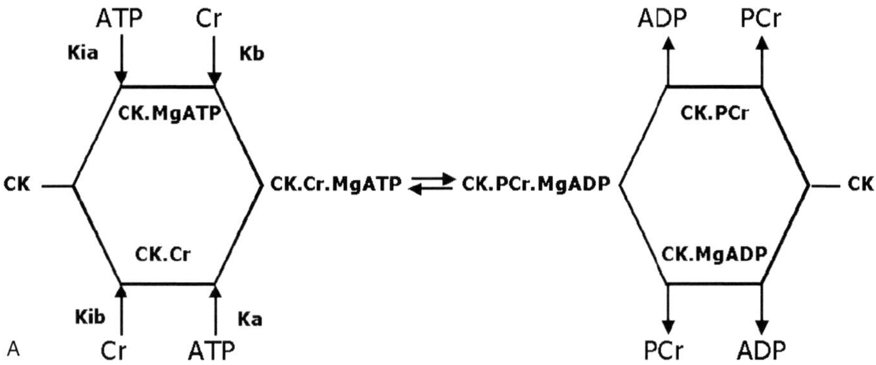

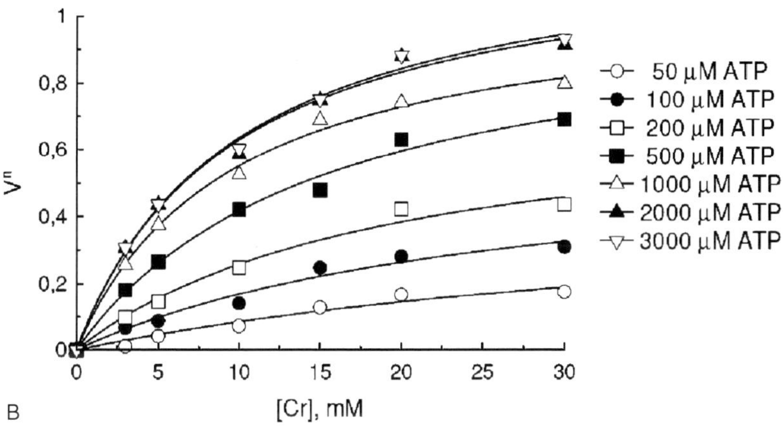

Figure 5. CK kinetics without oxidative phosphorylation in isolated cardiac mitochondria. (A) The complete kinetic scheme of the CK reaction. (B) Initial rates of the MtCK reaction without oxidative phosphorylation as a function of Cr concentration with different fixed ATP concentrations. (C) Secondary analysis of CK kinetics for fixed ATP concentrations, for finding values of i_o and slopes for calculation of dissociation constants K_b and K_{ib}. (D) Initial rates of the MtCK reaction without oxidative phosphorylation as a function of ATP concentration, when Cr concentrations are fixed. (E) Secondary analysis of CK kinetics for fixed Cr concentrations, for finding values of i_o and slopes for calculation of dissociation constants K_{ia} and K_a. (F) Determination of kinetic constants for ATP – K_a and K_{ia} – by secondary analysis of data obtained both in the absence (Figure 5E) and presence (not shown) of oxidative phosphorylation. (G) Determination of kinetic constants for Cr – K_b and K_{ib} – by secondary analysis of data obtained both in the absence (Figure 5C) and presence (not shown) of oxidative phosphorylation.

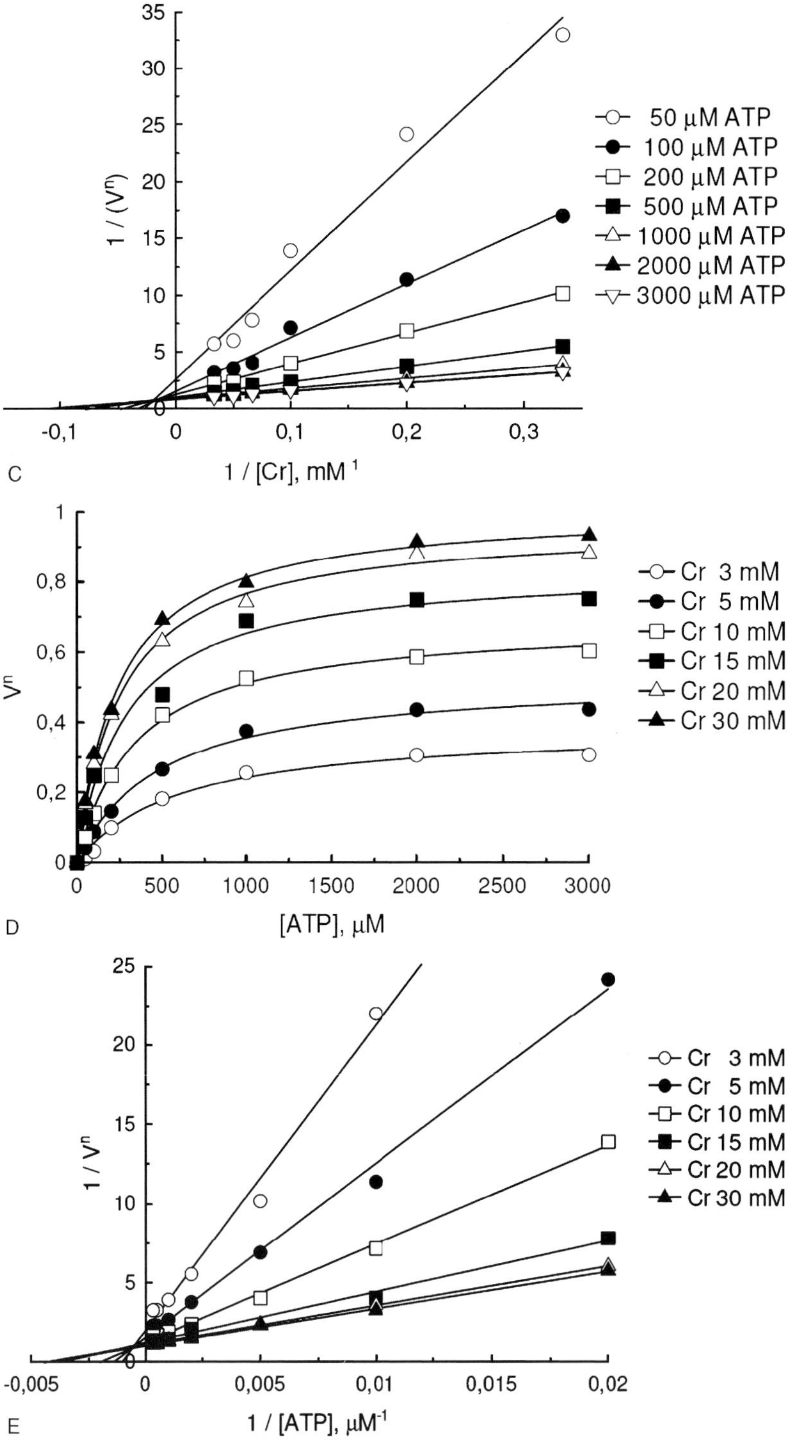

Figure 5. (continued)

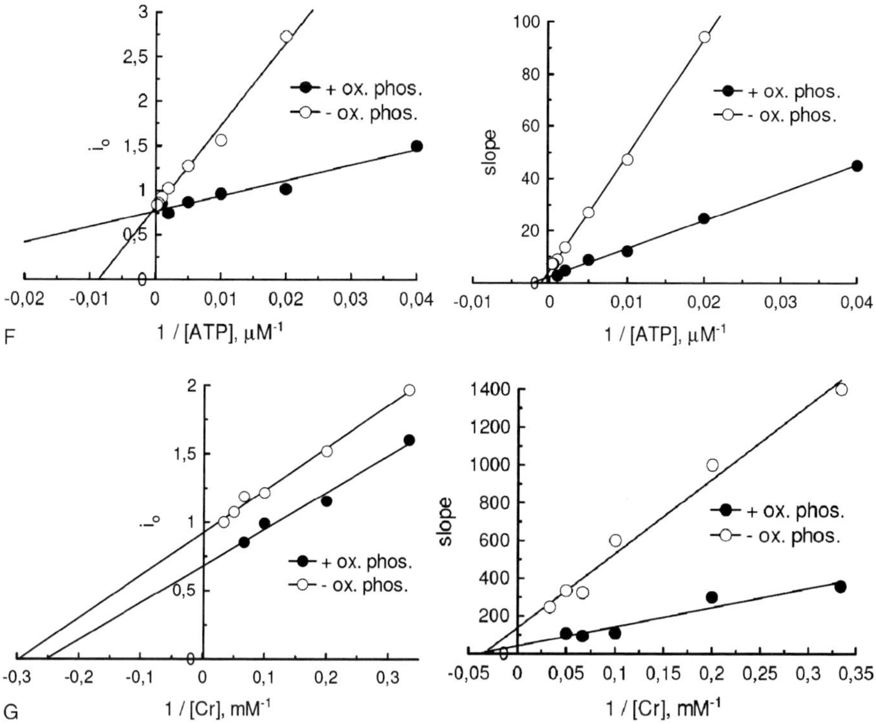

Figure 5. (continued)

effective channelling of ADP from MtCK to the ANT (Gellerich *et al.*, 1982). Gellerich's group has explained these latter data by the hypothesis of dynamic compartmentation of adenine nucleotides in the intermembrane space, meaning that there is some control of the permeability of the outer mitochondrial membrane, which would result in ADP and ATP concentration gradients across this membrane (Gellerich *et al.*, 1987, 1994, 1998, 2002). This represents an alternative hypothetical mechanism of coupling between MtCK and ANT without a need for direct transfer of the substrates between ANT and MtCK. Interestingly, this hypothesis focused attention on the role of the mitochondrial outer membrane in the control of mitochondrial function, and foresaw many important aspects of the control of mitochondrial function *in vivo*, but appeared to be insufficient to explain quantitatively the functional coupling between MtCK and ANT.

3.1.2. Mathematical modelling of the functional coupling between MtCK and ANT

Mathematical modelling of the coupling of the MtCK reaction with ANT was used to test both hypotheses – dynamic compartmentation described above and direct transfer of ATP from ANT to MtCK (Figure 6) – to explain the shifts in kinetic constants of the MtCK reaction caused by oxidative phosphorylation, as shown in Figure 5F and G (Jacobus and Saks, 1982).

The simplest kinetic scheme that was proposed to explain the kinetic data on MtCK-ANT coupling is based on dynamic compartmentation of adenine nucleotides as referred to above (Gellerich *et al.*, 1987). According to this hypothesis, there are metabolite gradients between the solution surrounding isolated mitochondria and the vicinity of the mitochondrial inner membrane in the intermembrane space. Due to the close vicinity of MtCK and ANT at and within the inner mitochondrial membrane, respectively, ATP and ADP are exchanged between MtCK and ANT. The kinetic properties of MtCK and ANT are not changed during interaction, and all the changes in apparent kinetic constants measured in the experiment are due to inability of experimentalists to measure local ATP and ADP concentrations next to these proteins. Since the gradients of ATP and ADP between the surrounding solution and the vicinity of the mitochondrial inner membrane are dependent on the activity of the mitochondria (i.e. are changing), this hypothesis is usually referred to as "dynamic compartmentation" hypothesis.

To check whether the dynamic compartmentation hypothesis is in correspondence with the data, one could compose a simple mathematical model and compare the model solutions with the experimental data (Vendelin *et al.*, 2004a). To find out whether a set of parameter values exists that can reproduce both dissociation constants for ATP, altered by oxidative phosphorylation at the same time, a number of parameters were varied across the entire allowable range: (a) ATP and ADP exchange coefficients across the outer mitochondrial membrane, (b) ATPase activity, and (c) ANT activity. Regardless of the selected parameter set, the model could not reproduce quantitatively the experimental data obtained in the presence of oxidative phosphorylation (Figure 7A). Thus, it is clear that the dynamic compartmentation hypothesis is not sufficient to reproduce the measurements, and a more complex mechanism of interaction between MtCK and ANT should be used to reproduce the data (Vendelin *et al.*, 2004a). Therefore, the alternative direct local transfer of ATP – generated by oxidative phosphorylation – to MtCK should be considered as a main mechanism of acceleration of aerobic PCr production in heart, skeletal muscle and brain mitochondria.

In mitochondria, ANT transports ADP from the mitochondrial intermembrane space into the matrix space. The stoichiometry of imported ADP to exported ATP is normally one. In mitochondria, the number of MtCK dimers is approximately equal to the number of ANT tetramers (Kuznetsov and Saks, 1986). Irrespective of the exact stoichiometry of ANT/MtCK complexes, the juxtaposition of these molecular entities creates a structural basis for direct transfer (channelling) of metabolites from one entity to the other without dissociation of metabolites into the intermembrane space of mitochondria. ANT in the inner mitochondrial membrane forms tight complexes with negatively charged cardiolipin in a ratio of 1 to 6 (Beyer and Klingenberg, 1985; Beyer and Nuscher, 1996). It has been shown that positively charged MtCK is fixed to this cluster by electrostatic forces due to three C-terminal lysine residues which strongly interact with the negatively

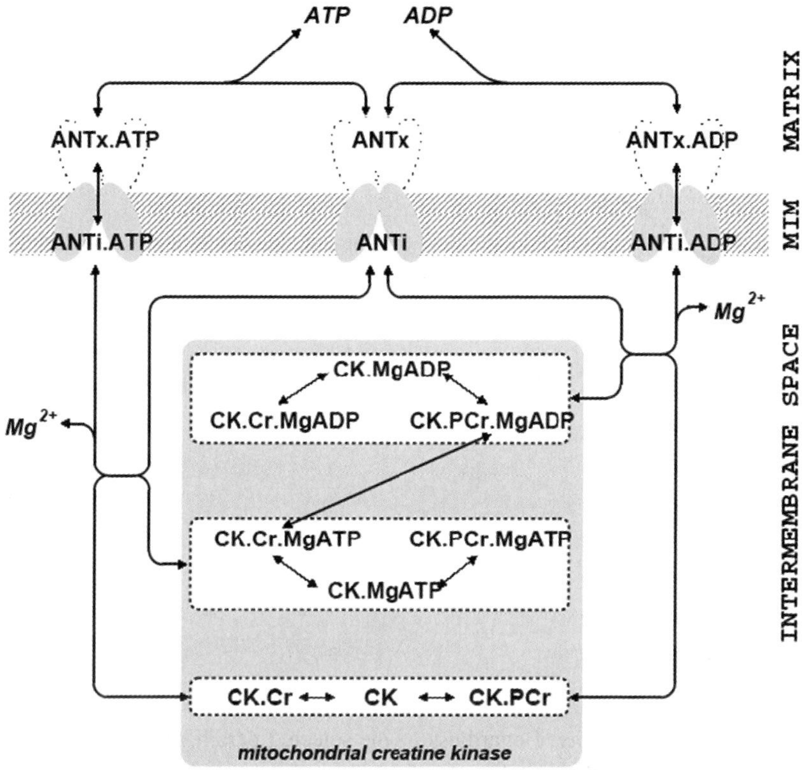

Figure 6. Scheme of interactions between mitochondrial CK (MtCK) and adenine nucleotide translocase (ANT) according to the direct transfer hypothesis. Interaction between the proteins is considered as a sum of two interaction modes: ATP and ADP are transferred through solution or directly channelled between the proteins. In this scheme, only the direct transfer of ATP and ADP between proteins is reflected. In direct transfer mode, ATP is transferred from ANT to MtCK without leaving the two-protein complex into solution. Since MtCK has only one binding site for ATP and ADP, such transfer is possible only if this site is free, i.e. MtCK is either free ("CK") or has only Cr or PCr bound ("CK.Cr" and "CK.PCr"). In the scheme, we grouped the states of MtCK according to whether ATP or ADP is bound to enzyme or not (white boxes in the scheme with three states of MtCK in each group). During direct transfer of ATP from ANT to MtCK, MtCK is transferred from states CK, CK.Cr, and CK.PCr to states CK.MgATP, CK.Cr.MgATP, and CK.PCr.MgATP, respectively. In the scheme, this transfer is shown as a link between ANTi.ATP and two corresponding groups of MtCK states. Next, after the MtCK reaction (link between states CK.Cr.MgATP and CK.PCr.MgADP in the scheme), ADP is transferred directly to ANT. Note that MtCK operates with Mg-bound ATP and ADP and ANT requires Mg-free ATP and ADP forms. Thus, during direct transfer between MtCK and ANT, Mg is either bound or released, as it is shown in the scheme. This complete scheme was the basis for thermodynamically consistent analysis of coupling (see the text). This figure is reproduced from Vendelin *et al.* (2004a) with permission from the Biophysical Society, USA.

A

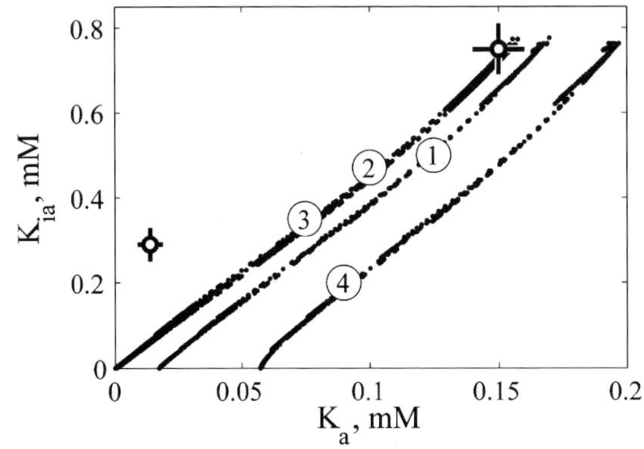

B

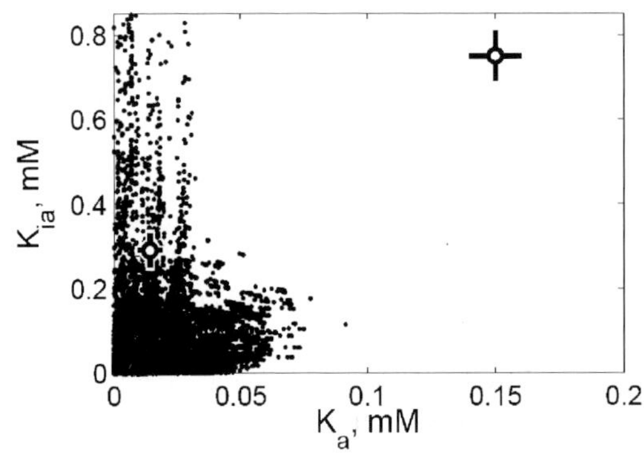

Figure 7. Analyses of the interaction between mitochondrial CK (MtCK) and adenine nucleotide translocase (ANT). (A) Calculated apparent dissociation constants K_a and K_{ia} of the MtCK reaction in the presence of oxidative phosphorylation, with coupling between MtCK and oxidative phosphorylation being modelled according to the dynamic compartmentation hypothesis. The dynamic compartmentation hypothesis assumes that 'interaction' between the proteins is the result of large ATP and ADP gradients between the vicinity of the mitochondrial inner membrane and the solution surrounding the mitochondria. Apparent dissociation constants K_a and K_{ia} (represented by small dots in the figure) were computed for different combinations of ATPase activities (v_{ATPase}) and exchange constants (D_{ATP} and D_{ADP}). The measured values are shown by open circles in the upper right corner (no oxidative phosphorylation) and lower left corner (with oxidative phosphorylation). When using the default kinetic constants of ANT transport, all combinations of computed K_a and K_{ia} values are aligned along line 1. By increasing the maximal activity of ANT 10- or 100-fold, the line can be shifted to the left (lines 2 and 3, respectively). When, instead of increasing the maximal activity of ANT, the apparent dissociation constant for ATP and ADP is increased, the line shifts to the right (line 4). Note that regardless of the used values of ANT kinetic constants, all computed combinations of K_a and K_{ia} were considerably adrift from the measured values in the presence of oxidative phosphorylation. Reproduced from Vendelin *et al.* (2004a) with permission from the Biophysical Society, USA.

(B) Calculated apparent dissociation constants K_a and K_{ia} of MgATP in the MtCK reaction in the presence of oxidative phosphorylation, with coupling between MtCK and oxidative phosphorylation being modelled assuming direct transfer of metabolites between ANT and MtCK. Apparent dissociation constants K_a and K_{ia} (represented by small dots in the figure) were computed for different combinations of free energies of MtCK-ANT complex states and free energies of transition. The measured values are charged cardiolipin in complex with ANT (Figures 2 and 3) (Muller *et al.*, 1985; Khuchua *et al.*, 1998; Schlattner *et al.*, 2004). The structure of the ANT was recently solved at 2.2 Å resolution by Brandolin's group (Pebay-Peyroula *et al.*, 2003). The translocation of both ATP^{4-} and ADP^{3-} in the Mg-free anionic forms is brought about by conformational changes of pore-forming ANT protomers (Pebay-Peyroula *et al.*, 2003). Klingenberg's group suggested that both protomers within an ANT dimer are involved in an alternating manner in the translocation, one accepting for example ATP from the matrix and translocating it towards the intermembrane space, while the second protomer only releases ADP translocated in the previous cycle ("half-site reactivity") (Gropp *et al.*, 1999; Huber *et al.*, 1999; Huang *et al.*, 2001; Brustovetsky *et al.*, 1996). This conformational change ("pore") mechanism leads in its simplest version to a ping-pong reaction mechanism of transport (Huang *et al.*, 2001). On the other hand, the kinetics of ATP-ADP exchange conform to a sequential mechanism involving binding of nucleotides on both sides before transport (Duyckaerts *et al.*, 1980). The structural data of Brandolin's group and the kinetics of ATP-ADP exchange by ANT are fitting well with the hypothesis that the dimers with alternatively activated protomers function in a coordinated manner in oligomeric ANT clusters, where the export of ATP from mitochondria by one protomer in a dimer occurs simultaneously with import of ADP by another protomer in another dimer (Kramer and Palmieri, 1992; Aliev and Saks, 2003; Nuri *et al.*, 2005, 2006).

The functional coupling between MtCK and ANT can be considered as the sum of two interaction modes. In the first mode, ATP and ADP are liberated into the intermembrane space and then bound to MtCK or ANT. This mode corresponds to an 'interaction' between MtCK and ANT as two separate proteins without any coupling (see Figure 6). In this case, ATP as well as all other substrates are in fast equilibrium with MtCK, and the reaction would follow a random Bi-Bi type mechanism. In the second mode, ATP and ADP would be directly channelled between the proteins. Such channelling is possible if the acceptor protein (ANT or MtCK) has no bound ATP or ADP molecule, i.e. it can accept ATP or ADP from the other protein. Additionally, we assume that when MtCK accepts ATP or ADP from ANT directly, bound ATP or ADP cannot be in fast equilibrium with the surrounding solution. Thus, the equilibration of the MtCK binding site for ATP and ADP with the surrounding solution is prevented and the system may have different kinetics if compared with the kinetics of isolated MtCK in solution.

With this model, we were able to reproduce the measured values of apparent kinetic constants of the MtCK reaction. During this test, the free energies of the states of the coupled system were varied as well as the free energies of activation.

Figure 7. shown by open circles in the upper right corner (no oxidative phosphorylation) and lower left corner (with oxidative phosphorylation). Note that the range of computed K_a-K_{ia} combinations covers the area near the measured values of these constants in the presence of oxidative phosphorylation. The figure is reproduced from Vendelin *et al.* (2004b) with permission from the Biophysical Society, USA.

The simulation results are shown in Figure 7B. Note that the region with the measured values of K_a and K_{ia} is covered by the model solutions, and it is possible to find model parameters which lead to the measured combination of K_a and K_{ia} values under conditions of oxidative phosphorylation. One can also reproduce the following experiments with the same set of model parameters: (a) changes in apparent kinetic properties of the MtCK reaction when coupled to oxidative phosphorylation (Jacobus and Saks, 1982; Saks *et al.*, 1985); (b) competition between MtCK-activated mitochondrial respiration and a separate ATP-regenerating system (Gellerich and Saks, 1982); and (c) studies on radioactively labelled adenine nucleotide uptake by mitochondria in the presence of MtCK activity (Barbour *et al.*, 1984). As a result, a free energy profile of the coupled MtCK-ANT system was proposed (Vendelin *et al.*, 2004a). According to our analysis, the main difference in free energy profiles between uncoupled and coupled MtCK-ANT reaction is the free energy change during reaction. In the coupled reaction, due to an increase in the free energy of the ANT_i.ATP state (Figure 6), the free energy decreases when ATP bound to ANT is used to synthesize PCr in the MtCK reaction. If compared with the models of the direct transfer using the probability approach, the proposed mechanism does not require strong changes in the free energy profile of the reaction of phosphate transfer itself (Vendelin *et al.*, 2004a). The proposed detailed kinetic scheme (Figure 6) can be used for further analysis of the kinetics of the MtCK-ANT interaction. In particular, the possible mechanism of an increase in the free energy level of the ANT_i.ATP complex when coupled to oxidative phosphorylation is intriguing and requires further experimental study.

Thus, both structural and functional data available now show convincingly that oxidative phosphorylation controls, via ANT, the MtCK reaction and forces it to produce PCr in spite of unfavourable kinetic and thermodynamic characteristics for this reaction. At the same time, MtCK plays the same role for ANT and oxidative phosphorylation, by channelling back ADP and, thus, directly controlling the rate of respiration.

Experimentally, the role of functional coupling between MtCK and ANT was verified recently in studies on energy metabolism of the heart of MtCK knock-out mice: as predicted by the model described above, these hearts had lower levels of PCr and reduced post-ischemic recovery (Spindler *et al.*, 2002, 2004). A new important role of MtCK-mediated control of ANT is the prevention of opening of the mitochondrial permeability transition pore, as recently discovered by Dolder *et al.* (2003), thereby inhibiting apoptosis and necrosis and, thus, preventing cell death. This illustrates once again the vital importance of the functional coupling phenomenon. Most important new data have recently been published by Meyer *et al.* (2006) showing that the functional coupling between MtCK and ANT strongly decreases production of reactive oxygen species (ROS) in mitochondria. Taking into account that ROS production is now considered a main reason for many age-related diseases and ageing itself (Jezek and Hlavata, 2005), it is difficult to overestimate the importance of these findings.

It is important to stress that the structural and functional coupling between MtCK and ANT does not prevent its participants from working in completely independent modes under some conditions. For example, it is well known that in a medium with only ADP, mitochondria can carry out oxidative phosphorylation without any limitation, despite the structural association of ANT with MtCK. On the other hand, inhibition of oxidative phosphorylation does not result in inhibition of MtCK but only alters its apparent kinetic behaviour (Saks *et al.*, 1975). These facts clearly indicate that the structural association of ANT with MtCK is rather dynamic and does not result in formation of a completely isolated space within these complexes. The substrates and products can leave this space, but can be arrested within to realize functional coupling between the partners, ANT and MtCK.

3.1.3. Heterogeneity of intracellular diffusion of ADP and the possible role of the outer mitochondrial membrane

Unusually high values of apparent K_m for exogenous ADP in regulation of the rate of mitochondrial respiration in permeabilized cardiac cells (up to 300–400 μM), as compared to isolated mitochondria (10–20 μM), have been found in many laboratories since 1988 (Saks *et al.*, 1989, 1991, 1993, 1994; Veksler *et al.*, 1995; Kuznetsov *et al.*, 1996; Kay *et al.*, 1997, 2000; Kummel *et al.*, 1988; Anflous *et al.*, 2001; Liobikas *et al.*, 2001; Boudina *et al.*, 2002; Burelle and Hochachka, 2002). Similarly high values of this parameter were found in several other oxidative muscles (Kay *et al.*, 1997; Kaasik *et al.*, 2001), in hepatocytes (Fontaine *et al.*, 1995), but not in fast-twitch skeletal muscles (Veksler *et al.*, 1995; Kuznetsov *et al.*, 1996; Burelle and Hochachka, 2002). Thus, this phenomenon is tissue specific. Figure 8 illustrates these results for isolated adult rat cardiomyocytes. Figure 8A and B show that in isolated mitochondria, respiration is rapidly activated by exogenous ADP in micromolar concentrations, with an apparent $K_m = 17$ μM (Appaix *et al.*, 2003). When permeabilized cardiomyocytes (Figure 8C) are used, activation of respiration requires addition of exogenous ADP in millimolar concentrations (Figure 8D), with an apparent K_m for ADP of 339 μM. However, in the presence of Cr (20 mM), these kinetics are changed, and the apparent K_m for ADP decreased 3–4-fold due to local production of ADP in the MtCK reaction (Appaix *et al.*, 2003). Figure 8E demonstrates most clearly the strong control of respiration by the MtCK reaction. In these experiments, respiration was activated in permeabilized cardiomyocytes by addition of exogenous MgATP (2 mM) in the presence of 3 mM PEP, and then PK was added in increasing amounts which decreased effectively the respiration rate due to trapping of ADP produced by ATPases. In the presence of this powerful PEP-PK system, stepwise addition of Cr rapidly and maximally activated respiration. Figure 8F shows that increases in PK activity up to 100 IU/ml were not able to suppress Cr-activated respiration. That means that ADP produced in the MtCK reaction coupled to the ANT was not accessible for the powerful ADP trapping system. Interestingly, however, when permeabilized cardiomyocytes were treated with trypsin (Figure 8F), Cr-activated respiration became sensitive to successive additions of PK (Figure 8G). Evidently, digestion by trypsin of some

A

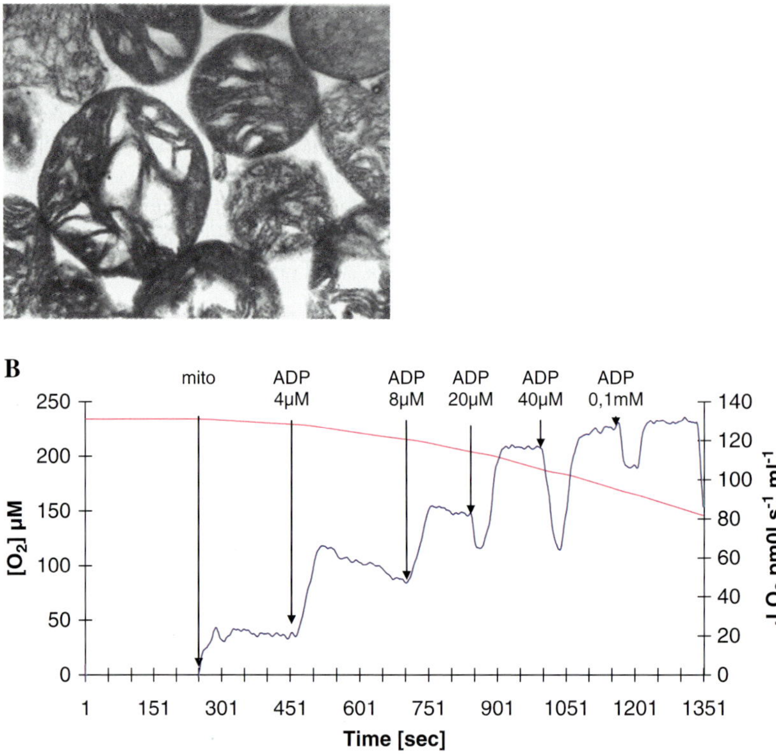

Figure 8. Regulation of respiration by exogenous ADP or ATP and Cr in rat heart mitochondria and permeabilized cardiomyocytes. Respiration rates were recorded using a two-channel high-resolution respirometer (Oroboros oxygraph-2k, Oroboros, Innsbruck, Austria). (A) Electron micrograph of isolated mitochondria. (B) Regulation of respiration of isolated cardiac mitochondria by exogenous ADP. JO_2 = rate of oxygen consumption. (C) Confocal microscopy of permeabilized cardiomyocytes: left: autofluorescence of NADH; right: preloaded with MitoTracker Green. (D) Regulation of respiration in permeabilized cardiomyocytes by exogenous ADP. At the end of the measurement, addition of cytochrome c (Cyt c, 8 μM) did not change respiration, indicating that the outer membrane was intact. Atractyloside (Atr, 30 μM) resulted in a decrease in respiration back to V_o due to inhibition of adenine nucleotide translocase. The respiratory control index in this experiment was 6.2. (E) Effective control of respiration by the CK system in permeabilized cardiomyocytes. Respiration was activated by addition of exogenous MgATP in the presence of 3 mM PEP, and then pyruvate kinase (PK) was added in increasing amounts up to 5 IU/ml. In the presence of this powerful PEP-PK system, stepwise addition of Cr maximally activated respiration. (F) Confocal microscopy image of the random arrangement of mitochondria in permeabilized cardiomyocytes preloaded with MitoTracker Red CMX Ros after treatment with trypsin; bar length = 10 μm; (G) Effect of the competing pyruvate kinase/phosphoenolpyruvate (PEP-PK) system on the relative respiration rate of isolated permeabilized cardiomyocytes initiated by endogenous ADP produced by 2 mM ATP and 20 mM Cr at optimal (0.4 μM) free $[Ca^{2+}]$ after trypsin treatment (4 °C, 5 min). Trypsin concentrations: 0 nM (•); 25 nM (▲); and 50 nM (■). (A)-(C) have been reproduced from Appaix *et al.* (2003) with kind permission from Experimental Physiology (The Physiological Society and Blackwell Publishing).

C

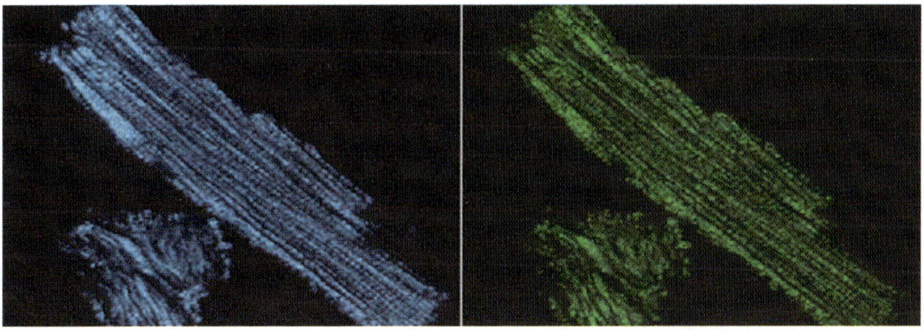

D

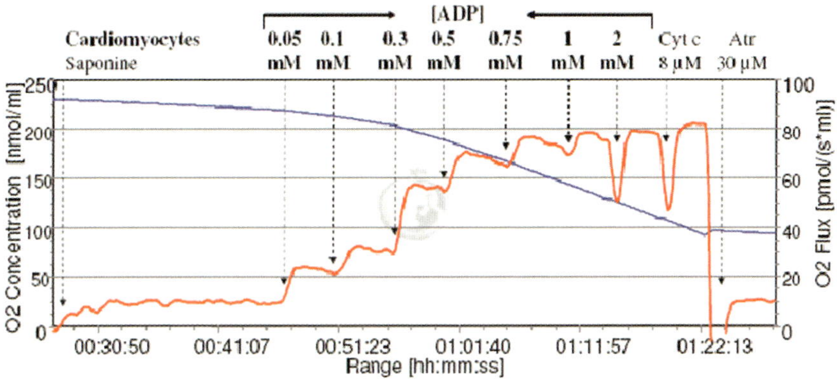

E

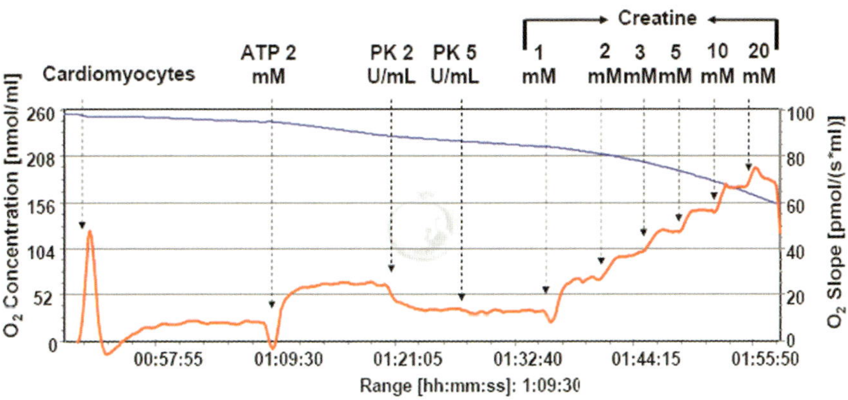

Figure 8. (continued)

F

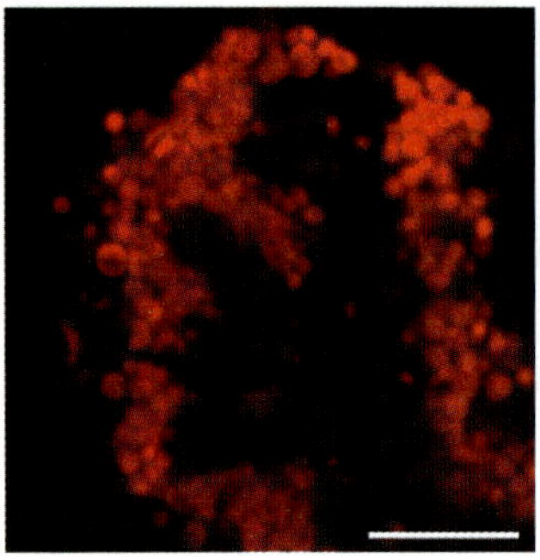

G

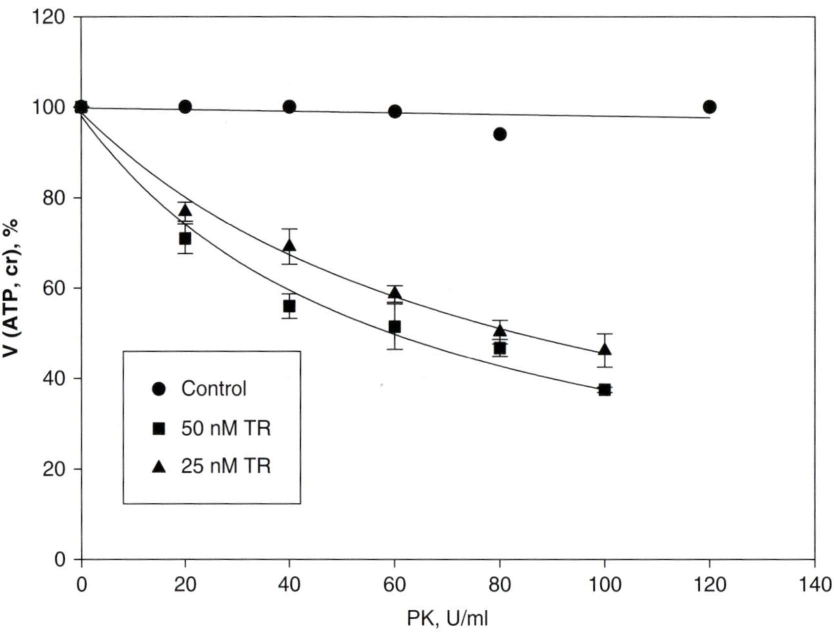

Figure 8. (continued)

cytoskeletal proteins that normally control mitochondrial positioning also opened the outer membrane for ADP, thus making it possible for some fraction of ADP to leave the intermembrane space and to be trapped by the PEP-PK system, showing that regulation of respiration is dependent on intracellular organization as well as on mitochondrial arrangement and/or integrity.

The high values of apparent K_m for exogenous ADP in permeabilized cardiac cells could be explained by heterogeneity of ADP diffusion inside cells, caused by contacts of mitochondria with the cytoskeleton and other cellular systems and, thus, by intracellular organization (Saks *et al.*, 2003; Vendelin *et al.*, 2004b). This conclusion was confirmed by results of mathematical modelling, which are described below.

On the way from the surrounding solution to the mitochondrial inner membrane, ATP and ADP molecules encounter two diffusion restrictions. The first one is on the level of the mitochondrial outer membrane and can be regulated by the state of VDAC (Colombini, 2004). The second one is due to macromolecular crowding in the cell. In addition, when simulating experiments performed on permeabilized fibers and cells, the endogenous ATPases have to be taken into account. The complete system used in the model is shown in Figure 9A. The aim of this particular model is to find a set of model parameters that can reproduce the experimental results on stimulation of respiration by exogenously added ADP, as well as on inhibition of respiration by an exogenously added ADP trapping system (PEP-PK). The model was used to determine possible values for two factors describing the restriction of ADP and ATP diffusion: the diffusion factor (*DF*) and the permeability factor (*PF*). *DF* describes the restriction of diffusion of adenine nucleotides within the cytoplasmic (extramitochondrial) space due to macromolecular crowding and cytoskeletal structures, and *PF* describes the decrease in permeability of the outer mitochondrial membrane due to the control of VDAC by some cytoskeletal proteins (Saks *et al.*, 1994, 1995; Capetanaki, 2002).

Diffusion restrictions between the mitochondrial inner membrane and the external solution influence the apparent K_m for exogenous ADP in regulation of respiration by shifting it to higher values, as referred to above. For example, if there are no considerable diffusion restrictions encountered by molecules between solution and ANT, then respiration should be stimulated by exogenous ADP as in isolated mitochondria. Indeed, when we assume diffusion coefficients equal to those measured in the bulk phase, assuming fast diffusion of nucleotides through the mitochondrial outer membrane, the computed respiration rate is very close to the values measured for isolated mitochondria (open squares in Figure 9B, simulations with large PF values). However, by increasing diffusion restriction imposed by the mitochondrial outer membrane (reduction in PF), the model managed to reproduce data on permeabilized cells and muscle fibers (closed circles in Figure 9B). Similar results can be obtained by reducing the apparent diffusion coefficient DF and by keeping PF at large values (Saks *et al.*, 2003). Thus, the effects of both diffusion restrictions on the model's solution are similar when the measurements of apparent K_m for exogenous ADP are simulated for permeabilized muscle fibers or cells. Therefore, by using only this approach, it is impossible to distinguish between the contributions of the two types of diffusion restrictions.

Determination of diffusion coefficients in muscle tissue yielded about two times lower values than in water (de Graaf *et al.*, 2000). The apparent diffusion coefficients for ADP obtained in the analysis of our data were lower by at least an order of magnitude (Saks *et al.*, 2003). Thus, it seems that there are some diffusion restrictions for ADP which are not visible in overall (average) diffusion coefficient measurements in muscle tissue, but which play an important role in permeabilized

A

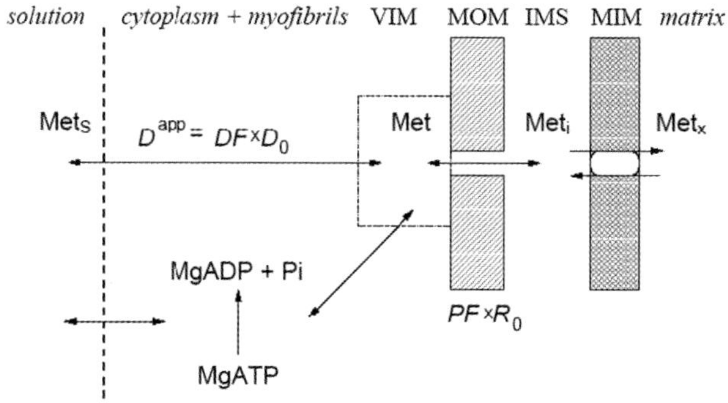

B

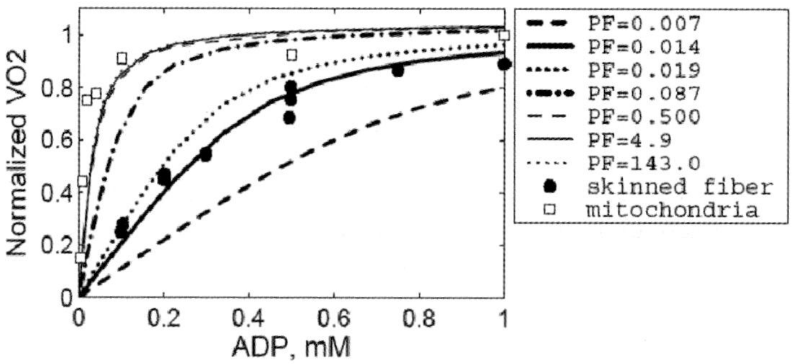

Figure 9. (A) Schematic representation of ADP (and ATP) diffusion pathways from solution into the mitochondrial matrix. Met_s, Met, Met_i, and Met_x are metabolite concentrations in the solution, in the vicinity of the mitochondria (VIM, inside ICEU), in the mitochondrial intermembrane space, and in the mitochondrial matrix, respectively. MOM, IMS, and MIM are mitochondrial outer membrane, intermembrane space, and inner membrane, respectively. D^{app}, D_0, and DF correspond to the apparent diffusion coefficient, the diffusion coefficient of the metabolite in bulk water phase, and the diffusion factor, respectively (see the text). R_0 is the permeability coefficient for passive diffusion of the metabolite across the outer mitochondrial membrane, and PF is the permeability factor for this metabolite. (B) Analysis of experimental data on regulation of respiration in permeabilized cardiac cells and fibers by mathematical modelling. Dependence of the calculated mitochondrial respiration rate in permeabilized cardiac fibers on the concentration of exogenous ADP computed with different mitochondrial outer membrane permeability factors (PF). The diffusion coefficients of the metabolites within the fiber were taken to be equal to the coefficients measured in bulk water phase of the cells, i.e. no diffusion restriction by macromolecular crowding was imposed. Simulations are compared with measurements of respiration in isolated mitochondria (open squares) and in skinned fibers (solid circles). Note that a reduction in outer membrane permeability increases the apparent K_m(ADP) of mitochondrial respiration. For permeabilized fibers, good fit with the experimental data is obtained for PF = 0.014. For isolated mitochondria, good fit is obtained for PF = 0.5 and higher. The figure was reproduced from Saks *et al.* (2003) with kind permission from the Biophysical Society, USA.

muscle fiber measurements when mitochondrial respiration rate is used as a probe for local ADP concentrations. Therefore, in addition to the value specifying diffusion restriction for ADP, DF, we tried to estimate the distribution of this restriction throughout the cell (Vendelin *et al.*, 2004b). These simulations indicate that the intracellular diffusion restrictions are not distributed uniformly, but are rather localized in certain areas in the cell (Vendelin *et al.*, 2004b). However, their precise localization is still unknown.

These results are consistent with the conclusions of strong localized diffusion restrictions for ADP and ATP in cardiac cells made by Abraham *et al.* (2002) and Selivanov *et al.* (2004). In particular, these authors studied the metabolic regulation of the sarcolemmal K_{ATP} channel, using both experimental methods and mathematical modelling, and concluded that the ATP diffusion coefficient is decreased in the subsarcolemmal area by several orders of magnitude. Earlier, a similar conclusion was drawn by Weiss and Lamp (1987). By using a microinjection technique, Bereiter-Hahn and Voth (1994) showed that the average diffusion coefficient of ATP in cells was decreased by an order of magnitude. Heterogeneity of ATP diffusion was also evidenced by using a targeted recombinant luciferase technique (Kennedy *et al.*, 1999).

To conclude, a model with localized diffusion restrictions for adenine nucleotides should be considered at present as a first approximation only, and insightful experiments are needed for measurement of local diffusion rates of ATP or ADP by modern experimental techniques. Under conditions of strong local restriction of diffusion of adenine nucleotides, compartmentalized energy transfer by the CK-PCr and other systems is likely to become vitally important for the cell.

3.2. Functionally Coupled MM-creatine Kinases

3.2.1. Myofibrillar creatine kinases

The function and roles of myofibrillar creatine kinase (MM-CK) have been studied and described extensively (Wallimann *et al.*, 1984; Ventura-Clapier *et al.*, 1987, 1998; Krause and Jacobus, 1992; Hornemann *et al.*, 2000, 2003). Wallimann's group has shown that MM-CK is bound specifically to the M-line (Wallimann *et al.*, 1984; Hornemann *et al.*, 2000, 2003), and that significant proportions of this isozyme are found in the space of the I-band of sarcomeres (Wegmann *et al.*, 1992). *In vitro*, the interactions of myosin with CK have been known for a long time (Yagi and Mase, 1962). Studies by Ventura-Clapier *et al.*, (1987, 1998) have shown that PCr accelerates the release of muscle from rigor tension in the presence of exogenous ATP, decreasing the necessary ATP concentration by an order of magnitude. Krause and Jacobus (1982) have shown close functional coupling between the actomyosin ATPase and the CK reaction in isolated rat heart myofibrils, seen as a decrease in the apparent K_m value for ATP from 79.9 ± 13.3 to 13.6 ± 1.4 μM after addition of 12.2 mM of phosphocreatine. In accordance with this finding, Sata *et al.* (1996) found that sliding velocity of fluorescently labeled actin on a cardiac myosin layer co-immobilized with CK showed a significantly lower apparent K_m value for MgATP than in the absence of CK. Ogut and Brozovich (2003) studied the kinetics of force development in skinned trabeculae from mouse hearts and found that in spite of the presence of 5 mM MgATP, the rate of force development depended on the concentration of PCr, and concluded that there is a direct functional link between the CK reaction and the actomyosin contraction cycle at the step of ADP release in myofibrils. Most probably, this efficient interaction occurs via small microcompartments of adenine nucleotides in myofibrils and is facilitated by anisotropy of their diffusion.

3.2.2. Membrane-bound creatine kinases

The role of MM-CK connected to the SR membrane and functionally coupled to the Ca,MgATP-dependent ATPase (SERCA) has been described in detail in many studies (Rossi *et al.*, 1990; Korge *et al.*, 1993; Korge and Campbell, 1994; Minajeva *et al.*, 1996). This coupling has been shown both for isolated SR vesicles and for intact SR in permeabilized cardiac fibers. Addition of PCr increased the rate of calcium uptake and the maximum SR Ca^{2+} content, while the exogenous ATP-regenerating PEP-PK system was less effective (Minajeva *et al.*, 1996). It was also shown in experiments with permeabilized cardiomyocytes that withdrawal of PCr from the medium reduced the frequency and amplitude, but increased the duration of spontaneous Ca^{2+} sparks (Yang and Steele, 2002). Thus, despite the presence of millimolar levels of cytosolic ATP, depletion of PCr impaired Ca^{2+} uptake (Korge *et al.*, 1993; Korge and Campbell, 1994; Minajeva *et al.*, 1996; Yang and Steele, 2002). All these data clearly show the importance of MM-CK, bound to the SR membrane, in rapid rephosphorylation of local MgADP produced in the Ca,MgATPase reaction, independently from the cytoplasmic environment, thus clearly demonstrating that SR-bound MM-CK is acting in a non-equilibrium manner.

An important player in the control of excitation-contraction coupling in the heart is the sarcolemmal membrane metabolic sensor complex. Its main part is the sarcolemmal ATP-sensitive K_{ATP} channel acting as an alarm system to adjust cell electrical activity to the metabolic state of the cell (Lederer and Nichols, 1989; Lorenz and Terzic, 1999; Abraham *et al.*, 2002). ATP closes the channel by interacting with its Kir6.2 subunit, but active membrane ATPases constantly reduce the local ATP concentration, which is distinct from that in the cytosol (Abraham *et al.*, 2002; Crawford *et al.*, 2002). Sarcolemmal MM-CK participates in this system by rephosphorylating local ADP and by maintaining a high ATP/ADP ratio in these microcompartments, with associated impacts on the coordination of membrane electrical activity with cellular metabolic status, and most notably with the PCr level. In this way, the PCr-CK network becomes the main intracellular regulatory pathway for cardiac cells, controlling electrical activity and cell excitability, calcium cycling, contraction and mitochondrial respiration. These energy transfer and control functions are shared by several systems including, besides CK, also the adenylate kinase (AK) and glycolytic systems, as it was seen in experiments with gene manipulation (Selivanov *et al.*, 2004; Carrasco *et al.*, 2001). MM-CK was first described in purified rat heart sarcolemmal preparations in 1977 (Saks *et al.*, 1977). Later, in experiments involving immunoprecipitation of guinea-pig cardiac membrane fractions with antibodies against the K_{ATP} channel's subunit SUR2, MM-CK was found to be physically associated with the cardiac K_{ATP} channel (Crawford *et al.*, 2002). Due to the sarcolemmal localization of MM-CK, the K_{ATP} channel's closed-open transitions are dependent upon PCr concentration at ATP concentrations higher than the threshold level for channel closure, as shown in experiments with isolated permeabilized cardiomyocytes for open cell-attached patch formation (Abraham *et al.*, 2002; Selivanov *et al.*, 2004). It was also concluded in these studies that strong local restrictions for ATP diffusion exist in the subsarcolemmal area of cardiac cells, which are by-passed by the CK system. This is in good concord with results of studies by Sasaki *et al.* (2001) showing that activation of mitochondrial ATP hydrolysis by uncouplers also activated sarcolemmal K_{ATP} channels in dependence of the activity of the CK system, which could be influenced by its inhibitor, 2,4-dinitrofluorobenzene. Similar functional coupling of CK with the K_{ATP} channel was described for pancreatic β-cells (Krippeit-Drews *et al.*, 2003).

Investigations in Wieringa's laboratory on genetic modification of CK and adenylate kinase (AK) provided firm evidence for the importance of this system: 'knock-outs' of the CK and AK genes result in significant adaptive changes in the cells such as structural remodelling (Janssen *et al.*, 2003; de Groof *et al.*, 2001; Ventura-Clapier *et al.*, 2004; Novatova *et al.*, 2006). A remarkable change associated with a M-CK knock-out was a manifold increase in the volume of the SR system, to compensate for the loss of efficiency of calcium uptake due to the absence of MM-CK (Steeghs *et al.*, 1997).

4. MATHEMATICAL MODELS OF THE PCr CIRCUIT IN HEART CELLS

4.1. Modelling Feedback Metabolic Regulation of Mitochondrial Respiration

The mitochondrial respiration rate *in vivo* may vary 20-fold, from 8–10 μmol min^{-1} g^{-1} dry mass in resting (KCl-arrested) aerobic hearts to at least 170 μmol min^{-1} g^{-1} dry mass in beating rat hearts (Williamson *et al.*, 1976). As shown by Neely *et al.* (1967, 1972), Williamson *et al.* (1976) and Balaban *et al.* (1986), oxygen consumption of the heart muscle is linearly dependent on the workload under conditions of metabolic stability of the heart. Parallel activation of energy-producing and consuming processes is not required to explain these observations in the heart muscle if CK compartmentation and the functional coupling mechanisms described above are taken into account. A model of compartmentalized energy transfer in heart muscle cells was initially developed by Aliev and Saks (1997) and later adapted by Vendelin and co-workers (Saks *et al.*, 2003; Vendelin *et al.*, 2000).

Mathematical modelling studies on energy exchange in working cardiac cells have been performed to gain insight into fundamental questions of cellular energetics: 1) is the cellular CK reaction always in equilibrium state during *in vivo* steady-state contractions of cardiac muscle? If so, ADP levels in the cytoplasm could be predicted from measured intracellular metabolite levels. 2) Is mitochondrial high-energy phosphate export *in vivo* mediated by ATP, according to the classical concept, or in the form of PCr, according to the concept of the CK phosphotransfer pathway? 3) Do metabolite levels in the myoplasm oscillate during cardiac contractions *in vivo*? 4) How can the cell maintain its metabolic stability at times when linearly interrelated manifold increases in cardiac work and oxygen consumption take place at practically constant metabolite levels and PCr/ATP ratios in the cells? In the absence of experimental methods for direct monitoring of cellular ADP levels, mathematical modelling of dynamic events in the cellular cytoplasm remains the only option.

4.2. Description of a Model

A new class of dynamic mathematical models of intracellular compartmentalized energy transport in cardiac cells was constructed, leading ultimately to the formulation of the "Intracellular Energetic Unit" (ICEU) concept (Saks *et al.*, 2001; Seppet *et al.*, 2001). These models are based mainly on the principles of chemical kinetics and the mass action law (Aliev and Saks, 1997; Dos Santos *et al.*, 2000). In addition, they consider the time dynamics of basic events of cellular energetics: ATP hydrolysis by actomyosin ATPase during a contraction cycle; diffusional exchange of metabolites between myofibrillar and mitochondrial compartments; VDAC-restricted diffusion of ATP and ADP across the mitochondrial outer membrane; mitochondrial synthesis of ATP by ATP synthase; P_i and ADP transport into the mitochondrial matrix controlled by the mitochondrial membrane potential, which is a function of ΔpH and $\Delta\Psi$; and PCr production in the coupled MtCK reaction and its utilization by cytoplasmic CKs. These factors are considered in a system consisting of a myofibril with a radius of 1 μm, a mitochondrion, and a thin layer of cytoplasm interposed between them (Aliev and Saks, 1997; Dos Santos *et al.*, 2000). The computations of diffusion and chemical events were performed for every 0.1-μm segment of the chosen diffusion path at each 0.01 ms time step (Aliev and Saks, 1997). This allowed simulation of space-dependent changes throughout the entire cardiac cycle. This system, with adjacent ADP-producing systems in myofibrils and in the SR, is supposed to represent the basic pattern of organization of muscle cell energy metabolism, the ICEU of cardiac cell energetics (see above).

In mitochondria, the ANT and the P_i carrier control the matrix concentrations of ATP, ADP and P_i available for ATP synthase. These carriers establish constant positive (i.e., higher concentration inside) ADP and P_i gradients between the mitochondrial matrix and intermembrane space. In the model, the ATP/ADP ratios in the matrix and the activity of ATP synthase are dependent on $\Delta\Psi$, the electric component of the mitochondrial membrane potential. The model also employs a complete mathematical description of the P_i carrier based on a probability approach, allowing prediction of the dynamics of P_i accumulation in the matrix in exchange for matrix OH^- ions, i.e. at the expense of mitochondrial proton-motive force, ΔpH. The model considers CK compartmentation as discussed above. In particular, cytoplasmic CK (MM-CK) molecules, representing 69% of total CK activity, are freely distributed in the myofibrillar and cytoplasmic spaces. The intrinsic thermodynamic parameters of MM-CK favour its functioning in the reverse direction of the CK reaction to transphosphorylate ADP to ATP at the expense of PCr utilization. The remaining part of cellular CK, i.e. the MtCK isoenzyme, is localized in the mitochondrial compartment. In mitochondria, MtCK is tightly anchored by cardiolipin molecules to the ANT at the outer surface of the inner mitochondrial membrane. The resulting close proximity of MtCK and ANT allows direct channelling of adenine nucleotides between their adjacent active sites; this channelling is the actual base for shifting the MtCK reaction towards synthesis of PCr from translocase-supplied ATP, even at high levels of ATP in the myoplasm of *in vivo* heart cells.

Mathematical modelling of the cellular CK circuit system was developed further by (i) more sophisticated modelling of the kinetics of mitochondrial ANT by a probability approach and (ii) a simplified modelling approach for the functional coupling between ANT and MtCK (see below). In both versions of the model (Aliev and Saks, 1997; Dos Santos *et al.*, 2000), functional coupling of MtCK to ANT was simulated by means of dynamically changing high local ATP concentrations in a 10-nm narrow space (microcompartment) between coupled molecules. This simplified approach – coupling by local dynamic compartmentation – was used because of a high demand for computing power in the original probability model of coupling. The probability model was used to check the validity of calculations in this simplified approach (Aliev and Saks, 1997).

A distinctive feature of this modelling approach is that we avoided, as much as possible, the formal description of chemical phenomena by adjustable mathematical terms. The living cell is a self-regulating chemical machine; therefore, relying on the principles of chemical and enzyme kinetics decreases the probability of errors during mathematical modelling. Proper choice of maximum rates of enzyme activities, taken from *in vivo* and biochemical data (Aliev and Saks, 1997), also served this goal. All details of modelling can be found in our publications (Aliev and Saks, 1997; Dos Santos *et al.*, 2000).

4.3. Main Results: Mathematical Evidence for the Phosphocreatine Circuit

Modelling revealed oscillations of all metabolite levels in the cytoplasm of the working heart (Figure 10A). Activation of contraction results in a small spike-like, transient (40 ms) decrease in ATP concentration and a symmetrical increase in ADP levels in the systole, followed by more pronounced and longer-lasting changes in PCr (transient decrease) and Cr (transient increase) concentrations. P_i concentrations change in a similar manner as free Cr concentrations (Aliev and Saks, 1997; Dos Santos *et al.*, 2000).

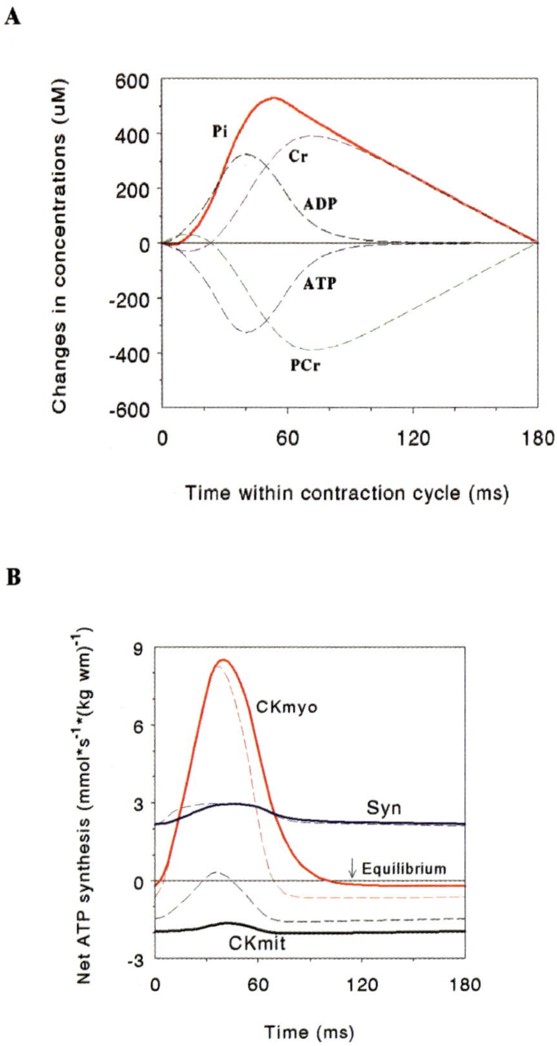

Figure 10. (A) Phasic changes in metabolite concentrations in a myofibril's core during a cardiac contraction cycle. Modelling for a high workload corresponding to a rate of oxygen consumption of 46 μatoms of O·(g wm)$^{-1}$·min^{-1}. (B) Modelling results of non-equilibrium behaviour of ATP-synthase (Syn), myofibrillar and mitochondrial CK (CKmyo and CKmit, respectively) under conditions with (solid lines) or without (dotted lines) restrictions for adenine nucleotide diffusion through the mitochondrial outer membrane. An arrow indicates the equilibrium position, when net ATP production is equal to zero. Reproduced from Aliev and Saks (1997) with kind permission from the Biophysical Society, USA.

These metabolite changes are the basis for respective changes in the corresponding chemical reactions. At very high workload, corresponding to a rate of oxygen consumption of 46 μ atoms of O·(g wet mass)$^{-1}$·min^{-1}, an increase in myoplasmic ADP concentration from basal diastolic levels (383 vs. 58 μM) is responsible for activation of net ATP synthesis by MM-CK from –0.2 to 8.5 mmol ATP·s^{-1}·(kg wet mass)$^{-1}$ (Aliev and Saks, 1997) (Figure 10B).

The simulation data indicate that myoplasmic MM-CK is clearly out of equilibrium in cyclically contracting cells (Figure 10B). Non-equilibrium behaviour of MM-CK is caused by cyclic increases in myoplasmic ADP levels during the systole of the cell. These oscillations cannot be dampened completely even when the activity of MM-CK is increased artificially by 10-fold (Saks and Aliev, 1996).

In such a system, based on published experimental data, regeneration of consumed myoplasmic PCr takes place mostly in mitochondria, as evidenced by a permanent shift of the MtCK reaction towards net PCr and ADP synthesis (Aliev and Saks, 1997) (Figure 10B). The mean value of net PCr synthesis in the mitochondrial compartment is 1.91 mmol PCr·s^{-1}·(kg wet mass)$^{-1}$. The sustained shift of the MtCK reaction towards PCr synthesis results both from local coupling of MtCK to ANT and from dynamic compartmentation imposed by restrictions for ADP diffusion through the mitochondrial outer membrane. The relative contribution of the former mechanism, i.e. local coupling, to this phenomenon is about 54%: in the same system, but without restrictions for ADP diffusion (Aliev and Saks, 1997), the mean value of net PCr synthesis by mitochondria drops to 1.03 mmol PCr·s^{-1}·(kg wet mass)$^{-1}$ (Figure 11).

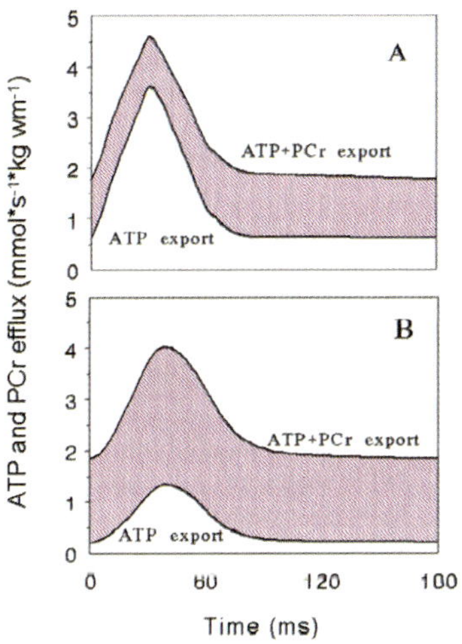

Figure 11. Modelling results of diffusional ATP and PCr export through the mitochondrial outer membrane under conditions without (A) or with (B) restrictions for adenine nucleotide diffusion through the mitochondrial outer membrane. The shaded area indicates PCr efflux. Reproduced from Aliev and Saks (1997) with kind permission from the Biophysical Society, USA.

As a whole, in a system with compartmentalized CK, the functional local coupling of CK to ANT leads to complete separation of functional roles of cellular CK isoenzymes: MM-CK becomes responsible for ATP regeneration in the myoplasm during the systole at the expense of PCr breakdown, while regeneration of myoplasmic Cr to PCr takes place in mitochondria in the coupled MtCK reaction throughout the contraction-relaxation cycle (Figure 11). Such a separation of functions is important for realization of metabolic stability of the working heart (Dos Santos *et al.*, 2000; Aliev *et al.*, 2003)

ACKNOWLEDGEMENTS

This work was supported by INSERM, France, and by grants from the Estonian Science Foundation (no. 7117 and 6142 to V.S.), the Swiss National Science Foundation (grant 3100AO-102075 to T.W. and U.S.), the Swiss Heart Foundation (to T.W. and U.S.) and the Russian Foundation for Basic Science (grant 060448620 to M.A.). Due to space limitations, as well as restriction of cited references, we apologize to all those colleagues and researchers in the field whose work is not directly cited here.

REFERENCES

Abraham, M.R., Selivanov, V.A., Hodgson, D.M., Pucar, D., Zingman, L.V., Wieringa, B., Dzeja P. Alekseev, A.E., and Terzic, A., 2002, Coupling of cell energetics with membrane metabolic sensing. Integrative signaling through creatine kinase phosphotransfer disrupted by M-CK gene knock-out. *J. Biol. Chem.* **277:** 24427–24434.

Aliev, M.K., and Saks, V.A., 1997, Compartmentalised energy transfer in cardiomyocytes. Use of mathematical modeling for analysis of in vivo regulation of respiration. *Biophys. J.* **73:** 428–445.

Aliev, M.K., and Saks, V.A., 2003, Analysis of mechanism of work of mitochondrial adenine nucleotide translocase using mathematical models. *Biofizika (Russian)* **48:** 1075–1085.

Aliev, M.K., Dos Santos, P., and Saks, V.A., 2003, Mathematical modeling of regulation of oxidative phosphorylation in cardiomyocytes. In: Kekelidze, T., and Holtzman, D. (eds.), Creatine Kinase and Brain Energy Metabolism. Function and Disease. IOS Press, Amsterdam, NATO Science Series: Life and Behavioral Sciences, Volume 342: 59–79.

Ames, A.3rd, 2000, CNS energy metabolism as related to function. *Brain Res. Rev.* **34:** 42–68.

Anflous, K., Armstrong, D.D., and Craigen, W.J., 2001, Altered mitochondrial sensitivity for ADP and maintenance of creatine-stimulated respiration in oxidative striated muscles from VDAC1-deficient mice. *J. Biol. Chem.* **276:** 1954–1960.

Appaix, F., Kuznetsov, A.V., Usson, Y., Andrienko, T., Olivares, J., Kaambre, T., Sikk, P., Margreiter, R., and Saks, V., 2003, Possible role of cytoskeleton in intracellular arrangement and regulation of mitochondria. *Exp. Physiol.* **88:** 175–190.

Balaban, R.S., Kantor, H.L., Katz, L.A., and Briggs, R.W., 1986, Relation between work and phosphate metabolite in the in vivo paced mammalian heart. *Science* **232:** 1121–1123.

Barbour, R.L., Ribaudo, J., and Chan, S.H.P., 1984, Effect of creatine kinase activity on mitochondrial ADP/ATP transport. Evidence for functional interaction. *J. Biol. Chem.* **259:** 8246–8251.

Belitzer, V.A., and Tsybakova, E.T., 1939, Sur le mécanisme des phosphorylations couplées avec la respiration. *Biochimia (Russian)* **4:** 516–535.

Bereiter-Hahn, J., and Voth, M., 1994, Dynamics of mitochondria in living cells: shape changes, dislocations, fusion, and fission of mitochondria. *Microsc. Res. Tech.* **27:** 198–219.

Bessman, S.P., and Fonyo, A., 1966, The possible role of the mitochondrial bound creatine kinase in regulation of mitochondrial respiration. *Biochem. Biophys. Res. Commun.* **22:** 597–602.

Bessman, S.P., and Geiger, P., 1981, Transport of energy in muscle: the phosphocreatine shuttle. *Science* **211:** 448–452.

Beyer, K., and Klingenberg, M., 1985, ADP/ATP carrier protein from beef heart mitochondria has high amounts of tightly bound cardiolipin, as revealed by ^{31}P nuclear magnetic resonance. *Biochemistry* **24:** 3821–3826.

Beyer, K., and Nuscher, B., 1996, Specific cardiolipin binding interferes with labeling of sulfhydryl residues in the adenosine diphosphate/adenosine triphosphate carrier protein from beef heart mitochondria. *Biochemistry* **35:** 15784–15790.

Boudina, S., Laclau, M.N., Tariosse, L., Daret, D., Gouverneur, G., Boron-Adele, S., Saks, V.A., and Dos Santos, P., 2002, Alteration of mitochondrial function in a model of chronic ischemia in vivo in rat heart. *Am. J. Physiol.* **282:** H821–H831.

Brustovetsky, N., Becker, A., Klingenberg, M., and Bamberg, E., 1996, Electrical currents associated with nucleotide transport by the reconstituted mitochondrial ADP/ATP carrier. *Proc. Natl. Acad. Sci. U.S.A.* **93:** 664–668.

Burelle, Y., and Hochachka, P.W., 2002, Endurance training induces muscle-specific changes in mitochondrial function in skinned muscle fibers. *J. Appl. Physiol.* **92:** 2429–2438.

Burklen, T.S., Schlattner, U., Homayouni, R., Gough, K., Rak, M., Szeghalmi, A., and Wallimann, T., 2006, The creatine kinase/creatine connection to Alzheimer's disease: CK-inactivation, APP-CK complexes and focal creatine deposits. *J. Biomed. Biotechnol.* **2006:** 1–11.

Capetanaki, Y., 2002, Desmin cytoskeleton: a potential regulator of muscle mitochondrial behavior and function. *Trends Cardiovasc. Med.* **12:** 339–348.

Carrasco, A.J., Dzeja, P.P., Alekseev, A.E., Pucar, D., Zingman, L.V., Abraham, M.R., Hodgson, D., Bienengraeber, M., Puceat, M., Janssen, E., Wieringa, B., and Terzic, A., 2001, Adenylate kinase phosphotransfer communicates cellular energetic signals to ATP-sensitive potassium channels. *Proc. Natl. Acad. Sci. U.S.A.* **98:** 7623–7628.

Cleland, W.W., 1963, The kinetics of enzyme-catalyzed reactions with two or more substrates or products. I. Nomenclature and rate equations. *Biochim. Biophys. Acta* **67:** 104–137.

Colombini, M., 2004, VDAC: the channel at the interface between mitochondria and the cytosol. *Mol. Cell. Biochem.* **256:** 107–115.

Crawford, R.M., Ranki, H.J., Botting, C.H., Budas, G.R., and Jovanovic, A., 2002, Creatine kinase is physically associated with the cardiac ATP-sensitive K^+ channel *in vivo*. *FASEB J.* **16:** 102–104.

De Furia, R.A., Ingwall, J.S., Fossel, E., and Dygert, M., 1980, The integration of isoenzymes for energy distribution. In: Jacobus, W.E., and Ingwall, J.S. (eds.), Heart creatine kinase, Williams & Wilkins, Baltimore-London, pp. 135–142.

Dolder, M., Walzel, B., Speer, O., Schlattner, U., and Wallimann, T., 2003, Inhibition of the mitochondrial permeability transition by creatine kinase substrates. Requirement for microcompartmentation. *J. Biol. Chem.* **278:** 17760–17766.

Dos Santos, P., Aliev, M.K., Diolez, P., Duclos, F., Bonoron-Adele, S., Besse, P., Canioni, P., Sikk, P., and Saks, V.A., 2000, Metabolic control of contractile performance in isolated perfused rat heart. Analysis of experimental data by reaction:diffusion mathematical model. *J. Mol. Cell. Cardiol.* **32:** 1703–1734.

Duyckaerts, C., Sluse-Coffart, C.M., Fux, J.P., Sluse, F.E., and Liebecq, C., 1980, Kinetic mechanism of the exchanges catalysed by the adenine nucleotide carrier. *Eur. J. Biochem.* **106:** 1–6.

Dzeja, P.P., Zeleznikar, R.J., and Goldberg, N.D., 1998, Adenylate kinase: kinetic behaviour in intact cells indicates it is integral to multiple cellular processes. *Mol. Cell. Biochem.* **184:** 169–182.

Dzeja, P., and Terzic, A., 2003, Phosphotransfer networks and cellular energetics. *J. Exp. Biol.* **206:** 2039–2047.

Epand, R.F., Tokarska-Schlattner, M., Schlattner, U., Wallimann, T., and Epand, R.M., 2007, Cardiolipin clusters and membrane domain formation induced by mitochondrial proteins. *J. Mol. Biol.* **365:** 968–980.

Epand, R.F., Schlattner, U., Wallimann, T., Lacombe, M.L., and Epand, R.M., 2006, Novel lipid transfer property of two mitochondrial proteins that bridge the inner and outer membranes. *Biophys. J.* **92:** 126–137.

Eppenberger, H.M., Dawson, D.M., and Kaplan, N.O., 1967, The comparative enzymology of creatine kinases. *J. Biol. Chem.* **242:** 204–209.

Fontaine, E.M., Keriel, C., Lantuejoul, S., Rigoulet, M., Leverve, X.M., and Saks, V.A., 1995, Cytoplasmic cellular structures control permeability of outer mitochondrial membrane for ADP and oxidative phosphorylation in rat liver cells. *Biochem. Biophys. Res. Commun.* **213:** 138–146.

Fritz-Wolf, K., Schnyder, T., Wallimann, T., and Kabsch, W., 1996, Structure of mitochondrial creatine kinase. *Nature* **381:** 341–345.

Gellerich, F., and Saks, V.A., 1982, Control of heart mitochondrial oxygen consumption by creatine kinase: the importance of enzyme localization. *Biochem. Biophys. Res. Commun.* **105:** 1473–1481.

Gellerich, F.N., Schlame, M., Bohnensack, R., and Kunz, W., 1987, Dynamic compartmentation of adenine nucleotides in the mitochondrial intermembrane space of rat-heart mitochondria. *Biochim. Biophys. Acta* **890:** 117–126.

Gellerich, F.N., Kapischke, M., Kunz, W., Neumann, W., Kuznetsov, A., Brdiczka, D., and Nicolay, K., 1994, The influence of the cytosolic oncotic pressure on the permeability of the mitochondrial outer membrane for ADP: implications for the kinetic properties of mitochondrial creatine kinase and for ADP channeling into the intermembrane space. *Mol. Cell. Biochem.* **133/134:** 85–104.

Gellerich, F.N., Laterveer, F.D., Korzeniewski, B., Zierz, S., and Nicolay, K., 1998, Dextran strongly increases the Michaelis constants of oxidative phosphorylation and of mitochondrial creatine kinase in heart mitochondria. *Eur. J. Biochem.* **254:** 172–180.

Gellerich, F.N., Laterveer, F.D., Zierz, S., and Nicolay, K., 2002, The quantitation of ADP diffusion gradients across the outer membrane of heart mitochondria in the presence of macromolecules. *Biochim. Biophys. Acta* **1554:** 48–56.

de Graaf, R.A., van Kranenburg, A., and Nicolay, K., 2000, In vivo ^{31}P-NMR diffusion spectroscopy of ATP and phosphocreatine in rat skeletal muscle. *Biophys J.* **78:** 1657–1664.

de Groof, A.J., Smeets, B., Groot Koerkamp, M.J., Mul, A.N., Janssen, E.E., Tabak, H.F., and Wieringa, B., 2001, Changes in mRNA expression profile underlie phenotypic adaptations in creatine kinase-deficient muscles. *FEBS Lett.* **506:** 73–78.

Gropp, T., Brustovetsky, N., Klingenberg, M., Müller, V., Fendler, K., and Bamberg, E., 1999, Kinetics of electrogenic transport by the ADP/ATP carrier. *Biophys. J.* **77:** 714–726.

Haas, R.C., and Strauss, A.W., 1990, Separate nuclear genes encode sarcomere-specific and ubiquitous human mitochondrial creatine kinase isoenzymes. *J. Biol. Chem.* **265:** 6921–6927.

Hornemann, T., Stolz, M., and Wallimann, T., 2000, Isoenzyme-specific interaction of muscle-type creatine kinase with the sarcomeric M-line is mediated by NH_2-terminal lysine charge-clamps. *J. Cell Biol.* **149:** 1225–1234.

Hornemann, T., Kempa, S., Himmel, M., Hayess, K., Furst, D.O., and Wallimann, T., 2003, Muscle-type creatine kinase interacts with central domains of the M-band proteins myomesin and M-protein. *J. Mol. Biol.* **332:** 877–887.

Huang, S.G., Odoy, S., and Klingenberg, M., 2001, Chimers of two fused ADP/ATP carrier monomers indicate a single channel for ADP/ATP transport. *Arch. Biochem. Biophys.* **394:** 67–75.

Huber, T., Klingenberg, M., and Beyer, K., 1999, Binding of nucleotides by mitochondrial ADP/ATP carrier as studied by ^{1}H nuclear magnetic resonance spectroscopy. *Biochemistry* **38:** 762–769.

Jacobus, W.E., and Lehninger, A.L., 1973, Creatine kinase of rat mitochondria. Coupling of creatine phosphorylation to electron transport. *J. Biol. Chem.* **248:** 4803–4810.

Jacobus, W.E., and Saks, V.A., 1982, Creatine kinase of heart mitochondria: changes in its kinetic properties induced by coupling to oxidative phosphorylation. *Arch. Biochem. Biophys.* **219:** 167–178.

Janssen, E., Terzic, A., Wieringa, B., and Dzeja, P., 2003, Impaired intracellular energy communication in muscles from creatine kinase and adenylate kinase (M-CK/AK1) double knock-out mice. *J. Biol. Chem.* **278:** 30441–30449.

Jezek, P., and Hlavata, L., 2005, Mitochondria in homeostasis of reactive oxygen species in cell, tissues and organism. *Intl. J. Biochem.* **37:** 2478–2503.

Kaasik, A., Veksler, V., Boehm, E., Novotova, M., Minajeva, A., and Ventura-Clapier, R., 2001, Energetic crosstalk between organelles. Architectural integration of energy production and utilization. *Circ. Res.* **89:** 153–159.

Kay, L., Li, Z., Mericskay, M., Olivares, J., Tranqui, L., Fontaine, E., Tiivel, T., Sikk, P., Kaambre, T., Samuel, J.L., Rappaport, L., Usson, Y., Leverve, X., Paulin, D., and Saks, V.A., 1997, Study of regulation of mitochondrial respiration in vivo. An analysis of influence of ADP diffusion and possible role of cytoskeleton. *Biochim. Biophys. Acta* **1322:** 41–59.

Kay, L., Nicolay, K., Wieringa, B., Saks, V., and Wallimann, T., 2000, Direct evidence for the control of mitochondrial respiration by mitochondrial creatine kinase in oxidative muscle cells in situ. *J. Biol. Chem.* **275:** 6937–6944.

Kennedy, H.J., Pouli, A.E., Ainscow, E.K., Jouaville, L.S., Rizzuto, R., and Rutter, G.A., 1999, Glucose generates sub-plasma membrane ATP microdomains in single islet beta-cells. Potential role for strategically located mitochondria. *J. Biol. Chem.* **274:** 13281–13291.

Khuchua, Z.A., Qin, W., Boero, J., Cheng, J., Payne, R.M., Saks, V.A., and Strauss, A.W., 1998, Octamer formation by cardiac sarcomeric mitochondrial creatine kinase is mediated by charged N-terminal residues. *J. Biol. Chem.* **273:** 22990–22996.

Kim, I.H., and Lee, H.J., 1987, Oxidative phosphorylation of creatine by respiring pig heart mitochondria in the absence of added adenine nucleotides. *Biochem. Int.* **14:** 103–110.

Klingenberg, M., 1964, Muskelmitochondrien. In: Kramer, K., Krayer, O., Lehnartz, E., Muralt, A., and Weber, H.H. (eds.), Ergebnisse der Physiologie, Biologischen Chemie und Experimentellen Pharmakologie, Springer Verlag, Berlin-Göttingen-Heidelberg-New York, Vol. 55, pp. 131–189.

Korge, P., Byrd, S.K., and Campbell, K.B., 1993, Functional coupling between sarcoplasmic-reticulum-bound creatine kinase and Ca^{2+} ATPase. *Eur. J. Biochem.* **213:** 973–980.

Korge, P., and Campbell, K.B., 1994, Local ATP regeneration is important for sarcoplasmic reticulum Ca^{2+} pump function. *Am. J. Physiol.* **267:** C357–366.

Kramer, R., and Palmieri, F., 1992, Metabolic carriers in mitochondria. In: *Molecular Mechanisms in Bioenergetics* (Ernster, L., ed.), Elsevier Science Publishers, pp. 359–384.

Krause, S.M., and Jacobus, W.E., 1992, Specific enhancement of the cardiac myofibrillar ATPase activity by bound creatine kinase. *J. Biol. Chem.* **267:** 2480–2486.

Krippeit-Drews, P., Backer, M., Dufer, M., and Drews, G., 2003, Phosphocreatine as a determinant of K_{ATP} channel activity in pancreatic β-cells. *Pflugers Arch.* **445:** 556–562.

Kummel, L., 1988, Ca,MgATPase activity of permeabilized rat heart cells and its functional coupling to oxidative phosphorylation in the cells. *Cardiovasc. Res.* **22:** 359–367.

Kuznetsov, A.V., and Saks, V.A., 1986, Affinity modification of creatine kinase and ATP-ADP translocase in heart mitochondria: determination of their molar stoichiometry. *Biochem. Biophys. Res. Commun.* **134:** 359–366.

Kuznetsov, A.V., Tiivel, T., Sikk, P., Kaambre, T., Kay, L., Daneshrad, Z., Rossi, A., Kadaja, L., Peet, N., Seppet, E., and Saks, V.A., 1996, Striking difference between slow and fast twitch muscles in the kinetics of regulation of respiration by ADP in the cells in vivo. *Eur. J. Biochem.* **241:** 909–915.

Lederer, W.J., and Nichols, C.G., 1989, Nucleotide modulation of the activity of rat heart ATP-sensitive K^+ channels in isolated membrane patches. *J. Physiol.* **419:** 193–211.

Liobikas, J., Kopustinskiene, D.M., and Toleikis, A., 2001, What controls the outer mitochondrial membrane permeability for ADP: facts for and against the oncotic pressure. *Biochim. Biophys. Acta* **1505:** 220–225.

Lorenz, E., and Terzic, A., 1999, Physical association between recombinant cardiac ATP-sensitive K^+ channel subunits Kir.6 and SUR2A. *J. Mol. Cell. Cardiol.* **31:** 425–434.

McLeish, M.J., and Kenyon, G.L., 2005, Relating structure to mechanism in creatine kinase. *Crit. Rev. Biochem. Mol. Biol.* **40**: 1–20.

Meyer, L.E., Machado, L.B., Santiago, A.P.S.A., da-Silva, S., De Felice, F.G., Holub, O., Oliviera, M., and Galina, A., 2006, Mitochondrial creatine kinase activity prevents reactive oxygen species generation: Antioxidant role of mitochondrial kinase-dependent ADP re-cycling activity. *J. Biol. Chem.* **281:** 29916–29928.

Minajeva, A., Ventura-Clapier, R., and Veksler, V., 1996, Ca^{2+} uptake by cardiac sarcoplasmic reticulum ATPase in situ strongly depends on bound creatine kinase. *Pflugers Arch.* **432:** 904–912.

Muller, M., Moser, R., Cheneval, D., and Carafoli, E., 1985, Cardiolipin is the membrane receptor for mitochondrial creatine phosphokinase. *J. Biol. Chem.* **260:** 3839–3843.

Neely, J.R., Liebermeister, H., Battersby, E.J., and Morgan, H.E., 1967, Effect of pressure development on oxygen consumption by isolated rat heart. *Am. J. Physiol.* **212:** 804–814.

Neely, J.R., Denton, R.M., England, P.J., and Randle, P.J., 1972, The effects of increased heart work on the tricarboxylate cycle and its interactions with glycolysis in the perfused rat heart. *Biochem. J.* **128:** 147–159.

Neely, J.R., and Morgan, H.E., 1974, Relationship between carbohydrate and lipid metabolism and the energy balance of heart muscle. *Annu. Rev. Physiol.* **63:** 413–459.

Novotova, M., Pavlovieova, M., Veksler, V., Ventura-Clapier, R., and Zahradnik, I., 2006, Ultrastructural remodeling of fast skeletal muscle fibers induced by invalidation of creatine kinase. *Am. J. Physiol. Cell Physiol.* **291:** C1279–1285.

Nury, H., Dahout-Gonzalez, C., Trezeguet, V., Lauquin, G.J., Brandolin, G., and Pebay-Peyroula, E., 2006, Relations between structure and function of the mitochondrial ADP/ATP carrier. *Annu. Rev. Biochem.* **75:** 713–741.

Nury, H., Dahout-Gonzalez, C., Trezeguet, V., Lauquin, G., Brandolin, G., and Pebay-Peyroula, E., 2005, Structural basis for lipid-mediated interactions between mitochondrial ADP/ATP carrier monomers. *FEBS Lett.* **579:** 6031–6036.

Ogut, O., and Brozovich, F.V., 2003, Creatine phosphate consumption and the actomyosin crossbridge cycle in cardiac muscles. *Circ. Res.* **93:** 54–60.

Opie, L.H., 1998, *The Heart. Physiology, from Cell to Circulation.* Lippincott-Raven Publishers, Philadelphia, USA, pp. 43–63.

Pebay-Peyroula, E., Dahout-Gonzalez, C., Trézéguet, V., Lauquin, G., and Brandolin, G., 2003, Structure of mitochondrial ADP/ATP carrier in complex with carboxyatractyloside. *Nature* **426:** 39–44.

Rao, J.K., Bujacz, G., and Wlodawer, A., 1998, Crystal structure of rabbit muscle creatine kinase. *FEBS Lett.* **439**: 133–137.

Qin, W., Khuchua, Z., Cheng, J., Boero, J., Payne, R.M, and Strauss, A.W., 1998, Molecular characterization of the creatine kinases and some historical perspectives. *Mol. Cell. Biochem.* **184:** 153–167.

Rossi, A.M., Eppenberger, H.M., Volpe, P., Cotrufo, R., and Wallimann, T., 1990, Muscle-type MM creatine kinase is specifically bound to sarcoplasmic reticulum and can support Ca^{2+} uptake and regulate local ATP/ADP ratios. *J. Biol. Chem.* **265:** 5258–5266.

Saks, V.A., Chernousova, G.B., Voronkov, U.I., Smirnov, V.N., and Chazov, E.I., 1974, Study of energy transport mechanism in myocardial cells. *Circ. Res.* **35:** 138–149.

Saks, V.A., Chernousova, G.B., Gukovsky, D.E., Smirnov, V.N., and Chazov, E.I., 1975, Studies of energy transport in heart cells. Mitochondrial isoenzyme of creatine phosphokinase: kinetic properties and regulatory action of Mg^{2+} ions. *Eur. J. Biochem.* **57:** 273–290.

Saks, V.A., Lipina, N.V., Sharov, V.G., Smirnov, V.N., Chazov, E.I., and Grosse, R., 1977, The localization of the MM isoenzyme of creatine phosphokinase on the surface membrane of myocardial cells and its functional coupling to ouabain-inhibited (Na^+, K^+)-ATPase. *Biochim. Biophys. Acta* **465:** 550–558.

Saks, V.A., Kuznetsov, A.V., Kupriyanov, V.V., Miceli, M.V., and Jacobus, W.E., 1985, Creatine kinase of rat heart mitochondria. The demostration of functional coupling to oxidative phosphorylation in an inner membrane-matrix preparation. *J. Biol. Chem.* **260:** 7757–7764.

Saks, V.A., Kapelko, V.I., Kupriyanov, V.V., Kuznetsov, A.V., Lakomkin, V.L., Veksler, V.I., Sharov, V.G., Javadov, S.A., Seppet, E.K., and Kairane, C., 1989, Quantitative evaluation of relationship between cardiac energy metabolism and post-ischemic recovery of contractile function. *J. Mol. Cell. Cardiol.* **21:** 67–78.

Saks, V.A., Belikova, Y.O., and Kuznetsov, A.V., 1991, In vivo regulation of mitochondrial respiration in cardiomyocytes: specific restrictions for intracellular diffusion of ADP. *Biochim. Biophys. Acta* **1074:** 302–311.

Saks, V.A., Vasilyeva, E., Belikova, Yu.O., Kuznetsov, A.V., Lyapina, S., Petrova, L., and Perov, N.A., 1993, Retarded diffusion of ADP in cardiomyocytes: possible role of mitochondrial outer membrane and creatine kinase in cellular regulation of oxidative phosphorylation. *Biochim. Biophys. Acta* **1144:** 134–148.

Saks, V.A., Khuchua, Z.A., Vasilyeva, E.V., Belikova, Yu.O., and Kuznetsov, A., 1994, Metabolic compartmentation and substrate channeling in muscle cells. Role of coupled creatine kinases in vivo regulation of cellular respiration - a synthesis. *Mol. Cell. Biochem.* **133/134:** 155–192.

Saks, V.A., Kuznetsov, A.V., Khuchua, Z.A., Vasilyeva, E.V., Belikova, J.O., Kesvatera, T., and Tiivel, T., 1995, Control of cellular respiration in vivo by mitochondrial outer membrane and by creatine kinase. A new speculative hypothesis: possible involvement of mitochondrial-cytoskeleton interactions. *J. Mol. Cell. Cardiol.* **27:** 625–645.

Saks, V.A., and Aliev, M.K., 1996, Is there the creatine kinase equilibrium in working heart cells? *Biochem. Biophys. Res. Commun.* **227:** 360–367.

Saks, V.A., Kaambre, T., Sikk, P., Eimre, M., Orlova, E., Paju, K., Piirsoo, A., Appaix, F., Kay, L., Regiz-Zagrosek, V., Fleck, E., and Seppet, E., 2001, Intracellular energetic units in red muscle cells. *Biochem. J.* **356:** 643–657.

Saks, V., Kuznetsov, A.V., Andrienko, T., Usson, Y., Appaix, F., Guerrero, K., Kaambre, T., Sikk, P., Lemba, M., and Vendelin, M., 2003, Heterogeneity of ADP diffusion and regulation of respiration in cardiac cells. *Biophys. J.* **84:** 3436–3456.

Saks, V.A., Vendelin, M., Aliev, M.K., Kekelidze, T., and Engelbrecht, J., 2006a, Mechanisms and modeling of energy transfer between intracellular compartments. In: *Handbook of Neurochemistry and Molecular Neurobiology*, 3*rd* edition, vol. 5, Neuronal Energy Utilization (Dienel, G., and Gibson, G., eds.), Springer-Verlag, Berlin-Heidelberg, pp. 1–46.

Saks, V., Dzeja, P., Schlattner, U., Vendelin, M., Terzic, A., and Wallimann, T., 2006b, Cardiac system bioenergetics: metabolic basis of Frank-Starling law. *J. Physiol.* **571:** 253–273.

Sasaki, N., Sato, T., Marban, E., and O'Rourke, B., 2001, ATP consumption by uncoupled mitochondria activates sarcolemmal K_{ATP} channels in cardiac myocytes. *Am. J. Physiol.* **280:** H1882–H1888.

Sata, M., Sugiura, S., Yamashita, H., Momomura, S.I., and Serizawa, T., 1996, Coupling between myosin ATPase cycle and creatine kinase cycle facilitates cardiac actomyosin sliding in vitro: a clue to mechanical dysfunction during myocardial ischemia. *Circulation* **93:** 310–317.

Schlattner, U., Forstner, M., Eder, M., Stachowiak, O., Fritz-Wolf, K., and Wallimann, T., 1998, Functional aspects of the X-ray structure of mitochondrial creatine kinase: a molecular physiology approach. *Mol. Cell. Biochem.* **184:** 125–140.

Schlattner, U., Eder, M., Dolder, M., Khuchua, Z.A., Strauss, A.W., and Wallimann, T., 2000, Divergent enzyme kinetics and structural properties of the two human mitochondrial creatine kinase isoenzymes. *Biol. Chem.* **381:** 1063–1070.

Schlattner, U., Gehring, F., Vernoux, N., Tokarska-Schlattner, M., Neumann, D., Marcillat, O., Vial, C., and Wallimann, T., 2004, C-terminal lysines determine phospholipid interaction of sarcomeric mitochondrial creatine kinase. *J. Biol. Chem.* **279:** 24334–24342.

Schlattner, U., Tokarska-Schlattner, M., and Wallimann, T., 2006, Mitochondrial creatine kinase in human health and disease. *Biochim. Biophys. Acta* **1762:** 164–180.

Schlattner, U., and Wallimann, T., 2000, Octamers of mitochondrial creatine kinase differ in stability and membrane binding. *J. Biol. Chem.* **275:** 17314–17320.

Schlattner, U., Dolder, M., Wallimann, T., and Tokarska-Schlattner, M., 2001, Mitochondrial creatine kinase and mitochondrial outer membrane porin show a direct interaction that is modulated by calcium. *J. Biol. Chem.* **276:** 48027–48030.

Schlattner, U., and Wallimann T., 2004, Metabolite channeling: creatine kinase microcompartments. In: Encyclopedia of Biological Chemistry (Lennarz, W.J., and Lane, M.D., eds.), Academic Press, New York, USA, pp. 646–651.

Schlattner, U., and Wallimann, T., 2006, Molecular structure and function of mitochondrial creatine kinases. In: Creatine kinase – biochemistry, physiology, structure and function (Uversky, V.N., ed.), Nova Science Publishers, New York, USA, pp. 123–170.

Stachowiak, O., Schlattner, U., Dolder, M., and Wallimann, T., 1998, Oligomeric state and membrane binding behaviour of creatine kinase isoenzymes: implications for cellular function and mitochondrial structure. *Mol. Cell. Biochem.* **184:** 141–151.

Schlegel, J., Wyss, M., Schurch, U., Schnyder, T., Quest, A., Wegmann, G., Eppenberger, H.M., and Wallimann, T., 1988a, Mitochondrial creatine kinase from cardiac muscle and brain are two distinct isoenzymes but both form octameric molecules. *J. Biol.Chem.* **263:** 16963–16969.

Schlegel, J., Zurbriggen, B., Wegmann, G., Wyss, M., Eppenberger, H.M., and Wallimann, T., 1988b, Native mitochondrial creatine kinase forms octameric structures. I. Isolation of two interconvertible mitochondrial creatine kinase forms, dimeric and octameric mitochondrial creatine kinase: characterization, localization, and structure-function relationships. *J. Biol. Chem.* **263:** 16942–16953.

Schnyder, T., Rojo, M., Furter, R., and Wallimann, T., 1994, The structure of mitochondrial creatine kinase and its membrane binding properties. *Mol. Cell. Biochem.* **133/134:** 115–123.

Scholte, H.R., 1973, On the triple localization of creatine kinase in heart and skeletal muscle cells of the rat: evidence for the existence of myofibrillar and mitochondrial isoenzymes. *Biochim. Biophys. Acta* **305:** 413–427.

Selivanov, V.A., Alekseev, A.E., Hodgson, D.M., Dzeja, P.P., and Terzic, A., 2004, Nucleotide-gated K_{ATP} channels integrated with creatine and adenylate kinases: Amplification, tuning and sensing of energetic signals in the compartmentalized cellular environment. *Mol. Cell. Biochem.* **256/257:** 243–256.

Seppet, E., Kaambre, T., Sikk, P., Tiivel, T., Vija, H., Kay, L., Appaix, F., Tonkonogi, M., Sahlin, K., and Saks, V.A., 2001, Functional complexes of mitochondria with MgATPases of myofibrils and sarcoplasmic reticulum in muscle cells. *Biochim. Biophys. Acta* **1504:** 379–395.

Sharov, V.G., Saks, V.A., Smirnov, V. N., and Chazov, E.I., 1977, An electron microscopic histochemical investigation of creatine phosphokinase in heart cells. *Biochim. Biophys. Acta* **468:** 495–501.

Soboll, S., Conrad, A., and Hebisch, S., 1994, Influence of mitochondrial creatine kinase on the mitochondrial/extramitochondrial distribution of high energy phosphates in muscle tissue: evidence for the leak in the creatine shuttle. *Mol. Cell. Biochem.* **133/134:** 105–115.

Spindler, M., Niebler, R., Remkes, H., Horn, M., Lanz, T., and Neubauer, S., 2002, Mitochondrial creatine kinase is critically necessary for normal myocardial high-energy phosphate metabolism. *Am. J. Physiol.***283:** H680–H687.

Spindler, M., Meyer, K., Stromer, H., Leupold, A., Boehm, E., Wagner, H., and Neubauer, S., 2004, Creatine kinase-deficient hearts exhibit increased susceptibility to ischemia-reperfusion injury and impaired calcium homeostasis. *Am. J. Physiol.* **287:** H1039–H1045.

Stanley, W.C., Recchia, F.A., and Lopaschuk, G.D., 2005, Myocardial substrate metabolism in the normal and failing heart. *Physiol. Rev.* **85:** 1093–1129.

Steeghs, K., Benders, A., Oerlemans, F., de Haan, A., Heerschap, A., Ruitenbeek, W., Jost, C., van Deursen, J., Perryman, B., Pette, D., Bruckwilder, M., Koudijs, J., Jap, P., Veerkamp, J., and Wieringa, B., 1997, Altered Ca^{2+} responses in muscles with combined mitochondrial and cytosolic creatine kinase deficiencies. *Cell* **89:** 93–103.

Taegtmeyer, H., Wilson, C.R., Razeghi, P., and Sharma, S., 2005, Metabolic energetics and genetics in the heart. *Ann. N. Y. Acad. Sci.* **1047:** 208–218.

Veksler, V.I., Kuznetsov, A.V., Anflous, K., Mateo, P., van Deursen, J., Wieringa, B., and Ventura-Clapier, R., 1995, Muscle creatine-kinase deficient mice. II Cardiac and skeletal muscles exhibit tissue-specific adaptation of the mitochondrial function. *J. Biol. Chem.* **270:** 19921–19929.

Vendelin, M., Kongas, O., and Saks, V., 2000, Regulation of mitochondrial respiration in heart cells analyzed by reaction-diffusion model of energy transfer. *Am. J. Physiol. Cell Physiol.* **278:** C747–C764.

Vendelin, M., Lemba, M., and Saks, V.A., 2004a, Analysis of functional coupling: mitochondrial creatine kinase and adenine nucleotide translocase. *Biophys. J.* **87:** 696–713.

Vendelin, M., Eimre, M., Seppet, E., Peet, N., Andrienko, T., Lemba, M., Engelbrecht, J., Seppet, E.K., and Saks, V.A., 2004b, Intracellular diffusion of adenosine phosphates is locally restricted in cardiac muscle. *Mol. Cell. Biochem.* **256/257:** 229–241.

Vendelin, M., Beraud, N., Guerrero, K., Andrienko, T., Kuznetsov, A.V, Olivares, J., Kay, L., and Saks, V.A., 2005, Mitochondrial regular arrangement in muscle cells: a "crystal-like" pattern. *Am. J. Physiol. Cell Physiol.* **288:** C757–C767.

Ventura-Clapier, R., Mekhfi, H., and Vassort, G., 1987, Role of creatine kinase in force development in chemically skinned rat cardiac muscle. *J. Gen. Physiol.* **89:** 815–837.

Ventura-Clapier, R., Kuznetsov, A., Veksler, V., Boehm, E., and Anflous, K., 1998, Functional coupling of creatine kinases in muscles: species and tissue specificity. *Mol. Cell. Biochem.* **184:** 231–247.

Ventura-Clapier, R., Kaasik, A., and Veksler, V., 2004, Structural and functional adaptations of striated muscles to CK deficiency. *Mol. Cell. Biochem.* **256:** 29–41.

Vial, C., Godinot, C., and Gautheron, D., 1972, Membranes: creatine kinase (E.C.2.7.3.2.) in pig heart mitochondria. Properties and role in phosphate potential regulation. *Biochimie* **54:** 843–852.

Wallimann, T., Schlosser, T., and Eppenberger, H., 1984, Function of M-line-bound creatine kinase as intramyofibrillar ATP regenerator at the receiving end of the phosphorylcreatine shuttle in muscle. *J. Biol. Chem.* **259:** 5238–5246.

Wallimann, T., Wyss, M., Brdiczka, D., Nicolay, K., and Eppenberger, H.M., 1992, Intracellular compartmentation, structure and function of creatine kinase isoenzymes in tissues with high and fluctuating energy demands: the 'phosphocreatine circuit' for cellular energy homeostasis. *Biochem. J.* **281:** 21–40.

Wallimann, T., and Hemmer, W., 1994, Creatine kinase in non-muscle tissues and cells. *Mol. Cell. Biochem.* **133/134:** 193–220.

Wallimann, T., Dolder, M., Schlattner, U., Eder, M., Hornemann, T., O'Gorman, E., Ruck, A., and Brdiczka, D., 1998, Some new aspects of creatine kinase (CK): compartmentation, structure, function and regulation for cellular and mitochondrial bioenergetics and physiology. *Biofactors* **8:** 229–234.

Wegmann, G., Zanolla, E., Eppenberger, H.M., and Wallimann, T., 1992, In situ compartmentation of creatine kinase in intact sarcomeric muscle: the acto-myosin overlap zone as a molecular sieve. *J. Muscle Res. Cell Motil.* **13:** 420–435.

Weiss, J.N., and Lamp, S.T., 1987, Glycolysis preferentially inhibits ATP-sensitive K^+ channels in isolated guinea pig cardiac myocytes. *Science* **238:** 67–69.

Weiss, J.N., Ling, Y., and Qu, Z., 2006, Network perspectives of cardiovascular metabolism. *J. Lipid Res.* **47:** 2355–2366.

Williamson, J.R., Ford, C., Illingworth, J., and Safer, B., 1976, Coordination of citric acid cycle activity with electron transport flux. *Circ. Res.* **38:** 39–51.

Williamson, J.R., 1979, Mitochondrial function in the heart. *Annu. Rev. Physiol.* **41:** 485-506.

Wyss, M., and Kaddurah-Daouk, R., 2000, Creatine and creatinine metabolism. *Physiol. Rev.* **80:** 1107–1213.

Yagi, K., and Mase, R., 1962, Coupled reaction of creatine kinase and myosin A-adenosine triphosphatase. *J. Biol. Chem.* **237:** 397–403.

Yang, Z., and Steele, D.S., 2002, Effects of phosphocreatine on SR regulation in isolated saponin-permeabilised rat cardiac myocytes. *J. Physiol.* **539:** 767–777.

Zhou, G., Somasundaram, T., Blanc, E., Parthasarathy, G., Ellington, W.R., and Chapman, M.S., 1998, Transition state structure of arginine kinase: implications for catalysis of bimolecular reactions. *Proc. Natl. Acad. Sci. U.S.A.* **95:** 8449–8454.

CHAPTER 4

EXPRESSION AND FUNCTION OF AGAT, GAMT AND CT1 IN THE MAMMALIAN BRAIN

OLIVIER BRAISSANT, CLAUDE BACHMANN AND HUGUES HENRY

Clinical Chemistry Laboratory, Centre Hospitalier Universitaire Vaudois and University of Lausanne, CH-1011 Lausanne, Switzerland

Abstract: In mammals, creatine is taken up from the diet and can be synthesized endogenously by a two-step mechanism involving the enzymes arginine:glycine amidinotransferase (AGAT) and guanidinoacetate methyltransferase (GAMT). Creatine (Cr) is taken up by cells through a specific transporter, CT1. While the major part of endogenous synthesis of Cr is thought to occur in kidney, pancreas and liver, the brain widely expresses *AGAT*, *GAMT* and *CT1*, both during development and in adulthood. The adult central nervous system (CNS) has a limited capacity to take up Cr from periphery, and seems to rely more on its endogenous Cr synthesis. In contrast, the embryonic CNS might be more dependent on Cr supply from periphery than on endogenous synthesis. This review will focus on the expression and function of AGAT, GAMT and CT1 in the mammalian CNS, both during development and in adulthood. Emphasis will also be placed on their specific roles in the different cell types of the brain, to analyze which brain cells are responsible for the CNS capacity of (i) endogenous Cr synthesis and (ii) Cr uptake from the periphery, and which brain cells are the main Cr consumers. The potential role of CT1 as guanidinoacetate transporter between "AGAT-only" and "GAMT-only" expressing cells will also be explored

1. INTRODUCTION

The creatine (Cr)/phosphocreatine (PCr)/creatine kinase (CK) system is essential for the buffering and transport of high-energy phosphates. In the central nervous system (CNS), Cr has been demonstrated to play a role in the migration of growth cones and the growth of dendrites and axons, in the activity of Na^+/K^+-ATPase, in the release of different neurotransmitters, in the maintenance of the membrane potential, in Ca^{2+} homeostasis or in the restoration of ion gradients (Wallimann *et al.*, 1992; Wyss and Kaddurah-Daouk, 2000). Cr was also recently hypothesized to act as a central neuromodulator (Almeida *et al.*, 2006).

G.S. Salomons and M. Wyss (eds.), Creatine and Creatine Kinase in Health and Disease, 67–81.

In mammals, Cr is taken up from the diet, and can be synthesized endogenously by a two-step mechanism involving the two enzymes arginine:glycine amidinotransferase (AGAT) and guanidinoacetate methyltransferase (GAMT). Cr is distributed through the blood and is taken up by cells with high energy demands through a specific Cr transporter, CT1, which belongs to the Na^+ and Cl^- dependent neurotransmitter transporter family (for a review, see Wyss and Kaddurah-Daouk, 2000). In adult mammals, *AGAT* is mainly expressed in kidney and pancreas, and *GAMT* in liver and pancreas. In addition, both enzymes are expressed in various other tissues, albeit at lower levels. The highest expression of *CT1* is found in kidney (see Wyss and Kaddurah-Daouk, 2000, and references therein).

The mammalian brain, as well as primary cultures of brain cells and nerve cell lines, are also able to synthesize Cr (Pisano *et al.*, 1963; Van Pilsum *et al.*, 1972; Daly, 1985; Dringen *et al.*, 1998; Braissant *et al.*, 2002). In the adult CNS, *AGAT* and *GAMT* genes are expressed by all the main cell types of the brain (Lee *et al.*, 1998; Braissant *et al.*, 2001b). Primary brain cell cultures, either neuronal, glial or mixed, and neuroblastoma cell cultures, have Cr transport activity (Daly, 1985; Möller and Hamprecht, 1989; Braissant *et al.*, 2002). In the adult mammalian brain, the *CT1* gene is expressed in neurons and oligodendrocytes while astrocytes, including those at the blood brain barrier (BBB), do not express it (Guimbal and Kilimann, 1993; Schloss *et al.*, 1994; Happe and Murrin, 1995; Saltarelli *et al.*, 1996; Braissant *et al.*, 2001b; Ohtsuki *et al.*, 2002; Tachikawa *et al.*, 2004). The brain capillary endothelial cells, however, which establish the BBB, express *CT1* (Braissant *et al.*, 2001b) and are able to import Cr (Ohtsuki *et al.*, 2002).

Until recently, the materno-fetal transport of Cr had been demonstrated (Koszalka *et al.*, 1975; Davis *et al.*, 1978; Walker, 1979), but little information was available on *AGAT*, *GAMT* and *CT1* gene expression during embryonic development, particularly for the CNS. *AGAT* and *GAMT* were found in whole extracts of the developing mouse embryo (Sandell *et al.*, 2003; Schmidt *et al.*, 2004), and *CT1* has been demonstrated in the neuraxis of the rat embryo (Schloss *et al.*, 1994). Our recent detailed analysis of *AGAT*, *GAMT* and *CT1* expression during the development of the rat embryo has revealed that *AGAT* and *CT1* are widely expressed as soon as E12.5, while *GAMT* expression could only be demontrated at a later stage (E18.5) in the developing rat brain (Braissant *et al.*, 2005b).

The CNS is the main organ affected in patients suffering from Cr deficiency syndromes due to either AGAT, GAMT or CT1 deficiency (Stöckler *et al.*, 1994; Salomons *et al.*, 2001; Item *et al.*, 2001). These patients present neurological symptoms in infancy and have mental retardation (see chapter 8; Stockler *et al.*, 2007; (Schulze *et al.*, 1997; Battini *et al.*, 2002; DeGrauw *et al.*, 2002)). All three deficiencies are characterized by an absence, or a severe decrease, of Cr in CNS (Stromberger *et al.*, 2003; Sykut-Cegielska *et al.*, 2004). AGAT and GAMT deficient patients can be treated with oral Cr supplementation. Although very high doses of Cr are being used, the replenishment of cerebral Cr takes months and results only in partial restoration of the cerebral Cr pool (Stöckler *et al.*, 1996;

Ganesan *et al.*, 1997; Schulze *et al.*, 1998; Item *et al.*, 2001; Battini *et al.*, 2002). Cr supplementation of CT1 deficient patients does not restore cerebral Cr levels (Cecil *et al.*, 2001; DeGrauw *et al.*, 2002; Bizzi *et al.*, 2002; Poo-Arguelles *et al.*, 2006).

The expression of *AGAT* and *GAMT* in the mammalian brain, the absence of *CT1* in the astrocytes lining the BBB (Braissant *et al.*, 2001b), and the fact that Cr poorly crosses the BBB of rodents (Ohtsuki *et al.*, 2002; Perasso *et al.*, 2003) as well as that of AGAT and GAMT deficient patients (Stromberger *et al.*, 2003; Sykut-Cegielska *et al.*, 2004) has led us to hypothesize that the postnatal and adult CNS might depend, at least in part, on its own Cr synthesis (Braissant *et al.*, 2001b). However, this question remains open, as CT1 deficient patients, who are expected to express *AGAT* and *GAMT* in their CNS, are nevertheless also depleted in intracerebral Cr (Salomons *et al.*, 2003). In contrast, our studies on develpmental biology of the rat brain suggest that in the early embryo, Cr uptake by the CNS from the periphery must predominate, with endogenous synthesis of Cr appearing only late in the embryonic development of the brain (Braissant *et al.*, 2005b).

2. EXPRESSION OF AGAT, GAMT AND CT1 IN THE ADULT BRAIN

It has long been thought that most, if not all, of the Cr necessary for the brain was of peripheral origin, be it taken from the diet or synthesized endogenously through the AGAT and GAMT activities in kidney, pancreas and liver (Wyss and Kaddurah-Daouk, 2000). However, since a long time it is also known that the mammalian brain is able to synthesize Cr (Pisano *et al.*, 1963; Van Pilsum *et al.*, 1972), which is also true for primary cultures of brain cells and nerve cell lines (Daly, 1985; Dringen *et al.*, 1998; Braissant *et al.*, 2002). While the highest expression of *AGAT* and *GAMT* is indeed found in kidney, pancreas and liver, these genes are also expressed, at lower levels, in various other tissues, including CNS (Lee *et al.*, 1998; Wyss and Kaddurah-Daouk, 2000; Braissant *et al.*, 2001b; Schmidt *et al.*, 2004; Tachikawa *et al.*, 2004; Nakashima *et al.*, 2005).

AGAT is expressed throughout the adult rat brain, shown both at the mRNA and protein levels in all the main CNS structures, with particularly high levels in the telencephalon and cerebellum (Braissant *et al.*, 2001b, 2005a). *AGAT* mRNA is also expressed in the rat retina (Nakashima *et al.*, 2005). The adult rat CNS expresses *AGAT* in all its main cell types, namely neurons, astrocytes and oligodendrocytes (Table 1 and Figure 1; Braissant *et al.*, 2001b). In the structures regulating the exchanges between the periphery and the CNS as well as between the brain parenchyma and cerebrospinal fluid (CSF), *AGAT* is expressed in microcapillary endothelial cells (MCEC) and the astrocytes contacting them (at the BBB), as well as in the choroid plexus and ependymal epithelia (Table 1 and Figure 1; Braissant *et al.*, 2001b).

GAMT is also expressed throughout the main structures of the adult mammalian brain, as shown in rat, mouse and human both at the mRNA and protein levels (Braissant *et al.*, 2001b, 2005a; Schmidt *et al.*, 2004; Tachikawa *et al.*, 2004).

Table 1. Expression of AGAT, GAMT and CT1 mRNAs and proteins in the different cell types of the adult brain, including blood brain barrier and blood CSF barrier. Presence (+) or absence (−) in the different cell types. For detailed structures of the brain, see Braissant *et al.*, 2001b; Ohtsuki *et al.*, 2002; Schmidt *et al.*, 2004; Tachikawa *et al.*, 2004).

	AGAT	GAMT	CTI
Neurons	+	+	+
Oligodendrocytes	+	+	+
Astrocytes	+	+	−
Microcapillary endothelial cells	+	−	+
Choroid plexus	+	+	+
Ependymal epithelium	+	+	+

Particularly high levels were identified in telencephalon (corpus callosum and hippocampus in particular), pons nuclei and cerebellum. Furthermore, *GAMT* is expressed in neurons, astrocytes and oligodendrocytes, with higher levels found in both glial cell types (Table 1 and Figure 1; Braissant *et al.*, 2001b; Schmidt *et al.*, 2004; Tachikawa *et al.*, 2004; Nakashima *et al.*, 2005). *GAMT* is not expressed in MCEC but is present in the astrocytes contacting them (at the BBB), as well as in the choroid plexus and ependymal epithelia (Table 1 and Figure 1; Braissant *et al.*, 2001b; Tachikawa *et al.*, 2004).

Organotypic rat cortical cultures, primary brain cell cultures – either neuronal, glial or mixed – and neuroblastoma cell cultures present Cr transport activity (Daly, 1985; Möller and Hamprecht, 1989; Braissant *et al.*, 2002; Almeida

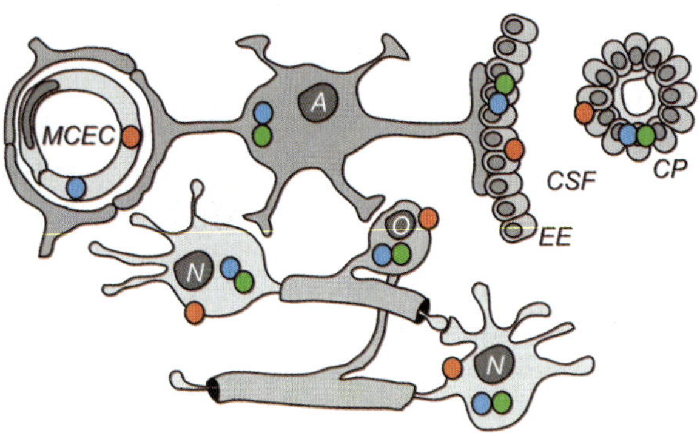

Figure 1. Expression of AGAT (•), GAMT (•) and CT1 (•) in the different cell types of the adult brain. AGAT and GAMT are expressed by neurons (N), astrocytes (A) and oligodendrocytes (O), as well as in choroid plexus (CP) and ependymal (EE) epithelia. AGAT is also found in microcapillary endothelial cells (MCEC). CT1 is expressed by MCEC but not by astrocytes sheathing them, and is also found in neurons and oligodendrocytes, as well as in choroid plexus and ependymal epithelia. CSF: cerebrospinal fluid; MC: microcapillary.

et al., 2006). *In vivo*, mouse and rat CNS are able to take up Cr from the blood against its concentration gradient, but this uptake of Cr through the BBB seems relatively inefficient (Ohtsuki *et al.*, 2002; Perasso *et al.*, 2003). The *CT1* gene is expressed throughout the main regions of the adult mammalian brain (Guimbal and Kilimann, 1993; Schloss *et al.*, 1994; Happe and Murrin, 1995; Saltarelli *et al.*, 1996; Braissant *et al.*, 2001b). In rat and mouse, *CT1* is found – both at the mRNA and protein levels – in neurons and oligodendrocytes, but, in contrast to AGAT and GAMT, cannot be detected in astrocytes (Table 1 and Figure 1; Braissant *et al.*, 2001b, 2005a; Ohtsuki *et al.*, 2002; Tachikawa *et al.*, 2004). This holds true also for the retina, where *CT1* is expressed in retinal neurons, but not in astrocytes (Nakashima *et al.*, 2004; Acosta *et al.*, 2005). In contrast to the absence of *CT1* in astrocytes lining microcapillaries, MCEC which form the BBB and the blood-retina barrier do express *CT1* (Table 1 and Figure 1; Braissant *et al.*, 2001b; Ohtsuki *et al.*, 2002; Tachikawa *et al.*, 2004; Nakashima *et al.*, 2004; Acosta *et al.*, 2005), and are able to take up Cr (Ohtsuki *et al.*, 2002). *CT1* is also expressed by choroid plexus and the ependymal epithelia (Table 1 and Figure 1; Braissant *et al.*, 2001b).

3. EXPRESSION OF AGAT, GAMT AND CT1 IN THE DEVELOPING BRAIN

The Cr/PCr/CK system plays essential roles in energy homeostasis during vertebrate embryonic development, in particular in tissues such as developing CNS and muscles (Wallimann *et al.*, 1992). Many structures of the vertebrate embryo express the CK genes at very early stages (Lyons *et al.*, 1991; Dickmeis *et al.*, 2001), and Cr concentrations between 5 and 8 mmol/kg wet weight have been measured in the fetal brain of rat and human, depending on the gestational stage (Miller *et al.*, 2000; Kreis *et al.*, 2002). Parts of the developmental needs of the CNS for Cr can be fulfilled by transport of Cr from the mother to the embryo (Koszalka *et al.*, 1975; Davis *et al.*, 1978; Walker, 1979). However, until recently, little information was available on *AGAT*, *GAMT* and *CT1* gene expression in the developing embryonic mammalian CNS: *AGAT* and *GAMT* had been found in whole extracts of the developing mouse embryo (Sandell *et al.*, 2003; Schmidt *et al.*, 2004), and *CT1* had been demonstrated in the neuraxis of the rat embryo (Schloss *et al.*, 1994). So far, it is not known whether alterations in Cr metabolism, as found in AGAT, GAMT or CT1 deficiencies, impair the development of the embryo, and particularly its CNS. The only functional indication that Cr metabolism genes are important for the mammalian embryo comes from GAMT knock-out mice, which present an increase in perinatal mortality (Schmidt *et al.*, 2004). Our recent detailed analysis, both at the mRNA and protein level, during the development of the rat embryo, has shed new light on AGAT, GAMT and CT1 expression and function in the developing mammalian CNS (Braissant *et al.*, 2005b).

We have shown that *AGAT* and *GAMT* are expressed in the hepatic primordium as soon as day 12.5, and then progressively acquire their adult pattern of expression, with high levels of *AGAT* in kidney and pancreas, and high levels of *GAMT* in liver and pancreas. As in adulthood, however, many other embryonic tissues also express *AGAT* and/or *GAMT*, including CNS. As soon as E12.5 in the rat, *AGAT* is expressed in the whole nervous system parenchyma, and remains expressed in CNS until birth (Table 2). In contrast, *GAMT* appears later, and only in scarce regions of the developing brain (Table 2): at E15.5, *GAMT* is detected only in striatum and pons, while at E18.5, it is found in neocortex, hippocampus, striatum, pallidum and spinal cord. In CNS, *CT1* is found expressed throughout development in all the main regions of the brain, with a similar pattern as that found for *AGAT*. An interesting pattern of expression was observed in the two structures responsible for metabolite exchange between the periphery and the CNS, i.e. the choroid plexus, and MCEC which build the BBB. *CT1* is found highly expressed in the choroid plexus (E15.5 and E18.5), but is absent from MCEC (E18.5), a situation reversed as compared to adulthood (see above, and Braissant *et al.*, 2001b). *AGAT* is absent from choroid plexus but appears in MCEC at E18.5, while *GAMT* is absent from both structures during the whole embryonic development (Table 2).

Table 2. Expression of AGAT, GAMT and CT1 mRNA and protein in the main structures of the developing embryonic brain, including the blood-brain and blood-CSF barriers. Presence (+) or absence (−) in the different cell types is indicated. E12.5, E15.5, E18.5: embryonic days 12.5, 15.5 and 18.5; S: striatum; P: pons nuclei. For detailed structures of the brain, see Braissant *et al.* (2005b).

	E12.5	E15.5	E18.5
AGAT			
Forebrain	+	+	+
Midbrain	+	+	+
Hindbrain	+	+	+
Spinal cord		+	+
Choroid plexus		−	−
Microcapillary endothelial cells			+
GAMT			
Forebrain	−	± (S)	+
Midbrain	−	−	−
Hindbrain	−	± (P)	−
Spinal cord		−	+
Choroid plexus		−	−
Microcapillary endothelial cells			−
CT1			
Forebrain	+	+	+
Midbrain	+	+	+
Hindbrain	+	+	+
Spinal cord		+	+
Choroid plexus		+	+
Microcapillary endothelial cells			−

AGAT, GAMT and CT1 deficiencies induce neurological symptoms in infancy and can lead to severe neurodevelopmental delay (Stromberger *et al.*, 2003; Salomons *et al.*, 2003). Despite developmental improvement upon Cr supplementation and partial recovery of the cerebral Cr pool in AGAT and GAMT deficient patients, sequelae of the developmental delay and mental retardation remain (Stromberger *et al.*, 2003; Schulze, 2003; Sykut-Cegielska *et al.*, 2004). In the case of GAMT deficiency, this may be partly due to toxicity of guanidinoacetic acid (GAA) accumulating in the CNS (Schulze *et al.*, 2001; Neu *et al.*, 2002). Despite the fact that most patients with AGAT, GAMT or CT1 deficiencies are diagnosed during infancy, and that significant damage to their brain occurs postnatally, the patterns of expression of *AGAT*, *GAMT* and *CT1* during CNS embryogenesis suggest that some of the irreversible damage observed in Cr deficient patients, while pre-symptomatic, may already occur *in utero* or in early post-natal development (Braissant *et al.*, 2005b). This is coherent with recent studies on Cr deficient neonates who, already at birth, present (i) a decreased Cr concentration together with either a decreased (AGAT deficiency) or increased (GAMT deficiency) GAA concentration in plasma and urine, and (ii) an absence or significant decrease of cerebral Cr levels. In addition, pre-symptomatic treatment of neonates by Cr supplementation prevents the phenotypic expression of AGAT and GAMT deficiencies (Battini *et al.*, 2006; Schulze *et al.*, 2006).

4. FUNCTION OF AGAT, GAMT AND CT1 IN THE MAMMALIAN BRAIN

The total level of Cr (Cr + PCr) and the CK activity are well correlated in the mammalian brain (Wyss and Kaddurah-Daouk, 2000), and their highest levels are reached in brain cells with particularly high and fluctuating energy demands, where *AGAT*, *GAMT* and *CT1* are also expressed (Holtzman *et al.*, 1989; Wallimann and Hemmer, 1994; Hemmer *et al.*, 1994; Whittingham *et al.*, 1995; Kaldis *et al.*, 1996; Wang and Li, 1998; Braissant *et al.*, 2001b, 2005b; Ohtsuki *et al.*, 2002; Tachikawa *et al.*, 2004; Acosta *et al.*, 2005).

The presence of AGAT and GAMT in every cell type of the brain (see above) suggests that the adult CNS must be able to synthesize Cr and thus secure part, if not all, of its Cr needs (Braissant *et al.*, 2001b, 2005b; Ohtsuki *et al.*, 2002; Schmidt *et al.*, 2004; Tachikawa *et al.*, 2004). The presence of *CT1* in neurons and oligodendrocytes (but not in astrocytes, see above) suggests transport of Cr between cells synthesizing Cr and cells with high energy requests (Braissant *et al.*, 2001b; Ohtsuki *et al.*, 2002).

Higher amounts of CK and PCr have been found in glial cells compared to neurons (Holtzman *et al.*, 1989; Hemmer *et al.*, 1994). Brain cell primary 3D cultures composed of mixed oligodendrocytes, astrocytes and neurons synthesize 10 times more Cr than their neuron-enriched counterparts (Braissant *et al.*, 2002), and primary astrocytes are able to synthesize Cr (Dringen *et al.*, 1998). Müller glial cells in the retina synthesize their own Cr (Nakashima *et al.*, 2005), and the GAMT protein is enriched in glial cells (oligodendrocytes and astrocytes) as compared to neurons in the mouse brain (Tachikawa *et al.*, 2004). These data suggest that glial cells may actually not only cover their own Cr needs but also partially those of the neurons.

Interestingly, we have observed the presence of *GAMT* and *CT1* mRNAs in neuronal processes (axons and dendrites), suggesting that *GAMT* and *CT1* transcripts might be transported along these processes in order to allow brain cells to translate these proteins at the sites where they are needed (Braissant *et al.*, 2001b). This would allow the cell to respond rapidly to immediate peripheral needs in, or recycling of, Cr. This process is of primary importance 1) in CNS development during synaptogenesis or growth cone migration which has been shown to be coupled directly to CK (mammals) or arginine kinase (insects) activity (Wang et al., 1998), and 2) in adulthood for the recently described role of Cr as central neuromodulator (see chapter 8; Stockler *et al.*, 2007; Almeida *et al.*, 2006). The transcript of the neuronal cationic amino acid transporter CAT3, which allows the uptake of arginine, the limiting substrate for Cr synthesis, is also found in neuronal processes (Braissant *et al.*, 1999).

5. ENDOGENOUS SYNTHESIS *VERSUS* UPTAKE OF CREATINE BY THE BRAIN

The long-term treatment of AGAT and GAMT deficient patients with very high doses of Cr replenishes their brain Cr pools very slowly and only partially (Stromberger *et al.*, 2003; Sykut-Cegielska *et al.*, 2004). Moreover, the blood to brain transport of Cr through the BBB has been demonstrated, but is relatively inefficient (Ohtsuki *et al.*, 2002; Perasso *et al.*, 2003). These *in vivo* data suggest that in the adult brain, the BBB has a limited permeability for Cr, which might be explained – despite the expression of *CT1* by MCEC and their capacity to import Cr – by the absence of *CT1* from astrocytes sheathing microcapillaries (Figure 2; Braissant *et al.*, 2001b; Ohtsuki *et al.*, 2002; Tachikawa *et al.*, 2004; Nakashima *et al.*, 2004; Acosta *et al.*, 2005).

One possibility for the restricted entry of Cr into the brain parenchyma, without going through astrocytes, could be the use of the limited surface of CNS capillary endothelium that is free of astrocytic endings (Figure 2; Virgintino *et al.*, 1997; Ohtsuki *et al.*, 2004). Thus, under normal physiological conditions, the adult mammalian brain might depend more on its own Cr synthesis, through the expression of *AGAT* and *GAMT*, than on Cr supply from the blood (Braissant *et al.*, 2001b). The brain capacity for Cr synthesis would thus depend on the efficient supply of arginine, the limiting substrate for Cr synthesis, from blood to CNS, and then also on the local trafficking of arginine between brain cells. We have shown that the cationic amino acid transporters might fulfill these roles in the adult rat brain, as *CAT1* is expressed at the BBB as well as ubiquitously in neuronal and glial cells, as *CAT2(B)* is present in neurons and oligodendrocytes, and as *CAT3* is restricted to neurons (Braissant *et al.*, 1999, 2001a).

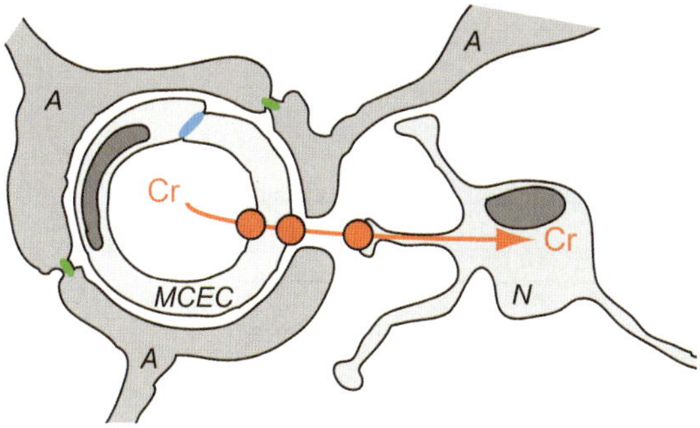

Figure 2. Expression of CT1 (•) in the adult BBB. CT1 is expressed on luminal and basal sides of microcapillary endothelial cells (MCEC), but not by astrocytes (A). CT1 is expressed by neurons (N) and oligodendrocytes. The restricted transport of creatine (Cr) from blood to brain through the BBB might be due to the limited surface of microcapillaries that are not covered by astrocytic endings. •: tight junctions; •: gap junctions.

The developing brain, in particular during embryogenesis, presents a different situation. While *AGAT* is expressed as soon as E12.5 in the whole CNS parenchyma and increases towards E18.5, *GAMT* expression is delayed in the developing brain, and remains at low levels (Table 2). Thus, in contrast to the adult rat CNS which appears able to synthesize its own Cr by expressing *AGAT* and *GAMT* in most brain cell types (Braissant *et al.*, 2001b), endogenous synthesis of Cr in CNS might appear only at the end of embryogenesis, and be restricted to discrete regions of the brain (Braissant *et al.*, 2005b). In consequence, the embryonic mammalian brain might depend mainly on extra-CNS supply of Cr, be it of maternal origin or synthesized in other tissues of the embryo, for the main part of its development time. This is supported by the high expression of *CT1* in the whole embryonic CNS. Interestingly, *CT1* is expressed in the choroid plexus of the rat, both in embryos and adults (Figure 3; Braissant *et al.*, 2001b, 2005b). In contrast, CT1 is absent from embryonic MCEC, while these cells express it in the adult rat CNS (Figure 3; Braissant *et al.*, 2001b, 2005b; Ohtsuki *et al.*, 2002). As choroid plexus differentiates earlier than brain capillaries and participates in early trophic supply for CNS (Dziegielewska *et al.*, 2001; Engelhardt, 2003), one might speculate that before angiogenesis occurs in CNS parenchyma, extra-CNS Cr is supplied from blood to CSF through the choroid plexus. Cr would then be available for the whole embryonic brain through CSF circulation (Segal, 2000) and through the observed high levels of *CT1* in the developing neuroepithelium, particularly in ependymal epithelium along ventricles (Figure 3; Braissant *et al.*, 2005b).

As the brain develops and enlarges and CNS angiogenesis progresses, the ratio of exchange surfaces in choroid plexus and CNS microcapillaries shifts to predominance of brain microcapillaries (Segal, 2000; Dziegielewska *et al.*, 2001; Engelhardt, 2003). Thus, at the end of embryonic development and then postnatally, Cr supply to the brain may occur preferentially at microcapillaries, with

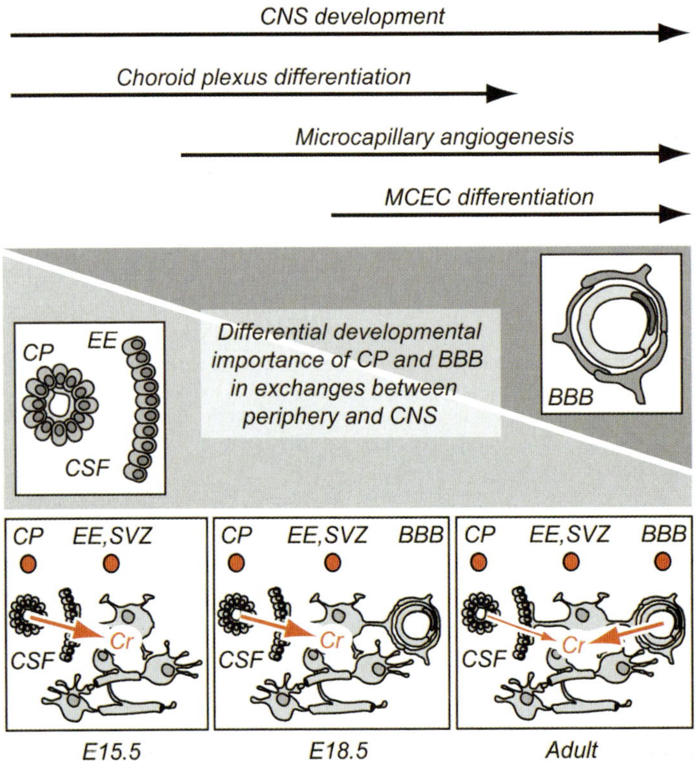

Figure 3. Expression of CT1 (•) in the developing blood-cerebrospinal fluid (CSF) and blood-brain (BBB) barriers of the rat. During the embryonic life, the brain uptake of Cr through CT1 probably occurs predominantly through choroid plexus (CP), while in adulthood, the ratio is reversed and Cr principally enters through the BBB. E15.5, E18.5: embryonic days 15.5 and 18.5; EE: ependymal epithelium, SVZ: cortical subventricular zone, MCEC: microcapillary endothelial cells.

CT1 being up-regulated in MCEC at the end of the fetal life and postnatally (Figure 3; Braissant *et al.*, 2001b; Ohtsuki *et al.*, 2002). It should be emphasized, however, that supply of Cr from blood to postnatal or adult brain is very likely of less quantitative importance than intra-cerebral Cr synthesis, as described above.

6. DEFICIENCIES IN AGAT, GAMT AND CT1: WHAT CAN WE LEARN FROM THEIR RESPECTIVE EXPRESSION?

The wide pattern of expression of *AGAT, GAMT* and *CT1* genes in the mammalian brain, which has been documented in every main region of rat (*AGAT*, *GAMT* and *CT*1), mouse (*GAMT* and *CT*1) and human (*GAMT*) CNS (Braissant *et al.*, 2001a, 2005b; Schmidt *et al.*, 2004; Tachikawa *et al.*, 2004; Galbraith *et al.*, 2006), probably accounts for the diverse phenotypic neurological spectrum observed in AGAT,

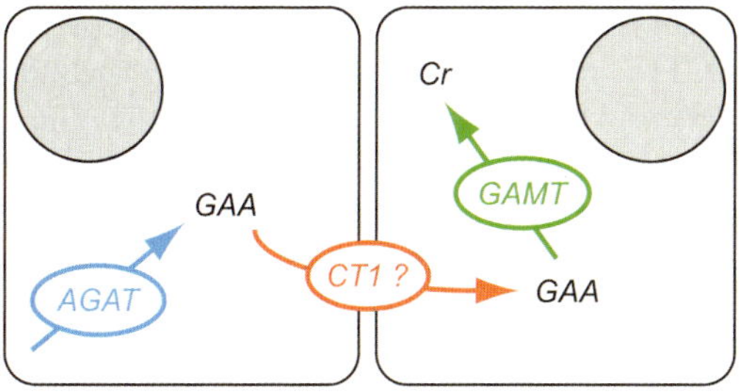

Figure 4. AGAT and GAMT expression in distinct brain cells. Is GAA transported through CT1, from AGAT expressing cells to GAMT expressing cells, to allow endogenous synthesis of Cr within the CNS?.

GAMT and CT1 deficient patients (Schulze, 2003; Battini *et al.*, 2006; Mercimek-Mahmutoglu *et al.*, 2006; Anselm *et al.*, 2006). The activation of $GABA_A$ receptors by GAA (Neu *et al.*, 2002) and the recently proposed role of Cr as co-transmitter on the widely distributed GABA postsynaptic receptors (Almeida *et al.*, 2006) might also contribute to this phenotypic diversity. The same is true, in case of GAMT deficiency and therefore accumulation of GAA, for the potential epileptogenic role of GAA (Schulze *et al.*, 2001), and its inhibitory effect on Na^+/K^+-ATPase and CK (Zugno *et al.*, 2006).

More particularly, specific features of *AGAT, GAMT* and *CT1* expression may contribute in the future to the understanding of the clinical characteristics of CT1 deficiency. The hypothesis of endogenous Cr synthesis in the brain might seem contradictory with the *in vivo* characteristics of CT1 deficiency. Despite the presence of *AGAT* and *GAMT* expression in the CNS of patients affected with CT1 deficiency, an absence or very low level of Cr in the brain is found, as in the case of AGAT or GAMT deficiency (Salomons *et al.*, 2003). This apparent contradiction might be explained by a recent observation we have made on the CNS expression of AGAT, GAMT and CT1: while AGAT and GAMT can be found in every cell type of the brain (Braissant *et al.*, 2001b), they rarely seem co-expressed within the same cell. This suggests that GAA, which is known to compete for Cr uptake through CT1 (Saltarelli *et al.*, 1996; Ohtsuki *et al.*, 2002), might be transported, through CT1, from AGAT to GAMT expressing cells, for Cr to be synthesized within the CNS (Figure 4; Braissant *et al.*, 2005a; Braissant and Henry, unpublished data). This could explain the absence of Cr synthesis in the brain of CT1 deficient patients.

7. CONCLUSIONS

AGAT, GAMT and *CT1* are widely expressed in the CNS, both during development and in adulthood. Most probably due to the absence of CT1 on the astrocytic feet sheathing microcapillaries, the adult brain has a limited capacity to take up Cr from the periphery, as has been shown *in vivo* both for mouse and human. Thus, the adult CNS seems to rely more on endogenous Cr synthesis than on Cr uptake from the blood. In contrast, *GAMT* is expressed only at the end of the embryonic brain development, while CT1 is present on the first exchange structures between CNS and periphery, namely choroid plexus and ependymal epithelia. Thus, the embryonic CNS might be more dependent on Cr supply from periphery than on endogenous synthesis, at least in its first stages of development. Among the questions remaining, more work is needed to understand the fine cell-to-cell and region-specific expression and function of AGAT, GAMT and CT1 in the brain. In particular, the potential role of CT1 as GAA transporter between "AGAT-only" and "GAMT-only" expressing cells has to be considered.

ACKNOWLEDGEMENTS

Our work is supported by the Swiss National Science Foundation, grants no. 3100-063892, 3100A0-100778 and 3100A0-116859.

REFERENCES

Acosta, M.L., Kalloniatis, M., and Christie, D.L., 2005, Creatine transporter localization in developing and adult retina: importance of creatine to retinal function. *Am. J. Physiol. Cell Physiol.* **289:** C1015–C1023.

Almeida, L.S., Salomons, G.S., Hogenboom, F., Jakobs, C., and Schoffelmeer, A.N., 2006, Exocytotic release of creatine in rat brain. *Synapse* **60:** 118–123.

Anselm, I.M., Alkuraya, F.S., Salomons, G.S., Jakobs, C., Fulton, A.B., Mazumdar, M., Rivkin, M., Frye, R., Poussaint, T.Y., and Marsden, D., 2006, X-linked creatine transporter defect: a report on two unrelated boys with a severe clinical phenotype. *J. Inherit. Metab. Dis.* **29:** 214–219.

Battini, R., Alessandri, M.G., Leuzzi, V., Moro, F., Tosetti, M., Bianchi, M.C., and Cioni, G., 2006, Arginine:glycine amidinotransferase (AGAT) deficiency in a newborn: early treatment can prevent phenotypic expression of the disease. *J. Pediatr.* **148:** 828–830.

Battini, R., Leuzzi, V., Carducci, C., Tosetti, M., Bianchi, M.C., Item, C.B., Stöckler-Ipsiroglu, S., and Cioni, G., 2002, Creatine depletion in a new case with AGAT deficiency: clinical and genetic study in a large pedigree. *Mol. Genet. Metab.* **77:** 326–331.

Bizzi, A., Bugiani, M., Salomons, G.S., Hunneman, D.H., Moroni, I., Estienne, M., Danesi, U., Jakobs, C., and Uziel, G., 2002, X-linked creatine deficiency syndrome: a novel mutation in creatine transporter gene SLC6A8. *Ann. Neurol.* **52:** 227–231.

Braissant, O., Villard, A., Henry, H., Speer, O., Wallimann, T., and Bachmann, C., 2005a, Synthesis and transport of creatine in the central nervous system. In Clinical and molecular aspects of defects in creatine and polyol metabolism, Jakobs, C., and Stöckler-Ipsiroglu, S., eds. (SPS Verlagsgesellschaft, Heilbronn, Germany), pp. 49–63.

Braissant, O., Gotoh T., Loup, M., Mori M., and Bachmann, C., 2001a, Differential expression of the cationic amino acid transporter 2(B) in the adult rat brain. *Mol. Brain Res.* **91:** 189–195.

Braissant, O., Gotoh, T., Loup, M., Mori, M., and Bachmann, C., 1999, L-arginine uptake, the citrulline-NO cycle and arginase II in the rat brain: an in situ hybridization study. *Mol. Brain Res.* **70:** 231–241.

Braissant, O., Henry, H., Loup, M., Eilers, B., and Bachmann, C., 2001b, Endogenous synthesis and transport of creatine in the rat brain: an in situ hybridization study. *Mol. Brain Res.* **86:** 193–201.

Braissant, O., Henry, H., Villard, A.M., Speer, O., Wallimann, T., and Bachmann, C., 2005b, Creatine synthesis and transport during rat embryogenesis: Spatiotemporal expression of AGAT, GAMT and CT1. *BMC Dev. Biol.* **5:** 9.

Braissant, O., Henry, H., Villard, A.M., Zurich, M.G., Loup, M., Eilers, B., Parlascino, G., Matter, E., Boulat, O., Honegger, P., and Bachmann, C., 2002, Ammonium-induced impairment of axonal growth is prevented through glial creatine. *J. Neurosci.* **22:** 9810–9820.

Cecil, K.M., Salomons, G.S., Ball, W.S., Jr., Wong, B., Chuck, G., Verhoeven, N.M., Jakobs, C., and DeGrauw, T.J., 2001, Irreversible brain creatine deficiency with elevated serum and urine creatine: a creatine transporter defect? *Ann. Neurol.* **49:** 401–404.

Daly, M.M., 1985, Guanidinoacetate methyltransferase activity in tissues and cultured cells. *Arch. Biochem. Biophys.* **236:** 576–584.

Davis, B.M., Miller, R.K., Brent, R.L., and Koszalka, T.R., 1978, Materno-fetal transport of creatine in the rat. *Biol. Neonate* **33:** 43–54.

DeGrauw, T.J., Salomons, G.S., Cecil, K.M., Chuck, G., Newmeyer, A., Schapiro, M.B., and Jakobs, C., 2002, Congenital creatine transporter deficiency. *Neuropediatrics* **33:** 232–238.

Dickmeis, T., Rastegar, S., Aanstad, P., Clark, M., Fischer, N., Plessy, C., Rosa, F., Korzh, V., and Strahle, U., 2001, Expression of brain subtype creatine kinase in the zebrafish embryo. *Mech. Dev.* **109:** 409–412.

Dringen, R., Verleysdonk, S., Hamprecht, B., Willker, W., Leibfritz, D., and Brand, A., 1998, Metabolism of glycine in primary astroglial cells: synthesis of creatine, serine, and glutathione. *J. Neurochem.* **70:** 835–840.

Dziegielewska, K.M., Ek, J., Habgood, M.D., and Saunders, N.R., 2001, Development of the choroid plexus. *Microsc. Res. Tech.* **52:** 5–20.

Engelhardt, B., 2003, Development of the blood-brain barrier. *Cell Tissue Res.* **314:** 119–129.

Galbraith, R.A., Furukawa, M., and Li, M., 2006, Possible role of creatine concentrations in the brain in regulating appetite and weight. *Brain Res.* **1101:** 85–91.

Ganesan, V., Johnson, A., Connelly, A., Eckhardt, S., and Surtees, R.A., 1997, Guanidinoacetate methyltransferase deficiency: new clinical features. *Pediatr. Neurol.* **17:** 155–157.

Guimbal, C. and Kilimann, M.W., 1993, A Na^+-dependent creatine transporter in rabbit brain, muscle, heart, and kidney. cDNA cloning and functional expression. *J. Biol. Chem.* **268:** 8418–8421.

Happe, H.K. and Murrin, L.C., 1995, In situ hybridization analysis of CHOT1, a creatine transporter, in the rat central nervous system. *J. Comp. Neurol.* **351:** 94–103.

Hemmer, W., Zanolla, E., Furter-Graves, E.M., Eppenberger, H.M., and Wallimann, T., 1994, Creatine kinase isoenzymes in chicken cerebellum: specific localization of brain-type creatine kinase in Bergmann glial cells and muscle-type creatine kinase in Purkinje neurons. *Eur. J. Neurosci.* **6:** 538–549.

Holtzman, D., McFarland, E., Moerland, T., Koutcher, J., Kushmerick, M.J., and Neuringer, L.J., 1989, Brain creatine phosphate and creatine kinase in mice fed an analogue of creatine. *Brain Res.* **483:** 68–77.

Item, C.B., Stöckler-Ipsiroglu, S., Stromberger, C., Muhl, A., Alessandri, M.G., Bianchi, M.C., Tosetti, M., Fornai, F., and Cioni, G., 2001, Arginine:glycine amidinotransferase deficiency: the third inborn error of creatine metabolism in humans. *Am. J. Hum. Genet.* **69:** 1127–1133.

Kaldis, P., Hemmer, W., Zanolla, E., Holtzman, D., and Wallimann, T., 1996, 'Hot spots' of creatine kinase localization in brain: cerebellum, hippocampus and choroid plexus. *Dev. Neurosci.* **18:** 542–554.

Koszalka, T.R., Jensh, R.P., and Brent, R.L., 1975, Placental transport of creatine in the rat. *Proc. Soc. Exp. Biol. Med.* **148:** 864–869.

Kreis, R., Hofmann, L., Kuhlmann, B., Boesch, C., Bossi, E., and Huppi, P.S., 2002, Brain metabolite composition during early human brain development as measured by quantitative in vivo 1H magnetic resonance spectroscopy. *Magn. Reson. Med.* **48:** 949–958.

Lee, H., Kim, J.H., Chae, Y.J., Ogawa, H., Lee, M.H., and Gerton, G.L., 1998, Creatine synthesis and transport systems in the male rat reproductive tract. *Biol. Reprod.* **58:** 1437–1444.

Lyons, G.E., Muhlebach, S., Moser, A., Masood, R., Paterson, B.M., Buckingham, M.E., and Perriard, J.C., 1991, Developmental regulation of creatine kinase gene expression by myogenic factors in embryonic mouse and chick skeletal muscle. *Development* **113:** 1017–1029.

Mercimek-Mahmutoglu, S., Stoeckler-Ipsiroglu, S., Adami, A., Appleton, R., Araujo, H.C., Duran, M., Ensenauer, R., Fernandez-Alvarez, E., Garcia, P., Grolik, C., Item, C.B., Leuzzi, V., Marquardt, I., Muhl, A., Saelke-Kellermann, R.A., Salomons, G.S., Schulze, A., Surtees, R., van der Knaap, M.S., Vasconcelos, R., Verhoeven, N.M., Vilarinho, L., Wilichowski, E., and Jakobs, C., 2006, GAMT deficiency: features, treatment, and outcome in an inborn error of creatine synthesis. *Neurology* **67:** 480–484.

Miller, T.J., Hanson, R.D., and Yancey, P.H., 2000, Developmental changes in organic osmolytes in prenatal and postnatal rat tissues. *Comp. Biochem. Physiol. A Mol. Integr. Physiol.* **125:** 45–56.

Möller, A. and Hamprecht, B., 1989, Creatine transport in cultured cells of rat and mouse brain. *J. Neurochem.* **52:** 544–550.

Nakashima, T., Tomi, M., Katayama, K., Tachikawa, M., Watanabe, M., Terasaki, T., and Hosoya, K., 2004, Blood-to-retina transport of creatine via creatine transporter (CRT) at the rat inner blood-retinal barrier. *J. Neurochem.* **89:** 1454–1461.

Nakashima, T., Tomi, M., Tachikawa, M., Watanabe, M., Terasaki, T., and Hosoya, K., 2005, Evidence for creatine biosynthesis in Muller glia. *GLIA* **52:** 47–52.

Neu, A., Neuhoff, H., Trube, G., Fehr, S., Ullrich, K., Roeper, J., and Isbrandt, D., 2002, Activation of $GABA_A$ receptors by guanidinoacetate: a novel pathophysiological mechanism. *Neurobiol. Dis.* **11:** 298–307.

Ohtsuki, S., Tachikawa, M., Takanaga, H., Shimizu, H., Watanabe, M., Hosoya, K., and Terasaki, T., 2002, The blood-brain barrier creatine transporter is a major pathway for supplying creatine to the brain. *J. Cereb. Blood Flow Metab.* **22:** 1327–1335.

Perasso, L., Cupello, A., Lunardi, G.L., Principato, C., Gandolfo, C., and Balestrino, M., 2003, Kinetics of creatine in blood and brain after intraperitoneal injection in the rat. *Brain Res.* **974:** 37–42.

Pisano, J.J., Abraham, D., and Udenfriend, S., 1963, Biosynthesis and disposition of γ-guanidinobutyric acid in mammalian tissues. *Arch. Biochem. Biophys.* **100:** 323–329.

Poo-Arguelles, P., Arias, A., Vilaseca, M.A., Ribes, A., Artuch, R., Sans-Fito, A., Moreno, A., Jakobs, C., and Salomons, G., 2006, X-Linked creatine transporter deficiency in two patients with severe mental retardation and autism. *J. Inherit. Metab. Dis.* **29:** 220–223.

Salomons, G.S., van Dooren, S.J., Verhoeven, N.M., Cecil, K.M., Ball, W.S., Degrauw, TJ, and Jakobs, C., 2001, X-linked creatine-transporter gene (SLC6A8) defect: a new creatine-deficiency syndrome. *Am. J. Hum. Genet.* **68:** 1497–1500.

Salomons, G.S., van Dooren, S.J., Verhoeven, N.M., Marsden, D., Schwartz, C., Cecil, K.M., DeGrauw, T.J., and Jakobs, C., 2003, X-linked creatine transporter defect: an overview. *J. Inherit. Metab. Dis.* **26:** 309–318.

Saltarelli, M.D., Bauman, A.L., Moore, K.R., Bradley, C.C., and Blakely, R.D., 1996, Expression of the rat brain creatine transporter in situ and in transfected HeLa cells. *Dev. Neurosci.* **18:** 524–534.

Sandell, L.L., Guan, X.J., Ingram, R., and Tilghman, S.M., 2003, Gatm, a creatine synthesis enzyme, is imprinted in mouse placenta. *Proc. Natl. Acad. Sci. U. S. A.* **100:** 4622–4627.

Schloss, P., Mayser, W., and Betz, H., 1994, The putative rat choline transporter CHOT1 transports creatine and is highly expressed in neural and muscle-rich tissues. *Biochem. Biophys. Res. Commun.* **198:** 637–645.

Schmidt, A., Marescau, B., Boehm, E.A., Renema, W.K., Peco, R., Das, A., Steinfeld, R., Chan, S., Wallis, J., Davidoff, M., Ullrich, K., Waldschutz, R., Heerschap, A., De Deyn, P.P., Neubauer, S., and Isbrandt, D., 2004, Severely altered guanidino compound levels, disturbed body weight homeostasis and impaired fertility in a mouse model of guanidinoacetate N-methyltransferase (GAMT) deficiency. *Hum. Mol. Genet.* **13:** 905–921.

Schulze, A., 2003, Creatine deficiency syndromes. *Mol. Cell. Biochem.* **244:** 143–150.

Schulze, A., Ebinger, F., Rating, D., and Mayatepek, E., 2001, Improving treatment of guanidinoacetate methyltransferase deficiency: reduction of guanidinoacetic acid in body fluids by arginine restriction and ornithine supplementation. *Mol. Genet. Metab.* **74:** 413–419.

Schulze, A., Hess, T., Wevers, R., Mayatepek, E., Bachert, P., Marescau, B., Knopp, M.V., De Deyn, P.P., Bremer, H.J., and Rating, D., 1997, Creatine deficiency syndrome caused by guanidinoacetate methyltransferase deficiency: diagnostic tools for a new inborn error of metabolism. *J. Pediatr.* **131:** 626–631.

Schulze, A., Hoffmann, G.F., Bachert, P., Kirsch, S., Salomons, G.S., Verhoeven, N.M., and Mayatepek, E., 2006, Presymptomatic treatment of neonatal guanidinoacetate methyltransferase deficiency 1. *Neurology* **67:** 719–721.

Schulze, A., Mayatepek, E., Bachert, P., Marescau, B., De Deyn, P.P., and Rating, D., 1998, Therapeutic trial of arginine restriction in creatine deficiency syndrome. *Eur. J. Pediatr.* **157:** 606–607.

Segal, M.B., 2000, The choroid plexuses and the barriers between the blood and the cerebrospinal fluid. *Cell. Mol. Neurobiol.* **20:** 183–196.

Stöckler, S., Holzbach, U., Hanefeld, F., Marquardt, I., Helms, G., Requart, M., Hänicke, W., and Frahm, J., 1994, Creatine deficiency in the brain: a new, treatable inborn error of metabolism. *Pediatr. Res.* **36:** 409–413.

Stöckler, S., Isbrandt, D., Hanefeld, F., Schmidt, B., and Von Figura, K., 1996, Guanidinoacetate methyltransferase deficiency: the first inborn error of creatine metabolism in man. *Am. J. Hum. Genet.* **58:** 914–922.

Stockler, S., Schutz, P.W., and Salomons, G.S., 2007, Cerebral creatine deficiency syndromes: clinical aspects, treatment and pathophysiology. *Subcell. Biochem.* **46:** 149–166.

Stromberger, C., Bodamer, O.A., and Stöckler-Ipsiroglu, S., 2003, Clinical characteristics and diagnostic clues in inborn errors of creatine metabolism. *J. Inherit. Metab. Dis.* **26:** 299–308.

Sykut-Cegielska, J., Gradowska, W., Mercimek-Mahmutoglu, S., and Stockler-Ipsiroglu, S., 2004, Biochemical and clinical characteristics of creatine deficiency syndromes. *Acta Biochim. Pol.* **51:** 875–882.

Tachikawa, M., Fukaya, M., Terasaki, T., Ohtsuki, S., and Watanabe, M., 2004, Distinct cellular expressions of creatine synthetic enzyme GAMT and creatine kinases uCK-Mi and CK-B suggest a novel neuron-glial relationship for brain energy homeostasis. *Eur. J. Neurosci.* **20:** 144–160.

Van Pilsum, J.F., Stephens, G.C., and Taylor, D., 1972, Distribution of creatine, guanidinoacetate and enzymes for their biosynthesis in the animal kingdom. Implications for phylogeny. *Biochem. J.* **126:** 325–345.

Walker, J.B., 1979, Creatine: biosynthesis, regulation, and function. *Adv. Enzymol. Relat. Areas Mol. Biol.* **50:** 177–242.

Wallimann, T. and Hemmer, W., 1994, Creatine kinase in non-muscle tissues and cells. *Mol. Cell. Biochem.* **133–134:** 193–220.

Wallimann, T., Wyss, M., Brdiczka, D., Nicolay, K., and Eppenberger, H.M., 1992, Intracellular compartmentation, structure and function of creatine kinase isoenzymes in tissues with high and fluctuating energy demands: the 'phosphocreatine circuit' for cellular energy homeostasis. *Biochem. J.* **281:** 21–40.

Wang, Y. and Li, S.J., 1998, Differentiation of metabolic concentrations between gray matter and white matter of human brain by in vivo 1H magnetic resonance spectroscopy. *Magn. Reson. Med.* **39:** 28–33.

Wang, Y.E., Esbensen, P., and Bentley, D., 1998, Arginine kinase expression and localization in growth cone migration. *J. Neurosci.* **18:** 987–998.

Whittingham, T.S., Douglas, A., and Holtzman, D., 1995, Creatine and nucleoside triphosphates in rat cerebral gray and white matter. *Metab. Brain Dis.* **10:** 347–352.

Wyss, M. and Kaddurah-Daouk, R., 2000, Creatine and creatinine metabolism. *Physiol. Rev.* **80:** 1107–1213.

Zugno, A.I., Scherer, E.B., Schuck, P.F., Oliveira, D.L., Wofchuk, S., Wannmacher, C.M., Wajner, M., and Wyse, A.T., 2006, Intrastriatal administration of guanidinoacetate inhibits Na^+, K^+-ATPase and creatine kinase activities in rat striatum. *Metab. Brain Dis.* **21:** 41–50.

CHAPTER 5

A NOVEL RELATIONSHIP BETWEEN CREATINE TRANSPORT AT THE BLOOD-BRAIN AND BLOOD-RETINAL BARRIERS, CREATINE BIOSYNTHESIS, AND ITS USE FOR BRAIN AND RETINAL ENERGY HOMEOSTASIS

MASANORI TACHIKAWA[1], KEN-ICHI HOSOYA[1], SUMIO OHTSUKI[2,3] AND TETSUYA TERASAKI[2,3]

[1] *Graduate School of Medicine and Pharmaceutical Sciences, University of Toyama, Toyama 930-0194, Japan*
[2] *Graduate School of Pharmaceutical Sciences, Tohoku University, Sendai 980-8578, Japan*
[3] *SORST, Japan Science and Technology Agency, Japan*

Abstract: Evidence is increasing that the creatine/phosphocreatine shuttle system plays an essential role in energy homeostasis in the brain and retina to ensure proper development and function. Thus, our understanding of the mechanism of creatine supply and creatine usage in the brain and retina and of creatine supplementation in patients with creatine deficiency syndromes is an important step towards improved therapeutic strategies for brain and retinal disorders. Our recent research provides novel molecular-anatomical evidence that (i) at the blood-brain barrier and the inner blood-retinal barrier, the creatine transporter (CRT/SLC6A8) functions as a major pathway for supplying creatine to the brain and retina, and that (ii) local creatine is preferentially synthesized in the glial cells, e.g., oligodendrocytes, astrocytes, and Müller cells, in the brain and retina. Thus, the blood-brain barrier and inner blood-retinal barrier play important roles not only in supplying energy sources (glucose and lactate), but also in supplying an energy 'buffer' (creatine). These findings lead to the novel insight that the creatine/phosphocreatine shuttle system is based on an intricate relationship between the blood-brain barrier, inner blood-retinal barrier, glia, and neurons (photoreceptor cells) to maintain and ensure energy homeostasis in the brain and retina

1. INTRODUCTION

The creatine/phosphocreatine shuttle system plays an important role in energy homeostasis in the brain and retina to ensure proper development and function. Brain cells have only a few seconds supply of ATP which, in part, accounts for

G.S. Salomons and M. Wyss (eds.), Creatine and Creatine Kinase in Health and Disease, 83–98.

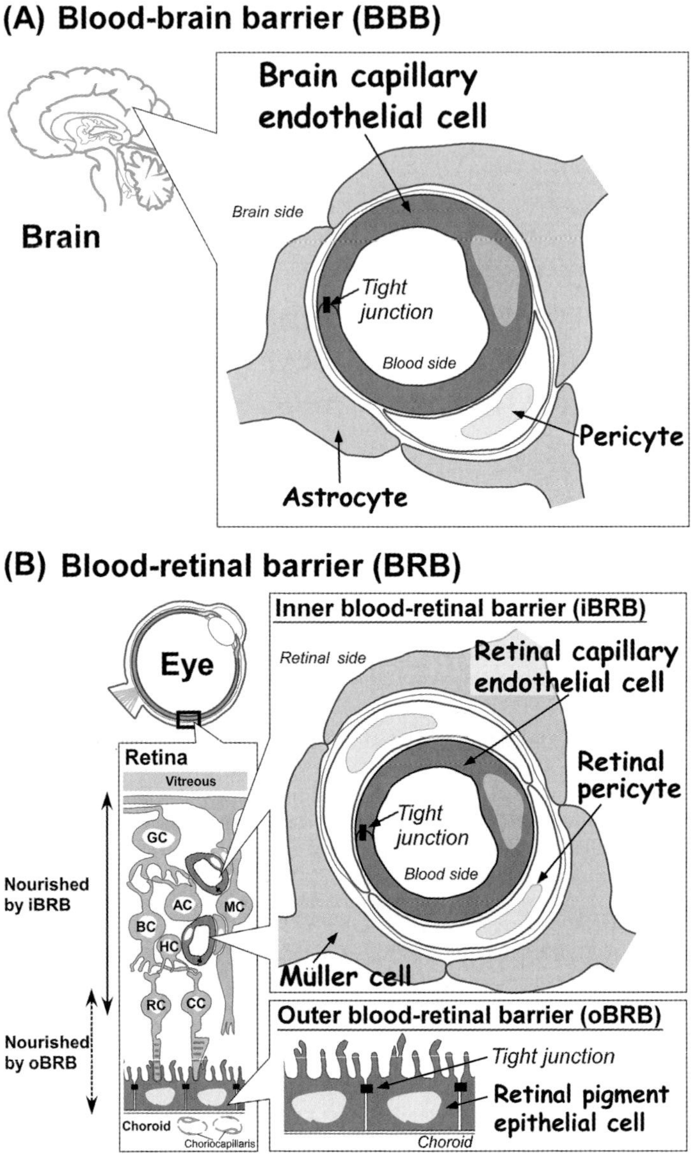

Figure 1. The structures of the blood-brain barrier (BBB) (A) and the blood-retinal barrier (BRB) (B). The BBB forms complex tight-junctions of brain capillary endothelial cells. The BRB forms complex tight junctions of retinal capillary endothelial cells (inner BRB, iBRB) and retinal pigment epithelial cells (outer BRB, oBRB). The inner two thirds of the human retina are nourished by the iBRB and the remainder is covered by the choriocapillaris via the oBRB. GC, ganglion cell; AC, amacrine cell; BC, bipolar cell; HC, horizontal cell; MC, Müller cell; RC, photoreceptor cell (rod cell); CC, photoreceptor cell (cone cell).

the rapid deterioration of brain tissue following oxygen deprivation (Voat and Voat, 1995). The phosphate group of ATP is transferred to creatine and 'stored' as phosphocreatine which can later-on be used to regenerate ATP from ADP. Indeed, *in vitro* studies of the hippocampus have demonstrated that exogenous supply of creatine increases the neuronal phosphocreatine store and protects neurons from hypoxic damage, glutamate excitotoxicity and β-amyloid-induced toxicity (Balestrino *et al.*, 1999; Brewer and Wallimann, 2000). However, there is only limited information available about the mechanisms underlying the constant supply and use of creatine in the brain and retina.

Recently, a model of the functional coupling between the blood-brain barrier (BBB) and inner blood-retinal barrier (iBRB), glia, and neurons (photoreceptor cells) has been proposed to explain the maintenance of constant energy supply in the brain and retina (Magistretti, 2006; Poitry-Yamate *et al.*, 1995). In particular, the BBB and iBRB (Figure 1) play a key role in the constant supply of energy sources (glucose and lactate) from the circulating blood to the brain and retina via facilitative glucose transporter 1 (GLUT1) and monocarboxylate transporter 1 (MCT1) (Enerson and Drewes, 2003; Gerhart *et al.*, 1999; Ohtsuki *et al.*, 2006, Pardridge *et al.*, 1990; Takata *et al.*, 1992). Therefore, it is conceivable that the functional relationships between the BBB, iBRB, glia, and neurons (photoreceptor cells) are also essential for the creatine/phosphocreatine shuttle system in the brain and retina. In this chapter, we describe the recent advances in our research on the cellular and molecular mechanisms of creatine transport at the BBB and iBRB, creatine biosynthesis, and the use of creatine in the brain and retina.

2. CREATINE TURNOVER IN BRAIN AND RETINA, IN HEALTH AND DISEASE

The creatine/phosphocreatine shuttle system operates in particular in tissues with high and fluctuating energy demands, such as striated muscle, brain, and retina (Wyss and Kaddurah-Daouk, 2000). In the human brain, the creatine concentration is 180-fold greater than in plasma (Marescau *et al.*, 1992). Indeed, the brain is one of the main targets of creatine deficiency syndromes, which are characterized by the absence or severe reduction of creatine in the brain. Patients exhibit mental retardation, delayed speech and language, epilepsy, extra-pyramidal signs and autistic behavior (see chapter 8; Almeida *et al.*, 2004; Stockler *et al.*, 2007). Creatine deficiency syndromes are caused by defects in the creatine biosynthesis enzymes, L-arginine:glycine amidinotransferase (AGAT) (Bianchi *et al.*, 2000; Item *et al.*, 2001) and S-adenosyl-L-methionine:guanidinoacetate N-methyltransferase (GAMT) (Stöckler *et al.*, 1994), or in the creatine transporter (CRT/SLC6A8) (Salomons *et al.*, 2001). Although the recent advances in creatine research make it clear that creatine plays an important role in brain function and development, the mechanisms controlling the maintenance of cerebral creatine levels remain uncertain. Furthermore, higher levels of creatine in the chicken retina (3 mM) and in photoreceptor cells (10–15 mM) suggest that the creatine/phosphocreatine system also plays a significant role in the retina, where photoreceptor cells require a large amount of metabolic energy for phototransduction maintained by ionic gradients across the plasma membrane (Hall and Kühn, 1986; Sather and Detwiler 1987; Wallimann *et al.*, 1986). In fact, it has been shown that gyrate atrophy of the choroid and retina is characterized by hyperornithinemia and hypocreatinemia, implying that excessive ornithine leads to chorioretinal degeneration through suspension of creatine synthesis (Sipilä *et al.*, 1992). Therefore, it is important to gain new insights into how creatine is supplied to the brain and retina.

Creatine levels in the body are maintained by biosynthesis, which occurs mainly in the kidney and in the liver, and by dietary supplementation. In the brain, the creatine supply from the circulating blood is believed to be limited due to the presence of the BBB, since oral administration of 20 g creatine per day for 4 weeks produces only a 9% increase in total creatine in human brain (Dechent *et al.*, 1999). Braissant *et al.* (2001) also reported that the signal for CRT mRNA was absent in astrocytes, particularly those in contact with capillary endothelial cells, suggesting that the rat brain is dependent – at least in part – on endogenous creatine synthesis (see also chapter 4; Braissant *et al.*, 2007). In contrast, it has been reported that patients with CRT deficiency failed to improve their neurological symptoms following oral creatine supplementation (Bizzi *et al.*, 2002; Cecil *et al.*, 2001; de Grauw *et al.*, 2002; Poo-Arguelles *et al.*, 2006), despite the presence of creatine biosynthesis enzymes and normal levels of plasma creatine. On the other hand, in patients with AGAT and GAMT deficiency, oral administration of creatine increased the creatine level in the brain (Stöckler *et al.*, 1994) and improved their neurologic symptoms (Bianchi *et al.*, 2000; Stöckler *et al.*, 1996). Oral administration of creatine has been reported to protect the neurons in animal models of amyotrophic lateral sclerosis (ALS), Huntington's disease and Parkinson's disease (Klivenyi *et al.*, 1999; Matthews *et al.*, 1998, 1999). Creatine has a net positive charge and an estimated log partition coefficient of -2.7 (Persky and Brazeau, 2001), which does not allow it to diffuse through plasma membranes. This body of evidence prompted us to hypothesize on a major contribution from a specific creatine transport system at the BBB, and/or on a neuron-glial interplay for local creatine biosynthesis and supply in the brain. Therefore, the processes of creatine transport are important for understanding the mechanisms governing the supply of creatine to the brain and retina and could help in designing improved regimens of oral creatine supplementation for the treatment of neurodegenerative diseases and gyrate atrophy of the choroid and retina.

3. MOLECULAR MECHANISMS OF CREATINE TRANSPORT AT THE BBB AND iBRB: A CONTRIBUTION FROM CRT

3.1. Transport of Creatine Across the BBB and BRB

The *in vivo* blood-to-brain and blood-to-retina transport of creatine across the BBB and BRB, including both the iBRB and the outer BRB (oBRB, see Figure 1), was evaluated by integration plot analysis after intravenous injection of [^{14}C]creatine in rats (Ohtsuki *et al.*, 2002; Nakashima *et al.*, 2004). This approach allows determination of the apparent brain and retinal creatine uptake, or clearance (CL_{BBB}, CL_{BRB}; in μL/(min·g tissue)) even if there is only low permeability across the BBB and BRB. As an index of the brain and retinal distribution characteristics of [^{14}C]creatine at a particular time point, the apparent brain-to-plasma or retina-to-plasma concentration ratio [$K_{p,app}$(t)] is used. This value [$K_{p,app}$(t); in mL/(g brain or retina)] is defined as the amount of [^{14}C]creatine per gram brain or retina divided by the amount of [^{14}C]creatine per milliliter plasma, calculated for a particular, defined time point (t). The CL_{BBB} or CL_{BRB} can be described by the following equation:

$$K_{p,app}(t) = (CL_{BBB} \ or \ CL_{BRB}) \ x \ AUC(t)/Cp(t) + V_i$$

where AUC(t) (dpm · min/mL), Cp(t) (dpm/mL), and V_i (mL/(g brain or retina)) represent the area under the plasma concentration time curve of [^{14}C]creatine from time 0 to t, the plasma [^{14}C]creatine concentration at time t, and the rapidly equilibrated distribution volume of [^{14}C]creatine in the brain or retina, respectively. V_i is usually comparable to the vascular volume of the brain or retina. The CL_{BBB} or CL_{BRB} can be obtained from the initial slope of a plot of $K_{p,app}$(t) versus AUC(t)/Cp(t), as an integration plot.

Integration plot analysis shows that creatine is transported from the circulating blood to the brain and retina. The CL_{BBB} and CL_{BRB} of [^{14}C]creatine through the BBB and BRB were 1.61 and 10.7 μL/(min·g tissue), respectively (Figure 2A,B). The CL_{BBB} and CL_{BRB} are about 6-fold and 40-fold greater than those of [^{14}C]sucrose in rats [0.29 and 0.26 μL/(min·g tissue)], respectively. [^{14}C]Sucrose

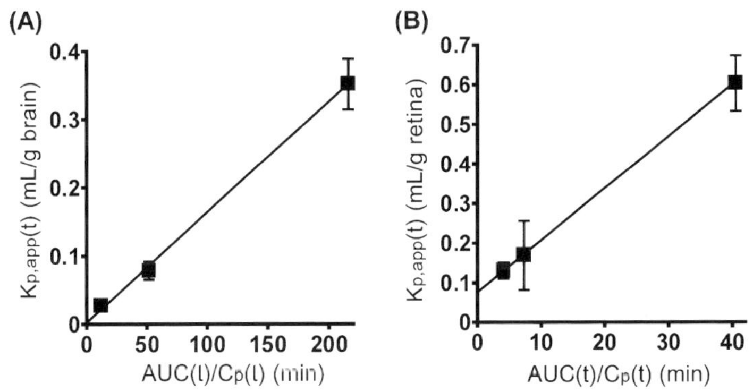

Figure 2. In vivo transport of [^{14}C]creatine from the circulating blood to the brain (A) and retina (B). The solid line represents the apparent brain or retinal uptake clearance of [^{14}C]creatine (CL_{BBB}, or CL_{BRB}) that can be obtained from the initial slope of a plot of $K_{p,app}$(t) versus AUC(t)/Cp(t). Each point represents the mean ± SEM (n = 3-4). From Ohtsuki *et al.*, 2002 (A) and Nakashima *et al.*, 2004 (B) with kind permission from Nature Publishing Group and Blackwell Publishing.

is used as a non-permeable paracellular marker (Lightman *et al.*, 1987). Furthermore, the apparent cerebrum-to-plasma concentration ratio ($K_{p,app}$) of [^{14}C]creatine was 30.8 mL/g tissue 24 hours after exogenous administration of [^{14}C]creatine to mice. This value is consistent with the endogenous brain-to-serum concentration ratio of creatine (Marescau *et al.*, 1992), supporting the proposal that the BBB is a major pathway for supplying creatine to the brain. Such evidence strongly suggests that creatine is transported via a carrier-mediated transport system at the BBB and BRB, rather than by passive diffusion, from the circulating blood to the brain and retina against its concentration gradient.

3.2. Characteristics of Creatine Transport Across the BBB and iBRB

The molecular mechanisms responsible for creatine transport across the BBB and iBRB have been elucidated by using our newly established conditionally immortalized mouse brain capillary endothelial cell line (TM-BBB) and rat retinal capillary endothelial cell line (TR-iBRB). These cell lines retain the *in vivo* expression and function of several transporters and are a suitable *in vitro* model for the BBB and iBRB, respectively (Hosoya and Tomi, 2005; Terasaki *et al.*, 2003). Using TM-BBB and TR-iBRB cells, a number of transporters expressed at the BBB and iBRB have already been identified, which are involved in nutrient supply and drug distribution to the brain and retina (see recent reviews: Hosoya and Tomi, 2005; Ohtsuki, 2004; Terasaki and Ohtsuki, 2005). The characteristics of [^{14}C]creatine uptake by TM-BBB and TR-iBRB cells are summarized in Table 1 (Nakashima *et al.*, 2004; Ohtsuki *et al.*, 2002), supporting the notion that CRT is involved in creatine transport across the BBB and iBRB. [^{14}C]Creatine uptake by TM-BBB and TR-iBRB cells takes place in a Na^+-, Cl^-- and concentration-dependent manner with a Michaelis-Menten constant (K_m) for creatine of 16.2 and 14.9 μM, respectively. The maximal rate (V_{max}) of [^{14}C]creatine uptake by TM-BBB and TR-iBRB cells is 105 and 49.3 pmol/(min·mg protein), respectively. The BBB clearance of creatine per gram brain ($CL_{BBB,\ in\ vitro}$) has been estimated to be 2.1 μL/(min·g brain) by using the above kinetic parameters of *in vitro* [^{14}C]creatine uptake by TM-BBB cells (Ohtsuki *et al.*, 2002). This is in good agreement with the $CL_{BBB,\ in\ vivo}$ value (1.6 μL/(min·g brain)), suggesting that the *in vitro* uptake study using TM-BBB cells reflects the *in vivo* situation. The corresponding K_m values are consistent with an apparent K_m for creatine of 15 and 29 μM for human and rat CRT, respectively (Saltarelli *et al.*, 1996, Sora *et al.*, 1994). These K_m values are 10- to 40-fold lower than the plasma concentrations of creatine (140–600 μM) in the mouse and rat (Marescau *et al.*, 1986); this suggests that the blood-to-brain and blood-to-retina transport of creatine is more than 90% saturated by endogenous plasma creatine and that the creatine transport system at the BBB and iBRB plays a role in continuously supplying creatine from the circulating blood to the brain and retina at a rate close to V_{max}. This physiological role of the BBB and iBRB may be important for a better understanding of creatine turnover in the brain and retina. Indeed, oral administration of a relatively large amount of creatine for 4 weeks produces only about

Table 1. Characteristics of [^{14}C]creatine uptake by TM-BBB and TR-iBRB cells. (A) Na^+- and Cl^--dependent [^{14}C]creatine uptake by TM-BBB and TR-iBRB cells at 37 °C. Each value represents the mean ± SEM (n = 3-4). (B) Kinetic parameters of [^{14}C]creatine uptake by TM-BBB and TR-iBRB cells at 37 °C. (C) Effect of several compounds (1 mM) on [^{14}C]creatine uptake by TM-BBB and TR-iBRB cells at 37 °C. Each value represents the mean ± SEM (n = 3-4). $*p < 0.001$, $**p < 0.01$: significantly different from control. ND, not determined. From Ohtsuki *et al.* (2002) and Nakashima *et al.* (2004) with kind permission from Nature Publishing Group and Blackwell Publishing .

	TM-BBB	TR-iBRB
(A) Na^+- and Cl^--dependence	Relative uptake (% of control)	
Control	100 ± 4	100 ± 9
Na^+-free condition	3.60 ± 0.13*	9.16 ± 0.58**
Cl^--free condition	8.83 ± 0.75*	18.8 ± 0.8**
(B) Kinetic parameters		
K_m μM	16.2	14.9
V_{max} in pmol/(min.mg protein)	105	49.3
(C) Inhibitory effects (1 mM)	Relative uptake (% of control)	
Control	100 ± 6.5	100 ± 2
Creatine	6.75 ± 0.34**	5.97 ± 0.56**
β-Guanidinopropionate	4.52 ± 0.88**	8.74 ± 1.01**
Guanidinoacetate	30.2 ± 3.8**	ND
γ-Guanidinobutyrate	ND	27.7 ± 3.0**
Phosphocreatine	67.5 ± 6.7	57.7 ± 3.2**
Creatinine	82.7 ± 2.2	91.1 ± 1.9
L-Arginine	79.9 ± 4.4	86.1 ± 3.9
Glycine	81.1 ± 4.3	ND
γ-Aminobutyric acid	88.7 ± 1.3	93.9 ± 2.2
Choline	75.0 ± 11	ND

a 9% increase in total creatine in human brain (Dechent *et al.*, 1999). Therefore, oral creatine treatment for patients with ALS or gyrate atrophy may depend on the transport mechanism at the BBB and iBRB.

In vitro inhibition studies further suggest that creatine is mainly transported via CRT, since β-guanidinopropionate produces marked inhibition of [^{14}C]creatine uptake by TM-BBB and TR-iBRB cells, as reported elsewhere (Guimbal and Kilimann, 1993), whereas precursors (glycine and L-arginine) and metabolites of creatine (creatinine) do not. RT-PCR and Western blot analyses revealed that CRT mRNA and protein are expressed in TM-BBB and TR-iBRB cells. Furthermore, Bélanger *et al.* (2007) recently reported that mRNA expression and function of CRT in TM-BBB cells are increased under hyperammonemic conditions. This evidence suggests that the CRT at the BBB and iBRB plays a major role in supplying creatine to the brain and retina under pathophysiological conditions.

3.3. Localization of CRT in the Brain and Retinal Capillary Endothelial Cells

Immunohistochemical studies have shown that CRT is localized at both the luminal and abluminal sides of adult mouse brain (Figure 3A,B) and rat retinal (Figure 3C) capillary endothelial cells. In the adult mouse brain, CRT is also found to be expressed in neurons, suggesting that creatine is transported into neuronal cells by CRT following BBB transport. CRT mRNA can also be detected in rat neurons by *in situ* hybridization (Braissant *et al.*, 2001). Regarding expression of CRT in retinal capillaries, our evidence is in good agreement with a recent report by Acosta *et al.* (2005) showing that CRT is expressed in the cow retinal inner blood vessels. CRT is also expressed in photoreceptor inner segments, inner neuronal cells, and retinal pigment epithelial cells (Acosta *et al.*, 2005). This evidence suggests that creatine is transported into photoreceptor and/or neuronal cells by CRT following iBRB transport in the retina. The role of CRT in the abluminal membrane of the BBB and iBRB and the contribution of oBRB to the supply of creatine to the retina remain unclear and are the subject of ongoing investigations.

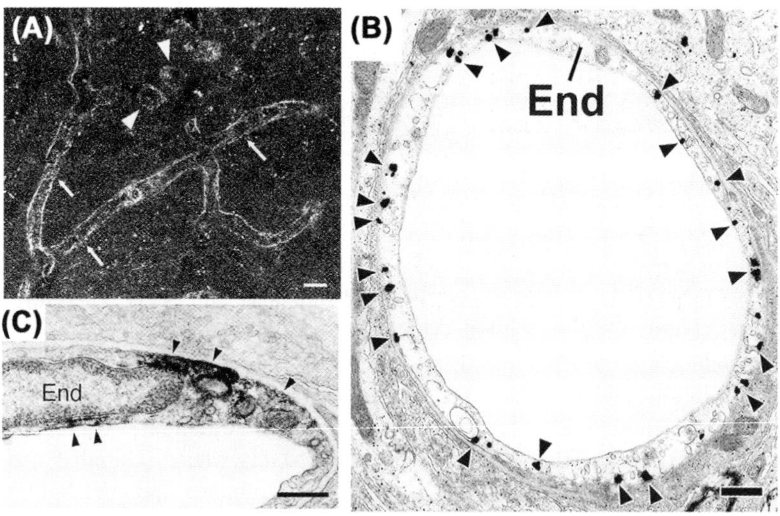

Figure 3. Localization of CRT in the brain (A, B) and retinal (C) capillary endothelial cells. (A) Abundant immunofluorescence of brain capillaries branching in all cortical layers (arrows) with moderate staining in neuronal perikarya (arrowheads). (B) Immunogold labeling of CRT in a cortical capillary. Immunogold particles (arrowheads) are associated with both the luminal and abluminal cell membranes of mouse brain capillary endothelial cells (End). (C) Immunoperoxidase labeling (arrowheads) was detected on the surface of both the luminal and abluminal cell membranes of rat retinal capillary endothelial cells (End). Scale bars: (A) 10 μm, (B, C) 0.5 μm. From Ohtsuki *et al.*, 2002 (A), Tachikawa *et al.*, 2004 (B) and Nakashima *et al.*, 2004 (C) with kind permission from Nature Publishing Group and Blackwell Publishing.

4. CELLULAR SYSTEMS OF CREATINE BIOSYNTHESIS AND USE IN THE BRAIN AND RETINA

4.1. Differential Cellular Expression of the Creatine Biosynthesis Enzyme, GAMT, and Creatine Kinases, uCK-Mi and CK-B, in the Adult Mouse Brain

The cellular system of creatine biosynthesis and its function in the adult mouse brain have been investigated by immunohistochemical localization studies for GAMT, ubiquitous mitochondrial creatine kinase (uCK-Mi) and brain-type cytoplasmic creatine kinase (CK-B) (Tachikawa *et al.*, 2004). GAMT is highly expressed in oligodendrocytes (Figure 4A, arrow) and olfactory ensheathing glia and moderately in astrocytes (Figure 4A, arrowhead), whereas the level of GAMT is very low in neurons and microglia and below the detection threshold in capillaries.

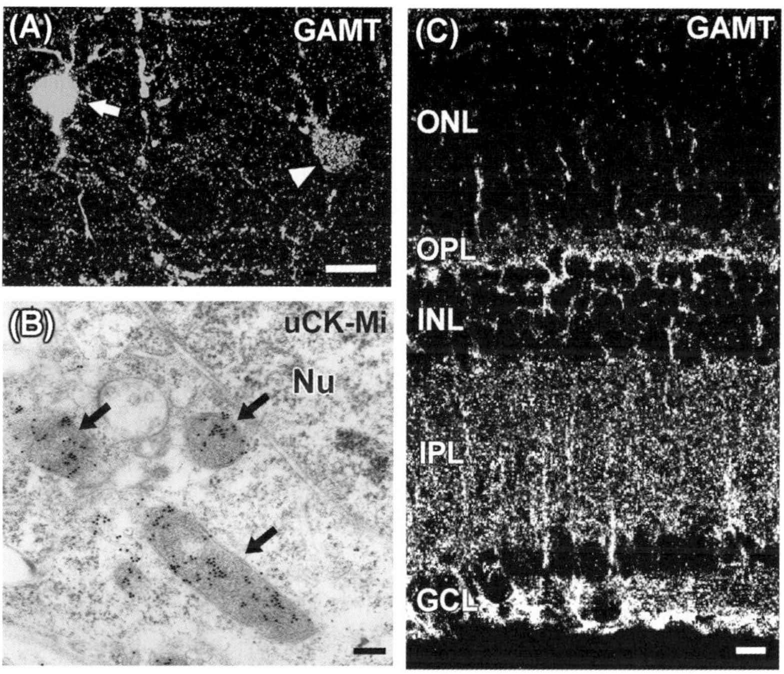

Figure 4. Cellular expression of GAMT (A) and uCK-Mi (B) in the mouse brain, and of GAMT (C) in the rat retina. (A) Confocal immunofluorescence microscope image of GAMT localization in the mouse cerebral cortex. Cells expressing GAMT strongly or moderately are indicated by arrows (oligo dendrocyte) and arrowheads (astrocyte). (B) Selective localization of uCK-Mi in neuronal mitochondria (arrows) as revealed by postembedding immunogold electron microscopy. Nu, nucleus. (C) Confocal immunofluorescence microscope image of GAMT localization in the rat retina. ONL, outer nuclear layer; OPL, outer plexiform layer; INL, inner nuclear layer; IPL, inner plexiform layer; GCL, ganglion cell layer. Scale bars: (A) 10 μm, (B) 0.1 μm, (C) 10 μm. From Tachikawa *et al.*, 2004 (A, B) and Nakashima *et al.*, 2005 (C) with kind permission from Blackwell Publishing and John Wiley & Sons.

The ability to synthesize creatine from glycine has been confirmed for astrocytes: when [^{13}C]glycine was added to the culture medium, radioactive creatine was detected in cell extracts of astrocyte-rich primary cultures (Dringen *et al.*, 1998). It remains unknown whether oligodendrocytes and olfactory ensheathing glia can synthesize creatine at higher rates than astrocytes.

uCK-Mi is expressed selectively in neurons and localized in mitochondria within cell bodies (Figure 4B, arrows), dendrites, axons, and terminals, suggesting that the creatine/phosphocreatine shuttle system is coupled with ATP production in neuronal mitochondria, but not in glial mitochondria. Indeed, this shuttle system has been reported to be involved in neuronal growth cone activity and axonal elongation (Wang *et al.*, 1998). The preferential glial expression of GAMT and selective neuronal expression of uCK-Mi support the notion that the neuron-glial relationship is essential for the functioning of this shuttle system. High GAMT expression in oligodendrocytes and olfactory ensheathing glia may suggest, in particular, preferential supply of creatine to axonal mitochondria. In this regard, neurological symptoms in patients with GAMT deficiency, such as epilepsy, bilateral myelination delay and extra-pyramidal movement disorder (Kahler and Fahey, 2003, Schulze, 2003, Stöckler *et al.*, 1994), can thus be ascribed not only to a reduced creatine supply from the circulating blood (Ohtsuki *et al.*, 2002), but also to a disrupted supply from local glial cells.

Immunohistochemical and immunoelectron microscopy investigations have shown that CK-B is expressed selectively in astrocytes among glial populations, and expressed exclusively in inhibitory neurons among neuronal populations. In contrast, CK-B immunoreactivity is virtually absent in excitatory neurons. Interestingly, the cell types with high CK-B immunoreactivity are known to be highly resistant to acute energy loss, such as in hypoxia or hypoglycemia.

Considering that phosphocreatine regenerates ATP much faster than the processes of glycolysis and oxidative phosphorylation (Wallimann *et al.*, 1992), the highly regulated cellular expression of creatine biosynthesis and metabolic enzymes suggests that the creatine/phosphocreatine shuttle system plays a role in brain energy homeostasis through a novel neuron-glial relationship.

4.2. Creatine is Preferentially Synthesized in Müller Glial Cells of the Retina

RT-PCR analysis has shown that AGAT and GAMT mRNAs are expressed in the retina and the Müller glial cell line, TR-MUL (Nakashima *et al.*, 2005). TR-MUL is a conditionally immortalized rat Müller cell line with properties of normal retinal Müller cells (Hosoya and Tomi, 2005). Western blot analysis confirmed the expression of GAMT protein in the rat retina and TR-MUL cells. Confocal immunofluorescence microscopy of rat retinal sections demonstrated that GAMT is preferentially localized in glutamine synthetase-positive Müller cells (Figure 4C). In contrast, there is little immunoreactivity in photoreceptor cells located in the outer nuclear layer. Furthermore, [^{14}C]creatine was detected in the isolated rat retina and TR-MUL cells after a 24 h-incubation with [^{14}C]glycine added to the culture medium, suggesting creatine biosynthesis in Müller glia (Nakashima *et al.*, 2005). Interestingly, creatine kinases are more concentrated in photoreceptor cells compared with other cells in the chicken retina (Wallimann *et al.*, 1986). Mitochondrial CK is present in the inner segments of bovine rod and cone cells, while the cytoplasmic brain isoform of creatine kinase is also located in the rod outer segments. The localization of creatine kinase isoforms supports the existence of a creatine/phosphocreatine shuttle in the highly polar photoreceptor cells. Although the function and destination of creatine synthesized in the retinal Müller cells remain unknown, it appears likely that synthesized creatine is not only used by Müller cells, but also supplied to other cells, most likely photoreceptor cells, as do lactate and amino acids which are assumed to be shuttled between Müller cells and photoreceptor cells (Poitry-Yamate *et al.*, 1995, Rauen and Wiessner, 2000). Bearing in mind the fact that creatine is also supplied from the circulating blood via CRT at the iBRB (Nakashima *et al.*, 2004), the creatine concentration in photoreceptor cells may be controlled by a dual system of creatine supply from the local retinal glial cells and from capillaries. Chorioretinal degeneration in patients with gyrate atrophy can be ascribed both to a reduced creatine supply from the circulating blood and a disrupted supply from the local Müller glia due to inhibition of creatine biosynthesis by hyperornithinemia (Sipilä *et al.*, 1980).

5. CONCLUSIONS

Our recent research provides novel molecular-anatomical evidence that (i) CRT at the BBB and iBRB is involved in regulating the creatine concentration in the brain and retina and that (ii) local creatine is preferentially synthesized in the glial cells in the brain and retina. These findings provide important information that will increase our understanding of the mechanism of creatine supply and creatine use in the brain and retina and of creatine supplementation in patients with creatine deficiency syndromes. In particular, the BBB and iBRB play essential roles in maintaining energy homeostasis not only in terms of supplying energy sources (glucose and lactate), but also for supplying an energy 'buffer' (creatine). Although oral creatine supplementation is reported to be an effective treatment for AGAT- and GAMT-deficient patients (Stöckler *et al.*, 1994, 1996), high doses of creatine given over a long period of time only partially replenish brain creatine pools. In this regard, the creatine concentration in the brain and retina appears to be dependent on BBB and iBRB functions.

Figure 5 summarizes the putative cellular mechanisms of creatine transport, biosynthesis and use in the brain and retina. This leads to a novel insight that the creatine/phosphocreatine shuttle system is based on an intricate relationship between the BBB, iBRB, glia, and neurons (photoreceptor cells) to maintain and ensure energy homeostasis in the brain and retina.

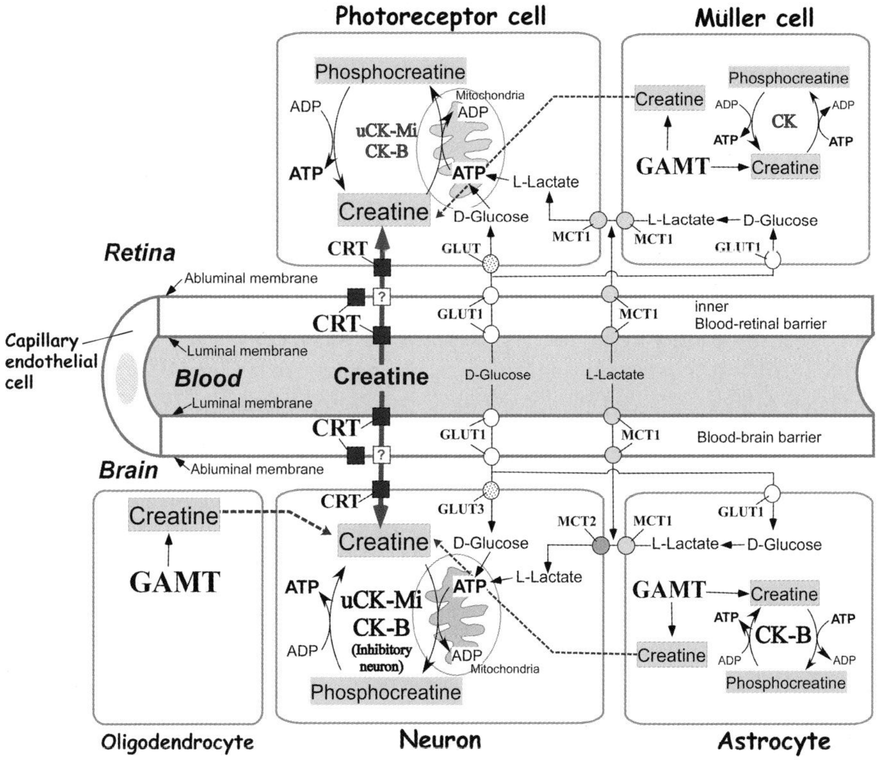

Figure 5. Schematic diagram of the functional relationships between creatine transport at the BBB and iBRB, creatine biosynthesis, and its use for brain and retinal energy homeostasis. The role of CRT in the abluminal membrane of the BBB and iBRB (isolated black squares) in the supply of creatine to the brain and retina remains unclear. CK, creatine kinase; CRT, creatine transporter; GAMT, S-adenosyl-L-methionine:guanidinoacetate N-methyltransferase; GLUT, glucose transporter; MCT, monocarboxylate transporter.

6. RESEARCH OUTLOOK

Multidisciplinary approaches to creatine research will lead to new insights into the functional relationships between the BBB and iBRB, glia, and neurons (photoreceptor cells) for brain and retinal energy homeostasis and will offer improved therapeutic strategies for brain and retinal disorders.

Genetic manipulation of BBB and iBRB function in mice or rats, e.g. by BBB-, iBRB- or glia-selective gene transfer and silencing, will be a future key method. Using such techniques will be a particularly important issue to investigate the relationships between creatine transport dysfunction at the BBB and iBRB and mental retardation, speech problems, language delay, and epilepsy. We recently produced a transgenic rat harboring the mouse Tie2 promoter/enhancer linked green fluorescent protein (GFP) gene (Ohtsuki *et al.*, 2005). The mouse Tie2 promoter/enhancer induces selective gene expression in vascular endothelial cells in the brain and retina of transgenic rats. Therefore, use of the Tie2 promoter/enhancer for the induction or silencing of the CRT gene selectively in the BBB and iBRB will be an important strategy for analyzing the relevance of CRT at the BBB and iBRB under physiological and pathophysiological conditions.

Oral administration of creatine is a potentially promising treatment for neurodegenerative diseases owing to the neuroprotective effects found in animal models of ALS, Huntington's disease and Parkinson's disease. However, the blood-to-brain transport of creatine is apparently limited due to the almost complete saturation of CRT at the BBB. Indeed, it has been reported that the effect of creatine supplementation in patients with ALS or gyrate atrophy appears to be limited (Shefner *et al.*, 2004, Vannas-Sulonen *et al.*, 1985). Therefore, the regulatory mechanisms of CRT expression and function need to be further clarified for a more rational creatine therapy of neurodegenerative diseases and gyrate atrophy of the choroid and retina. This may lead to the discovery and development of drugs which increase the density of CRT on the plasma membrane and/or CRT transport activity at the BBB and iBRB.

ACKNOWLEDGEMENTS

The authors gratefully acknowledge Drs. M. Watanabe, S. Hori, M. Tomi, and H. Takanaga for their insights and contributions to ongoing research in the authors' laboratories. This study is supported, in part, by a Grant-in Aid for Scientific Research and a 21st Century Center of Excellence (COE) Program Grant from the Ministry of Education, Science, Sports, and Culture, and by a grant for Research on Sensory and Communicative Disorders by the Ministry of Health, Labor, and Welfare, and by Research Fellowships of the Japan Society for the Promotion of Science for Young Scientists, Japan.

REFERENCES

Acosta, M.L., Kalloniatis, M., and Christie, D.L., 2005, Creatine transporter localization in developing and adult retina: importance of creatine to retinal function. *Am. J. Physiol. Cell Physiol.* **289:** C1015–C1023.

Almeida, L.S., Verhoeven, N.M., Roos, B., Valongo, C., Cardoso, M.L., Vilarinho, L., Salomons, G.S., and Jakobs, C., 2004, Creatine and guanidinoacetate: diagnostic markers for inborn errors in creatine biosynthesis and transport. *Mol. Genet. Metab.* **82:** 214–219.

Balestrino, M., Rebaudo, R., and Lunardi, G., 1999, Exogenous creatine delays anoxic depolarization and protects from hypoxic damage: dose-effect relationship. *Brain Res.* **816:** 124–130.

Belanger, M., Asashima, T., Ohtsuki, S., Yamaguchi, H., Ito, S., and Terasaki, T., 2007, Hyperammonemia induces transport of taurine and creatine and suppresses claudin-12 gene expression in brain capillary endothelial cells in vitro. *Neurochem. Int.* **50:** 95–101.

Bianchi, M.C., Tosetti, M., Fornai, F., Alessandri, M.G., Cipriani, P., De Vito, G., and Canapicchi, R., 2000, Reversible brain creatine deficiency in two sisters with normal blood creatine level. *Ann. Neurol.* **47:** 511–513.

Bizzi, A., Bugiani, M., Salomons, G.S., Hunneman, D.H., Moroni, I., Estienne, M., Danesi, U., Jakobs, C., and Uziel, G., 2002, X-linked creatine deficiency syndrome: a novel mutation in creatine transporter gene SLC6A8. *Ann. Neurol.* **52:** 227–231.

Braissant, O., Henry, H., Loup, M., Eilers, B., and Bachmann, C., 2001, Endogenous synthesis and transport of creatine in the rat brain: an in situ hybridization study. *Brain Res. Mol. Brain Res.* **86:** 193–201.

Braissant, O., Bachmann, C., and Henry, H., 2007, Expression and function of AGAT, GAMT and CT1 in the mammalian brain. *Subcell. Biochem.* **46:** 67–81.

Brewer, G.J., and Wallimann, T.W., 2000, Protective effect of the energy precursor creatine against toxicity of glutamate and beta-amyloid in rat hippocampal neurons. *J. Neurochem.* **74:** 1968–1978.

Cecil, K.M., Salomons, G.S., Ball, W.S., Jr., Wong, B., Chuck, G., Verhoeven, N.M., Jakobs, C., and DeGrauw, T.J., 2001, Irreversible brain creatine deficiency with elevated serum and urine creatine: a creatine transporter defect? *Ann. Neurol.* **49:** 401–404.

Dechent, P., Pouwels, P.J., Wilken, B., Hanefeld, F., and Frahm, J., 1999, Increase of total creatine in human brain after oral supplementation of creatine-monohydrate. *Am. J. Physiol.* **277:** R698–R704.

deGrauw, T.J., Salomons, G.S., Cecil, K.M., Chuck, G., Newmeyer, A., Schapiro, M.B., and Jakobs, C., 2002, Congenital creatine transporter deficiency. *Neuropediatrics* **33:** 232–238.

Dringen, R., Verleysdonk, S., Hamprecht, B., Willker, W., Leibfritz, D., and Brand, A., 1998, Metabolism of glycine in primary astroglial cells: synthesis of creatine, serine, and glutathione. *J. Neurochem.* **70:** 835–840.

Enerson, B.E., and Drewes, L.R., 2003, Molecular features, regulation, and function of monocarboxylate transporters: implications for drug delivery. *J. Pharm. Sci.* **92:** 1531–1544.

Gerhart, D.Z., Leino, R.L., and Drewes, L.R., 1999, Distribution of monocarboxylate transporters MCT1 and MCT2 in rat retina. *Neuroscience* **92:** 367–375.

Guimbal, C., and Kilimann, M.W., 1993, A Na^+-dependent creatine transporter in rabbit brain, muscle, heart, and kidney. cDNA cloning and functional expression. *J. Biol. Chem.* **268:** 8418–8421.

Hall, S.W., and Kühn, H., 1986, Purification and properties of guanylate kinase from bovine retinas and rod outer segments. *Eur. J. Biochem.* **161:** 551–556.

Hosoya, K., and Tomi, M., 2005, Advances in the cell biology of transport via the inner blood-retinal barrier: establishment of cell lines and transport functions. *Biol. Pharm. Bull.* **28:** 1–8.

Item, C.B., Stöckler-Ipsiroglu, S., Stromberger, C., Muhl, A., Alessandri, M.G., Bianchi, M.C., Tosetti, M., Fornai, F., and Cioni, G., 2001, Arginine:glycine amidinotransferase deficiency: the third inborn error of creatine metabolism in humans. *Am. J. Hum. Genet.* **69:** 1127–1133.

Kahler, S.G., and Fahey, M.C., 2003, Metabolic disorders and mental retardation. *Am. J. Med. Genet. C Semin. Med. Genet.* **117:** 31–41.

Klivenyi, P., Ferrante, R.J., Matthews, R.T., Bogdanov, M.B., Klein, A.M., Andreassen, O.A., Mueller, G., Wermer, M., Kaddurah-Daouk, R., and Beal, M.F., 1999, Neuroprotective effects of creatine in a transgenic animal model of amyotrophic lateral sclerosis. *Nat. Med.* **5:** 347–350.

Lightman, S.L., Palestine, A.G., Rapoport, S.I., and Rechthand, E., 1987, Quantitative assessment of the permeability of the rat blood-retinal barrier to small water-soluble non-electrolytes. *J. Physiol.* **389:** 483–490.

Magistretti, P.J., 2006, Neuron-glia metabolic coupling and plasticity. *J. Exp. Biol.* **209:** 2304–2311.

Marescau, B., De Deyn, P., Wiechert, P., Van Gorp, L., and Lowenthal, A., 1986, Comparative study of guanidino compounds in serum and brain of mouse, rat, rabbit, and man. *J. Neurochem.* **46:** 717–720.

Marescau, B., Deshmukh, D.R., Kockx, M., Possemiers, I., Qureshi, I.A., Wiechert, P., and De Deyn, P.P., 1992, Guanidino compounds in serum, urine, liver, kidney, and brain of man and some ureotelic animals. *Metabolism* **41:** 526–532.

Matthews, R.T., Ferrante, R.J., Klivenyi, P., Yang, L., Klein, A.M., Mueller, G., Kaddurah-Daouk, R., and Beal, M.F., 1999, Creatine and cyclocreatine attenuate MPTP neurotoxicity. *Exp. Neurol.* **157:** 142–149.

Matthews, R.T., Yang, L., Jenkins, B.G., Ferrante, R.J., Rosen, B.R., Kaddurah-Daouk, R., and Beal, M.F., 1998, Neuroprotective effects of creatine and cyclocreatine in animal models of Huntington's disease. *J. Neurosci.* **18:** 156–163.

Nakashima, T., Tomi, M., Katayama, K., Tachikawa, M., Watanabe, M., Terasaki, T., and Hosoya, K., 2004, Blood-to-retina transport of creatine via creatine transporter (CRT) at the rat inner blood-retinal barrier. *J. Neurochem.* **89:** 1454–1461.

Nakashima, T., Tomi, M., Tachikawa, M., Watanabe, M., Terasaki, T., and Hosoya, K.I., 2005, Evidence for creatine biosynthesis in Müller glia. *Glia* **52:** 47–52.

Ohtsuki, S., 2004, New aspects of the blood-brain barrier transporters; its physiological roles in the central nervous system. *Biol. Pharm. Bull.* **27:** 1489–1496.

Ohtsuki, S., Kamiya, N., Hori, S., and Terasaki, T., 2005, Vascular endothelium-selective gene induction by Tie2 promoter/enhancer in the brain and retina of a transgenic rat. *Pharm. Res.* **22:** 852–857.

Ohtsuki, S., Kikkawa, T., Hori, S., and Terasaki, T., 2006, Modulation and compensation of the mRNA expression of energy related transporters in the brain of glucose transporter 1-deficient mice. *Biol. Pharm. Bull.* **29:** 1587–1591.

Ohtsuki, S., Tachikawa, M., Takanaga, H., Shimizu, H., Watanabe, M., Hosoya, K., and Terasaki, T., 2002, The blood-brain barrier creatine transporter is a major pathway for supplying creatine to the brain. *J. Cereb. Blood Flow Metab.* **22:** 1327–1335.

Pardridge, W.M., Boado, R.J., and Farrell, C.R., 1990, Brain-type glucose transporter (GLUT-1) is selectively localized to the blood-brain barrier. Studies with quantitative western blotting and in situ hybridization. *J. Biol. Chem.* **265:** 18035–18040.

Persky, A.M., and Brazeau, G.A., 2001, Clinical pharmacology of the dietary supplement creatine monohydrate. *Pharmacol. Rev.* **53:** 161–176.

Poitry-Yamate, C.L., Poitry, S., and Tsacopoulos, M., 1995, Lactate released by Müller glial cells is metabolized by photoreceptors from mammalian retina. *J. Neurosci.* **15:** 5179–5191.

Poo-Arguelles, P., Arias, A., Vilaseca, M.A., Ribes, A., Artuch, R., Sans-Fito, A., Moreno, A., Jakobs, C., and Salomons, G., 2006, X-Linked creatine transporter deficiency in two patients with severe mental retardation and autism. *J. Inherit. Metab. Dis.* **29:** 220–223.

Rauen, T., and Wiessner, M., 2000, Fine tuning of glutamate uptake and degradation in glial cells: common transcriptional regulation of GLAST1 and GS. *Neurochem. Int.* **37:** 179–189.

Salomons, G.S., van Dooren, S.J., Verhoeven, N.M., Cecil, K.M., Ball, W.S., Degrauw, T.J., and Jakobs, C., 2001, X-linked creatine-transporter gene (SLC6A8) defect: a new creatine-deficiency syndrome. *Am. J. Hum. Genet.* **68:** 1497–1500.

Saltarelli, M.D., Bauman, A.L., Moore, K.R., Bradley, C.C., and Blakely, R.D., 1996, Expression of the rat brain creatine transporter in situ and in transfected HeLa cells. *Dev. Neurosci.* **18:** 524–534.

Sather, W.A., and Detwiler, P.B., 1987, Intracellular biochemical manipulation of phototransduction in detached rod outer segments. *Proc. Natl. Acad. Sci. USA* **84:** 9290–9294.

Schulze, A., 2003, Creatine deficiency syndromes. *Mol. Cell. Biochem.* **244:** 143–150.

Shefner, J.M., Cudkowicz, M.E., Schoenfeld, D., Conrad, T., Taft, J., Chilton, M., Urbinelli, L., Qureshi, M., Zhang, H., Pestronk, A., Caress, J., Donofrio, P., Sorenson, E., Bradley, W., Lomen-Hoerth, C., Pioro, E., Rezania, K., Ross, M., Pascuzzi, R., Heiman-Patterson, T., Tandan, R., Mitsumoto, H., Rothstein, J., Smith-Palmer, T., MacDonald, D., and Burke, D., 2004, A clinical trial of creatine in ALS. *Neurology* **63:** 1656–1661.

Sipilä, I., Simell, O., and Arjomaa, P., 1980, Gyrate atrophy of the choroid and retina with hyperornithinemia. Deficient formation of guanidinoacetic acid from arginine. *J. Clin. Invest.* **66:** 684–687.

Sipilä, I., Valle, D., and Brusilow, S., 1992, Low guanidinoacetic acid and creatine concentrations in gyrate atrophy of choroids and retina (GA). In: Guanidino Compounds in Biology and Medicine (De Deyn, P.P., Marescau, B., Stalon, V., and Qureshi, I., eds.), John Libbey Ltd, London, pp 379–383.

Sora, I., Richman, J., Santoro, G., Wei, H., Wang, Y., Vanderah, T., Horvath, R., Nguyen, M., Waite, S., Roeske, W.R., and *et al.*, 1994, The cloning and expression of a human creatine transporter. *Biochem. Biophys. Res. Commun.* **204:** 419–427.

Stöckler, S., Hanefeld, F., and Frahm, J., 1996, Creatine replacement therapy in guanidinoacetate methyltransferase deficiency, a novel inborn error of metabolism. *Lancet* **348:** 789–790.

Stöckler, S., Holzbach, U., Hanefeld, F., Marquardt, I., Helms, G., Requart, M., Hanicke, W., and Frahm, J., 1994, Creatine deficiency in the brain: a new, treatable inborn error of metabolism. *Pediatr. Res.* **36:** 409–413.

Stockler, S., Schutz, P.W., and Salomons, G.S., 2007, Cerebral creatine deficiency syndromes: clinical aspects, treatment and pathophysiology. *Subcell. Biochem.* **46:** 149–166.

Tachikawa, M., Fukaya, M., Terasaki, T., Ohtsuki, S., and Watanabe, M., 2004, Distinct cellular expressions of creatine synthetic enzyme GAMT and creatine kinases uCK-Mi and CK-B suggest a novel neuron-glial relationship for brain energy homeostasis. *Eur. J. Neurosci.* **20:** 144–160.

Takata, K., Kasahara, T., Kasahara, M., Ezaki, O., and Hirano, H., 1992, Ultracytochemical localization of the erythrocyte/HepG2-type glucose transporter (GLUT1) in cells of the blood-retinal barrier in the rat. *Invest. Ophthalmol. Vis. Sci.* **33:** 377–383.

Terasaki, T., and Ohtsuki, S., 2005, Brain-to-blood transporters for endogenous substrates and xenobiotics at the blood-brain barrier: an overview of biology and methodology. *NeuroRx* **2:** 63–72.

Terasaki, T., Ohtsuki, S., Hori, S., Takanaga, H., Nakashima, E., and Hosoya, K., 2003, New approaches to in vitro models of blood-brain barrier drug transport. *Drug Discov. Today* **8:** 944–954.

Vannas-Sulonen, K., Sipila, I., Vannas, A., Simell, O., and Rapola, J., 1985, Gyrate atrophy of the choroid and retina. A five-year follow-up of creatine supplementation. *Ophthalmology* **92:** 1719–1727.

Voat, D., and Voat, J.G., 1995, *Biochemistry*, 2nd edn. John Wiley & Sons, Inc, Toronto, Canada.

Wallimann, T., Wegmann, G., Moser, H., Huber, R., and Eppenberger, H.M., 1986, High content of creatine kinase in chicken retina: compartmentalized localization of creatine kinase isoenzymes in photoreceptor cells. *Proc. Natl. Acad. Sci. USA* **83:** 3816–3819.

Wallimann, T., Wyss, M., Brdiczka, D., Nicolay, K., and Eppenberger, H.M., 1992, Intracellular compartmentation, structure and function of creatine kinase isoenzymes in tissues with high and fluctuating energy demands: the 'phosphocreatine circuit' for cellular energy homeostasis. *Biochem. J.* **281:** 21–40.

Wang, Y.E., Esbensen, P., and Bentley, D., 1998, Arginine kinase expression and localization in growth cone migration. *J. Neurosci.* **18:** 987–998.

Wyss, M., and Kaddurah-Daouk, R., 2000, Creatine and creatinine metabolism. *Physiol. Rev.* **80:** 1107–1213.

CHAPTER 6

FUNCTIONAL INSIGHTS INTO THE CREATINE TRANSPORTER

DAVID L. CHRISTIE

Molecular, Cell and Developmental Biology Section, School of Biological Sciences, University of Auckland, Private Bag 92019, Auckland, New Zealand

Abstract: Creatine and phosphocreatine provide an intracellular, high-energy phosphate buffering system, essential to maintain ATP levels in tissues with high energy demands. A specific plasma membrane creatine transporter (CRT) is required for the cellular uptake of creatine. This transporter is related to the γ-aminobutyric acid (GAT) and norepinephrine (NET) transporters and is part of a large gene family of Na^+- and Cl^--dependent neurotransmitter transporters, now known as solute carrier family 6 (SLC6). CRT is essential for normal brain function as mutations in the CRT gene (*SLC6A8*) result in X-linked mental retardation, associated with the almost complete lack of creatine in the brain, severe speech and language delay, epilepsy, and autistic behaviour. Insight into the structure and function of the CRT has come from studies of creatine transport by tissues and cells, *in vitro* studies of CRT mutations, identification of mutations associated with CRT deficiency, and from the recent high resolution structure of a prokaryotic homologue of the SLC6 transporters. CRT antibodies have been developed enabling the localization of creatine uptake sites in the brain, retina, muscle and other tissues. These tools in conjunction with the use of appropriate cell models should allow further progress in our knowledge on the regulation and cellular trafficking of the CRT. Development of suitable mouse models may allow improved understanding of the importance of the CRT for normal brain function and how the transporter is regulated *in vivo*

1. INTRODUCTION

The physiological importance of creatine transport was first appreciated for muscle where high concentrations of phosphocreatine are required for muscle contractions. Creatine biosynthesis occurs mostly in other tissues, so ~2 g/day of creatine must be taken up by skeletal muscle to compensate for losses due to its irreversible conversion to creatinine. Uptake of [^{14}C]creatine in isolated rat muscle tissue was shown to be a saturable transport process (Fitch and Shields, 1966). Uptake was inhibited by creatine analogues such as β-guandinopropionate and guanidinoacetate, but not by amino acids

G.S. Salomons and M. Wyss (eds.), Creatine and Creatine Kinase in Health and Disease, 99–118.

or other compounds lacking an amidino group (Fitch *et al.*, 1968). The kinetics and specificity of creatine transport were subsequently studied in cultured cell preparations (Daly and Seifter, 1980), human monocytes and macrophages (Loike *et al.*, 1986) and astroglial-rich cultures from neonatal rats and mice (Moller and Hamprecht, 1989). The creatine uptake system was confirmed to be saturable, sodium-dependent and highly specific. The properties of the CRT have been confirmed and extended through the functional expression of the cloned creatine transporter from a variety of species.

2. MOLECULAR CHARACTERIZATION OF THE CRT

2.1. The CRT is a Member of the Family of Neurotransmitter Transporters

Molecular cloning showed the CRT to be a member of solute carrier family 6, a large family of membrane transporters that mediate the translocation of a range of solutes across plasma membranes, through the co-transport of sodium and chloride down their electrochemical gradients (Chen *et al.*, 2004). The γ-aminobutryic acid (GABA) and norepinephrine transporters (Guastella *et al.*, 1990; Paczkowski *et al.*, 1999) were the first to be discovered, followed rapidly by identification of transporters for dopamine and serotonin, accounting for reference to this family as Na^+/Cl^--dependent neurotransmitter transporters or neurotransmitter:sodium symporters (NSS) (Saier, 1999). Subsequently, homologous transporters for solutes other than neurotransmitters were discovered, e.g. for taurine, betaine and creatine. The human genome encodes twenty SLC6 transporters, a large sub-branch of which are Na^+-dependent amino acid transporters (Broer, 2006). A role in amino acid transport is likely to account for the large number of SLC6 homologues in bacteria (Androutsellis-Theotokis *et al.*, 2003; Quick *et al.*, 2006; Saier, 1999). SLC6 transporters are absent in yeast, fungi and plants (Nelson, 1998).

2.2. The CRT Gene (*SLC6A8*)

The CRT gene (*SLC6A8*) is located on human chromosome Xq28, contains 13 exons and spans ~8.5 kb of genomic DNA (Gregor *et al.*, 1995; Sandoval *et al.*, 1996). Mutations in *SLC6A8* were first discovered in a boy with mental retardation who had severe speech and language delay, cerebral creatine deficiency and an increased urinary creatine/creatinine ratio (see chapter 8; Stockler *et al.*, 2007; Cecil *et al.*, 2003; Salomons *et al.*, 2001). Subsequently, *SLC6A8* was shown to be an X-linked mental retardation gene (Hahn *et al.*, 2002). Three studies have reported the presence of a second CRT gene (*SLC6A10*) on chromosome 16 (Eichler *et al.*, 1996; Iyer *et al.*, 1996; Xu *et al.*, 1997). The physiological significance of the second CRT gene is unclear. Evidence has been obtained supporting the expression of *SLC6A10* transcripts specific to the testis (Iyer *et al.*, 1996). However, it has also been stated that *SLC6A10* is a pseudogene as the predicted amino acid sequence indicated the presence of a premature stop codon in exon 4 (Eichler *et al.*, 1996). There appears to be no correlate of *SLC6A10* in the mouse (Broer, 2006).

2.3. Molecular Cloning of the CRT cDNA

Our understanding of the CRT has progressed in a similar way to other members of the neurotransmitter transporter family. The cDNAs from a variety of tissues and species were cloned and characterized by functional expression in *Xenopus* oocytes or mammalian cell lines. The first structural and functional characterization of the CRT came with the cloning of cDNAs from rabbit brain and muscle (Guimbal and Kilimann, 1993). Functional expression of the cDNA in COS-7 cells demonstrated Na^+- and Cl^--dependent creatine uptake. The cDNA was 98% identical to a cDNA isolated from rat brain, first described as a Na^+-dependent choline transporter and later confirmed to be a creatine transporter (Mayser *et al.*, 1992; Schloss *et al.*, 1994). Subsequently cDNAs encoding CRTs from human were identified (Nash *et al.*, 1994; Saltarelli *et al.*, 1996; Sora *et al.*, 1994). Several non-mammalian CRTs have also been cloned. The CRT from the electric lobe of the ray, *Torpedo marmorata*, was found to be 64% identical to the rabbit CRT (Guimbal and Kilimann, 1994). More recently, the sequence of the CRT has been deduced from genome sequencing of zebrafish, *Danio rerio* (accession no. XP_695932) and the pufferfish, *Tetraodon nigroviridis* (accession no. CAF95116). The functional properties of these fish CRTs have not been investigated so far.

3. SUBSTRATE SPECIFICITY AND TRANSPORT MECHANISM OF THE CRT

The CRT is highly specific for creatine. Neither creatinine nor phosphocreatine are substrates. The key features for substrate specificity are a carboxyl group and a guanidino group, separated by no more than 2–3 carbon atoms (Guimbal and Kilimann, 1993; Moller and Hamprecht, 1989; Saltarelli *et al.*, 1996; Sora *et al.*, 1994). A benzyl ester derivative of creatine which no longer contains a free carboxyl group is not a substrate for the transporter although it can be taken up by brain slices in a non-transporter dependent manner (Lunardi *et al.*, 2006). β-Guanidinopropionate is a better competitive inhibitor than guanidinoacetate or 4-guanidinobutyrate. 2-Amino-3-guandinopropionate competes more efficiently with creatine for rabbit than electric ray CRT, suggesting slightly altered binding sites in different species (Guimbal and Kilimann, 1993, 1994). Arginine, citrulline, carnitine, GABA and choline are all without a significant effect on creatine transport.

Many studies have shown that the CRT is Na^+-dependent. Sodium cannot be replaced by other cations such as lithium, choline or N-methylglucamine (Dodd *et al.*, 1999; Guimbal and Kilimann, 1993, 1994; Moller and Hamprecht, 1989; Saltarelli *et al.*, 1996; Sora *et al.*, 1994). The creatine transporter is also Cl^--dependent (Dai *et al.*, 1999; Guimbal and Kilimann, 1993; Walzel *et al.*, 2002). This is an important consideration as not all SLC6 transporters are dependent on chloride. The mammalian neutral amino acid transporter B°AT1 (SLC6A19) is not Cl^--dependent (Broer *et al.*, 2004; Broer, 2006). It appears that the CRT has a coupling ratio of 2 Na^+:1 Cl^-:1 creatine molecule based on studies on the expression of human CRT in *Xenopus* oocytes (Dai *et al.*, 1999) and creatine transport by brush-border membranes of rat kidney cortex (Garcia-Delgado *et al.*, 2001), rat jejunal enterocytes (Tosco *et al.*, 2004) and chicken enterocytes (Peral *et al.*, 2002). These studies make it clear that the CRT in apical membranes of intestinal enterocytes and proximal kidney epithelial cells has an important role in absorption and reabsorption of creatine, respectively. However, kinetic studies and consideration of electrogenicity make it unlikely that the CRT plays a role in the export of creatine from the enterocyte across the basolateral membrane (Orsenigo *et al.*, 2005).

The transport of creatine through the CRT is saturable. K_m values for the rat (Garcia-Delgado *et al.*, 2001; Moller and Hamprecht, 1989; Saltarelli *et al.*, 1996; Tosco *et al.*, 2004), human (Dai *et al.*, 1999; Loike *et al.*, 1986; Nash *et al.*, 1994; Sora *et al.*, 1994), rabbit (Guimbal and Kilimann, 1993) and bovine CRT (Dodd *et al.*, 1999) have been determined as 15–46 μM, 15–77 μM, 35 μM and 188 μM, respectively. By comparison, the serum creatine concentrations for rats, rabbits and humans have been estimated as 528, 149 and 58 μM, respectively (Marescau *et al.*, 1986). It should be noted that plasma creatine levels may decrease with age. In humans, plasma creatine concentrations were found to range from 17–109 μM and from 6–50 μM at <10 and > 10 years, respectively (Almeida *et al.*, 2004). It is apparent, however, from the concentration of creatine in plasma that CRTs work close to saturation, suggesting that the amount of transporter in the cell membrane may limit creatine accumulation.

4. STRUCTURE AND FUNCTION OF THE CRT

4.1. Properties of the CRT Protein

Mammalian CRTs from mouse, rat, rabbit, cow and man contain 635 amino acids and are ~96% identical at the amino acid level. Each is predicted to contain 12 trans-membrane spanning domains (TMs) with the amino- and carboxy-terminal regions facing the cytoplasmic side of the membrane. Two potential sites for N-linked glycosylation are found in the extracellular loop between TM3 and TM4. There is another predicted glycosylation site located in the extracellular loop between TMs 11 and 12 (Snow and Murphy, 2001; Sora *et al.*, 1994). However, this is a very short loop and it may not be accessible for glycosylation. The lectin, wheat germ agglutinin (specificity for N-acetylglucosamine), was used in the purification of bovine CRT from membranes of HEK293 cells expressing high levels of the transporter (West *et al.*, 2005). N-glycosidase F digestion reduced the mass for the purified CRT (70–80 kDa) to ~50 kDa. The mass for deglycosylated CRT is less than predicted from the amino acid sequence (70.68 kDa) suggesting that this hydrophobic protein may run faster on SDS polyacrylamide gels than expected from its molecular mass. The mobility of the CRT on SDS gels varies if other detergents are present. Western blotting of HEK293-CRT cell membranes solubilized directly into SDS gel loading buffer demonstrated a 70–80 kDa CRT-immunoreactive protein. However, cell lysates, prepared first in a buffer containing 1% Triton X-100, 1% sodium deoxycholate and 0.1% SDS, gave 80–90 kDa immunoreactive protein bands (Dodd and Christie, 2001;

West *et al.*, 2005). This apparent size agrees with the ~80 kDa immunoreactive protein band seen for the expression of the related norepinephrine transporter in HEK293 cells (Burton *et al.*, 1998; Galli *et al.*, 1995). It is difficult to reconcile this data with a recent report indicating a molecular mass of 58 kDa for the glycosylated form of haemagglutinin-tagged rat CRT expressed in HEK293 cells (Straumann *et al.*, 2006).

4.2. Structure of LeuT$_{Aa}$, a Bacterial Homologue of SLC6 Transporters

The key question for the CRT and other transporters is how they are able to recognize their substrates and translocate them across the lipid bilayer. Determination of the crystal structure of a bacterial leucine transporter (LeuT$_{Aa}$) has provided the very first structure for a member of the SLC6 transporter family (Yamashita *et al.*, 2005). Remarkably, for a membrane protein, the structure of LeuT$_{Aa}$ from *Aquifex aeolicus* was solved by Gouaux and coworkers at extraordinarily high resolution (1.65 Å). This structure not only revealed the protein folds of the transporter but also provided clear details of the substrate binding site, with leucine and two Na^+ ions bound. The importance of the LeuT$_{Aa}$ structure in relation to the structure and function of SLC6 transporters has been reviewed recently (Broer, 2006; Gether *et al.*, 2006; Henry *et al.*, 2006; Kanner, 2005).

As predicted from hydropathy models of mammalian SLC6 transporters (Guastella *et al.*, 1990; Paczkowski *et al.*, 1999), LeuT$_{Aa}$ contains 12 TM domains with the N- and C-termini facing the cytoplasm. The core structure shows the substrate binding site for leucine buried within a unique fold. An unexpected structural repeat was found in the first ten TMs. Amino acid residues in TMs 1–5 can be superimposed on those from TMs 6–10 by rotation through a pseudo-twofold axis of symmetry in the plane of the membrane. TMs 1 and 6 are highly conserved in SLC6 transporters and are orientated anti-parallel to each other. Both the TM1 and TM6 helices are partially unwound in the middle of the lipid bilayer. The exposed main chain carbonyl oxygen and nitrogen atoms are involved in direct interactions with the substrate, leucine. The two Na^+ ions also interact with the unwound portions of these helices. LeuT$_{Aa}$ is not Cl^--dependent, so its structure could not provide information on the Cl^- binding site. TMs 3 and 8 also relate to each other through the two-fold symmetry axis. These are strongly tilted (~50°) and contribute a few key residues located near the unwound portions of TMs 1 and 6 that are also involved in substrate binding.

The structure of LeuT$_{Aa}$ supports the two gate theory for transporters. In this model, exposure of the binding site to the outside of the cell enables the substrate to bind. Then, closure of the external gate, traps the substrate within the membrane. Subsequent opening of an internal gate, on the other side of the membrane, allows the release of the substrate into the cell. In the structure of LeuT$_{Aa}$ the external and inner gates are closed and leucine and Na^+ are bound within the binding site formed by residues from TMs 1, 3, 6 and 8. The transport process was suggested to require major movements of TM1 and TM6 relative to TM3 and TM8 (Yamashita *et al.*, 2005). A few residues from extracellular loop 4 (ECL4) between TMs 7 and 8 form the external gate and cover the substrate binding site. The intracellular gate extends from the substrate binding site to the cytoplasmic face of the membrane, spanning ~20 Å, and consists of ordered protein structure from regions of TMs 1, 6 and 8.

4.3. Is $LeuT_{Aa}$ a Suitable Model for the CRT?

$LeuT_{Aa}$ is only 20–25% identical to the mammalian SLC6 transporters. However, the conservation of residues in SLC6 transporters at key positions and the topology of $LeuT_{Aa}$ suggest that this bacterial SLC6 homologue will prove to be a valuable model for the CRT and other SLC6 transporters. A schematic diagram for the topology of the CRT based on $LeuT_{Aa}$ is shown in Figure 1.

It is interesting to review studies of *in vitro* mutagenesis and mutations associated with CRT deficiency in relation to a structural template based on $LeuT_{Aa}$. Comparison of the phylogenetically distant *Torpedo marmorata* and mammalian CRTs indicated that only a few residues were better conserved in CRTs than other members of the neurotransmitter transporter family (Guimbal and Kilimann, 1994). One of the 'CRT-specific' residues was Cys-144 in TM3. The importance of this residue was investigated by chemical modification with methanethiosulfonates (MTS), membrane impermeable sulphydryl-modifying reagents, and by site-directed mutagenesis (Dodd and Christie, 2001). Low concentrations of 2-aminoethyl methanethiosulfonate (MTSEA) rapidly inactivated the transporter. A Cys144Ser mutant retained 70% of activity and was completely resistant to MTSEA while a Cys144Leu mutant had a much higher K_m for creatine. Creatine and its close analogues, but not arginine or GABA, protected Cys144 from modification. It was concluded that Cys144 may be close to a binding site or part of a permeation channel for creatine. A very good way of identifying channels and pores in membrane proteins is the substituted cysteine accessibility method (SCAM). The functionally active and MTS-resistant p.Cys144Ser mutant was used as a background to replace 22 residues in TM3, one at a time, with cysteine (Dodd

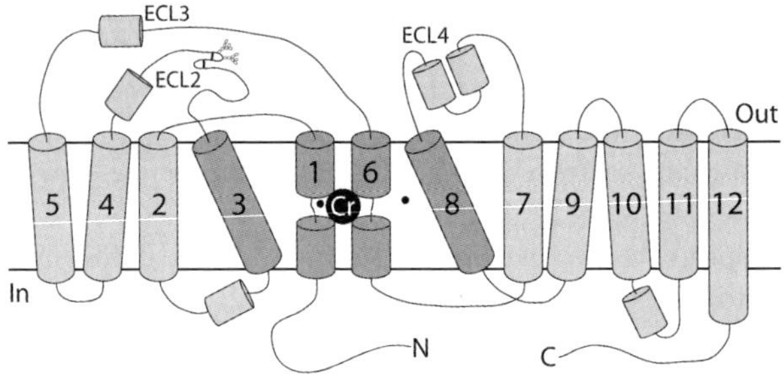

Figure 1. Schematic representation of transmembrane topology of the CRT. The topology of the CRT is based on $LeuT_{Aa}$ (Yamashita *et al.*, 2005). TMs 1-5 are related to TMs 6-10 by rotation through a pseudo-twofold axis of symmetry in the plane of the membrane. The CRT, as indicated, has much longer N- and C-terminal domains than $LeuT_{Aa}$. The figure also shows two predicted disulphide bonds and two N-linked carbohydrate groups in extracellular loop 2 (ECL2). The predicted binding site for creatine (Cr) is indicated. The small black dots indicate bound Na^+. TMs 1, 3, 6 and 8, involved in substrate binding are shaded dark grey.

and Christie, 2005). The activity of each Cys substitution mutant was tested before and after reaction with MTS reagents to determine substituted cysteine accessibility. Inactive and MTS-sensitive Cys mutants corresponded to positions Trp154, Val151, Tyr148, Tyr147, Cys144 and Ile140. These were aligned 3–4 amino acids apart along one face of the predicted α-helix of TM3. It was suggested that these residues face the substrate binding pocket and that three residues, Ile140, Cys144 and Tyr147, each protected from MTSEA inactivation by creatine, may be close to the substrate binding site. Residues equivalent to these positions in the CRT have also been shown to be involved in substrate specificity through SCAM analysis of the human serotonin transporter (Chen and Rudnick, 2000; Chen *et al.*, 1997). Thus, biochemical analysis of the creatine and serotonin transporters, indicating the involvement of TM3 in substrate binding, are in excellent agreement with the $LeuT_{Aa}$ structural model (Yamashita *et al.*, 2005). This gives us confidence that $LeuT_{Aa}$ is an appropriate structural template for mammalian SLC6 transporters.

We have attempted to obtain a preliminary model of the CRT based on the coordinates of $LeuT_{Aa}$ (Yamashita *et al.*, 2005). In the model of the creatine binding site, Cys144 and Tyr148 in TM3 are opposite to Phe315 and Leu321 in TM6 and Tyr68 in TM1 (Figure 2). This model agrees well with previous biochemical studies of the binding site of the CRT. A close-up view of the predicted substrate binding site is shown in Figure 3. Interestingly, of the 12 residues involved in the binding of Na^+ and leucine in $LeuT_{Aa}$, only 4 of the CRT and the GABA transporter (GAT-1) residues differ at the equivalent positions (Guastella *et al.*, 1990; Yamashita *et al.*, 2005). Phe68, Cys144, Ala318 and Gly421 are found in CRT and Tyr, Leu, Gly and Thr in GAT-1 in place of Asn, Val, Ser and Ile in $LeuT_{Aa}$, respectively (Yamashita *et al.*, 2005). One

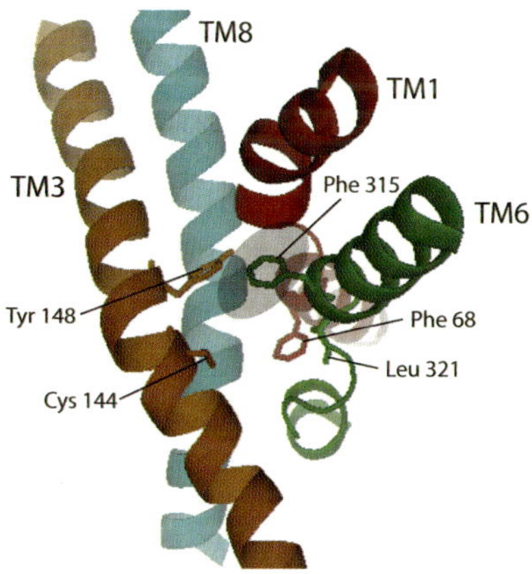

Figure 2. Model of CRT substrate binding site. An approximate model of the bovine CRT binding site was obtained by exchanging the sequences of TMs 1, 3, 6 and 8 in $LeuT_{Aa}$ with the sequences of the equivalent regions of the bovine CRT. The modified $LeuT_{Aa}$/CRT sequence was fitted to the coordinates of $LeuT_{Aa}$ (PDB, ID 2A65) using SWISS-MODEL (http://swissmodel.expasy.org/). Residues thought to be involved in substrate binding are shown by stick representation. The figure was generated using PyMOL v0.99. The shaded oval indicates the predicted binding site for creatine.

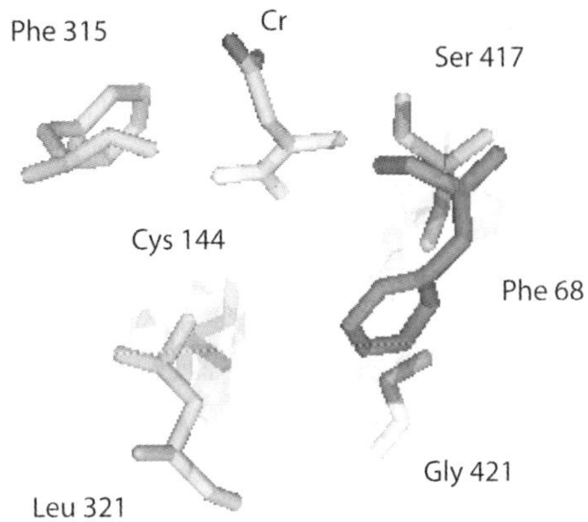

Figure 3. Predicted hydrophobic CRT binding pocket. The predicted hydrophobic pocket of the CRT based on $LeuT_{Aa}$ and the model shown in Figure 6.2. The side chains of Phe68 in TM1, Cys144 in TM3, Phe315 and Leu321 in TM6 and Ser417 and Gly421 in TM8 are indicated by semitransparent van der Waals surface spheres, coloured red, orange, green and blue, respectively. A stick representation of creatine is shown (oxygen, carbon and nitrogen atoms are shown in red, yellow and blue, respectively). The position of creatine was aligned with respect to the carboxyl group and C_α atom of leucine in $LeuT_{Aa}$. Some residues predicted to be involved in substrate binding (CRT residues Leu72 and Gly73 in TM1, Tyr148 in TM3, as well as Phe314 and Ala318) were omitted for clarity.

can imagine that small changes can alter the shape of the binding site to accommodate different substrates. We have also tested the model by asking the following question: if the model for the substrate binding site is correct, can we alter the substrate specificity of the CRT? We find that substitution of just three of the four above described amino acids results in the complete loss of creatine transport and the gain of specific GABA transport function (Dodd and Christie, 2007).

4.4. How do Pathogenic *SLC6A8* Mutations Relate to the Proposed CRT Structural Model?

Mutations in *SLC6A8* are found in families with (X-linked) mental retardation. In order to assess the frequency of *SLC6A8* mutations, males with no known cause for mental retardation were screened by sequencing genomic DNA (Clark *et al.*, 2006). It was concluded that *SLC6A8* mutations occur in $\sim$1% of males with mental retardation of unknown etiology. This study identified six novel mutations and summarized known mutations in the *SLC6A8* gene. In total 18 mutations, from 20 unrelated families, have been reported and occur throughout the gene. Criteria used to classify the variants as pathogenic included absence of creatine in the brain, a high creatine:creatinine ratio in the urine and in some cases, demonstration of impaired uptake of creatine by fibroblasts from affected males (Cecil *et al.*, 2003; Clark *et al.*, 2006; Salomons *et al.*, 2001, 2003). It is not surprising that major changes (e.g. large genomic deletion, frameshift, splice error, nonsense mutation (i.e. premature stop codon)) would lead to inactive CRT. Analysis of missense mutations (i.e. one amino acid substitution by another amino acid) and perhaps also one amino acid deletions may provide greater insight into regions critical for the function of the CRT.

The missense mutation p.Gly87Arg is part of a highly conserved sequence (Asn-Gly-Gly-Gly-Ala/Val-Phe) found in all animal and bacterial SLC6 transporters. Insertion of the larger, positively charged Arg residue is expected to disrupt this loop, possibly interfering with the orientation of TM1, one of the 4 TMs known to be important for substrate binding in $LeuT_{Aa}$.

Interestingly, other pathogenic mutations have also been found in loops. The mutation p.Cys337Trp is at the end of TM7 that faces the cytoplasm. $LeuT_{Aa}$ and GAT-1 have Ile and Val at this position, respectively, and Trp may be too large and bulky to be tolerated. Three other missense mutations, p.Gly381Arg, p.Pro390Leu and p.Arg391Trp are located in ECL4. This is an important loop that is likely to participate in conformational changes during transport. This loop forms a lid that covers the substrate binding site, extends into the centre of the transporter and interacts with residues in TM1. Evidence supporting conformational changes have come from engineering of Zn^{2+} binding sites and accessibility studies in the dopamine, serotonin and GABA transporters (Loland *et al.*, 1999; MacAulay *et al.*, 2001; Mitchell *et al.*, 2004; Norregaard *et al.*, 2003). The ECL4 loop contains two short helices separated by a sharp bend (Yamashita *et al.*, 2005). Gly-381 is adjacent to the second short helix and substitution by Arg would have a major affect on this loop. *In vitro* studies in the author's laboratory have shown that p.Gly381Arg-containing bovine CRT was inactive and failed to be inserted into the membrane of HEK293 cells, suggesting a problem with the folding of the transporter protein. Overexpression of a p.Gly381Cys mutation, on the other hand, was expressed in the membrane but remained inactive (Lisa Tolich, Joanna Dodd and David Christie, unpublished data). The mutations (p.Pro390Leu and p.Arg391Trp) were found in the second short helix. It is likely that each of these mutations disrupts this functionally important segment of secondary structure.

The pathogenic mutation p.Pro554Leu was attributed to the loop between TMs 11 and 12 based on the predicted topology of mammalian SLC6 transporters (Rosenberg *et al.*, 2004). However, revision of the topology based on $LeuT_{Aa}$ (Figure 1) suggests that this conserved proline may be part of the helix of TM11 (Yamashita *et al.*, 2005). Proline residues are known to introduce twists in TM helices and, for this reason, leucine may not be tolerated at this position.

$LeuT_{Aa}$ has very short N- and C-terminal domains so its structure does not help with interpretation on the effect of mutations in the much longer N-terminal and C-terminal domains of CRT. While these domains are not highly conserved in mammalian SLC6 transporters, they do have some functional importance. Syntaxin 1A regulates the activity of the GABA and noradrenaline transporters by binding to their N-terminal domains (Deken *et al.*, 2000; Sung *et al.*, 2003). The unclassified variant p.Lys4Arg, discovered through sequencing of CRT genomic sequences, may represent a rare polymorphism. Although Lys-4 is conserved in mammalian CRTs (Rosenberg *et al.*, 2004), an Arg residue is found at a similar position in the electric ray CRT (Guimbal and Kilimann, 1994). The p.Gly26Arg mutation is also unclassified as no biological material was available from the patient. A glycine is found at this position in mammalian CRTs and proline, another neutral amino acid, is present in the electric ray CRT. The effect of introduction of the positively charged Arg residue at this position needs investigation through *in vitro* studies. There have been no reported studies of the functional importance of the C-terminal region of the CRT. However the sequence of the C-terminal four amino acids (-Glu-Ser-Val-Met) may constitute a binding motif recognized by PDZ-domain proteins which are known to regulate many channel and transporter proteins (Brone and Eggermont, 2005; Sheng and Sala, 2001). A similar sequence (-Glu-Thr-Val-Met) at the C-terminus of the parathyroid hormone receptor is recognized by the PDZ domain protein Na^+/H^+ regulatory factor 2 (Mahon *et al.*, 2002). The unclassified variation p.Val629Ile is likely to represent a polymorphism as Ile is found in the marble electric ray (Guimbal and Kilimann, 1994; Rosenberg *et al.*, 2004) and the zebrafish CRTs (accession no. XP_695932).

5. TISSUE DISTRIBUTION OF THE CRT

5.1. Localization of CRT mRNA

The distribution of the CRT mRNA was determined soon after cloning of the cDNA. Initially, the CRT cDNA was thought to encode a choline transporter. However, localization of CRT mRNA indicated that the transporter was present in many tissues outside the nervous system, being most abundant in tissues with high energy demands such as skeletal muscle, heart, brain and retina, or with absorptive functions such as kidney and intestine (Gonzalez and Uhl, 1994; Guimbal and Kilimann, 1993; Peral *et al.*, 2002; Saltarelli *et al.*, 1996; Schloss *et al.*, 1994). *In situ* hybridization has revealed further details of the distribution of the CRT mRNA in brain (Happe and Murrin, 1995; Saltarelli *et al.*, 1996; Schloss *et al.*, 1994). Transporter expression was high in the cerebellum (Purkinje and granule cell layers), hippocampus (pyramidal cell layer) and some brain stem nuclei with moderate levels detected in the cortex, globus pallidus and most white matter tracts. The striatum, nucleus accumbens, hippocampus molecular layer, and cerebellar molecular layer showed very low levels of CRT mRNA (Happe and Murrin, 1995). In the rat auditory brain stem, most auditory nuclei expressed CRT mRNA although the amounts varied among auditory nuclei, possibly reflecting differences in high-energy phosphate metabolism (Hiel *et al.*, 1996).

5.2. Localization of CRT Protein

Localization of the CRT protein identifies sites of creatine uptake and thus has functional significance. Unfortunately, studies of the localization of the CRT protein have been scarce due to the limited availability and specificity of CRT antibodies.

Several antibodies are available that appear to give reliable identification of the CRT by immunocytochemistry (Acosta *et al.*, 2005; Murphy *et al.*, 2001; Ohtsuki *et al.*, 2002; Peral *et al.*, 2002). However, the development of antibodies with adequate specificity and sensitivity to detect the CRT in tissue extracts by Western blotting has been more problematic. The CRT does not appear to be an abundant component of plasma membranes, even in tissues containing high levels of creatine and phosphocreatine. Also, depending on conditions and possibly different glycosylation states, the CRT may run with different apparent molecular masses on SDS polyacrylamide gels. The CRT may also aggregate once solubilized from membranes.

Antibodies to the CRT, developed against peptide sequences corresponding to the N- and C-terminal regions of the CRT (Guerrero-Ontiveros and Wallimann, 1998) have been used to localize the CRT in rat skeletal muscle (Murphy *et al.*, 2001). Immunohistochemical analysis indicated that the CRT was predominantly associated with the sarcolemmal membrane in both type I and II fibres. The CRT was localized with the N-terminally directed CRT antibody to the apical membrane of cells lining the villus from rat, chicken and human ileum (Peral *et al.*, 2002). The immunolocalization of the CRT in muscle and intestinal enterocytes is entirely consistent with its known role for creatine transport in these tissues. However, some caution is required when considering the results for studies in which these antibodies were used to quantify up- or down-regulation of the CRT protein, especially if analyzed by Western blotting. The N- and C-terminal CRT anti-peptide antibodies cross-react with 55- and 70 kDa species also present in mitochondria, and the C-terminal antibody has been shown to recognize E2 components of mitochondrial dehydrogenases (Speer *et al.*, 2004).

An affinity-purified antibody directed to the C-terminal 21 amino acids of the CRT (Dodd and Christie, 2001) was used to localize the CRT in adult vertebrate retinas and the retina of the developing mouse (Acosta *et al.*, 2005). Strong labeling for the CRT was seen for inner segments and a variety of inner neuronal cells. CRT localization revealed the sites of creatine transport into the retina at the retinal pigment epithelium, inner retinal blood vessels and perivascular astrocytes. This antibody detected a ~65 kDa protein in membranes from bovine retina. Detection of this band was prevented when the antibody was blocked with the fusion protein used to generate the antibody. Based on the known CRT content of HEK293-CRT cells, it was estimated that the CRT content in retina and choriocapillaris membranes represents less than 0.03% of retina membrane protein, indicating that the CRT is not abundant, even in tissues like retina with known requirements for high levels of creatine and phosphocreatine (Hemmer *et al.*, 1993). A C-terminal antibody prepared by a different group detected 85- and 71 kDa CRT species in rat brain and retina (Nakashima *et al.*, 2004; Ohtsuki *et al.*, 2002). CRT was detected at both the luminal and abluminal membranes from rat retinal capillary cells by immunoperoxidase electron microscopy (Nakashima *et al.*, 2004). The CRT was also shown to be expressed in mouse brain capillaries, indicating an important role in supplying creatine to the brain via the blood-brain barrier (Ohtsuki *et al.*, 2002). CRT antibodies have also localized the CRT to the plasma membrane of keratinocytes of human skin (Lenz *et al.*, 2005; Schlattner *et al.*, 2002).

6. REGULATION OF THE CRT

The prime function of the CRT is for the cellular accumulation of creatine which, in turn, is crucial for the maintenance of ATP in tissues with high energy demands like brain and muscle. Also, creatine is neuroprotective in many animal models of neurological diseases (Wyss and Schulze, 2002). Consequently, the regulation of the CRT is an important topic. It is essential to know whether creatine supplementation down-regulates the transporter. Alternatively, if we can learn how to up-regulate the transporter, then it may be possible to enhance the uptake of creatine into the brain, facilitating its use as a neuroprotective agent. Efforts have been made to understand how the CRT is regulated in relation to tissue creatine levels, particularly for muscle. However, deficiencies in the specificity of some of the available CRT antibodies (discussed in section 5.2 above) have so far limited our understanding of the regulation of CRT activity at the cellular level. Understanding of the regulation of the CRT lags well behind that of regulation for other members or the SLC6 neurotransmitter transporter family.

CRT homologues are known to be regulated by changes in their distribution between intracellular and plasma membrane pools (see section 6.2, below). It will be important to consider this mode of regulation for the CRT. The K_m values determined for CRTs are lower than the concentration of creatine in the blood, meaning that the transporter normally works close to saturation. If the creatine uptake rate is close to its maximal capacity, then it follows that regulation of the level of CRT in the plasma membrane would alter creatine uptake.

6.1. Relationship of Skeletal Muscle and Heart Creatine Levels and CRT Activity

The relationship between creatine content of muscle and the levels and activity of the CRT has been studied following dietary supplementation with creatine or its analogues. Feeding rats with 3% creatine for 6 weeks resulted in a 30% decrease in the V_{max} of creatine transport activity of isolated perfused hearts. In contrast, feeding rats with β-guanidinopropionate (β-GPA), an analogue of creatine that depleted intracellular creatine, decreased the K_m and increased the V_{max} (~70%) for creatine uptake (Boehm *et al.*, 2003). Plasma membrane CRT (~60- and 75-kDa immunoreactive bands) decreased 30% in cardiac myocytes from the creatine-fed group, and increased ~5-fold in β-GPA-fed rats. Interestingly, neither treatment changed CRT content (72- and 52 kDa bands) of whole cell extracts of cardiac myocytes. It was concluded that in heart, creatine transport is determined by the plasma membrane content of the CRT and not by its total cellular content.

In a separate study, rats were fed 1% creatine or 1% β-GPA for up to 7 weeks, followed by hindlimb perfusion to assess [^{14}C]creatine accumulation. A rabbit polyclonal antiserum, raised against a bacterial fusion protein containing a 50 amino acid, non-conserved region of the CRT, was used to determine CRT content by Western blotting (Brault *et al.*, 2003). In white gastrocnemius muscle creatine supplementation significantly decreased the creatine uptake rate but did not alter the CRT content (55- and 58-kDa bands). Creatine depletion from β-GPA supplementation increased creatine uptake capacity in the soleus and red gastrocnemius muscle, a change accompanied by 70–150% increases in CRT protein. However, the creatine uptake rate of white gastrocnemius muscle was unchanged, even though its CRT content increased by 230%, suggesting that total CRT content is unrelated to creatine uptake activity. It should be noted that unlike the previous study, this study determined the total rather than the plasma membrane content of the CRT. The relationship between creatine content and the levels of CRT remains unclear. In a human study, moderate supplementation with creatine increased creatine content of skeletal muscle without significant changes in levels of CRT protein or mRNA (Tarnopolsky *et al.*, 2003). Disease states also affect the CRT content of skeletal and heart muscle. In humans, the CRT content of skeletal muscle was lower in inflammatory myopathy, mitochondrial myopathy and muscular dystrophy/congenital myopathy (Tarnopolsky *et al.*, 2001). In a rat model, heart failure led to a 30% decrease in intracellular creatine and a 26% decrease in creatine uptake capacity (Ten Hove *et al.*, 2005), making the CRT a potential therapeutic target. Growth hormone administration was able to significantly increase CRT mRNA in a rat model of myocardial infarction (Omerovic *et al.*, 2003). However, too much CRT may be harmful. Overexpression of the CRT in transgenic mice resulted in supranormal levels of myocardial creatine and phosphocreatine. However, the mice developed substantial left ventricular dysfunction and hypertrophy. It was suggested that these defects resulted from increased free ADP levels from the need to phosphorylate the abnormally high intracellular creatine pool (Wallis *et al.*, 2005).

6.2. Cellular Mechanisms for the Regulation of the CRT

The prime function of the CRT is for the cellular accumulation of creatine. This role appears distinct from the role of 'true' neurotransmitter transporters to rapidly clear transiently high concentrations of a neurotransmitter from the synapse. However, this assumption may need some revision, as a recent paper has shown that central neurons release creatine by exocytosis in a similar manner to neurotransmitters (Almeida *et al.*, 2006). Neurotransmitter transporters are known to be regulated by acute changes in their rate of endocytosis from the plasma membrane to a pool of recycling endosomes (Buckley *et al.*, 2000; Melikian, 2004). Exposure of cultured neurons and glial cells to GABA increased GABA transport activity rapidly by increasing the expression of the transporter at the plasma membrane (Bernstein and Quick, 1999), an effect mediated by substrate-induced phosphorylation of the transporter (Law *et al.*, 2000; Whitworth and Quick, 2001). While the CRT does not normally need to respond to rapid changes in extracellular creatine, there is growing evidence to suggest that mechanisms similar to those identified for other SLC6 transporters may be involved in regulating the CRT.

Creatine accumulation by G8 cultured skeletal myoblasts is increased by a range of hormones, including insulin, insulin-like growth factor I, 3,3',5-triiodothyronine, amylin and a β-receptor agonist, isoproterenol (Odoom *et al.*, 1996). It was suggested that some of these agents may work by increasing the Na^+ concentration gradient, thus stimulating Na^+/creatine co-transport. However, it appears that the CRT also becomes activated through cell signaling pathways. Phorbol esters, activators of protein kinase C, are known to decrease the activity of the creatine transporter in the same manner that has been observed for other SLC6 transporters (Dai *et al.*, 1999). It is likely that activation of protein kinase C promotes the internalization of the CRT by endocytosis. A decrease in cell surface expression was shown to be responsible for the reduction in creatine uptake through the recombinant human CRT in *Xenopus* oocytes and the endogenous CRT in cultured C_2C_{12} muscle cells in response to the immunosuppressive drug, cyclosporine A (Tran *et al.*, 2000). The CRT has been shown to be upregulated by serum and glucocorticoid inducible kinase, isoforms SGK1 and SGK3, in *Xenopus* oocytes (Shojaiefard *et al.*, 2005). SGK1 is known to regulate a number of carriers and channels (Lang *et al.*, 2003). While the CRT does not contain sites for SGK phosphorylation, SGK may prolong CRT half-life at the cell surface by phosphorylating Nedd4-2, which decreases the affinity of this ubiquitin ligase for the CRT. The activity of SGK1 is stimulated by IGF-1, so SGK1 could also participate in the regulation of the CRT by growth hormone (Omerovic *et al.*, 2003). The mammalian target of rapamycin, mTOR, has also been shown to stimulate creatine transport in *Xenopus* oocytes (Shojaiefard *et al.*, 2006). Both SGK1 and mTOR could be involved in the increased CRT expression in C_2C_{12} muscle cells following exposure to hyperosmotic stress (Alfieri *et al.*, 2006). A significant increase in tyrosine phosphorylation of the CRT, possibly through its association with c-Src tyrosine kinase, occurred following caecal ligation and puncture, a surgical rat model for polymicrobial sepsis. This was associated with an increase in the creatine content of fast-twitch gastrocnemius muscle (Wang *et al.*, 2002). In a recent study, electric field stimulation of isolated mouse skeletal muscle was used as a model to study the effect of contractions on creatine uptake (Derave *et al.*, 2006). Interestingly, it was not muscle contractions but electrolysis, a process often overlooked in field stimulation experiments, that stimulated both creatine transport and surface expression of the CRT. The mechanism is likely to involve the production of reactive oxygen species due to electrolysis in the highly oxygenated medium used to incubate the isolated muscle.

There is evidence that the plasma membrane levels of the CRT may be regulated in response to extracellular creatine levels. In rat L6 myoblasts and human myoblasts, extracellular creatine was found to down-regulate the CRT (Loike *et al.*, 1988). This down-regulation was reduced if protein synthesis inhibitors were present, suggesting that a protein may inhibit the CRT or be involved in its removal from the plasma membrane. Exposure to creatine-free medium resulted in a time-dependent increase in creatine transport capacity. To date, changes in the plasma membrane expression of the CRT in response to extracellular creatine have not been confirmed by Western blotting. The CRT may also be regulated in response to nutrient availability. Total creatine concentrations were found to be increased in gastrocnemius muscles in rats that were starved for 4 days to model severe nutrient deprivation commonly seen in surgical patients. A combined approach of immunoprecipitation and Western blotting indicated that the increase in CRT activity was associated with a 30% reduction in the phosphorylation of the CRT on serine residues (Zhao *et al.*, 2002).

7. FUTURE DIRECTIONS

The importance of the CRT for brain function and the growing interest in creatine as a neuropotective agent are likely to stimulate more detailed studies of the regulation of the CRT. We need a much better understanding of the membrane trafficking of the CRT. To achieve this, the CRT needs to be studied in suitable cell models, both in cell lines overexpressing the CRT and expressing the endogenous transporter. We need to know whether, like other SLC6 transporters (Deken *et al.*, 2003; Loder and Melikian, 2003), the CRT cycles between the cytosol and the plasma membrane. If so, can this flux be altered by levels of creatine, growth factors and other regulatory factors? What is the role of phosphorylation of the CRT or interacting proteins on the membrane turnover of the CRT? What is the fate of the CRT internalized from the plasma membrane?

We also know very little about the targeting of the CRT to defined regions of plasma membranes. The CRT works at the apical surface of intestinal and kidney epithelial cells but what is the nature of its sorting determinants? Is the CRT localized to axonal or dendritic surfaces of neurons?

Transgenic mice overexpressing the CRT, despite their limitations (Wallis *et al.*, 2005), may enable testing of the hypothesis that creatine is neuroprotective *in vivo*. It would be important to know if these transgenic mice exhibit greater resistance to paradigms that mimic neurological disease. Furthermore, fusions between CRT and green fluorescent protein (GFP) could be used to generate a strain of knock-in mice that produce GFP-CRT in place of the wild-type CRT as has been done for the GABA transporter (Chiu *et al.*, 2002). Cells expressing low and high levels of CRT could be identified from GFP fluorescence. Questions could be asked whether cells with low CRT were more susceptible to ischemic or anoxic damage. Also, the effects of creatine supplementation or deficiency and growth factors on CRT levels could be studied more easily.

ACKNOWLEDGEMENTS

I wish to thank Joanna Dodd for help with the preparation of the figures.

REFERENCES

Acosta, M.L., Kalloniatis, M., and Christie, D.L., 2005, Creatine transporter localization in developing and adult retina: importance of creatine to retinal function. *Am. J. Physiol. Cell Physiol.* **289:** C1015–C1023.

Alfieri, R.R., Bonelli, M.A., Cavazzoni, A., Briggotti, M., Fumarola, C., Sestili, P., Mozzoni, P., G, D.E.P., Mutti, A., Carnicelli, D., Vacondio, F., Silva, C., Borghetti, A.F., Wheeler, K.P., and Petronini, P.G., 2006, Creatine as a compatible osmolyte in muscle cells exposed to hypertonic stress. *J. Physiol.* **576:** 391–401.

Almeida, L.S., Salomons, G.S., Hogenboom, F., Jakobs, C., and Schoffelmeer, A.N., 2006, Exocytotic release of creatine in rat brain. *Synapse* **60:** 118–123.

Almeida, L.S., Verhoeven, N.M., Roos, B., Valongo, C., Cardoso, M.L., Vilarinho, L., Salomons, G.S., and Jakobs, C., 2004, Creatine and guanidinoacetate: diagnostic markers for inborn errors in creatine biosynthesis and transport. *Mol. Genet. Metab.* **82:** 214–219.

Androutsellis-Theotokis, A., Goldberg, N.R., Ueda, K., Beppu, T., Beckman, M.L., Das, S., Javitch, J.A., and Rudnick, G., 2003, Characterization of a functional bacterial homologue of sodium-dependent neurotransmitter transporters. *J. Biol. Chem.* **278:** 12703–12709.

Bernstein, E.M., and Quick, M.W., 1999, Regulation of gamma-aminobutyric acid (GABA) transporters by extracellular GABA. *J. Biol. Chem.* **274:** 889–895.

Boehm, E., Chan, S., Monfared, M., Wallimann, T., Clarke, K., and Neubauer, S., 2003, Creatine transporter activity and content in the rat heart supplemented by and depleted of creatine. *Am. J. Physiol. Endocrinol. Metab.* **284:** E399–E406.

Brault, J.J., Abraham, K.A., and Terjung, R.L., 2003, Muscle creatine uptake and creatine transporter expression in response to creatine supplementation and depletion. *J. Appl. Physiol.* **94:** 2173–2180.

Broer, A., Klingel, K., Kowalczuk, S., Rasko, J.E., Cavanaugh, J., and Broer, S., 2004, Molecular cloning of mouse amino acid transport system B0, a neutral amino acid transporter related to Hartnup disorder. *J. Biol. Chem.* **279:** 24467–24476.

Broer, S., 2006, The SLC6 orphans are forming a family of amino acid transporters. *Neurochem. Int.* **48:** 559–567.

Brone, B., and Eggermont, J., 2005, PDZ proteins retain and regulate membrane transporters in polarized epithelial cell membranes. *Am. J. Physiol. Cell Physiol.* **288:** C20–C29.

Buckley, K.M., Melikian, H.E., Provoda, C.J., and Waring, M.T., 2000, Regulation of neuronal function by protein trafficking: a role for the endosomal pathway. *J. Physiol.* **525:** 11–19.

Burton, L.D., Kippenberger, A.G., Lingen, B., Bruss, M., Bonisch, H., and Christie, D.L., 1998, A variant of the bovine noradrenaline transporter reveals the importance of the C-terminal region for correct targeting to the membrane and functional expression. *Biochem. J.* **330:** 909–914.

Cecil, K.M., DeGrauw, T.J., Salomons, G.S., Jakobs, C., Egelhoff, J.C., and Clark, J.F., 2003, Magnetic resonance spectroscopy in a 9-day-old heterozygous female child with creatine transporter deficiency. *J. Comput. Assist. Tomogr.* **27:** 44–47.

Chen, J.G., and Rudnick, G., 2000, Permeation and gating residues in serotonin transporter. *Proc. Natl. Acad. Sci. USA* **97:** 1044–1049.

Chen, J.G., Sachpatzidis, A., and Rudnick, G., 1997, The third transmembrane domain of the serotonin transporter contains residues associated with substrate and cocaine binding. *J. Biol. Chem.* **272:** 28321–28327.

Chen, N.H., Reith, M.E., and Quick, M.W., 2004, Synaptic uptake and beyond: the sodium- and chloride-dependent neurotransmitter transporter family SLC6. *Pflugers Arch.* **447:** 519–531.

Chiu, C.S., Jensen, K., Sokolova, I., Wang, D., Li, M., Deshpande, P., Davidson, N., Mody, I., Quick, M.W., Quake, S.R., and Lester, H.A., 2002, Number, density, and surface/cytoplasmic distribution of GABA transporters at presynaptic structures of knock-in mice carrying GABA transporter subtype 1-green fluorescent protein fusions. *J. Neurosci.* **22:** 10251–10266.

Clark, A.J., Rosenberg, E.H., Almeida, L.S., Wood, T.C., Jakobs, C., Stevenson, R.E., Schwartz, C.E., and Salomons, G.S., 2006, X-linked creatine transporter (SLC6A8) mutations in about 1% of males with mental retardation of unknown etiology. *Hum. Genet.* **119:** 604–610.

Dai, W., Vinnakota, S., Qian, X., Kunze, D.L., and Sarkar, H.K., 1999, Molecular characterization of the human CRT-1 creatine transporter expressed in Xenopus oocytes. *Arch. Biochem. Biophys.* **361:** 75–84.

Daly, M.M., and Seifter, S., 1980, Uptake of creatine by cultured cells. *Arch. Biochem. Biophys.* **203:** 317–324.

Deken, S.L., Beckman, M.L., Boos, L., and Quick, M.W., 2000, Transport rates of GABA transporters: regulation by the N-terminal domain and syntaxin 1A. *Nat. Neurosci.* **3:** 998–1003.

Deken, S.L., Wang, D., and Quick, M.W., 2003, Plasma membrane GABA transporters reside on distinct vesicles and undergo rapid regulated recycling. *J. Neurosci.* **23:** 1563–1568.

Derave, W., Straumann, N., Olek, R.A., and Hespel, P., 2006, Electrolysis stimulates creatine transport and transporter cell surface expression in incubated mouse skeletal muscle: potential role of ROS. *Am. J. Physiol. Endocrinol. Metab.* **291:** E1250–E1257.

Dodd, J.R., Zheng, T., and Christie, D.L., 1999, Creatine accumulation and exchange by HEK293 cells stably expressing high levels of a creatine transporter. *Biochim. Biophys. Acta* **1472:** 128–136.

Dodd, J.R., and Christie, D.L., 2001, Cysteine 144 in the third transmembrane domain of the creatine transporter is located close to a substrate-binding site. *J. Biol. Chem.* **276:** 46983–46988.

Dodd, J.R., and Christie, D.L., 2005, Substituted cysteine accessibility of the third transmembrane domain of the creatine transporter: defining a transport pathway. *J. Biol. Chem.* **280:** 32649–32654.

Dodd, J.R., and Christie, D.L., 2007, Selective amino acid substitutions convert the creatine transporter to a gamma -aminobutyric acid transporter. *J. Biol. Chem.* **282:** 15528–15533.

Eichler, E.E., Lu, F., Shen, Y., Antonacci, R., Jurecic, V., Doggett, N.A., Moyzis, R.K., Baldini, A., Gibbs, R.A., and Nelson, D.L., 1996, Duplication of a gene-rich cluster between 16p11.1 and Xq28: a novel pericentromeric-directed mechanism for paralogous genome evolution. *Hum. Mol. Genet.* **5:** 899–912.

Fitch, C.D., and Shields, R.P., 1966, Creatine metabolism in skeletal muscle. I. Creatine movement across muscle membranes. *J. Biol. Chem.* **241:** 3611–3614.

Fitch, C.D., Shields, R.P., Payne, W.F., and Dacus, J.M., 1968, Creatine metabolism in skeletal muscle. 3. Specificity of the creatine entry process. *J. Biol. Chem.* **243:** 2024–2027.

Galli, A., DeFelice, L.J., Duke, B.J., Moore, K.R., and Blakely, R.D., 1995, Sodium-dependent norepinephrine-induced currents in norepinephrine- transporter-transfected HEK-293 cells blocked by cocaine and antidepressants. *J. Exp. Biol.* **198:** 2197–2212.

Garcia-Delgado, M., Peral, M.J., Cano, M., Calonge, M.L., and Ilundain, A.A., 2001, Creatine transport in brush-border membrane vesicles isolated from rat kidney cortex. *J. Am. Soc. Nephrol.* **12:** 1819–1825.

Gether, U., Andersen, P.H., Larsson, O.M., and Schousboe, A., 2006, Neurotransmitter transporters: molecular function of important drug targets. *Trends Pharmacol. Sci.* **27:** 375–383.

Gonzalez, A.M., and Uhl, G.R., 1994, 'Choline/orphan V8-2-1/creatine transporter' mRNA is expressed in nervous, renal and gastrointestinal systems. *Brain Res. Mol. Brain Res.* **23:** 266–270.

Gregor, P., Nash, S.R., Caron, M.G., Seldin, M.F., and Warren, S.T., 1995, Assignment of the creatine transporter gene (SLC6A8) to human chromosome Xq28 telomeric to G6PD. *Genomics* **25:** 332–333.

Guastella, J., Nelson, N., Nelson, H., Czyzyk, L., Keynan, S., Miedel, M.C., Davidson, N., Lester, H.A., and Kanner, B.I., 1990, Cloning and expression of a rat brain GABA transporter. *Science* **249:** 1303–1306.

Guerrero-Ontiveros, M.L., and Wallimann, T., 1998, Creatine supplementation in health and disease. Effects of chronic creatine ingestion in vivo: down-regulation of the expression of creatine transporter isoforms in skeletal muscle. *Mol. Cell. Biochem.* **184:** 427–437.

Guimbal, C., and Kilimann, M.W., 1993, A Na^+-dependent creatine transporter in rabbit brain, muscle, heart, and kidney. cDNA cloning and functional expression. *J. Biol. Chem.* **268:** 8418–8421.

Guimbal, C., and Kilimann, M.W., 1994, A creatine transporter cDNA from Torpedo illustrates structure/function relationships in the GABA/noradrenaline transporter family. *J. Mol. Biol.* **241:** 317–324.

Hahn, K.A., Salomons, G.S., Tackels-Horne, D., Wood, T.C., Taylor, H.A., Schroer, R.J., Lubs, H.A., Jakobs, C., Olson, R.L., Holden, K.R., Stevenson, R.E., and Schwartz, C.E., 2002, X-linked mental retardation with seizures and carrier manifestations is caused by a mutation in the creatine-transporter gene (SLC6A8) located in Xq28. *Am. J. Hum. Genet.* **70:** 1349–1356.

Happe, H.K., and Murrin, L.C., 1995, In situ hybridization analysis of CHOT1, a creatine transporter, in the rat central nervous system. *J. Comp. Neurol.* **351:** 94–103.

Hemmer, W., Riesinger, I., Wallimann, T., Eppenberger, H.M., and Quest, A.F., 1993, Brain-type creatine kinase in photoreceptor cell outer segments: role of a phosphocreatine circuit in outer segment energy metabolism and phototransduction. *J. Cell. Sci.* **106:** 671–683.

Henry, L.K., Defelice, L.J., and Blakely, R.D., 2006, Getting the message across: a recent transporter structure shows the way. *Neuron* **49:** 791–796.

Hiel, H., Happe, H.K., Warr, W.B., and Morley, B.J., 1996, Regional distribution of a creatine transporter in rat auditory brainstem: an in-situ hybridization study. *Hear. Res.* **98:** 29–37.

Iyer, G.S., Krahe, R., Goodwin, L.A., Doggett, N.A., Siciliano, M.J., Funanage, V.L., and Proujansky, R., 1996, Identification of a testis-expressed creatine transporter gene at 16p11.2 and confirmation of the X-linked locus to Xq28. *Genomics* **34:** 143–146.

Kanner, B.I., 2005, Molecular physiology: intimate contact enables transport. *Nature* **437:** 203–205.

Lang, F., Henke, G., Embark, H.M., Waldegger, S., Palmada, M., Bohmer, C., and Vallon, V., 2003, Regulation of channels by the serum and glucocorticoid-inducible kinase - implications for transport, excitability and cell proliferation. *Cell. Physiol. Biochem.* **13:** 41–50.

Law, R.M., Stafford, A., and Quick, M.W., 2000, Functional regulation of gamma-aminobutyric acid transporters by direct tyrosine phosphorylation. *J. Biol. Chem.* **275:** 23986–23991.

Lenz, H., Schmidt, M., Welge, V., Schlattner, U., Wallimann, T., Elsasser, H.P., Wittern, K.P., Wenck, H., Stab, F., and Blatt, T., 2005, The creatine kinase system in human skin: protective effects of creatine against oxidative and UV damage in vitro and in vivo. *J. Invest. Dermatol.* **124:** 443–452.

Loder, M.K., and Melikian, H.E., 2003, The dopamine transporter constitutively internalizes and recycles in a protein kinase C-regulated manner in stably transfected PC12 cell lines. *J. Biol. Chem.* **278:** 22168–22174.

Loike, J.D., Somes, M., and Silverstein, S.C., 1986, Creatine uptake, metabolism, and efflux in human monocytes and macrophages. *Am. J. Physiol.* **251:** C128–C135.

Loike, J.D., Zalutsky, D.L., Kaback, E., Miranda, A.F., and Silverstein, S.C., 1988, Extracellular creatine regulates creatine transport in rat and human muscle cells. *Proc. Natl. Acad. Sci. USA* **85:** 807–811.

Loland, C.J., Norregaard, L., and Gether, U., 1999, Defining proximity relationships in the tertiary structure of the dopamine transporter. Identification of a conserved glutamic acid as a third coordinate in the endogenous Zn^{2+}-binding site. *J. Biol. Chem.* **274:** 36928–36934.

Lunardi, G., Parodi, A., Perasso, L., Pohvozcheva, A.V., Scarrone, S., Adriano, E., Florio, T., Gandolfo, C., Cupello, A., Burov, S.V., and Balestrino, M., 2006, The creatine transporter mediates the uptake of creatine by brain tissue, but not the uptake of two creatine-derived compounds. *Neuroscience* **142:** 991–997.

MacAulay, N., Bendahan, A., Loland, C.J., Zeuthen, T., Kanner, B.I., and Gether, U., 2001, Engineered Zn^{2+} switches in the gamma-aminobutyric acid (GABA) transporter-1. Differential effects on GABA uptake and currents. *J. Biol. Chem.* **276:** 40476–40485.

Mahon, M.J., Donowitz, M., Yun, C.C., and Segre, G.V., 2002, Na^+/H^+ exchanger regulatory factor 2 directs parathyroid hormone 1 receptor signalling. *Nature* **417:** 858–861.

Marescau, B., De Deyn, P., Wiechert, P., Van Gorp, L., and Lowenthal, A., 1986, Comparative study of guanidino compounds in serum and brain of mouse, rat, rabbit, and man. *J. Neurochem.* **46:** 717–720.

Mayser, W., Schloss, P., and Betz, H., 1992, Primary structure and functional expression of a choline transporter expressed in the rat nervous system. *FEBS Lett.* **305:** 31–36.

Melikian, H.E., 2004, Neurotransmitter transporter trafficking: endocytosis, recycling, and regulation. *Pharmacol. Ther.* **104:** 17–27.

Mitchell, S.M., Lee, E., Garcia, M.L., and Stephan, M.M., 2004, Structure and function of extracellular loop 4 of the serotonin transporter as revealed by cysteine-scanning mutagenesis. *J. Biol. Chem.* **279:** 24089–24099.

Moller, A., and Hamprecht, B., 1989, Creatine transport in cultured cells of rat and mouse brain. *J. Neurochem.* **52:** 544–550.

Murphy, R., McConell, G., Cameron-Smith, D., Watt, K., Ackland, L., Walzel, B., Wallimann, T., and Snow, R., 2001, Creatine transporter protein content, localization, and gene expression in rat skeletal muscle. *Am. J. Physiol. Cell Physiol.* **280:** C415–C422.

Nakashima, T., Tomi, M., Katayama, K., Tachikawa, M., Watanabe, M., Terasaki, T., and Hosoya, K., 2004, Blood-to-retina transport of creatine via creatine transporter (CRT) at the rat inner blood-retinal barrier. *J. Neurochem.* **89:** 1454–1461.

Nash, S.R., Giros, B., Kingsmore, S.F., Rochelle, J.M., Suter, S.T., Gregor, P., Seldin, M.F., and Caron, M.G., 1994, Cloning, pharmacological characterization, and genomic localization of the human creatine transporter. *Receptors Channels* **2:** 165–174.

Nelson, N., 1998, The family of Na^+/Cl^- neurotransmitter transporters. *J. Neurochem.* **71:** 1785–1803.

Norregaard, L., Loland, C.J., and Gether, U., 2003, Evidence for distinct sodium-, dopamine-, and cocaine-dependent conformational changes in transmembrane segments 7 and 8 of the dopamine transporter. *J. Biol. Chem.* **278:** 30587–30596.

Odoom, J.E., Kemp, G.J., and Radda, G.K., 1996, The regulation of total creatine content in a myoblast cell line. *Mol. Cell. Biochem.* **158:** 179–188.

Ohtsuki, S., Tachikawa, M., Takanaga, H., Shimizu, H., Watanabe, M., Hosoya, K., and Terasaki, T., 2002, The blood-brain barrier creatine transporter is a major pathway for supplying creatine to the brain. *J. Cereb. Blood Flow Metab.* **22:** 1327–1335.

Omerovic, E., Bollano, E., Lorentzon, M., Walser, M., Mattsson-Hulten, L., and Isgaard, J., 2003, Growth hormone induces myocardial expression of creatine transporter and decreases plasma levels of IL-1beta in rats during early postinfarct cardiac remodeling. *Growth Horm. IGF Res.* **13:** 239–245.

Orsenigo, M.N., Faelli, A., De Biasi, S., Sironi, C., Laforenza, U., Paulmichl, M., and Tosco, M., 2005, Jejunal creatine absorption: what is the role of the basolateral membrane? *J. Membr. Biol.* **207:** 183–195.

Paczkowski, F.A., Bryan-Lluka, L.J., Porzgen, P., Bruss, M., and Bonisch, H., 1999, Comparison of the pharmacological properties of cloned rat, human, and bovine norepinephrine transporters. *J. Pharmacol. Exp. Ther.* **290:** 761–767.

Peral, M.J., Garcia-Delgado, M., Calonge, M.L., Duran, J.M., De La Horra, M.C., Wallimann, T., Speer, O., and Ilundain, A., 2002, Human, rat and chicken small intestinal Na^{+}-Cl^{-}-creatine transporter: functional, molecular characterization and localization. *J. Physiol.* **545:** 133–144.

Quick, M., Yano, H., Goldberg, N.R., Duan, L., Beuming, T., Shi, L., Weinstein, H., and Javitch, J.A., 2006, State-dependent conformations of the translocation pathway in the tyrosine transporter Tyt1, a novel neurotransmitter: sodium symporter from *Fusobacterium nucleatum*. *J. Biol. Chem.* **281:** 26444–26454.

Rosenberg, E.H., Almeida, L.S., Kleefstra, T., deGrauw, R.S., Yntema, H.G., Bahi, N., Moraine, C., Ropers, H.H., Fryns, J.P., deGrauw, T.J., Jakobs, C., and Salomons, G.S., 2004, High prevalence of SLC6A8 deficiency in X-linked mental retardation. *Am. J. Hum. Genet.* **75:** 97–105.

Saier, M.H., Jr., 1999, A functional-phylogenetic system for the classification of transport proteins. *J. Cell. Biochem.* **Suppl. 32–33:** 84–94.

Salomons, G.S., van Dooren, S.J., Verhoeven, N.M., Cecil, K.M., Ball, W.S., Degrauw, T.J., and Jakobs, C., 2001, X-linked creatine-transporter gene (SLC6A8) defect: a new creatine- deficiency syndrome. *Am. J. Hum. Genet.* **68:** 1497–1500.

Salomons, G.S., van Dooren, S.J., Verhoeven, N.M., Marsden, D., Schwartz, C., Cecil, K.M., DeGrauw, T.J., and Jakobs, C., 2003, X-linked creatine transporter defect: an overview. *J. Inherit. Metab. Dis.* **26:** 309–318.

Saltarelli, M.D., Bauman, A.L., Moore, K.R., Bradley, C.C., and Blakely, R.D., 1996, Expression of the rat brain creatine transporter in situ and in transfected HeLa cells. *Dev. Neurosci.* **18:** 524–534.

Sandoval, N., Bauer, D., Brenner, V., Coy, J.F., Drescher, B., Kioschis, P., Korn, B., Nyakatura, G., Poustka, A., Reichwald, K., Rosenthal, A., and Platzer, M., 1996, The genomic organization of a human creatine transporter (CRTR) gene located in Xq28. *Genomics* **35:** 383–385.

Schlattner, U., Mockli, N., Speer, O., Werner, S., and Wallimann, T., 2002, Creatine kinase and creatine transporter in normal, wounded, and diseased skin. *J. Invest. Dermatol.* **118:** 416–423.

Schloss, P., Mayser, W., and Betz, H., 1994, The putative rat choline transporter CHOT1 transports creatine and is highly expressed in neural and muscle-rich tissues. *Biochem. Biophys. Res. Commun.* **198:** 637–645.

Sheng, M., and Sala, C., 2001, PDZ domains and the organization of supramolecular complexes. *Annu. Rev. Neurosci.* **24:** 1–29.

Shojaiefard, M., Christie, D.L., and Lang, F., 2005, Stimulation of the creatine transporter SLC6A8 by the protein kinases SGK1 and SGK3. *Biochem. Biophys. Res. Commun.* **334:** 742–746.

Shojaiefard, M., Christie, D.L., and Lang, F., 2006, Stimulation of the creatine transporter SLC6A8 by the protein kinase mTOR. *Biochem. Biophys. Res. Commun.* **341:** 945–949.

Snow, R.J., and Murphy, R.M., 2001, Creatine and the creatine transporter: a review. *Mol. Cell. Biochem.* **224:** 169–181.

Sora, I., Richman, J., Santoro, G., Wei, H., Wang, Y., Vanderah, T., Horvath, R., Nguyen, M., Waite, S., Roeske, W.R., and et al., 1994, The cloning and expression of a human creatine transporter. *Biochem. Biophys. Res. Commun.* **204:** 419–427.

Speer, O., Neukomm, L.J., Murphy, R.M., Zanolla, E., Schlattner, U., Henry, H., Snow, R.J., and Wallimann, T., 2004, Creatine transporters: a reappraisal. *Mol. Cell. Biochem.* **256–257:** 407–424.

Stockler, S., Schutz, P.W., and Salomons, G.S., 2007, Cerebral creatine deficiency syndromes: clinical aspects, treatment and pathophysiology. *Subcell. Biochem.* **46:** 149–166.

Straumann, N., Wind, A., Leuenberger, T., and Wallimann, T., 2006, Effects of N-linked glycosylation on the creatine transporter. *Biochem. J.* **393:** 459–469.

Sung, U., Apparsundaram, S., Galli, A., Kahlig, K.M., Savchenko, V., Schroeter, S., Quick, M.W., and Blakely, R.D., 2003, A regulated interaction of syntaxin 1A with the antidepressant-sensitive norepinephrine transporter establishes catecholamine clearance capacity. *J. Neurosci.* **23:** 1697–1709.

Tarnopolsky, M., Parise, G., Fu, M.H., Brose, A., Parshad, A., Speer, O., and Wallimann, T., 2003, Acute and moderate-term creatine monohydrate supplementation does not affect creatine transporter mRNA or protein content in either young or elderly humans. *Mol. Cell. Biochem.* **244:** 159–166.

Tarnopolsky, M.A., Parshad, A., Walzel, B., Schlattner, U., and Wallimann, T., 2001, Creatine transporter and mitochondrial creatine kinase protein content in myopathies. *Muscle Nerve* **24:** 682–688.

Ten Hove, M., Chan, S., Lygate, C., Monfared, M., Boehm, E., Hulbert, K., Watkins, H., Clarke, K., and Neubauer, S., 2005, Mechanisms of creatine depletion in chronically failing rat heart. *J. Mol. Cell. Cardiol.* **38:** 309–313.

Tosco, M., Faelli, A., Sironi, C., Gastaldi, G., and Orsenigo, M.N., 2004, A creatine transporter is operative at the brush border level of the rat jejunal enterocyte. *J. Membr. Biol.* **202:** 85–95.

Tran, T.T., Dai, W., and Sarkar, H.K., 2000, Cyclosporin A inhibits creatine uptake by altering surface expression of the creatine transporter. *J. Biol. Chem.* **275:** 35708–35714.

Wallis, J., Lygate, C.A., Fischer, A., ten Hove, M., Schneider, J.E., Sebag-Montefiore, L., Dawson, D., Hulbert, K., Zhang, W., Zhang, M.H., Watkins, H., Clarke, K., and Neubauer, S., 2005, Supranormal myocardial creatine and phosphocreatine concentrations lead to cardiac hypertrophy and heart failure: insights from creatine transporter-overexpressing transgenic mice. *Circulation* **112:** 3131–3139.

Walzel, B., Speer, O., Boehm, E., Kristiansen, S., Chan, S., Clarke, K., Magyar, J.P., Richter, E.A., and Wallimann, T., 2002, New creatine transporter assay and identification of distinct creatine transporter isoforms in muscle. *Am. J. Physiol. Endocrinol. Metab.* **283:** E390–E401.

Wang, W., Jobst, M.A., Bell, B., Zhao, C.R., Shang, L.H., and Jacobs, D.O., 2002, Cr supplementation decreases tyrosine phosphorylation of the CreaT in skeletal muscle during sepsis. *Am. J. Physiol. Endocrinol. Metab.* **282:** E1046–E1054.

West, M., Park, D., Dodd, J., Kistler, J., and Christie, D., 2005, Purification and characterization of the creatine transporter expressed at high levels in HEK293 cells. *Prot. Expr. Purif.* **41:** 393–401.

Whitworth, T.L., and Quick, M.W., 2001, Substrate-induced regulation of gamma-aminobutyric acid transporter trafficking requires tyrosine phosphorylation. *J. Biol. Chem.* **276:** 42932–42937.

Wyss, M., and Schulze, A., 2002, Health implications of creatine: can oral creatine supplementation protect against neurological and atherosclerotic disease? *Neuroscience* **112:** 243–260.

Xu, W., Liu, L., Gorman, P.A., Sheer, D., and Emson, P.C., 1997, Assignment of the human creatine transporter type 2 (SLC6A10) to chromosome band 16p11.2 by in situ hybridization. *Cytogenet. Cell Genet.* **76:** 19.

Yamashita, A., Singh, S.K., Kawate, T., Jin, Y., and Gouaux, E., 2005, Crystal structure of a bacterial homologue of $Na+/Cl^-$-dependent neurotransmitter transporters. *Nature* **437:** 215–223.

Zhao, C.R., Shang, L., Wang, W., and Jacobs, D.O., 2002, Myocellular creatine and creatine transporter serine phosphorylation after starvation. *J. Surg. Res.* **105:** 10–16.

CHAPTER 7

IN VIVO MAGNETIC RESONANCE SPECTROSCOPY OF TRANSGENIC MICE WITH ALTERED EXPRESSION OF GUANIDINOACETATE METHYLTRANSFERASE AND CREATINE KINASE ISOENZYMES

AREND HEERSCHAP[1], HERMIEN E. KAN[1], CHRISTINE I.H.C. NABUURS[1], W. KLAASJAN RENEMA[1], DIRK ISBRANDT[2] AND BÉ WIERINGA[3]

[1]*Department of Radiology, Radboud University Nijmegen Medical Centre, 6500 HB Nijmegen, the Netherlands*
[2]*Centre for Molecular Neurobiology Hamburg (ZMNH), University of Hamburg, D-20246 Hamburg, Germany*
[3]*Department of Cell Biology, Nijmegan Centre for Molecular Life Science, Radboud University Nijmegen Medical Centre, 6500 HB Nijmegen, the Netherlands*

Abstract: Mice with an under- or over-expression of enzymes catalyzing phosphoryl transfer in high-energy supplying reactions are particulary attractive for *in vivo* magnetic resonance spectroscopy (MRS) studies as substrates of these enzymes are visible in MR spectra.

This chapter reviews results of *in vivo* MRS studies on transgenic mice with alterations in the expression of the enzymes creatine kinase and guanidinoacetate methyltransferase. The particular metabolic consequences of these enzyme deficiencies in skeletal muscle, brain, heart and liver are addressed. An overview is given of metabolite levels determined by *in vivo* MRS in skeletal muscle and brain of wild-type and transgenic mice.

MRS studies on mice lacking guanidinoacetate methyltransferase have demonstrated metabolic changes comparable to those found in the deficiency of this enzyme in humans, which are (partly) reversible upon creatine feeding. Apart from being a model for a creatine deficiency syndrome, these mice are also of interest to study fundamental aspects of the biological role of creatine.

MRS studies on transgenic mice lacking creatine kinase isoenzymes have contributed significantly to the view that the creatine kinase reaction together with other enzymatic steps involved in high-energy phosphate transfer builds a large metabolic energy network, which is highly versatile and can dynamically adapt to genotoxic or physiological challenges

G.S. Salomons and M. Wyss (eds.), Creatine and Creatine Kinase in Health and Disease, 119–148.

1. INTRODUCTION

Studies of genetically altered mice have proven to be important in our understanding of the biological role of specific enzymes or pathways involved in cell signaling, metabolism, or structural organization of organismal form and function. It is now widely recognized that non-invasive bio-imaging and mapping can significantly contribute to the proper phenotyping of these animals (Budinger *et al.*, 1999; MacLaren *et al.*, 2000; Ntziachristos *et al.*, 2003). A particularly useful tool to investigate metabolism *in vivo* and *in vitro* is MRS. The MR sensitive nuclei mostly used in MRS of living beings, ^{31}P, ^{1}H and ^{13}C, each offer unique information on distinct metabolic and physiologic processes or conditions. MRS of transgenic animals is most powerful when genetic modifications have an effect on enzymes that interact directly with MR-visible metabolites, such as creatine (Cr) and phosphocreatine (PCr). MRS can also be applied to tissues *ex vivo* and to isolated body fluid samples for obtaining detailed metabolic information. MRS approaches thus offer new opportunities in the area of metabolomics and metabonomics.

One of the first MRS studies on genetically altered mice – in this case a natural mutant model – was performed on the ReJ 129 dy/dy strain, a model for human myopathy (Heerschap *et al.*, 1988). Since then, MRS studies were conducted on classical transgenic mice (Koretsky *et al.*, 1990) and on metabolic knock-out mice with mutations created by embryonic stem cell technology (van Deursen *et al.*, 1993). This chapter focuses on animal models with alterations in the expression of the enzyme guanidinoacetate methyltransferase (GAMT), catalyzing the last step in the synthesis of creatine, and of isoenzymes of creatine kinase (CK), involved in the phosphorylation of creatine. Deficiencies in these enzymes either cause depletion of creatine or affect its use in energy supplying pathways. We first briefly introduce *in vivo* MRS. Then we will provide an overview of MRS studies on transgenic mouse models with altered GAMT and CK expression, emphazising findings in mouse skeletal muscle, brain, heart and liver.

2. *IN VIVO* MAGNETIC RESONANCE SPECTROSCOPY (MRS)

In vivo MRS has been applied extensively in studies of several tissues, including skeletal muscle, brain, heart and liver, in which it has opened new windows on metabolism, physiology and biochemistry in selected volumes of interest. Here only a brief introduction is provided in basic principles of MR which are relevant for the studies described in this chapter. A detailed review of the classical and quantum mechanical principles of nuclear magnetic resonance and its localized application in living tissues is beyond the scope of this chapter and can be found in several textbooks (e.g. de Graaf, 1998; Gadian, 1982; Macomber, 1998).

MR is based on a quantum mechanical property of atomic nuclei called spin. Nuclei most commonly used for biomedical MR are the spin ½ nuclei of hydrogen (^{1}H), phosphorus-31 (^{31}P), carbon-13 (^{13}C) and fluorine-19 (^{19}F), which naturally occur in bodily compounds or drugs. When placed in the field of the magnet of a MR machine, these spins become aligned parallel or anti-parallel to the direction of the main field. At body temperature, there is a small difference in the population of these two energy levels, which is exploited to generate an MR signal. The MR signal strength is proportional to this polarization and the number of spins. The magnetized state can be perturbed by applying a short radiofrequency field (RF) with an external coil. After this RF pulse the spin system induces a current in this coil with a specific frequency proportional to the magnetic field and nature of the nucleus. Simultaneously the spin system relaxes back to its equilibrium state. The position of a nucleus in a molecule results in a slightly different magnetic field, by which the resonance frequency of its signal is shifted, which is called the chemical shift. Thus nuclei at different positions in a molecule will have different chemical shifts and hence they will show signals at different resonance frequencies. This is commonly plotted in a spectrum in which the x-axis is expressed in parts per million (ppm), a measure for the chemical shift relative to a reference signal. The y-axis in a MR spectrum is expressed in arbitrary units, and the surface area under the signals is a measure for the concentration of the nuclei (see Figures 1 and 2).

At field strengths commonly employed for *in vivo* studies the direct MR visibility of molecules is generally restricted to those that are present at tissue concentrations in the millimolar range or higher. Furthermore, only molecular components that are small and/or mobile generate peaks with sufficiently small line widths to be visible by MRS. Thus, most of the compounds visible in an MR spectrum arise from small metabolites. The unique property of MRS is that these molecules can be assessed in a non-invasive and dynamic or longitudinal way. The sensitivity of MRS usually does not allow the direct assessment of tissue metabolites at a spatial resolution down to (sub-)cellular size, but prior knowledge on the compartmentalization of metabolites may be used to interpret data at this level. Localized measurements can be performed for single voxels or multiple voxels (metabolite mapping).

Of the nuclei described in this chapter, ^{1}H has the highest sensitivity and a natural abundance of nearly 100%, enabling MRS of relatively small volumes of interest or voxels (about a few μl) in mice or rats (in't Zandt *et al.*, 2004; Pfeuffer *et al.*, 1999; Renema *et al.*, 2003). As the high signal for water (~45 M) may perturb the signal contributions from other MR visible molecules (~0.1–15 mM), it is commonly suppressed. In ^{1}H MR spectra of brain, signals of up to about 30 different molecules may be detected of which those of creatine, N-acetyl aspartate (NAA) and choline compounds (Cho) are usually dominating the spectrum (Figure 1). ^{1}H MR spectra of skeletal muscle reveal mainly signals of creatine, taurine (Tau), and lipids (Figure 1). When acquiring ^{1}H MR spectra of mouse skeletal muscle, the orientation of the muscle fibers with respect to B_0 is important as the orientation of the fibers may result in splitting of the signals of Cr and of other molecules (in't Zandt *et al.*, 2000).

^{31}P has a natural abundance of 100%, but its sensitivity is lower than that of ^{1}H (~7%). Therefore, ^{31}P MR spectra are usually acquired for larger regions

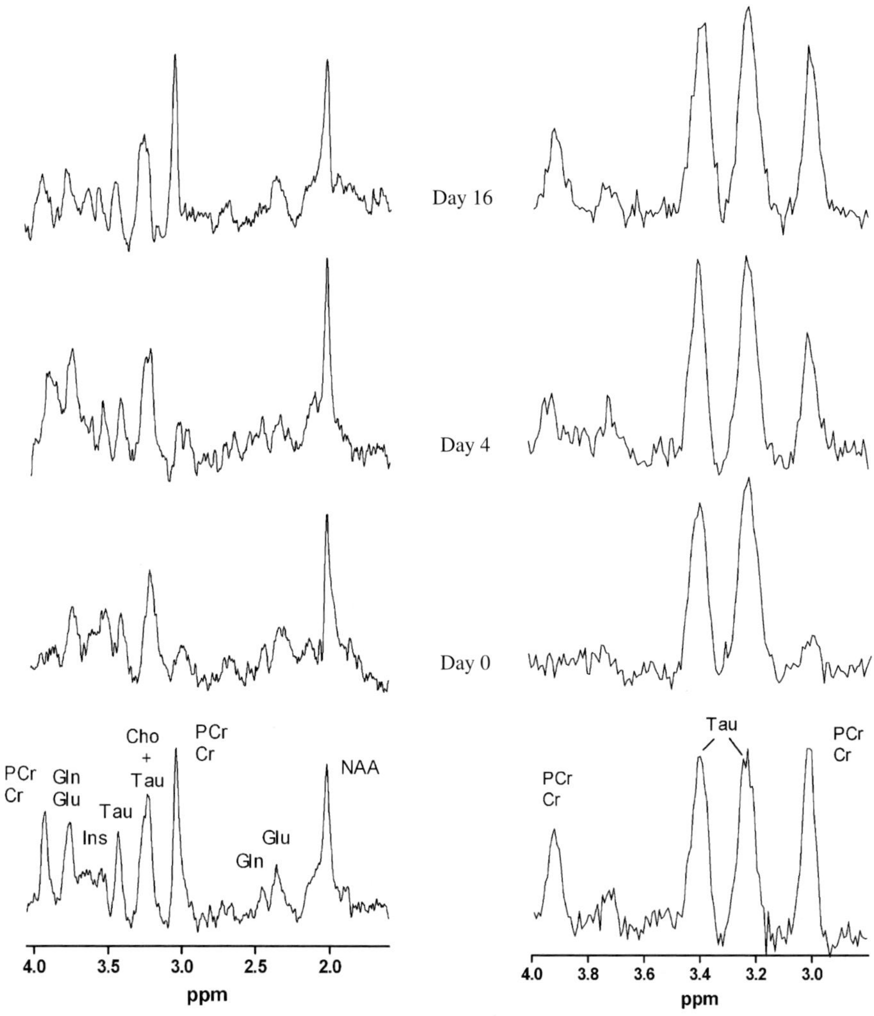

Figure 1. Left: [1]H MR spectra of a central region in the cerebrum of the mouse brain. Right: [1]H MR spectra of the triceps surae region in mouse hind leg skeletal muscle. The two bottom spectra are from a control animal. Signals for tCr (PCr and Cr) are clearly distinguishable at 3.0 (methylene group) and 3.9 ppm (methyl group). Methyl and methylene lipid protons resonate at about 0.8-1.5 ppm in the muscle spectra (right, not shown). Peak assignments: PCr = phosphocreatine; Cr = creatine; Gln = glutamine; Glu = glutamate; Ins = *myo*-inositol; Tau = taurine; Cho = choline; NAA = N-acetylaspartate. The spectra above were obtained from GAMT deficient mice during creatine supplementation. The number of days on supplementation is indicated

of interest as compared to those for ^{1}H MRS. In mouse hind leg, for example, ^{31}P MR spectra are usually acquired of the whole hind leg, while ^{1}H MR spectra can be localized in specific muscle groups. ^{31}P MRS has been applied extensively in the study of energy metabolism, as important phosphates in energy metabolism

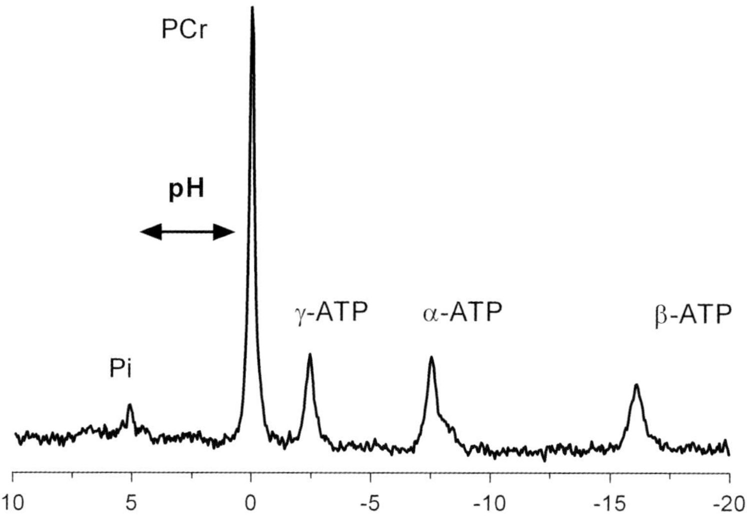

Figure 2. Typical ^{31}P MR spectrum of normal mouse skeletal muscle (hindleg) under resting conditions. The difference in resonance position between the PCr and Pi peak can be used to calculate tissue pH. Peak assignments: Pi = inorganic phosphate, PCr = phosphocreatine, γ-, α-, β-ATP = resonance frequencies of ATP

are clearly visible in an *in vivo* ^{31}P MR spectrum (Figure 2). Apart from signals for inorganic phosphate (Pi), PCr and ATP, also the intracellular tissue pH can be calculated from the shift in resonance position between Pi and PCr (Brindle, 1988; Moon and Richards, 1973; Taylor *et al.*, 1983). By a method called magnetization transfer, flux rates in some biochemical pathways can be determined by ^{31}P MRS (Brindle, 1988; Forsén and Hoffman, 1963; Rudin and Sauter, 1992). This technique can be applied to determine the flux rates in the CK reaction, involving the phosphorylation of Cr to PCr.

The ^{13}C nucleus has the lowest MR sensitivity of those nuclei described in this chapter (~0.02% of that of ^{1}H), and also a low natural abundance (1.1%). Therefore, a natural abundance ^{13}C MR spectrum of tissues only shows signals for a few compounds present at high concentrations, like long chain fatty acids, or glucosyl elements of glycogen in skeletal muscle. This low natural presence of ^{13}C enables the dynamic detection of several metabolites by infusing compounds labeled with ^{13}C. For instance, in brain studies, glucose labeled at the C1 position can be infused, and over time, the transfer of this label to other metabolites enables the calculation of flux rates (e.g. de Graaf *et al.*, 2003; Gruetter *et al.*, 2003). A similar approach has been applied in skeletal muscle studies, where the transfer of the label to lactate and glycogen can be monitored (Jucker *et al.*, 1997; Jue *et al.*, 1989). Cr can also be labeled with ^{13}C, and this enables the detection of the specific increase in Cr after supplementation. As in a ^{13}C MR spectrum, signals for both Cr and PCr can be observed separately, this technique further enables the determination of the phosphorylated fraction of the total Cr pool (in't Zandt *et al.*, 2003) (Figure 3).

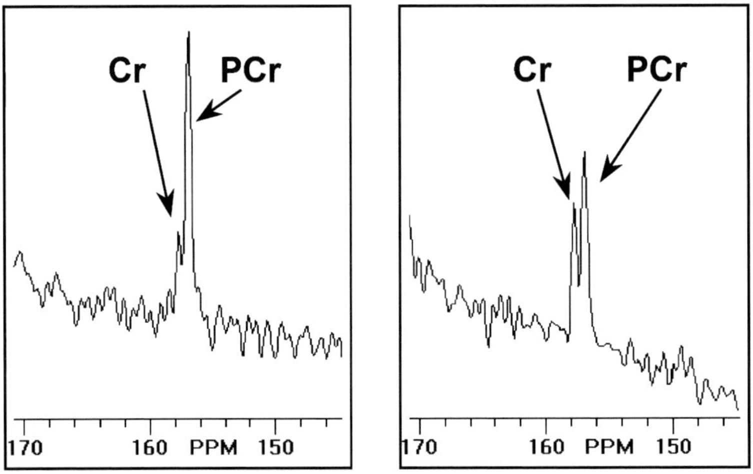

Figure 3. Selected region of ^{13}C MR spectra of adult hindlimb muscle after administration of ^{13}C-4 labeled creatine. M-CK/ScCKmit--/-- double knockout animals show uptake and phosphorylation of ^{13}C-labeled creatine (right) although PCr/tCr ratios are significantly lower than in wild-type animals (left). Adapted from in't Zandt *et al.* (2003) and reprinted with permission

3. METABOLIC REACTIONS CATALYZED BY GUANIDINOACETATE METHYLTRANSFERASE AND CREATINE KINASE

Creatine is a small metabolite that fulfills a central role in energy metabolism. The final step in its biosynthesis is controlled by the enzyme guanidinoacetate methyltransferase (GAMT), which catalyzes the methylation of guanidinoacetate to form Cr. In mammals, this biosynthesis is assumed to take place mainly in the pancreas and liver. Subsequently, Cr is exported into the blood and taken up by tissues requiring Cr, such as muscle and brain (Wyss and Kaddurah-Daouk, 2000). Intake of creatine-containing food may also increase the blood level of creatine, and tissues such as muscle and brain may have the enzymatic machinery to synthesize creatine themselves.

Tissues taking up creatine are generally characterized as having high and fluctuating energy demands. In these tissues, Cr is reversibly phosphorylated to phosphocreatine (PCr) in the near-equilibrium reaction Cr + ATP $\leftrightarrow$ PCr + ADP + H^+, which is catalyzed by the enzyme creatine kinase (CK). This reaction is thought to be important for the temporal buffering of the levels of so-called high-energy phosphates (~P) to balance (local) ATP/ADP ratios, and also for pH buffering (for more comprehensive reviews see Wallimann, 1994; Wallimann *et al.*, 1992; Wyss and Kaddurah-Daouk, 2000; Wyss *et al.*, 1992). Over the years, discussion has arisen concerning the role of the Cr/CK system in the distribution and intracellular transport of high-energy phosphates, the importance of which is related to cellular diffusion distances. The presence and partitioning of different iso-enzymes of CK, apparently specifically associated with loci of energy production and consumption in the cell, has stimulated the idea of a structured uni-directional high-energy phosphate transport system, the so-called CK shuttle. This shuttle concept was originally proposed in the late last century (Bessman and Fonyo, 1966; Bessman and Geiger, 1981) and implicates cellular compartmentation of the reactants of the CK reaction. However, the spatial transport role of the Cr/CK system can also be understood in a simpler way, as a specific case of facilitated diffusion not necessitating the structural organization proposed for shuttle-type systems (Meyer *et al.*, 1984).

In vertebrates, four CK genes have been identified, coding for altogether five different isoenzymes. Two genes for cytoplasmatic CK encode for muscle (M-CK) and brain (B-CK) isoforms, which assemble into active homo-dimeric (MM-CK, BB-CK) and hetero-dimeric (MB-CK) enzymes in the cytosol. MM-CK is expressed at high levels in muscle, representing between 85% and 95% of total CK activity depending on fiber type (Wallimann *et al.*, 1992) and stage of development. In embryonic or young mouse skeletal muscle, trace amounts of BB-CK also occur (in't Zandt *et al.*, 2003). MB-CK is a transitional isoform, which is present in immature stages of skeletal muscle. In heart, this isoform is permanently expressed at relatively low amounts. BB-CK is expressed at high levels in brain and various other tissues in adult mice. Expression of the other class of CKs, ubiquitous (UbCKmit) and sarcomeric mitochondrial CK (ScCKmit), is also cell-type dependent. UbCKmit is mostly co-expressed with BB-CK. Recent studies indicate that in the brain, BB-CK is preferentially expressed in astrocytes and inhibitory neurons, while mitochondrial CK seems to be restricted primarily to neurons (Tachikawa *et al.*, 2004). The expression of ScCKmit is confined to muscle. UbCKmit and ScCKmit isoforms occur as dimeric or octameric enzymes and reside between the inner and outer membranes of mitochondria (Wallimann *et al.*, 1992).

Another key enzyme in $\sim$P transfer that may be assessed by MRS is adenylate kinase (AK), which serves to balance ATP/ADP ratios by catalyzing the reaction 2 ADP $\leftrightarrow$ ATP + AMP. The CK and AK catalyzed reactions appear to be closely connected in a metabolic network (see, e.g., Dzeja and Terzic, 2003; Zeleznikar *et al.*, 1995).

4. TRANSGENIC MICE LACKING GUANIDINOACETATE METHYLTRANSFERASE (GAMT)

4.1. General Phenotype of GAMT Deficient Mice

Recently, GAMT knockout mice were generated by gene targeting in embryonic stem cells (Schmidt *et al.*, 2004), as a model for the first-discovered Cr deficiency syndrome in humans (Stöckler *et al.*, 1994). As these animals are not able to synthesize their own Cr, their tissues are completely deprived of Cr unless Cr is orally ingested. GAMT knockout mice have markedly increased guanidinoacetate (GAA) and reduced creatine and creatinine levels in brain, serum and urine, which are the typical abnormal metabolic findings also seen in GAMT deficient patients.

Biochemical alterations were highly reminiscent to those found in GAMT deficient patients and can be attributed to the overall similar GAMT expression patterns in human and mouse tissues. GAMT deficiency in mouse is associated with increased neonatal mortality, muscular hypotonia, decreased male fertility and a non-leptin-mediated life-long reduction in body weight due to reduced body fat mass. Apart from the lack of creatine, the genotoxic effects of this mutation may also be caused by accumulation of guanidinoacetate (Kan *et al.*, 2005; Schmidt *et al.*, 2004).

4.2. MRS Studies on GAMT Deficient Mice

Localized ^{1}H MRS of mice lacking the GAMT enzyme (GAMT-/-) showed reduced levels of Cr in muscle and brain although residual Cr may still be present due to oral intake (Renema *et al.*, 2003). In ^{1}H MR spectra of skeletal muscle, a broad signal of GAA is present, the immediate biosynthetic precursor of Cr, but this signal only becomes clearly visible when Cr is almost completely depleted. GAA is phosphorylated (PGAA) and metabolically active in skeletal muscle (Renema *et al.*, 2003). In ischemia-reperfusion experiments monitored by ^{31}P MRS, the PGAA signal decreased similarly to the PCr signal. The recovery of the PGAA signal, however, was significantly delayed (Kan *et al.*, 2004), similar to the recovery of the PCr signal in skeletal muscle lacking M-CK and AK (Renema, 2005), but the nature of this delay was different. As the recovery rate of PGAA did not change upon Cr feeding in GAMT-/- mice, it was concluded that the slower recovery is caused by a lower affinity of CK for GAA (Kan *et al.*, 2004). Accordingly, saturation transfer experiments showed negligible transfer of the γ-phosphate group of ATP to PGAA indicating slower enzymatic kinetics (Figure 4; Kan *et al.*, 2004).

Another interesting phenotype of GAMT-/- mice was an increased Pi signal in skeletal muscle during basal conditions which was similar as observed in M-CK/ScCKmit--/-- mice (Table 2; Kan *et al.*, 2004). This increased Pi was only detected when PCr was fully depleted for an extended time period. Upon (intentional) Cr feeding the Pi signal normalized, the PGAA signal decreased and a PCr signal appeared, quickly exceeding the PGAA signal in skeletal muscle (Kan *et al.*, 2004).

In ^{31}P MR spectra of the brain a PGAA signal could be observed in GAMT-deficient mice (Renema *et al.*, 2003; Schmidt *et al.*, 2004), but this signal had a much lower intensity than the corresponding signal in spectra of skeletal muscle. Large fluctuations of the PGAA signal, as seen in skeletal muscle, were not observed in the brain of different animals. Limited permeability of the blood-brain barrier for Cr (Wyss and Kaddurah-Daouk, 2000) could explain these observations. This limited permeability was already suspected from the slow recovery of the Cr signal after long-term feeding of high amounts of Cr to GAMT-deficient patients (von Figura *et al.*, 2001; Wyss and Kaddurah-Daouk, 2000). In a recent study, we indeed observed a slower uptake of Cr in brain as compared to skeletal muscle of GAMT deficient mice upon Cr feeding (Figure 1). However, the faster uptake in skeletal muscle only occurred during the first day of Cr supplementation, thereafter

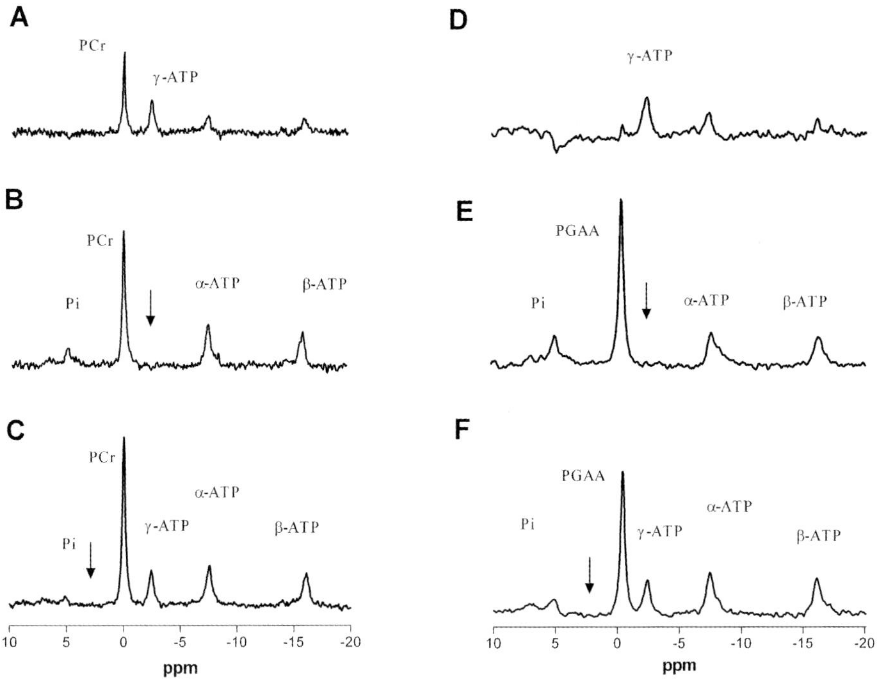

Figure 4. Typical saturation transfer spectra of a GAMT-/- (D,E,F; n = 6) and a wild-type (A,B,C; n = 3) mouse hindleg. The appropriate control spectra (C, F), a spectrum with a 5000 ms (B) or a 3000 ms (E) saturation pulse applied at the frequency of γ-ATP and the corresponding difference spectra (A, D). Arrows denote the frequency of the selective saturation pulse. Adapted from Kan *et al.* (2004) and reprinted with permission

the uptake in brain and muscle was similar (Kan *et al.*, 2007). This suggests a more subtle difference in Cr transport into these tissues of GAMT deficient mice. NAA levels in the brain of GAMT lacking mice were increased, but significantly decreased at higher Cr levels (see also section 5.2.2). At the end of the supplementation period, Cr levels were typically higher than in wild-type animals not supplemented with Cr.

Also in MR spectra of the heart, a signal for PGAA was observed (Figure 5) and, as for skeletal muscle, fluctuations in oral Cr ingestion seemed to play a role in the fluctuations of the PGAA signal as can be concluded from a comparison of three MRS studies (Schmidt *et al.*, 2004; Schneider *et al.*, 2004; ten Hove *et al.*, 2005). In one study (Schmidt *et al.*, 2004), similar PCr and PGAA concentrations were observed in GAMT-deficient animals; the PCr level was reduced by 67% compared to control PCr values. In contrast, in two other studies, PCr was undetectable in the knockout animals (Schneider *et al.*, 2004; ten Hove *et al.*, 2005) while PGAA concentrations were only reduced by 33% compared to PCr in control hearts. These differences in (P)Cr concentrations are most likely

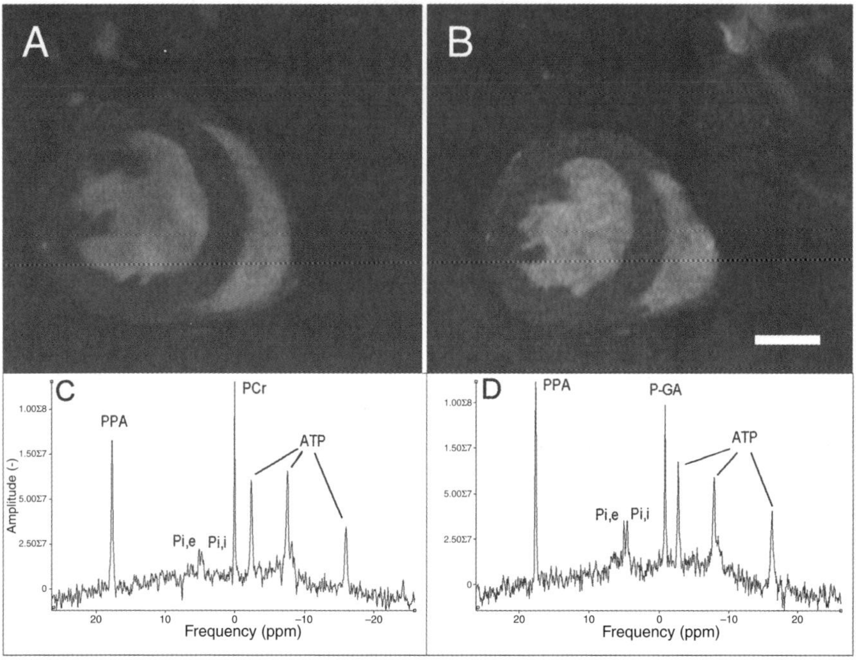

Figure 5. MR images and spectra of wild-type and GAMT-deficient heart. *In vivo* end-diastolic MR images from a wild-type (A) and a GAMT-/- (B) mouse (bar = 2 mm applies to both panels). Note that the GAMT-/- heart is smaller. Typical ^{31}P-MR spectra of a heart from a wild-type (C) and a GAMT-/- (D) mouse. In wild-type hearts, PCr and 3 peaks for ATP are visible. In GAMT-/- hearts, ATP is visible, but no PCr could be detected. Instead, PGAA (= P-GA in the figure) appears at –0.5 ppm. In both spectra, the phenylphosphonic acid (PPA) resonance originates from a PPA solution inside the intraventricular balloon. Reprinted from ten Hove *et al.* (2005) with permission

caused by coprophagia (Kan *et al.*, 2004; ten Hove *et al.*, 2005). Besides the fluctuations in PGAA content, another parallel between skeletal muscle and heart is the response in ischemia/reperfusion experiments, where a delayed recovery of PGAA compared to PCr in control animals is observed in both tissues (Kan *et al.*, 2004).

In general, MRS experiments on GAMT-/- mice showed that despite the lower affinity of CK for PGAA, the latter compound functions surprisingly well *in vivo* to match moderate energy demands. Only small amounts of PCr were necessary to prevent elevated levels of Pi in skeletal muscle, and increased levels of Pi itself may be an adaptation to stimulate glycolysis, yet another pathway for relatively fast energy production. To optimize treatment regimens in patients, further experiments may be performed on GAMT deficient mice that not only include Cr supplementation but also efforts to lower GAA levels.

5. MRS OF MICE TRANSGENIC FOR CREATINE KINASE ISOENZYMES

This section reviews the literature on MRS of mice in which CK isoenzymes are lacking, substituted or overexpressed, in skeletal muscle, brain, heart and/or liver. In many MRS studies that have been performed in the past on ~P transfer systems in these organs, the CK reaction has played an essential role as levels of key substrates in this reaction (PCr, ATP, H^+) can be determined non-invasively from ^{31}P MR spectra of living tissue. In addition, the kinetics of the CK reaction can be determined from magnetization transfer experiments.

Research on high-energy phosphoryl transfer in skeletal muscle benefited enormously from the availability of mice with null mutations in the M-CK gene (M-CK-/-; van Deursen *et al.*, 1993), the ScCKmit gene (ScCKmit-/-; Steeghs *et al.*, 1997b), or both mutations combined (M-CK/ScCKmit --/--; Steeghs *et al.*, 1997a). Also the availibility of animals with a leaky mutation in the M-CK gene (and graded levels of MM-CK activity) helped in these studies (van Deursen *et al.*, 1994b). Mice lacking either the cytosolic B-CK (B-CK-/- mice) (Jost *et al.*, 2002), the ubiquitous mitochondrial CK isoenzyme (UbCKmit-/- mice; Kekelidze *et al.*, 2001) or both enzymes (B-CK/UbCKmit--/-- mice; in't Zandt *et al.*, 2004) have been generated to study the role of the CK reaction in other tissues than muscle, most notably in brain.

5.1. Skeletal Muscle

5.1.1. *Single creatine kinase knock-outs*

Initially it came as a surprise that ^{31}P MRS studies of resting skeletal muscle of mice with reduced or absent M-CK expression showed no significant difference in PCr, Pi or ATP levels, nor in tissue pH, compared to wild-type (Table 1; (van Deursen *et al.*, 1993, 1994b)). Nevertheless, during a short ischemic period of 20 minutes, more pronounced acidification was observed in M-CK-/- compared to wild-type muscle (in't Zandt *et al.*, 1999), which may be due to diminished proton buffering. In wild-type animals, chemically determined PCr/ATP ratios were somewhat lower than PCr/ATP ratios determined by *in vivo* MRS (van Deursen *et al.*, 1993) and PCr levels determined chemically were higher in M-CK-/- mice than in control mice while MRS analysis showed similar levels. This discrepancy between *in vivo* MRS and *in vitro* chemical analysis of PCr levels can be ascribed to the hydrolysis of PCr during the isolation and freezing of tissues and/or tissue extraction (Ackerman *et al.*, 1980; Meyer *et al.*, 1985). Support for this view came from a study in which, despite rapid resection and quick freezing of the muscle, higher PCr levels were detected in rats treated with a CK inhibitor (Brault *et al.*, 2003). This illustrates the value of non-invasive assessment by MRS of metabolites involved in rapid biochemical conversions.

Skeletal muscles of mice deficient in cytosolic CK were unable to phosphorylate β-GPA (β-guanidinopropionic acid), demonstrating that M-CK is the only isoform

Table 1. Tissue levels and ratios of metabolites in mouse skeletal muscle of CK knockout mice

	in't Zandt *et al*., 2003		Steeghs *et al*., 1997a				van Deursen *et al*., 1993	
	wild-type	M-CK/ ScCKmit--/-- double k.o.	wild-type	ScCKmit-/-	M-CK-/-	M-CK/ ScCKmit--/-- double k.o.	wild-type	M-CK-/-
Ratios (by MRS)[a,b]								
PCr/ATP	2.9 ± 0.2	2.4 ± 0.2 *	3.0 ± 0.1[f]	3.0 ± 0.2[f]	3.4 ± 0.1[f]	2.1 ± 0.1[f]	2.8 ± 0.3[f]	2.9 ± 0.2[f]
Pi/ATP	–	–	0.36 ± 0.06[f]	0.31 ± 0.05[f]	0.32 ± 0.06[f]	0.51 ± 0.04*[f]	0.37 ± 0.11[f]	0.32 ± 0.06[f]
pH	7.22 ± 0.03	7.24 ± 0.04	7.29 ± 0.07	7.17 ± 0.05	7.24 ± 0.02	7.21 ± 0.05	7.18 ± 0.08	7.19 ± 0.09
Concentrations (mM)								
[tCr]	30.5 ± 2.0	31.4 ± 2.8	29.9 ± 1.0 [c,d]	29.9 ± 1.3[c,d]	29.9 ± 1.3[c,d]	29.9 ± 2.3[c,d]	27.0 ± 1.8[c]	30.2 ± 2.2[c]
[PCr]	22.5 ± 1.9	17.6 ± 2.2	19.7 ± 0.7[c]	20.6 ± 0.9[c]	21.8 ± 1.0*[c]	16.4 ± 1.7*[c]	17.2 ± 1.5[c]	23.0 ± 1.9[c]*
[ATP]	7.8 ± 0.8	7.5 ± 1.1	7.8 ± 1.5[c]	7.3 ± 0.6[c]	6.1 ± 0.7[c]	5.8 ± 0.8*[c]	8.4 ± 0.5[c]	7.4 ± 0.6[c]
[PCr]/[ATP][g]	2.9 ± 0.4	2.3 ± 0.5	2.5 ± 0.5	2.8 ± 0.3	3.6 ± 0.4	2.8 ± 0.5	2.0 ± 0.2	3.1 ± 0.4

(*continued*)

Table 1. (continued)

	in't Zandt *et al.*, 2003		Steeghs *et al.*, 1997a				van Deursen *et al.*, 1993	
	wild-type	M-CK / ScCKmit--/-- double k.o.	wild-type	ScCKmit-/-	M-CK-/-	M-CK/ ScCKmit--/-- double k.o.	wild-type	M-CK-/-
Other values								
[PCr]/[tCr]	0.74 ± 0.04[h]	0.56 ± 0.05[h,*]	0.66 ± 0.03	0.69 ± 0.04	0.73 ± 0.05	0.55 ± 0.07	0.64 ± 0.07	0.76 ± 0.1
[ADP][i]	0.037	(0.083) [e]	–	–	–	–	–	–

Data obtained *in vivo*, essentially from the hind limb GPS complex by quantitative MRS, except when indicated otherwise. Concentrations are expressed in mmol·(l tissue)$^{-1}$ (mean ± s.d.), except [ADP] which is expressed as mmol·(l intracellular water)$^{-1}$. Values that are expressed per l tissue can be converted to values per l intracellular water by using a factor of 0.73 for the intracellular water fraction.

[*] Significantly different from wt animals.

[a] Ratios are determined from fully relaxed MR spectra. Studies not using fully relaxed spectra to compare wt and transgenic animals were not included in this table.

[b] Unless stated otherwise, ratios are based on the β-ATP signal.

[c] Concentrations were determined by chemical methods in mmol·(g dry weight)$^{-1}$ and calculated to mM assuming the fraction dw:ww to be 0.3:0.7 and a tissue density of 1.06.

[d] Cr + PCr were normalized over all muscles.

[e] Equilibrium condition necessary for calculation of ADP may not be valid in this knockout mouse.

[f] Calculated from individual metabolite concentrations as presented in original paper.

[g] Calculated from biochemical concentrations for comparative reasons; not in original paper.

[h] Directly derived from ^{13}C MR spectra.

[i] As calculated from MR spectra and the CK equilibrium constant.

Table 2. Tissue levels and ratios of metabolites in mouse skeletal muscle of GAMT transgenic mice

	Renema *et al.*, 2003		Kan *et al.*, 2004a		
	wild-type	GAMT-/-[d]	wild-type	GAMT-/-	GAMT-/- [e] (suppl. 1 month)
Ratios (by MRS) [a,b]					
PCr/ATP	2.7 ± 0.1	0.8 ± 0.3*	3.2 ± 0.3	n.d.	2.9 ± 0.06
PGAA/ATP	–	1.8 ± 0.4*	n.d.	3.0 ± 0.2	n.d.
Pi/ATP	0.40 ± 0.06	0.40 ± 0.08	0.40 ± 0.1	0.65 ± 0.1*	0.26 ± 0.05♧
pH	7.24 ± 0.09	7.18 ± 0.06	7.20 ± 0.04	7.19 ± 0.11	–
Concentrations (mM)					
[tCr]	28.4 ± 2.6	8.9 ± 3.8*	–	–	–
[ATP]	–	–	5.0 ± 0.9 [c]	3.5 ± 0.9 [c]	–

Data obtained *in vivo*, essentially from the hind limb GPS complex by quantitative MRS, except when indicated otherwise. Concentrations are expressed in mmol·(l tissue)$^{-1}$ (mean ± s.d.). n.d. = not detectable.

* Significantly different from wt animals; §significantly different from value before supplementation.

[a] Ratios were determined from fully relaxed MR spectra. Studies not using not fully relaxed spectra to compare wt and transgenic animals were not included in this table.

[b] Unless stated otherwise, ratios are based on the β-ATP signal.

[c] Concentrations determined chemically in mmol·(g dry weight)$^{-1}$ and reverted to mM by assuming the fraction dw:ww to be 0.3:0.7 and a tissue density of 1.06.

[d] Some creatine is detectable in these GAMT deficient animals because of coprophagia.

[e] After 1 month of Cr supplementation.

that can phosphorylate this Cr analogue in muscle (van Deursen *et al.*, 1994a), as was previously observed in *in vitro* studies (Boehm *et al.*, 1996; Clark *et al.*, 1994).

Physiological experiments in which skeletal muscles of mice were electrically stimulated to contract unequivocally demonstrated lack of burst activity in M-CK-/- muscles. However, MRS analysis of electrical stimulation of muscle by 1 or 5 Hz (van Deursen *et al.*, 1993) or a challenge by ischemia-reperfusion (in't Zandt *et al.*, 1999) initially showed a similar breakdown of PCr and accumulation of Pi in wild-type and M-CK-/- muscle, within the time resolution of the experiments (~140 s per spectrum). If the duration of the assessment was shortened, however, as done in gated ^{31}P MRS experiments, a faster initial decrease of PCr was found during the first few seconds of 5 Hz stimulation in wild-type mice as compared to the mutant mice (Roman *et al.*, 2002). This latter observation is in agreement with an important role for PCr in burst activity. Impairment of the CK reaction in M-CK-/- mice was demonstrated earlier in inversion transfer (van Deursen *et al.*, 1993) and saturation transfer (van Dorsten *et al.*, 1997) experiments of skeletal muscles at rest. These studies revealed that rapid high-energy phosphate transfer from PCr to ATP or visa versa was not detectable in M-CK-/- mice. In fact, this apparent lack of activity was already observed in mice expressing 34% or less of wild-type MM-CK activity (van Deursen *et al.*, 1994b), indicating that the phosphate transfer rate was below the detection threshold of the MRS experiment. Interestingly, the fact that reciprocal PCr to Cr conversions occurs in the M-CK knockout animals demonstrates that ScCKmit can catalyze the CK reaction bi-directionally *in vivo*.

Similar to what was observed with M-CK-/- mice, the single mitochondrial CK knock-out mice (ScCKmit-/-) showed no difference in PCr, Pi and pH levels compared to wild-type mice at rest. Both inversion and saturation transfer experiments (Nicolay *et al.*, 1998; Steeghs *et al.*, 1997b) revealed no difference in the flux through the CK reaction. Small differences compared to wild-type animals were observed in ischemia experiments. Recovery of PCr and Pi levels appeared to be faster than in wild-type animals, and the hydrolysis of PCr was not completely compensated for by increases in Pi and phosphomonoesters (PME). This may be related to changes in the mitochondrial membrane microenvironment (in't Zandt *et al.*, 1999).

5.1.2. Mice lacking both creatine kinase isoenzymes in skeletal muscle

Inbreeding of M-CK-/- and ScCKmit-/- mice resulted in mice lacking both isoenzymes: M-CK/ScCKmit--/-- double knockout mice, which have virtually no CK activity in skeletal muscle (Steeghs *et al.*, 1997a). In the course of aging, the level of total Cr (tCr; i.e. the sum of Cr and PCr levels) rose in parallel in M-CK/ScCKmit--/-- and wild-type mice as shown by ^{1}H MRS and biochemical methods (in't Zandt *et al.*, 2003), both yielding the same tissue concentrations. This latter finding demonstrated that virtually all Cr is visible in ^{1}H MR spectra if measured under the proper conditions (in't Zandt *et al.*, 2000, 2003). Remarkably, the postnatal increase in PCr levels in these knockout mice was the same as in wild-type animals. We have attributed this to the presence of residual BB-CK activity, which is present in muscle of M-CK/ScCKmit--/-- mice during the first postnatal weeks (in't Zandt *et al.*, 2003). In muscle of mature M-CK/ScCKmit--/-- mice, PCr levels decreased with age, possibly reflecting decreasing deposition of CK from fusing satellite cells. When M-CK/ScCKmit--/-- mice were three weeks old, they still were able to fully hydrolyze PCr (in't Zandt *et al.*, 2003). At older age, this CK activity decreased to an undetectably low level and as a consequence, conversion through the CK reaction gradually decreased (in't Zandt *et al.*, 2003) until, in adult mice, the MRS-detectable flux through the CK reaction was fully absent (Nicolay *et al.*, 1998; Steeghs *et al.*, 1997a). The silencing of the CK reaction in adult animals was reflected by a decrease in ATP levels and a profound immediate decline of tissue pH during 20 minutes of ischemia (in't Zandt *et al.*, 1999).

The Cr signals in ^{1}H MRS spectra of skeletal muscle show so-called off-resonance magnetization transfer and dipolar coupling effects which are likely due to weak interactions between Cr and macromolecular structures, affecting the mobility of (P)Cr molecules to some extent. A potential candidate could be the interaction of Cr with CK isoenzymes. However, studies with single and double CK knockouts showed no differences in these effects (in't Zandt *et al.*, 2000; Kruiskamp *et al.*, 2000), implying that other physical interactions of (P)Cr in myofibers are responsible for these phenomena.

By using ^{13}C MRS and ^{13}C-4 labeled Cr, simultaneous observation of Cr and PCr became possible in mice. After injection with ^{13}C-labeled Cr in adult M-CK/ScCKmit--/-- mice, Cr uptake and phosphorylation was observed in muscle, demonstrating that very low CK activity is sufficient for this process (Figure 3; in't Zandt *et al.*, 2003). Using this ^{13}C MRS data, it was possible to calculate PCr/tCr ratios. This PCr/tCr ratio, together with the absolute tCr tissue content, obtained from ^{1}H MR spectra, and the metabolite ratios determined from ^{31}P MR spectra, enabled complete non-invasive determination of the tissue levels of ATP and PCr in skeletal muscle (Table 1; in't Zandt *et al.*, 2003), thus avoiding artifacts due to hydrolysis associated with freeze clamping as described above.

5.1.3. Creatine kinase overexpression or isoenzyme replacement

To test whether the proposed specific localization of MM-CK in muscle, which is close to the M-line and the sarcoplasmic reticulum, is important, M-CK-/- mice (van Deursen *et al.*, 1993) were crossed with mice expressing B-CK in muscle (Brosnan *et al.*, 1993) to generate mice in which M-CK has been 'replaced' by B-CK, which lacks the above-mentioned M-CK-specific subcellular localization (Roman *et al.*, 1997). *In vivo*, ^{31}P saturation transfer experiments showed a decrease in CK activity at rest to only 58% of that of wild-type animals. As the contractile phenotype of M-CK-/- mice returned to normal in mice containing B-CK instead, it was concluded that partitioning and compartmentation of CK to the myofibril is not an essential prerequisite for proper contractile function (Roman *et al.*, 1997). Another implication of these results is that the CK shuttle, as originally proposed by Bessman and co-workers (Bessman and Fonyo, 1966; Bessman and Geiger, 1981), is not of vital importance to skeletal muscle, at least under the conditions tested.

Interesting findings about the function of CK were not only obtained by knocking out the enzyme or enzyme swapping, but also by over-expression approaches. Ectopic over-expression of the B-CK isoform in muscle of transgenic mice resulted in the formation of the stable dimers MB-CK and BB-CK, and increased total muscle CK activity as assessed under maximal velocity conditions in tissue extracts (up to 150% of that of controls; Brosnan *et al.*, 1991b, 1993). Using saturation transfer techniques, the *in vivo* CK activity appeared to be two-fold up-regulated (Brosnan *et al.*, 1993). This up-regulation of CK had no effect on PCr/ATP or Pi/ATP ratios as measured by ^{31}P MRS, or the ATP or Cr concentrations as measured by HPLC, which is expected for the CK reaction being at (near) equilibrium under resting conditions. The above studies were initially performed in animals that were heterozygous for the transgene, but subsequent analysis of homozygous transgenics showed similar effects (Roman *et al.*, 1996). The only difference between wild-type and homozygous transgenic mice was a decrease in the rise time of force of a 5 s tetanic stimulation, which showed that only during severe metabolic demand, extra CK may increase performance (Roman *et al.*, 1996). It may be of note that adaptation in other phosphotransfer enzymes occurred as a result of overexpression, as AK activity appeared to be decreased by 9% (Roman *et al.*, 1996). This may reflect the close harmony in which CK and AK work in an integrated network (Dzeja and Terzic, 2003).

Altogether, MRS studies on skeletal muscle of transgenic mice lacking one or more CK isoenzymes, or with a replacement of M-CK by B-CK, or with over-expression of CK have provided the most direct evidence as yet under *in vivo* conditions for the 'classical' role of CK in buffering global ATP levels, in maintaining tissue pH and in the liberation of Pi. Moreover, these studies have revealed metabolic reasons for the critical role of CK in burst activity of skeletal muscle, but at the same time indicate that the CK-shuttle is not of vital importance for maintenance of (twitch) contractile function. It remains to be seen whether this is also true under more demanding conditions. Further conclusions from the MRS studies are that CK has little effect on cellular uptake of Cr and its physical state in the muscle cell. If the remaining PCr levels in the double knockout mice indeed reflect B-CK activity in satellite cells, this provides us with a new and non-invasive window on the role of satellite cells in muscle aging and repair.

5.2. MR Examinations of the Brain of Single and Double Creatine Kinase Knock-out Mice

5.2.1. General phenotype of brain CK deficiency

Using gene-knockout in embryonic stem cells, mouse strains were generated that carry null mutations for either B-CK (B-CK-/-) or UbCKmit (UbCKmit-/-), or for both enzymes (B-CK/UbCKmit--/--). The double mutant B-CK/UbCKmit--/-- mice, which were studied most intensively (Streijger *et al.*, 2005), appeared viable, but their development was affected, resulting in an overall smaller body posture. Furthermore, the complete lack of CK in brain in these animals severely impaired spatial learning and was coupled to abnormal startle responses and hearing problems. In contrast, the visual and motor functions, exploration behavior, and anxiety-related responses were not changed, suggesting no global deficit in sensorimotor function or motivation. Various effects on behavior were consistent with underlying anatomical changes in the hippocampus. Others can best be explained by loss of functional integrity that may occur in development. Parallel to the anatomical and behavioral abnormalities, B-CK/UbCKmit--/-- mice also had problems with thermoregulation, presumably due to an altered hypothalamic thermostat. Furthermore, their response to chemical seizure induction appeared suppressed.

5.2.2. MRS studies of brain CK isoenzyme deficiency

Localized ^{31}P MRS measurements of a major part of the brain of transgenic mice without cytosolic CK (B-CK-/- mice) showed normal PCr, Pi and ATP levels and pH values, but the flux of ~P transfer through the CK reaction measured by saturation transfer experiments was dramatically lowered, although still measurable (Jost *et al.*, 2002). Because the brain is an oxidative organ, the mitochondrial content and fractional contribution of mitochondrial CK activity (up to 30% of total CK activity; Bé Wieringa *et al.*, unpublished result) is much higher than in muscle. Unlike in muscle, mitochondrial CK in brain may therefore be present at large enough quantities to mediate a visible exchange between the high-energy phosphates of γ-ATP and PCr. As compared to muscle, the brain is composed of a much more heterogeneous mixture of cell types. Moreover, the cell-type distribution and co-expression of B-CK and UbCKmit isoforms in glial and neuronal cells appears to be rather variable and does not seem to support the co-existence of cytosolic and mitochondrial endpoints of the CK circuit in various brain regions (see, e.g., Streijger *et al.*, 2005 and references therein). Evidence from several studies showed close metabolic collaboration between glial and neuronal systems for brain energetics (Kasischke *et al.*, 2004; Pellerin and Magistretti, 2004), which makes it intrinsically difficult to know how the PCr shuttle – if present – is integrated into the energy network of brain. Recently, Cr and the CK reaction have been proposed as players in a neuron-glial shuttle (Tachikawa *et al.*, 2004).

UbCKmit-/- mice appeared to have rather normal basal PCr, Pi, ATP and pH levels (in't Zandt *et al.*, 2004), although a significantly lower PCr/ATP ratio in homozygous UbCKmit-/- mice was also reported in one study (Kekelidze *et al.*, 2001). During seizure induction by PTZ injection, UbCKmit-/- mice showed a more pronounced decrease in PCr and β-ATP signals while the PCr/β-ATP ratio remained lower than in wild-type mice (Kekelidze *et al.*, 2001).

PCr in brains of double-knockout B-CK/UbCKmit--/-- mice appeared to be fully depleted, while Pi and ATP levels and pH were virtually normal (Figure 6; in't Zandt *et al.*, 2004). This picture is different from that in muscle, where PCr is still present in double-knockout animals despite the absence of cytosolic and mitochondrial CK as outlined above. In the brain of B-CK/UbCKmit double-knockout mutant, tCr was decreased by 30% and levels of N-acetyl aspartate (NAA) increased (Figure 7, Table 3), which could point to an osmolar redistribution or other adaptation mechanisms (in't Zandt *et al.*, 2004). Interestingly, a similar inverse relationship between Cr and NAA was observed during Cr feeding in GAMT deficient mice (see section 4.2).

The role of the PCr/CK circuit in the brain is far less well understood than in muscle. Apparently the circuit is not vital to basic brain physiology, but behavioral studies have demonstrated structural, sensory and cognitive deficiencies in brain CK mutants (Streijger *et al.*, 2004, 2005). Testing brain function and metabolism under energy stress conditions in these mice is needed to uncover more specific functions of the circuit such as, for instance, in metabolic coupling between neuroactivation and hemodynamic response (Aubert *et al.*, 2007).

5.3. Studies of the Heart in Creatine Kinase Knockouts

Cardiac muscle cells possess a relatively high mitochondrial capacity as compared to skeletal muscle cells, and therefore have a more oxidative character with smaller diffusion distances between energy-producing and energy-consuming sites. Although the total CK activity (activity units per mg tissue) is only 10–30% of that in skeletal muscle, the relative contribution by mitochondrial ScCKmit activity

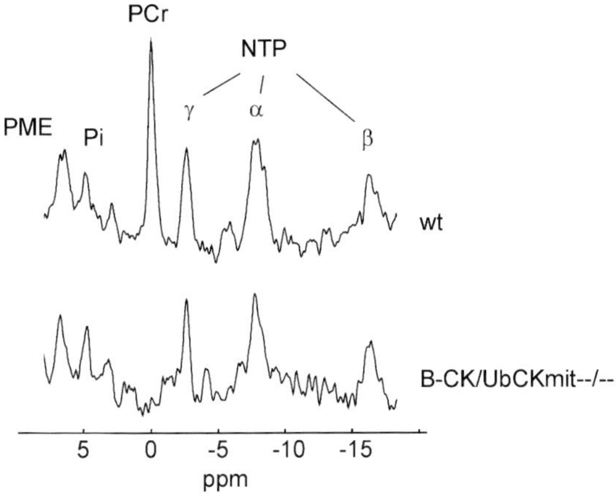

Figure 6. ^{31}P MR spectra of volumes covering the major part of the cerebrum of a wild-type (wt) mouse (top) and of a B-CK/UbCKmit--/-- mouse (bottom). The obvious difference is the absence of a signal for PCr in the mouse brain completely lacking creatine kinase. NTP is nucleoside triphosphate, mainly ATP; PME is phosphomonoester; Pi is inorganic phosphate; PCr is phosphocreatine

is much higher. Moreover, a small fraction of cytosolic CK activity stems from MB-CK (Veksler *et al.*, 1995).

^{31}P MRS revealed that M-CK-/- hearts displayed normal PCr/ATP ratios as well as normal absolute PCr and ATP levels, thus keeping the phosphorylation potential high (Saupe *et al.*, 1998; van Dorsten *et al.*, 1998). At different workloads (i.e. at different pacing frequencies), PCr and other ^{31}P MRS-detectable metabolites appeared to be equal in M-CK-/- and wild-type mice. However, differences in very fast events may have been missed in these studies as rather long acquisition times were used (~8-10 minutes per spectrum). Free ADP levels were also equal, or somewhat higher in transgenic mice at a workload of 600 beats per minute (Saupe *et al.*, 1998; van Dorsten *et al.*, 1998). In contrast to skeletal muscle lacking M-CK, saturation transfer experiments clearly demonstrated phosphoryl transfer from PCr to ATP in the heart. Again, this is most easily explained by the presence of relatively large amounts of mitochondrial CK (van Dorsten *et al.*, 1997, 1998). At higher workloads, however, CK flux increased less in M-CK-/- mice than in wild-type mice (van Dorsten *et al.*, 1998). From these studies, it was concluded that mitochondrial CK activity is sufficient to maintain most of the CK functions in heart. The authors of this work hypothesized that mitochondrial CK, although quantitatively present at lower concentrations, has a higher contribution to the total measured CK flux than M-CK in the wild-type heart (van Dorsten *et al.*, 1998).

This conclusion was further corroborated by ^{31}P MRS experiments performed on ScCKmit-/- hearts, which showed that mitochondrial CK was necessary to maintain normal high-energy phosphate levels (Spindler *et al.*, 2002). Mice lacking only mitochondrial CK showed 20-30% lower PCr levels in the heart (Spindler *et al.*, 2002, 2003), which is in contrast to the situation in skeletal muscle, where PCr

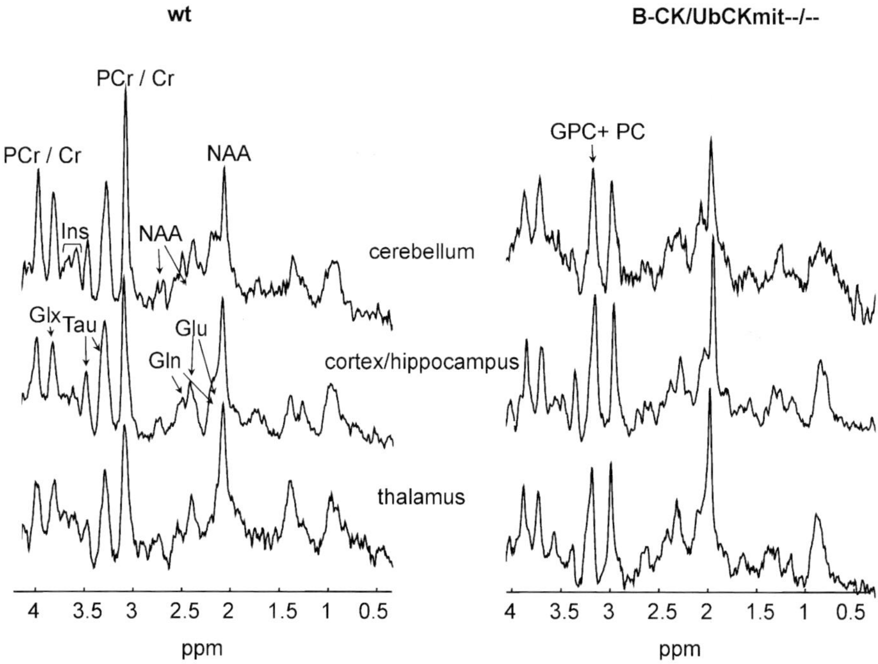

Figure 7. ^{1}H MR spectra of selected regions in the brain of wild type (wt) mice (left) and of double knockout (B-CK/UbCKmit--/--) mice (right). Peak assignments: PCr = phosphocreatine; Cr = creatine; Gln = glutamine; Glu = glutamate; Ins = *myo*-inositol; Tau = taurine; GPC = glycerolphosphocholine; PC= phosphocholine; NAA = N-acetylaspartate. Reprinted from in't Zandt *et al.* (2004) with permission

levels remained unaltered (Steeghs *et al.*, 1997b). Furthermore, ScCKmit-/- hearts showed two-fold elevated ADP levels, as calculated from the CK equilibrium, while Pi and ATP levels appeared normal (Spindler *et al.*, 2002). This remarkable finding underlines the role of ScCKmit in maintaining (local) ADP concentrations. At increased workload, the ScCKmit-/- mice were still able to hydrolyze PCr, although the decrease in PCr was less than in wild-type mice (24% vs. 37% of baseline PCr). When recovering from this higher workload, ScCKmit-/- mice resynthesized PCr. Therefore, it was concluded that bidirectional ∼P transfer is maintained even if CK is lacking in the inter-membrane compartment of mitochondria (Spindler *et al.*, 2002), similar to what was found for skeletal muscle lacking ScCKmit (in't Zandt *et al.*, 1999).

As demonstrated by ^{31}P MRS, intact hearts of M-CK/ScCKmit--/-- mice have a lower basal PCr level as compared to wild-type animals (Table 3; Saupe *et al.*, 1998). This lower PCr level resembles the situation in skeletal muscle. However, hearts of double knockout M-CK/ScCKmit--/-- mice are able to hydrolyze PCr at increased workload while maintaining ATP concentrations. This is in sharp contrast to an ischemic challenge of skeletal muscle of adult M-CK/ScCKmit--/-- mice where PCr was virtually metabolically inert and ATP levels decreased considerably (in't Zandt

Table 3. Tissue levels of brain metabolites in CK and GAMT knockout mice and their corresponding littermates measured by *in vivo* ^{1}H MRS

	wild-type (n = 6-8)	wild-type littermate (n = 5)	B-CK-/- (n = 5)	UbCKmit-/- (n = 6)	B-CK/ UbCKmit--/-- (n = 5)	wild-type littermate (n = 4)	GAMT-/- (n = 14)	GAMT-/- (1 month Cr suppl.) (n = 6)
Taurine	7.7 – 12.9	–	–	–	–	7.1 – 7.8	5.1 – 6.2	3.9 – 5.6
Total choline[a]	1.2 – 1.9	0.9 – 2.3	0.9 – 1.8	1.4 – 3.0	0.8 – 2.8	1.7 – 3.1	2.1 – 2.3	1.7 – 3.1
Total NAA[b]	8.2 – 10.5	6.3 – 7.3	5.9 – 6.5	6.3 – 7.0	8.1 – 8.8	8.6 – 9.6	11.7 – 11.8	9.5 – 10.4
Total creatine[c]	7.9 – 13.0	8.3 – 10.0	8.9 – 10.3	8.4 – 11.2	6.5 – 7.9	10.7 – 14.7	0.8 – 1.2	13.7 – 20.6
Myo-inositol	4.4 – 7.1	2.9 – 4.3	2.0 – 3.1	2.9 – 3.8	2.0 – 2.9	5.1 – 8.8	5.7 – 9.0	6.8 – 7.9
Glutamate	9.2 – 12.4	6.1 – 6.6	5.8 – 6.4	7.4 – 8.3	7.0 – 8.8	10.4 – 11.7	9.7 – 11.3	10.0 – 10.0
Glutamine	3.4 – 3.8	–	–	–	–	3.0 – 4.5	3.6 – 5.0	3.2 – 3.9
	Tkac *et al.*, 2004	in't Zandt *et al.*, 2004					Kan *et al.*, 2007	

All concentrations are represented in mM (average range over different brain regions).

[a] Choline + phosphocholine + glycero-phosphocholine;

[b] N-acetylaspartate + N-aspartylaspartateglutamate;

[c] creatine + phosphocreatine.

et al., 2003). This difference can be explained by the presence of BB-CK in cardiac muscle although the remaining CK activity is only approximately twice as high as that in skeletal muscle of CK double-knockouts (1.9 mM·s^{-1} vs. 0.5-1.0 mM·s^{-1}; in't Zandt *et al.*, 2003; Saupe *et al.*, 1998). Localization of the BB-CK enzyme and its substrates in different compartments may play a role here.

The amount of PCr hydrolyzed is larger in double CK knockouts than in M-CK-/- or wild-type hearts in response to the same workload (Ingwall, 1998; Saupe *et al.*, 1998). Since the tissue content of PCr is a net balance between Cr phosphorylation and PCr hydrolysis, this effect was attributed to impaired synthesis (normally primarily mediated by ScCKmit, now only by BB-CK) during cardiac work. As calculated from the CK equilibrium reaction (Veech *et al.*, 1979), [ADP] was increased by 95% in hearts of M-CK/ScCKmit--/-- mice, which may also be the result of the disrupted functioning of mitochondrial CK in balancing ATP/ADP levels. Interestingly, as ScCKmit-/- single knock-outs (relying on M-CK and B-CK) do not display altered PCr behavior at increased workload, the total CK activity is probably an important factor besides the presence of ScCKmit.

Altogether, these studies show that the different distribution of CK isoenzymes in the heart of mice as compared to that in skeletal muscle results in a different metabolic phenotype upon deletion of the enzymes. In particular, mitochondrial CK has a more prominent role, but as in skeletal muscle, different CK isoenzymes may behave in a compensatory way.

5.4. Transgenic Expression of Creatine Kinase in Liver

Liver is an ideal tissue model to study Cr metabolism and CK function in a null background, since it normally does not express CK. When CK is ectopically expressed from a transgene, differences from the normal situation in liver can provide direct information on the working mechanisms and importance of CK. Moreover, as the CK reaction in equilibrium enables to calculate free ADP levels, its presence opens a possibility to study this important parameter in liver.

One of the first genetic manipulations involving the CK system was by expressing the B subunit of CK in liver (Koretsky *et al.*, 1990). Livers were analyzed by ^{31}P MRS using a radiofrequency probe adjacent to a surgically exposed liver. The ability to phosphorylate Cr was only observed after Cr feeding (Figure 8), and the concentration of PCr measured correlated with the quantity of Cr ingested (Brosnan *et al.*, 1990). This represented the first use of MRS to detect a change in gene expression, and the idea to use CK as reporter for MRS detection of gene expression was suggested (Koretsky *et al.*, 1990). Subsequently, a number of papers have appeared, demonstrating the use of CK, or a related guanidino or arginine kinase, as reporters for MRS detection of gene expression (Walter *et al.*, 2000; Auricchio *et al.*, 2001). With CK expression in the liver, it became possible to calculate the amount of free ADP. Consistent with values in brain and heart, the free ADP levels in the liver were low, as might be expected for well-perfused tissue. Moreover, these levels were lower than total ADP

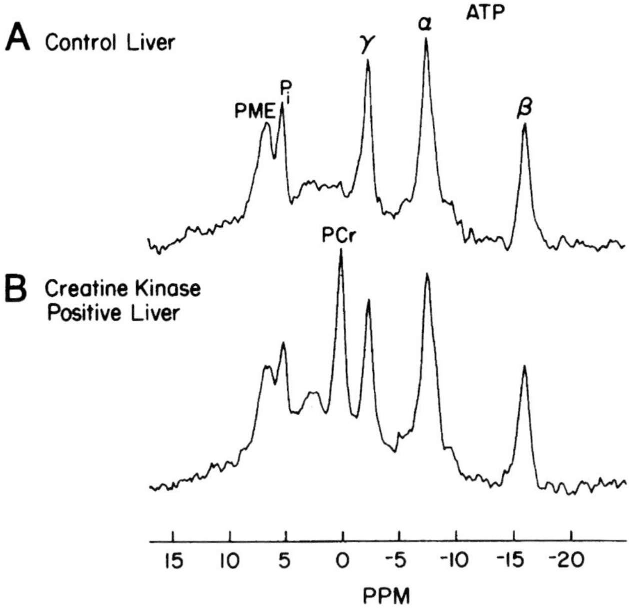

Figure 8. ^{31}P MR spectra obtained from wild-type liver (A) and CK-positive (B) liver. Reprinted from Koretsky *et al.* (1990) with permission

levels measured in tissue extracts, also consistent with results from heart and brain. Furthermore, a non-natural CK substrate, cyclocreatine, was fed to the animals, which resulted in high levels of phosphocyclocreatine. ADP levels were similar with cyclocreatine and Cr (Brosnan *et al.*, 1990, 1991b). These results all support a passive buffering role for the CK-catalyzed reaction in these transgenic livers.

In studies of the perfused liver during an ischemia protocol, ^{31}P MRS showed that upon introduction of CK in the liver, the dramatic decrease in ATP and pH observed in wild-type mice is prevented (Miller *et al.*, 1993). Thus CK rescues the liver from the adverse effects of the ischemic challenge. As in skeletal muscle studies, this unambiguously demonstrates the energetic and pH stabilization that CK imparts on tissues. As expected, Pi was increasing more during ischemia in CK-expressing livers than in wild-type livers. After reperfusion, Pi and ATP levels in the CK-expressing livers returned to pre-ischemia values within 30 minutes, in contrast to normal livers for which ATP remained reduced and Pi increased (Miller *et al.*, 1993). Besides protecting the liver from ischemia (where the tissue is deprived of nutrients and oxygen), the CK system could also protect the liver from only oxygen deprivation, depending on the amount of PCr present (Miller *et al.*, 1993). The protection from ischemia and hypoxia afforded by CK represents an as yet unexplored opportunity to engineer liver to be saved from the consequences of such events. The pH of the perfusate was also shown to have a large influence on the amount and rate of PCr formed with varying Cr concentrations in the buffer (Masson and Quistorff, 1994).

A fructose load is a severe challenge of energy metabolism in the liver due to the consumption of ATP and Pi in the conversion of fructose to fructose-1-phosphate. When fructose was injected intraperitoneally, ^{31}P MRS revealed that PME levels rose two- to three-fold, and Pi and ATP levels decreased by 60% and 50%, respectively, similarly in wild-type and in transgenic mice not fed with Cr. While Pi returned to resting levels in about 30 minutes, ATP and PME failed to do so even after one hour. Upon feeding of Cr, the fall in ATP levels was prevented, and only a small decrease in ATP was observed for relatively low PCr values (PCr/ATP = 1.0-1.5; Brosnan *et al.*, 1991a). In addition, the changes in Pi and PME levels were diminished in the presence of PCr. ADP levels as calculated from the CK reaction, however, changed independently of Cr feeding (Brosnan *et al.*, 1991a). Based on these findings, it was concluded that prevention of the decline in ATP must be due to the newly available source of high-energy phosphates from PCr (Brosnan *et al.*, 1991a). Interestingly, these results demonstrated for the first time that small amounts of PCr already have a significant role in the buffering of high-energy phosphates, a finding that was confirmed in later studies on Cr-deficient mice (see section 4.2). The results also explain why large losses in ATP in the normal liver under a fructose challenge is associated with a decrease in Pi rather than with an increase in ADP.

In addition to mice expressing cytosolic CK, mice have been generated that express the mitochondrial UbCKmit isoenzyme in liver (Miller *et al.*, 1997). These mice were able to synthesize PCr upon Cr feeding, and during ischemia of perfused livers PCr decreased, indicating that it is utilized (Miller *et al.*, 1997). Furthermore, mice were generated which express both cytosolic and mitochondrial CK in liver, in this way providing a series of mice expressing a range of CK activities (B-CK, UbCKmit or both, ranging in activity from 28-1635 μmol$\cdot$(g wet wt)$^{-1}\cdot$min^{-1}; Askenasy and Koretsky, 2002). Saturation transfer experiments on perfused livers showed a linear relation between the CK rate *in vivo* and the biochemically determined CK activity. Furthermore, it was demonstrated in this study that the forward and backward flux through the CK reaction were similar because k_{for}/k_{rev} of the simplified (pseudo first order) CK reaction PCr $\leftrightarrow$ ATP was equal to [ATP]/[PCr], even in mouse liver expressing only mitochondrial CK. These results point to the absence of any vectorial predeliction in the mitochondrial CK reaction, which would have been expected for common PCr shuttle models. Variable expression of CK also had a large effect on ADP levels: only in mice expressing very low levels of CK, ADP levels were relatively high. When a certain threshold in CK activity was reached, apparent ADP levels dropped significantly, which could be due to functional incorporation of the mitochondrial CK isoform in the livers of these transgenic mice (Askenasy and Koretsky, 2002). It is possible that this CK stabilizes the mitochondrial membrane in a way that enables increased efficiency of oxidative phosphorylation (Speer *et al.*, 2005).

6. CONCLUDING REMARKS

As described in this chapter, MRS provides unique contributions to the understanding of creatine-related bioenergetics and metabolism in tissues of transgenic animals. Typical properties of MRS, such as its non-destructive and non-invasive character, which enables dynamic and longitudinal studies, are major assets as compared to more invasive approaches in investigations of tissues in which rapid changes in bioenergetics and metabolism may occur. Because of its relatively crude spatial resolution, the use of results from MRS examinations in the interpretation of processes at the (sub)cellular level may not be trivial and will require careful consideration of morphological, physiological, and biochemical knowledge of the tissue being examined, or dedicated experimental approaches (e.g., Wiseman and Kushmerick, 1995). Cellular heterogeneity in tissues may be an issue when the tissue composition changes due to the mutation introduced. Phenotypic consequences may also not be 100% penetrant, as was shown for ventricle enlargement by MR imaging of B-CK and other knockout mice (Auerbach *et al.*, 2001; in't Zandt *et al.*, 2004). When only average values are evaluated, such changes may remain unnoticed.

Knockout mice often show a remarkable plasticity in coping with the effects of gene deficiency. Unexpected secondary adaptations may occur with wider metabolic consequences than the direct effect of gene deletion. For example, muscle of CK double knock-out mice showed a shift in fiber type characteristics, an upregulation of glycolytic and oxidative potential (Steeghs *et al.*, 1998) and lowered mitochondrial affinity for ADP (ter Veld *et al.*, 2005). In a complex biological system, interpretation of a protein's function may not be straightforward (Bailey, 1999), due to partial redundancy or overlap with functions of other proteins. However, although numerous adaptations can occur that may obscure the identification of the primary function of a (deleted) enzyme, the adaptations themselves provide valuable information on the flexibility of the cellular network in which the enzyme is involved. For CK, which was long considered an irreplaceable element in tissues with high energy demand, a more subtle and dynamic picture of the protein within such a network has emerged due to studies with genetic manipulation. Further work under different challenging conditions *in vivo* is needed to uncover more specific functions of ~P-metabolizing iso-enzymes and the role(s) of creatine, in particular in the brain.

Besides as a tool to gain insight in the biological significance of metabolic enzymes, MRS of transgenic mice that represent models for disorders caused by defects in these enzymes may have clinical relevance. A typical example of such use is the study of the mousemodel for GAMT deficiency, for which the previously unknown *in vivo* metabolic activity of PGAA in skeletal muscle was demonstrated by MRS. This particular model provides unique possibilities to study the potential reversibility of disease consequences, by studying the uptake of Cr in muscle and brain. Moreover, the effects of a genetic lesion causing deficiency in creatine synthesis can be studied without interfering developmental adaptations by suppressing these adaptations with Cr feeding until switching to a Cr-free diet.

MRS of tissues in living mice is certainly not without technical challenges, but the studies described in this paper and many other MRS studies on mice have demonstrated that it is feasible if proper conditions are met. New dedicated practical solutions for MRS measurements will make MRS of mice more widely applicable. In the future, the development of new MRS equipment (e.g., higher fields, better radiofrequency probes) and methods dedicated to mice are expected to uncover still more physiological details of energy transfer systems.

ACKNOWLEDGEMENT

This work was supported by NWO-ZONMW grants 901-01191 (BW) and 902-37-088 (AH), and NWO-ALW grant 834-04-007 (AH). The authors would like to thank Dr. Alan Koretsky for useful suggestions.

REFERENCES

Ackerman, J.J., Grove, T.H., Wong, G.G., Gadian, D.G., and Radda, G.K., 1980, Mapping of metabolites in whole animals by ^{31}P NMR using surface coils. *Nature* **283:** 167–170.

Askenasy, N., and Koretsky, A.P., 2002, Transgenic livers expressing mitochondrial and cytosolic CK: mitochondrial CK modulates free ADP levels. *Am. J. Physiol. Cell Physiol.* **282:** C338-C346.

Aubert, A., Pellerin, L., Magistretti, P.J., and Costalat, R., 2007, A coherent neurobiological framework for functional neuroimaging provided by a model integrating compartmentalized energy metabolism. *Proc. Natl. Acad. Sci. USA* **104:** 4188–4193.

Auerbach, W., Hurlbert, M.S., Hilditch-Maguire, P., Wadghiri, Y.Z., Wheeler, V.C., Cohen, S.I., Joyner, A.L., MacDonald, M.E., and Turnbull, D.H., 2001, The HD mutation causes progressive lethal neurological disease in mice expressing reduced levels of huntingtin. *Hum. Mol. Genet.* **10:** 2515–2523.

Auricchio, A., Zhou, R., Wilson, J.M., and Glickson, J.D., 2001, *In vivo* detection of gene expression in liver by ^{31}P nuclear magnetic resonance spectroscopy employing creatine kinase as a marker gene. *Proc. Natl. Acad. Sci. USA* **98:** 5205–5210.

Bailey, J.E., 1999, Lessons from metabolic engineering for functional genomics and drug discovery. *Nat. Biotechnol.* **17:** 616–618.

Bessman, S.P., and Fonyo, A., 1966, The possible role of the mitochondrial bound creatine kinase in regulation of mitochondrial respiration. *Biochem. Biophys. Res. Commun.* **22:** 597–602.

Bessman, S.P., and Geiger, P.J., 1981, Transport of energy in muscle: the phosphorylcreatine shuttle. *Science* **211:** 448–452.

Boehm, E.A., Radda, G.K., Tomlin, H., and Clark, J.F., 1996, The utilisation of creatine and its analogues by cytosolic and mitochondrial creatine kinase. *Biochim. Biophys. Acta* **1274:** 119–128.

Brault, J.J., Abraham, K.A., and Terjung, R.L., 2003, Phosphocreatine content of freeze-clamped muscle: influence of creatine kinase inhibition. *J. Appl. Physiol.* **94:** 1751–1756.

Brindle, K.M., 1988, NMR methods for measuring enzyme kinetics *in vivo*. *Progr. NMR Spectr.* **20:** 257–293.

Brosnan, M.J., Chen, L., Van Dyke, T.A., and Koretsky, A.P., 1990, Free ADP levels in transgenic mouse liver expressing creatine kinase. Effects of enzyme activity, phosphagen type, and substrate concentration. *J. Biol. Chem.* **265:** 20849–20855.

Brosnan, M.J., Chen, L.H., Wheeler, C.E., Van Dyke, T.A., and Koretsky, A.P., 1991a, Phosphocreatine protects ATP from a fructose load in transgenic mouse liver expressing creatine kinase. *Am. J. Physiol.* **260:** C1191–C1200.

Brosnan, M.J., Halow, J.M., and Koretsky, A.P., 1991b, Manipulating creatine kinase activity in transgenic mice to study control of energy metabolism. *Biochem. Soc. Trans.* **19:** 1010–1014.

Brosnan, M.J., Raman, S.P., Chen, L., and Koretsky, A.P., 1993, Altering creatine kinase isoenzymes in transgenic mouse muscle by overexpression of the B subunit. *Am. J. Physiol.* **264:** C151-C160.

Budinger, T.F., Benaron, D.A., and Koretsky, A.P., 1999, Imaging transgenic animals. *Annu. Rev. Biomed. Eng.* **1:** 611–648.

Clark, J.F., Khuchua, Z., Kuznetsov, A.V., Vassil'eva, E., Boehm, E., Radda, G.K., and Saks, V., 1994, Actions of the creatine analogue beta-guanidinopropionic acid on rat heart mitochondria. *Biochem. J.* **300:** 211–216.

de Graaf, R.A., 1998, *In vivo* NMR spectroscopy. Principles and techniques. John Wiley, Chichester.

de Graaf, R.A., Mason, G.F., Patel, A.B., Behar, K.L., and Rothman, D.L., 2003, *In vivo* ^{1}H-[^{13}C]-NMR spectroscopy of cerebral metabolism. *NMR Biomed.* **16:** 339–357.

Dzeja, P.P., and Terzic, A., 2003, Phosphotransfer networks and cellular energetics. *J. Exp. Biol.* **206:** 2039–2047.

Forsén, S., and Hoffman, R.A., 1963, Study of moderately rapid chemical exchange reactions by means of nuclear magnetic double resonance. *J. Chem. Phys.* **39:** 2892–2901.

Gadian, D.G., 1982, NMR and its application to living systems. Clarendon Press, Oxford.

Gruetter, R., Adriany, G., Choi, I.Y., Henry, P.G., Lei, H., and Oz, G., 2003, Localized *in vivo* ^{13}C NMR spectroscopy of the brain. *NMR Biomed.* **16:** 313–338.

Heerschap, A., Bergman, A.H., van Vaals, J.J., Wirtz, P., Loermans, H.M., and Veerkamp, J.H., 1988, Alterations in relative phosphocreatine concentrations in preclinical mouse muscular dystrophy revealed by *in vivo* NMR. *NMR Biomed.* **1:** 27–31.

in 't Zandt, H.J., Oerlemans, F., Wieringa, B., and Heerschap, A., 1999, Effects of ischemia on skeletal muscle energy metabolism in mice lacking creatine kinase monitored by *in vivo* ^{31}P nuclear magnetic resonance spectroscopy. *NMR Biomed.* **12:** 327–334.

in 't Zandt, H.J., Klomp, D.W., Oerlemans, F., Wieringa, B., Hilbers, C.W., and Heerschap, A., 2000, Proton MR spectroscopy of wild-type and creatine kinase deficient mouse skeletal muscle: dipole-dipole coupling effects and post-mortem changes. *Magn. Reson. Med.* **43:** 517–524.

in 't Zandt, H.J., Groof, A.J., Renema, W.K., Oerlemans, F.T., Klomp, D.W., Wieringa, B., and Heerschap, A., 2003, Presence of (phospho)creatine in developing and adult skeletal muscle of mice without mitochondrial and cytosolic muscle creatine kinase isoforms. *J. Physiol.* **548:** 847–858.

in 't Zandt, H.J.A., Renema, W.K.J., Streijger, F., Jost, C., Klomp, D.W.J., Oerlemans, F., van der Zee, C.E.E.M., Wieringa, B., and Heerschap, A., 2004, Cerebral creatine kinase deficiency influences metabolite levels and morphology in the mouse brain: a quantitative *in vivo* ^{1}H and ^{31}P magnetic resonance study. *J. Neurochem.* **90:** 1321–1330.

Ingwall, J.S., 1998, Creatine kinase knockout mice - what is the phenotype: heart. *Magma* **6:** 120–121.

Jost, C.R., Van Der Zee, C.E., In 't Zandt, H.J., Oerlemans, F., Verheij, M., Streijger, F., Fransen, J., Heerschap, A., Cools, A.R., and Wieringa, B., 2002, Creatine kinase B-driven energy transfer in the brain is important for habituation and spatial learning behaviour, mossy fibre field size and determination of seizure susceptibility. *Eur. J. Neurosci.* **15:** 1692–1706.

Jucker, B.M., Rennings, A.J., Cline, G.W., Petersen, K.F., and Shulman, G.I., 1997, *In vivo* NMR investigation of intramuscular glucose metabolism in conscious rats. *Am. J. Physiol.* **273:** E139–E148.

Jue, T., Rothman, D.L., Shulman, G.I., Tavitian, B.A., DeFronzo, R.A., and Shulman, R.G., 1989, Direct observation of glycogen synthesis in human muscle with ^{13}C NMR. *Proc. Natl. Acad. Sci. USA* **86:** 4489–4491.

Kan, H.E., Buse-Pot, T.E., Peco, R., Isbrandt, D., Heerschap, A., and de Haan, A., 2005, Lower force and impaired performance during high-intensity electrical stimulation in skeletal muscle of GAMT-deficient knockout mice. *Am. J. Physiol. Cell Physiol.* **289:** C113–C119.

Kan, H.E., Meeuwissen, E., van Asten, J.J., Veltien, A., Isbrandt, D., and Heerschap, A., 2007, Creatine uptake in brain and skeletal muscle of mice lacking guanidinoacetate methyltransferase assessed by magnetic resonance spectroscopy. *J. Appl. Physiol.*, **102:** 2121–2127.

Kan, H.E., Renema, W.K., Isbrandt, D., and Heerschap, A., 2004, Phosphorylated guanidinoacetate partly compensates for the lack of phosphocreatine in skeletal muscle of mice lacking guanidinoacetate methyltransferase. *J. Physiol.* **560:** 219–229.

Kasischke, K.A., Vishwasrao, H.D., Fisher, P.J., Zipfel, W.R., and Webb, W.W., 2004, Neural activity triggers neuronal oxidative metabolism followed by astrocytic glycolysis. *Science* **305:** 99–103.

Kekelidze, T., Khait, I., Togliatti, A., Benzecry, J.M., Wieringa, B., and Holtzman, D., 2001, Altered brain phosphocreatine and ATP regulation when mitochondrial creatine kinase is absent. *J. Neurosci. Res.* **66:** 866–872.

Koretsky, A.P., Brosnan, M.J., Chen, L., and Van Dyke, T., 1990, NMR detection of creatine kinase expressed in liver of transgenic mice: Determination of free ADP levels. *Proc. Natl. Acad. Sci. USA* **87:** 3112–3116.

Kruiskamp, M.J., van Vliet, G., and Nicolay, K., 2000, ^{1}H and ^{31}P magnetization transfer studies of hindleg muscle in wild-type and creatine kinase-deficient mice. *Magn. Reson. Med.* **43:** 657–664.

MacLaren, D.C., Toyokuni, T., Cherry, S.R., Barrio, J.R., Phelps, M.E., Herschman, H.R., and Gambhir, S.S., 2000, PET imaging of transgene expression. *Biol. Psychiatry* **48:** 337–348.

Macomber, R.S., 1998, A complete introduction to modern NMR spectroscopy. John Wiley & Sons, New York.

Masson, S., and Quistorff, B., 1994, Kinetics of creatine uptake in the perfused mouse liver: a ^{31}P-n.m.r. study of transgenic mice expressing creatine kinase (CKBB) in the liver. *Biochem. J.* **303:** 531–538.

Meyer, R.A., Brown, T.R., and Kushmerick, M.J., 1985, Phosphorus nuclear magnetic resonance of fast- and slow-twitch muscle. *Am. J. Physiol.* **248:** C279-C287.

Meyer, R.A., Sweeney, H.L., and Kushmerick, M.J., 1984, A simple analysis of the "phosphocreatine shuttle". *Am. J. Physiol.* **246:** C365-C377.

Miller, K., Halow, J., and Koretsky, A.P., 1993, Phosphocreatine protects transgenic mouse liver expressing creatine kinase from hypoxia and ischemia. *Am. J. Physiol.* **265:** C1544-C1551.

Miller, K., Sharer, K., Suhan, J., and Koretsky, A.P., 1997, Expression of functional mitochondrial creatine kinase in liver of transgenic mice. *Am. J. Physiol.* **272:** C1193-C1202.

Moon, R.B., and Richards, J.H., 1973, Determination of intracellular pH by ^{31}P magnetic resonance. *J. Biol. Chem.* **248:** 7276–7278.

Nicolay, K., van Dorsten, F.A., Reese, T., Kruiskamp, M.J., Gellerich, J.F., and van Echteld, C.J., 1998, *In situ* measurements of creatine kinase flux by NMR. The lessons from bioengineered mice. *Mol. Cell. Biochem.* **184:** 195–208.

Ntziachristos, V., Bremer, C., and Weissleder, R., 2003, Fluorescence imaging with near-infrared light: new technological advances that enable *in vivo* molecular imaging. *Eur. Radiol.* **13:** 195–208.

Pellerin, L., and Magistretti, P.J., 2004, Neuroscience. Let there be (NADH) light. *Science* **305:** 50–52.

Pfeuffer, J., Tkac, I., Provencher, S.W., and Gruetter, R., 1999, Toward an *in vivo* neurochemical profile: quantification of 18 metabolites in short-echo-time ^{1}H NMR spectra of the rat brain. *J. Magn. Reson.* **141:** 104–120.

Renema, W.K., Schmidt, A., Van Asten, J.J., Oerlemans, F., Ullrich, K., Wieringa, B., Isbrandt, D., and Heerschap, A., 2003, MR spectroscopy of muscle and brain in guanidinoacetate methyltransferase (GAMT)-deficient mice: Validation of an animal model to study creatine deficiency. *Magn. Reson. Med.* **50:** 936–943.

Renema, W.K.J., 2005, Magnetic resonance spectroscopy of creatine-related energy metabolism in skeletal muscle and brain. PhD Thesis, Radboud University, Nijmegen, The Netherlands.

Roman, B.B., Foley, J.M., Meyer, R.A., and Koretsky, A.P., 1996, Contractile and metabolic effects of increased creatine kinase activity in mouse skeletal muscle. *Am. J. Physiol.* **270:** C1236–C1245.

Roman, B.B., Meyer, R.A., and Wiseman, R.W., 2002, Phosphocreatine kinetics at the onset of contractions in skeletal muscle of MM creatine kinase knockout mice. *Am. J. Physiol. Cell Physiol.* **283:** C1776-C1783.

Roman, B.B., Wieringa, B., and Koretsky, A.P., 1997, Functional equivalence of creatine kinase isoforms in mouse skeletal muscle. *J. Biol. Chem.* **272:** 17790–17794.

Rudin, M., and Sauter, A., 1992, Measurement of reaction rates *in vivo* using magnetization transfer techniques. *NMR Basic Princ. Progr.* **27:** 257–293.

Saupe, K.W., Spindler, M., Tian, R., and Ingwall, J.S., 1998, Impaired cardiac energetics in mice lacking muscle-specific isoenzymes of creatine kinase. *Circ. Res.* **82:** 898–907.

Schmidt, A., Marescau, B., Boehm, E.A., Renema, W.K., Peco, R., Das, A., Steinfeld, R., Chan, S., Wallis, J., Davidoff, M., Ullrich, K., Waldschutz, R., Heerschap, A., De Deyn, P.P., Neubauer, S., and Isbrandt, D., 2004, Severely altered guanidino compound levels, disturbed body weight homeostasis and impaired fertility in a mouse model of guanidinoacetate N-methyltransferase (GAMT) deficiency. *Hum. Mol. Genet.* **13:** 905–921.

Schneider, J.E., Tyler, D.J., ten Hove, M., Sang, A.E., Cassidy, P.J., Fischer, A., Wallis, J., Sebag-Montefiore, L.M., Watkins, H., Isbrandt, D., Clarke, K., and Neubauer, S., 2004, *In vivo* cardiac ^{1}H-MRS in the mouse. *Magn. Reson. Med.* **52:** 1029–1035.

Speer, O., Back, N., Buerklen, T., Brdiczka, D., Koretsky, A., Wallimann, T., and Eriksson, O., 2005, Octameric mitochondrial creatine kinase induces and stabilizes contact sites between the inner and outer membrane. *Biochem. J.* **385:** 445–450.

Spindler, M., Engelhardt, S., Niebler, R., Wagner, H., Hein, L., Lohse, M.J., and Neubauer, S., 2003, Alterations in the myocardial creatine kinase system precede the development of contractile dysfunction in β_1-adrenergic receptor transgenic mice. *J. Mol. Cell. Cardiol.* **35:** 389–397.

Spindler, M., Niebler, R., Remkes, H., Horn, M., Lanz, T., and Neubauer, S., 2002, Mitochondrial creatine kinase is critically necessary for normal myocardial high-energy phosphate metabolism. *Am. J. Physiol. Heart Circ. Physiol.* **283:** H680-H687.

Steeghs, K., Benders, A., Oerlemans, F., de Haan, A., Heerschap, A., Ruitenbeek, W., Jost, C., van Deursen, J., Perryman, B., Pette, D., Bruckwilder, M., Koudijs, J., Jap, P., Veerkamp, J., and Wieringa, B., 1997a, Altered Ca^{2+} responses in muscles with combined mitochondrial and cytosolic creatine kinase deficiencies. *Cell* **89:** 93–103.

Steeghs, K., Heerschap, A., de Haan, A., Ruitenbeek, W., Oerlemans, F., van Deursen, J., Perryman, B., Pette, D., Brueckwilder, M., Koudijs, J., Jap, P., and Wieringa, B., 1997b, Use of gene targeting for compromising energy homeostasis in neuro-muscular tissues: The role of sarcomeric mitochondrial creatine kinase. *J. Neurosci. Meth.* **71:** 29–41.

Steeghs, K., Oerlemans, F., de Haan, A., Heerschap, A., Verdoodt, L., de Bie, M., Ruitenbeek, W., Benders, A., Jost, C., van Deursen, J., Tullson, P., Terjung, R., Jap, P., Jacob, W., Pette, D., and Wieringa, B., 1998, Cytoarchitectural and metabolic adaptations in muscles with mitochondrial and cytosolic creatine kinase deficiencies. *Mol. Cell. Biochem.* **184:** 183–194.

Stöckler, S., Holzbach, U., Hanefeld, F., Marquardt, I., Helms, G., Requart, M., Hanicke, W., and Frahm, J., 1994, Creatine deficiency in the brain: a new, treatable inborn error of metabolism. *Pediatr. Res.* **36:** 409–413.

Streijger, F., Jost, C.R., Oerlemans, F., Ellenbroek, B.A., Cools, A.R., Wieringa, B., and Van der Zee, C.E., 2004, Mice lacking the UbCKmit isoform of creatine kinase reveal slower spatial learning acquisition, diminished exploration and habituation, and reduced acoustic startle reflex responses. *Mol. Cell. Biochem.* **256–257:** 305–318.

Streijger, F., Oerlemans, F., Ellenbroek, B.A., Jost, C.R., Wieringa, B., and Van der Zee, C.E., 2005, Structural and behavioural consequences of double deficiency for creatine kinases BCK and UbCKmit. *Behav. Brain Res.* **157:** 219–234.

Tachikawa, M., Fukaya, M., Terasaki, T., Ohtsuki, S., and Watanabe, M., 2004, Distinct cellular expressions of creatine synthetic enzyme GAMT and creatine kinases uCK-Mi and CK-B suggest a novel neuron-glial relationship for brain energy homeostasis. *Eur. J. Neurosci.* **20:** 144–160.

Taylor, D.J., Bore, P.J., Styles, P., Gadian, D.G., and Radda, G.K., 1983, Bioenergetics of intact human muscle. A ^{31}P nuclear magnetic resonance study. *Mol. Biol. Med.* **1:** 77–94.

ten Hove, M., Lygate, C.A., Fischer, A., Schneider, J.E., Sang, A.E., Hulbert, K., Sebag-Montefiore, L., Watkins, H., Clarke, K., Isbrandt, D., Wallis, J., and Neubauer, S., 2005, Reduced inotropic reserve and increased susceptibility to cardiac ischemia/reperfusion injury in phosphocreatine-deficient guanidinoacetate-N-methyltransferase-knockout mice. *Circulation* **111:** 2477–2485.

ter Veld, F., Jeneson, J.A., and Nicolay, K., 2005, Mitochondrial affinity for ADP is twofold lower in creatine kinase knock-out muscles. Possible role in rescuing cellular energy homeostasis. *FEBS J.* **272:** 956–965.

Tkac, I., Henry, P.G., Andersen, P., Keene, C.D., Low, W.C., and Gruetter, R., 2004, Highly resolved *in vivo* ^{1}H NMR spectroscopy of the mouse brain at 9.4 T. *Magn. Reson. Med.* **52:** 478–484.

van Deursen, J., Heerschap, A., Oerlemans, F., Ruitenbeek, W., Jap, P., ter Laak, H., and Wieringa, B., 1993, Skeletal muscles of mice deficient in muscle creatine kinase lack burst activity. *Cell* **74:** 621–631.

van Deursen, J., Jap, P., Heerschap, A., ter Laak, H., Ruitenbeek, W., and Wieringa, B., 1994a, Effects of the creatine analogue beta-guanidinopropionic acid on skeletal muscles of mice deficient in muscle creatine kinase. *Biochim. Biophys. Acta* **1185:** 327–335.

van Deursen, J., Ruitenbeek, W., Heerschap, A., Jap, P., ter Laak, H., and Wieringa, B., 1994b, Creatine kinase (CK) in skeletal muscle energy metabolism: a study of mouse mutants with graded reduction in muscle CK expression. *Proc. Natl. Acad. Sci. USA* **91:** 9091–9095.

van Dorsten, F.A., Reese, T., Gellerich, J.F., van Echteld, C.J., Nederhoff, M.G., Muller, H.J., van Vliet, G., and Nicolay, K., 1997, Fluxes through cytosolic and mitochondrial creatine kinase, measured by P-31 NMR. *Mol. Cell. Biochem.* **174:** 33–42.

van Dorsten, F.A., Nederhoff, M.G., Nicolay, K., and Van Echteld, C.J., 1998, ^{31}P NMR studies of creatine kinase flux in M-creatine kinase-deficient mouse heart. *Am. J. Physiol.* **275:** H1191-H1199.

Veech, R.L., Lawson, J.W., Cornell, N.W., and Krebs, H.A., 1979, Cytosolic phosphorylation potential. *J. Biol. Chem.* **254:** 6538–6547.

Veksler, V.I., Kuznetsov, A.V., Anflous, K., Mateo, P., van Deursen, J., Wieringa, B., and Ventura-Clapier, R., 1995, Muscle creatine kinase-deficient mice. II. Cardiac and skeletal muscles exhibit tissue-specific adaptation of the mitochondrial function. *J. Biol. Chem.* **270:** 19921–19929.

von Figura, K., Hanefeld, F., Isbrandt, D., and Stöckler-Ipsiroglu, S., 2001, Guanidinoacetate methyltransferase deficiency. In: Scriver, C.R., Beaudet, A.L., Sly, W.S., Valle, D., and Childs, V., eds., The Metabolic and Molecular Bases of Inherited Disease, Chapter 84, pp. 3695–3724, McGraw-Hill, USA.

Wallimann, T., 1994, Bioenergetics. Dissecting the role of creatine kinase. *Curr. Biol.* **4:** 42–46.

Wallimann, T., Wyss, M., Brdiczka, D., Nicolay, K., and Eppenberger, H.M., 1992, Intracellular compartmentation, structure and function of creatine kinase isoenzymes in tissues with high and fluctuating energy demands: the phosphocreatine circuit for cellular energy homeostatis. *Biochem. J.* **281:** 21–40.

Walter, G., Barton, E.R., and Sweeney, H.L., 2000, Noninvasive measurement of gene expression in skeletal muscle. *Proc. Natl. Acad. Sci. USA* **97:** 5151–5155.

Wiseman, R.W., and Kushmerick, M.J., 1995, Creatine kinase equilibration follows solution thermodynamics in skeletal muscle. ^{31}P NMR studies using creatine analogs. *J. Biol. Chem.* **270:** 12428–12438.

Wyss, M., and Kaddurah-Daouk, R., 2000, Creatine and creatinine metabolism. *Physiol. Rev.* **80:** 1107–1213.

Wyss, M., Smeitink, J., Wevers, R.A., and Wallimann, T., 1992, Mitochondrial creatine kinase: a key enzyme of aerobic energy metabolism. *Biochim. Biophys. Acta* **1102:** 119–166.

Zeleznikar, R.J., Dzeja, P.P., and Goldberg, N.D., 1995, Adenylate kinase-catalyzed phosphoryl transfer couples ATP utilization with its generation by glycolysis in intact muscle. *J. Biol. Chem.* **270:** 7311–7319.

CHAPTER 8

CEREBRAL CREATINE DEFICIENCY SYNDROMES: CLINICAL ASPECTS, TREATMENT AND PATHOPHYSIOLOGY

SYLVIA STOCKLER[1], PETER W. SCHUTZ[2] AND GAJJA S. SALOMONS[3]

[1] *Department of Pediatrics, University of British Columbia, Division of Biochemical Diseases, British Columbia Children's Hospital, Vancouver, B.C., V6H 3V4, Canada*

[2] *Departments of Pediatrics, Pathology and Laboratory Medicine, University of British Columbia, Division of Biochemical Diseases, British Columbia Children's Hospital, Vancouver, B.C., V6H 3V4, Canada*

[3] *Department of Clinical Chemistry, Metabolic Unit, VU University Medical Center, Amsterdam, The Netherlands*

Abstract: Cerebral creatine deficiency syndromes (CCDSs) are a group of inborn errors of creatine metabolism comprising two autosomal recessive disorders that affect the biosynthesis of creatine – i.e. arginine:glycine amidinotransferase deficiency (AGAT; MIM 602360) and guanidinoacetate methyltransferase deficiency (GAMT; MIM 601240) – and an X-linked defect that affects the creatine transporter, SLC6A8 deficiency (SLC6A8; MIM 300036). The biochemical hallmarks of these disorders include cerebral creatine deficiency as detected *in vivo* by ^{1}H magnetic resonance spectroscopy (MRS) of the brain, and specific disturbances in metabolites of creatine metabolism in body fluids. In urine and plasma, abnormal guanidinoacetic acid (GAA) levels are found in AGAT deficiency (reduced GAA) and in GAMT deficiency (increased GAA). In urine of males with SLC6A8 deficiency, an increased creatine/creatinine ratio is detected. The common clinical presentation in CCDS includes mental retardation, expressive speech and language delay, autistic like behaviour and epilepsy. Treatment of the creatine biosynthesis defects has yielded clinical improvement, while for creatine transporter deficiency, successful treatment strategies still need to be discovered. CCDSs may be responsible for a considerable fraction of children and adults affected with mental retardation of unknown etiology. Thus, screening for this group of disorders should be included in the differential diagnosis of this population. In this review, also the importance of CCDSs for the unravelling of the (patho)physiology of cerebral creatine metabolism is discussed

G.S. Salomons and M. Wyss (eds.), Creatine and Creatine Kinase in Health and Disease, 149–166.

1. INTRODUCTION

Over the last decade, a novel group of inborn errors affecting proteins involved in creatine biosynthesis and its transport (Figure 1) has been identified, the cerebral creatine deficiency syndromes (CCDSs). Creatine is synthesized in a two-step process: 1) transfer of the amidino group from arginine to glycine, yielding guanidinoacetic acid (GAA) and ornithine. This reaction is catalyzed by L-arginine:glycine amidinotransferase (AGAT); 2) transfer of a methyl group from *S*-adenosylmethionine to GAA, yielding creatine and *S*-adenosylhomocysteine. This reaction is catalyzed by *S*-adenosyl-L-methionine:*N*-guanidinoacetate methyltransferase (GAMT) (Walker, 1979; Wyss and Kaddurah-Daouk, 2000). Creatine synthesis primarily occurs in the kidney and pancreas which have high AGAT activity, and in liver which has high GAMT activity. From these organs and from nutritional sources (e.g. meat and fish), creatine is distributed via the bloodstream to the organs of usage – mainly muscle and brain – and is taken up into these tissues by a Na^+- and Cl^--dependent creatine transporter (SLC6A8). The rate-limiting step in creatine biosynthesis is catalyzed by the AGAT enzyme. AGAT activity is repressed by high creatine concentrations at the pretranslational level. Additional allosteric inhibition is effected by high ornithine concentrations (Wyss and Kaddurah-Daouk, 2000). *In vitro*, GAMT activity is inhibited allosterically by high *S*-adenosylhomocysteine concentrations. However, no *in vivo* regulatory mechanism is known for GAMT activity. Understanding the regulation of the creatine transporter by investigating its expression and its activity is of significant importance, especially in the development or improvement of treatment strategies for CCDSs. Unfortunately, there is only limited information available so far on this topic, and particularly on creatine transporter regulation in brain. Early experiments suggested that creatine uptake is down-regulated in cultured rat myoblasts in the presence of high levels of extracellular creatine (Loike *et al.*, 1988). However, this has not been proven at the protein level due to the lack of specific antibodies against the creatine transporter (Speer *et al.*, 2004). In human skeletal muscle, creatine-monohydrate supplementation resulted in increased muscular creatine content without affecting the *SLC6A8* mRNA levels (Tarnopolsky *et al.*, 2003). Various factors may be involved in the activation and regulation of the creatine transporter, including signal transduction proteins, hormones and nutrition as discussed in chapter 6 of this book (Christie, 2007). Intracellularly, creatine is reversibly converted into the high-energy compound phosphocreatine by the action of creatine kinase (CK). Cytosolic CK transphosphorylates glycolytically generated ATP to phosphocreatine, or it uses phosphocreatine of either cytosolic or mitochondrial origin to regenerate ATP in the vicinity of energy-expending ATPases. Mitochondrial CK catalyses phosphate transfer from mitochondrial ATP to creatine in the mitochondrial intermembrane space, in a concerted manner with the ATP/ADP translocator of the inner mitochondrial membrane. The reversible transfer of high-energy phosphate groups to creatine as storage and carrier vehicle facilitates intracellular delivery of high-energy phosphates and provides additional energy resources during peak energy demands (Wyss and Kaddurah-Daouk, 2000).

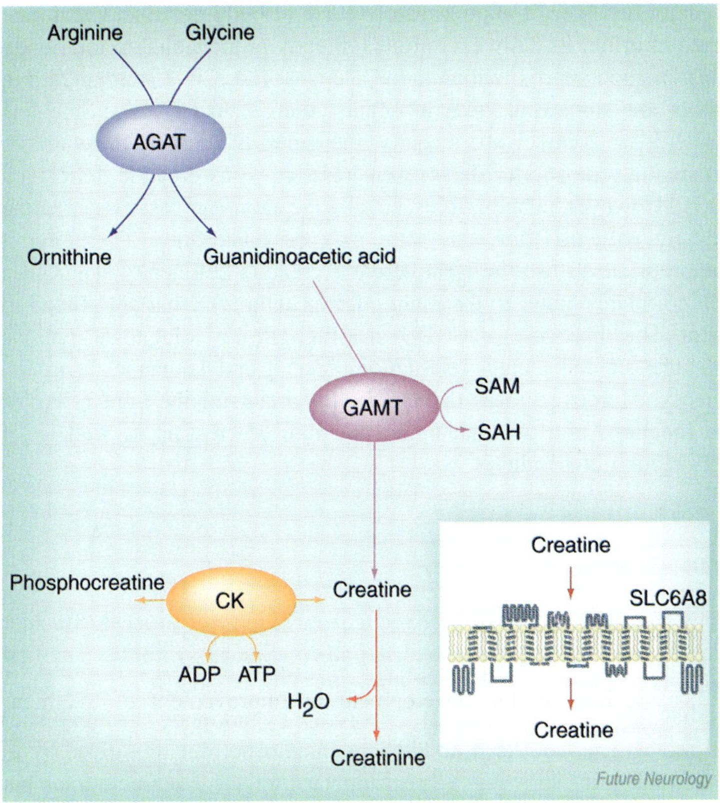

Figure 1. Schematic representation of creatine metabolism and transport, illustrating the proteins involved in the cerebral creatine deficiency syndromes. Creatine biosynthesis is a two-step process: L-arginine:glycine amidinotransferase (AGAT) catalyses the formation of guanidinoacetic acid from the amino acids arginine and glycine. If AGAT is impaired due to the presence of homozygous or compound heterozygous mutations, patients are affected with AGAT deficiency (MIM 602360). The second step involves the methylation of guanidinoacetic acid by *S*-adenosyl-L-methionine:*N*-guanidinoacetate methyltransferase (GAMT), which results in the formation of creatine and *S*-adenosylhomocysteine (SAH). *S*-Adenosylmethionine (SAM) functions as the donor of the methyl group. In case GAMT is impaired due to the presence of homozygous or compound heterozygous mutations, patients are affected with GAMT deficiency (MIM 601240). Creatine is distributed via the bloodstream and is subsequently taken up by cells via the creatine transporter (SLC6A8). The X-linked form of CCDSs – SLC6A8 deficiency (SLC6A8; MIM 300036) – is caused by hemizygous mutations in the *SLC6A8* gene in males, resulting in impaired creatine uptake. Due to skewed X-inactivation, the presence of a heterozygous *SLC6A8* mutation in females is associated with clinical symptoms of varying degrees. Creatine kinase (CK) catalyzes the (de)phosphorylation of creatine and phosphocreatine during ATP synthesis and usage. Reproduced from Almeida *et al.* (2006) with kind permission of Future Medicine Ltd.

Creatine and phosphocreatine are non-enzymatically converted into creatinine, with a constant daily turnover of approximately 1.5% of body creatine. Creatinine is mainly excreted in urine. Its daily excretion is directly proportional to total body creatine and, thus, in good approximation to muscle mass (i.e., 20–25 mg/kg/24 h in children and adults, and somewhat lower in infants younger than 2 years; Stöckler-Ipsiroglu and Salomons, 2006). Until recently, the group of creatine biosynthesis defects and the creatine transporter defect were referred to as creatine deficiency syndromes (CDSs). However, in body fluids, no creatine deficiency exists in creatine transporter deficient patients; thus, this term may be misleading. Therefore, it may be more appropriate to use the term cerebral creatine deficiency syndromes (CCDSs), which correlates better to the main clinical hallmarks that are related to CNS involvement. The discovery of CCDSs has brought new diagnostic options in patients with unexplained mental retardation, speech and language disorders, autism and epilepsy. Moreover, these defects are important for the unravelling of the physiologic functions and pharmacologic potential of creatine as well as of intermediates of creatine biosynthesis.

2. GAMT DEFICIENCY

GAMT deficiency was the first CCDS to be recognized in humans in 1994 (Stöckler *et al.*, 1994). The disorder affects the second step in creatine biosynthesis. The first patient, a German boy, seemed to develop normally until 4 months of age, when he was noted to have developmental arrest, hypotonia, hyperkinetic (hemiballistic) extra-pyramidal movements and head nodding. A diagnostic hint was given by *in vivo* ^{1}H MRS of the brain, showing a spectrum lacking the creatine signal but positive for a signal which finally was identified as GAA. The combination of deficient creatine with a high GAA signal was suggestive of GAMT deficiency. Subsequent studies confirmed the absence of GAMT activity in liver and identified the pathogenic mutations in the *GAMT* gene (Stöckler *et al.*, 1996b). The *GAMT* gene has been mapped to chromosome 15q15.3, thus confirming that GAMT deficiency is an autosomal recessive disorder. Oral supplementation with creatine monohydrate resulted in a rise of cerebral creatine levels up to 50% of normal within the first 3 months of treatment. After 24 months of treatment, cerebral creatine levels had reached up to 90% of normal levels. The restoration of cerebral creatine levels was accompanied by significant clinical improvement: extra-pyramidal movement disorder and head nodding resolved, and the patient was able to walk at the age of 4 years. The EEG, which had shown theta delta background activity with multifocal spikes, normalised, as did the bilateral pathologic signal intensities in the globus pallidus (Stöckler *et al.*, 1996a). Disappointingly, the long-term outcome revealed that the severe mental retardation remained, and there was no improvement in speech development and progressive autistic and aggressive behaviour disorder. One reason for the incomplete clinical improvement seems to be the accumulation of GAA which is not normalised by creatine supplementation (Stöckler *et al.*, 1997).

Shortly after the description of the first patient with GAMT deficiency, a second patient was described (Schulze *et al.*, 1998). In this patient, the authors could show convincingly that dietary restriction of arginine, the immediate precursor of GAA, results in significant reduction in GAA accumulation (Schulze *et al.*, 1998). Since the description of the first patient with GAMT deficiency, approximately 29 patients have been diagnosed worldwide (Almeida *et al.*, 2007; Mercimek-Mahmutoglu *et al.*, 2006). An overview of 27 cases shows that mental retardation and epilepsy are the most consistent clinical features (Mercimek-Mahmutoglu *et al.*, 2006). The severity may range from mild to severe mental retardation and from occasional to drug-resistant seizures. Additional extra-pyramidal movement disorder and pathologic signal intensities in the basal ganglia on brain MRI are observed in the most severe cases. The clinical onset of the disease is between an age of four months to three years. Treatment with creatine monohydrate consistently resulted in improvement of epilepsy and movement disorder. However, this treatment did not have an impact on the intellectual deficit of the patients. A few patients were treated with a combination of creatine monohydrate supplementation and dietary arginine restriction (i.e., protein restriction), resulting in reduction in GAA accumulation, but in all patients taking this diet, complete normalisation of GAA levels could not be achieved. On such a case-by-case basis, final conclusions about the clinical effects of additional dietary arginine restriction cannot be made. However, as GAA is considered neurotoxic, it is now a general consensus to treat all patients with the combined approach and to aim at reduction in GAA accumulation as much as possible. Besides dietary arginine restriction, ornithine and sodium benzoate supplementation were proposed as strategies for reducing GAA accumulation. Experience with the only patient diagnosed at birth and treated at a pre-symptomatic stage of the disease suggests that early treatment might widely prevent neurological manifestations (Schulze *et al.*, 2006; see also chapter 9; Schulze and Battini, 2007).

So far, 15 different GAMT mutations have been described (Almeida *et al.*, 2007; Mercimek-Mahmutoglu *et al.*, 2006), including nonsense and missense mutations, splice errors, insertions, deletions and frameshifts. There is no evidence for a hotspot region, although certain mutations appear to occur more frequently than others (c.59G > A; p.Trp20Ser and a mutation that results in erroneous splicing: c.327G > A). A relatively high carrier rate for the c.59G > A mutation has been detected in certain areas of Portugal, and 10 out of the 29 patients affected with GAMT deficiency are of Portuguese origin (Almeida *et al.*, 2007). Comparison of the type and severity of clinical presentation of the patients who are homozygous for any one of these mutations does not show any genotype-phenotype correlation yet (Mercimek-Mahmutoglu *et al.*, 2006).

3. AGAT DEFICIENCY

AGAT deficiency affects the first enzyme in creatine biosynthesis. This disorder was first described in two Italian sisters with unspecific developmental and speech delay, and occasional (fever-induced) seizures in one of them. *In vivo*

^{1}H MRS of the brain revealed creatine deficiency, but opposite to GAMT deficiency, GAA levels were low in body fluids. This combination pointed to a defect in GAA biosynthesis, and AGAT deficiency was finally confirmed by deficient enzyme activity and identification of a homozygous nonsense mutation in the *AGAT* gene (Item *et al.*, 2001). The *AGAT* gene has been mapped to chromosome 19p13.3, thus confirming that AGAT deficiency is an autosomal recessive disorder. Supplementation with creatine monohydrate resulted in almost complete normalisation of cerebral creatine levels and a catch-up in psychomental development (Bianchi *et al.*, 2000). Although, after 6 years of treatment, the 13- and 11-year-old sisters still have a moderate intellectual deficit, there was improvement and stabilisation of their condition. In a younger sibling, AGAT deficiency was diagnosed prenatally and creatine supplementation was started at the age of 4 months. This patient revealed normal development at the age of 18 months, while his sisters at this age already showed signs of retardation (Battini *et al.*, 2006; see also chapter 9; Schulze and Battini, 2007). Observations on this limited number of patients (n = 3) suggest that, compared to GAMT deficiency, AGAT deficiency is associated with a milder phenotype, and/or with a better response to creatine supplementation. More importantly, early neurological sequelae may be prevented by early treatment. Currently, five patients are known worldwide with this condition, of whom four are from one Italian family. Only two mutations have been found so far: a nonsense mutation (in the Italian family; Battini *et al.*, 2002; Item *et al.*, 2001), and a splice error mutation (Almeida *et al.*, 2006a). Both mutations are associated with impaired AGAT activity in lymphoblasts.

4. CREATINE TRANSPORTER (SLC6A8) DEFICIENCY

4.1. Creatine Transporter (SLC6A8) Deficiency in Males

SLC6A8 deficiency – a defect of the creatine transporter – was originally described in a 6-year-old American boy with a history of central hypotonia, one episode of status epilepticus, multifocal epileptiform discharges in his EEG recordings, delayed speech and language development, and creatine deficiency in the brain. In contrast to GAMT and AGAT deficiency, GAA levels were normal and oral creatine supplementation was not effective in restoring cerebral creatine levels (Cecil *et al.*, 2001). Due to these findings, and the X-linked pattern of inheritance in the family, cerebral creatine transporter deficiency was suspected. This hypothesis was confirmed by demonstration of deficient creatine uptake in fibroblasts and by identification of a hemizygous mutation in the *SLC6A8* gene mapped to the X chromosome (Salomons *et al.*, 2001).

Since the first description of SLC6A8 deficiency in 2001, more than 150 patients from 60 families have been diagnosed. In boys as well as adult males (age range 2–66 years), the diagnosis of SLC6A8 deficiency has been ascertained. The main biochemical hallmark is an increased urinary creatine to creatinine ratio in males (Stöckler-Ipsiroglu and Salomons, 2006). The clinical phenotype of males with SLC6A8 deficiency varies from mild to severe mental retardation associated with speech and language delay, expressive language disorder, epileptic seizures, behaviour disorders and gastrointestinal problems in adult patients (Kleefstra *et al.*, 2005). A neuropsychological profile was obtained for four affected Dutch boys from two unrelated families and revealed hyperactive impulsive attention deficit and a semantic-pragmatic language disorder with oral dyspraxia (Mancini *et al.*, 2005). Growth retardation, dysmorphic features, microcephaly, and brain atrophy have been described as accompanying structural characteristics in some, but not all patients (Kleefstra *et al.*, 2005). Brain atrophy may become more marked during the course of the disease, as may the behavioural symptoms (Degrauw *et al.*, 2002). Muscular hypotrophy and hypotonia and secondary mitochondriopathy have been reported as additional features in selected SLC6A8 deficient patients (Anselm *et al.*, 2006).

According to numerous reports, SLC6A8 deficiency is responsible for a substantial number of males with mental retardation of unknown cause. In a recently studied cohort of males presenting with risk symptoms, the urinary creatine to creatinine ratio was used to analyze SLC6A8 frequency. The frequency of SLC6A8 deficiency was confirmed by molecular and functional studies (i.e., creatine uptake in fibroblasts) to be 2.1% (2 out of 96 males; Mercimek-Mahmutoglu *et al.*, 2007). This frequency is consistent with results from other studies. SLC6A8 deficiency was detected in 5.4% (2 out of 37) and 2.1% (6 out of 288) of males with non-syndromic X-linked mental retardation (Lion-Francois *et al.*, 2006; Rosenberg *et al.*, 2004), in 2.2% (2 out of 92) of males with global developmental delay (IQ < 70) (Newmeyer *et al.*, 2005), as well as in 0.8% (4 out of 478) and 3.5% (4 out of 114) of males with mental retardation of unknown causes and negative for fragile X syndrome (Clark *et al.*, 2006; Lion-Francois *et al.*, 2006).

SLC6A8 deficiency appears not to be treatable by any of the approaches described above. Treatment of both males and females affected with SLC6A8 deficiency with creatine monohydrate has not proven to be successful (Anselm *et al.*, 2006; Poo-Arguelles *et al.*, 2006; Stöckler-Ipsiroglu and Salomons, 2006). Currently, supplementation with high doses of arginine and glycine, which are the primary substrates for creatine biosynthesis, combined with high doses of creatine monohydrate is being investigated. The rationale for this protocol is based on an increased cerebral uptake of both amino acids with the aim to enhance intracerebral creatine synthesis (Gracia M. Mancini, Marjo S. van der Knaap, and Gajja S. Salomons, unpublished data). In addition, alternative strategies may be developed that facilitate creatine transport into the brain either by modified transport via carrier molecules (e.g., peptides) or by supplementation with suitable creatine analogs.

Since the first description of SLC6A8 deficiency in 2001, 24 families have been described at the molecular level, and 20 different mutations have been identified. Four mutations have been identified in at least two unrelated families (c.321_323delCTT;p.Phe107del, c.1169C > T;p.Pro390Leu, c.1222_1224delTTC; p.Phe408del; c.1631C > T;p.Pro544Leu; Almeida *et al.*, 2006a; Stöckler-Ipsiroglu and Salomons, 2006). The most frequently identified type of mutations in *SLC6A8* are missense mutations (i.e., single amino acid substitutions) and single amino acid deletions located throughout the gene. Especially in case of novel missense mutations, the diagnosis should be confirmed by investigating creatine uptake in cultured fibroblasts or by ^{1}H-MRS of the brain. The pathogenicity of the mutation can be studied by overexpression of the mutant allele in SLC6A8-deficient cells, as recently described by Rosenberg *et al.* (2007). Approximately 10% of the mutations were shown to occur *de novo*, indicating that the mothers of children with such a *de novo* mutation will not have any clinical phenotype. Therefore, also in sporadic mental retardation, SLC6A8 deficiency should be considered in the differential diagnosis. Analyses of genotype-phenotype correlation have not been made so far. Characterization of a variety of missense mutations and one-amino acid deletions may provide insights into the structure and function of the protein.

4.2. Creatine Transporter (SLC6A8) Deficiency in Females

In addition to affected males who have a hemizygous mutation in *SLC6A8*, at least 80 female relatives as well as three index girls with a heterozygous mutation (carriers) have been identified so far; however, for the majority of them, no neuropsychological profiles have been obtained. Clinical symptoms of the index girls included learning disabilities and therapy-resistant epilepsy. Based on data on the first described females (Degrauw *et al.*, 2002, 2003), it is expected that approximately 50% of heterozygous females have learning and behavioural problems. Skewed X-inactivation may cause pronounced clinical manifestations in these cases that are similar to the male phenotype, whereas others remain seemingly asymptomatic. This phenomenon also applies to the biochemical hallmarks of the disease, which makes screening for a creatine transporter defect in females more difficult: 1) the urinary creatine to creatinine ratio is usually not informative; 2) the creatine signal in brain may be reduced only slightly; and 3) uptake studies in fibroblasts are also not informative and, thus, cannot be used as primary screening. Therefore, a combination of diagnostic tests may be needed, including molecular analysis of the *SLC6A8* gene. Treatment data is available for only one heterozygous female patient with learning disability and mildly decreased creatine concentration as revealed by brain ^{1}H MRS. This data showed mild improvement on neuropsychological testing after 18 weeks of treatment with creatine monohydrate (250–750 mg/kg/day; Cecil *et al.*, 2001).

5. BIOCHEMICAL PATHOLOGY AND LABORATORY DIAGNOSIS

CCDSs are characterized by an almost complete lack of creatine in the brain, which is shown *in vivo* by brain ^{1}H MRS. It should be noted that in females with a heterozygous mutation in the *SLC6A8* gene, a milder reduction in cerebral creatine is usually detected. However, cerebral creatine can vary in heterozygous females from being absent to normal levels, depending on the skewing in X-inactivation. Interestingly, in CCDS patients, creatine deficiency is much less severe in muscle than in brain. Although creatine levels in skeletal muscle have not been investigated extensively in CCDS patients and insights are based on case studies only, creatine content of skeletal muscle was measured *in vivo* by proton MRS and *in vitro* in a muscle biopsy of a SLC6A8-deficient patient (Pyne-Geithman *et al.*, 2004). This case report revealed normal creatine levels, suggesting that endogenous creatine biosynthesis is able to maintain proper creatine levels, and/or that another creatine transporter is able to take up creatine. It is worthy of note that phosphorus MRS of skeletal muscle of a GAMT-deficient patient showed reduced phosphocreatine levels (Schulze *et al.*, 2003). Alternatively, the high levels of GAA in GAMT deficiency may interfere with sufficient uptake of creatine via the creatine transporter. This data is in agreement with GAMT knock-out mice that show reduced creatine levels in muscle and high levels of GAA (Renema *et al.*, 2003; Schmidt *et al.*, 2004). No data are available on creatine levels in heart muscle of affected patients. Clinically, however, there is no evidence of cardiac dysfunction in the patients who underwent cardiologic evaluation.

GAA is the main intermediary product in creatine biosynthesis. In GAMT deficiency, GAA accumulates 2- to 30-fold in urine and plasma, and about 200-fold in CSF (Mercimek-Mahmutoglu *et al.*, 2006). The high GAA concentrations in GAMT deficiency are due to at least two mechanisms: firstly, defective GAMT activity causes an accumulation of GAA *per se*. Secondly, creatine deficiency and consequent absence of AGAT repression cause an increased rate of GAA biosynthesis. Creatine's regulatory feedback mechanism on GAA synthesis is confirmed in GAMT-deficient patients who show a significant decrease (but not normalisation) in GAA concentrations as a response to oral creatine supplementation (Mercimek-Mahmutoglu *et al.*, 2006; Stöckler *et al.*, 1997). In AGAT deficiency, GAA is low as a result of defective synthesis (Battini *et al.*, 2002; Item *et al.*, 2001). In SLC6A8 deficiency, the main site of metabolic disturbance is the intracellular creatine pool, whereas plasma GAA and creatine levels are in the normal range. However, the ratio of creatine to creatinine concentrations in urine of affected males is increased (Stöckler-Ipsiroglu and Salomons, 2006). This could be the result of an elevated creatine excretion due to impaired tubular reabsorption and a reduced creatinine excretion due to reduced intracellular creatine pools (e.g. in muscle and brain). Since the urinary creatine to creatinine ratio is usually only mildly increased, it is important to realize that there is an inverse relationship between this ratio and age (Almeida *et al.*, 2004). Moreover, as mentioned above, the creatine to creatinine ratio is usually not increased in urine of females with a heterozygous mutation. The biochemical profile of the individual CCDS determines its laboratory diagnosis. In practical terms, this means that determination of urinary GAA concentrations allows to differentiate between AGAT (low concentration) and GAMT deficiency (high concentration). The determination of the urinary creatine/creatinine ratio has been found to be a useful diagnostic marker of SLC6A8 deficiency in males. Laboratory diagnosis is confirmed by demonstration of the enzyme or transporter activity in either fibroblasts or lymphoblasts, and by DNA analysis of the relevant gene. Recently, we reviewed the biochemical changes in CCDSs (Stöckler-Ipsiroglu and Salomons, 2006).

6. PATHOPHYSIOLOGICAL CONSIDERATIONS

Present evidence suggests that each of the CCDSs has potential effects on several aspects of creatine physiology, including the intermediates of creatine biosynthesis and their biochemical reactions, the transfer of creatine from biosynthesis to usage compartments, intracellular creatine phosphorylation cycles, and all cellular functions normally supported by creatine in developing and mature brain.

6.1. Disrupted Creatine Transport

The concept of an organ-specific transfer of creatine from sites of biosynthesis in liver, pancreas and kidney to sites of usage in brain and muscle must be refined, at least with regard to the brain. Immunohistochemical and functional studies have established the presence of AGAT and GAMT in all types of brain cells *in vivo* (Braissant *et al.*, 2001) and the capacity of cultured astrocytes to synthesise creatine (Dringen *et al.*, 1998). This suggests a dual origin of cerebral creatine: from intracerebral biosynthesis and from transport across the blood-brain barrier. The creatine transporter, SLC6A8, is found in capillary endothelial cells, neurons, and oligodendrocytes (Braissant *et al.*, 2001). Based on these findings, creatine is thought to be transferred between different cell types within the brain, for instance from astrocytes to neurons (Tachikawa *et al.*, 2004), but also among the same cell type (Braissant *et al.*, 2005; see also chapters 4 and 5; Braissant *et al.*, 2007; Tachikawa *et al.*, 2007). Creatine transport across the blood-brain barrier and its transfer within the brain are the basis of our understanding of the pattern of cerebral creatine concentrations upon replacement therapy in the creatine biosynthesis deficiency syndromes, and of our understanding of the pathogenesis of SLC6A8 deficiency. Oral supplementation with creatine in patients with AGAT and GAMT deficiency results in a biphasic restoration of the cerebral creatine pool, with initially fast creatine accumulation followed by a slow, but continuous increase over several months to 80–90% of normal levels. Increasing the dosage or changing the time intervals between creatine administration do not result in any further increase in brain creatine levels (Stöckler *et al.*, 1997). The biphasic and incomplete restoration of the cerebral creatine pool suggests different intracerebral compartments of creatine metabolism, which may reflect creatine pools predominantly supplied by intracerebral creatine biosynthesis as opposed to those predominantly supplied by blood. In SLC6A8 deficiency, cerebral creatine levels are very low or even undetectable in *in vivo* ^{1}H MRS measurements. This indicates that the fraction of intracerebral creatine biosynthesis under these conditions is rather low. SLC6A8 may be indispensable for the intracerebral redistribution of locally synthesised GAA and/or creatine. GAA is a known competitive inhibitor of the creatine transporter and is thought to be taken up via the SLC6A8 transporter (Wyss and Kaddurah-Daouk, 2000). Therefore, due to the absence of a functional creatine transporter, GAA uptake might not be possible either. This may interfere with GAA methylation and, thus, creatine biosynthesis. In addition, low creatine levels may also result from disrupted intracerebral transport of creatine.

6.2. Intracellular Creatine Depletion in the Developing Brain

An important role for creatine during embryogenesis is suggested by the appearance of *AGAT* and *SLC6A8* transcripts in rat brain from as early as embryonal day 12.5 (Braissant *et al.*, 2005). Cell culture experiments on developing neurons suggest that creatine is needed for regular axonal growth (Braissant *et al.*, 2002; Wang *et al.*, 1998). Taken together, these findings indicate that creatine deficiency during brain development may cause impoverished axonal networks and reduced synaptic density. Microstructural anomalies of dendrites and neuronal networks are thought to be a morphological hallmark of mental retardation in Down syndrome and fragile X syndrome (Volpe, 2001). These considerations support the view that microstructural defects may be responsible, at least in part, for those clinical symptoms that are not reversible upon restoration of creatine content after a CCDS has manifested itself. Detailed morphological studies will be necessary to evaluate this hypothesis. Should these mechanisms indeed contribute to the clinical phenotype, pre-symptomatic treatment may be obligatory.

6.3. Intracellular Creatine Depletion in the Mature Brain

Creatine deficiency may cause a disruption of cellular energy homeostasis. The absence of creatine and phosphocreatine in the cytosol is expected to reduce the buffering capacity for peak ATP demands and the transport capacity for high-energy phosphate compounds. Since phosphocreatine is in part responsible for high-energy phosphate transport from mitochondria to ATPases elsewhere in the cell, its lack may resemble mitochondrial dysfunction. Its lack may also result in chronic energy failure. Such energy failure may explain pathologic signal intensities in the globus pallidus as observed in severe cases of GAMT deficiency (Mercimek-Mahmutoglu *et al.*, 2006). Comparable changes are found in patients with mitochondrial encephalopathies (e.g., cytochrome C oxidase deficiency), and in other cases of disrupted maintenance of cerebral energy state such as in hypoxic-ischemic events in newborns and in carbon monoxide intoxication. Recent observations suggest that oxidative phosphorylation may be impaired in patients affected by GAMT or SLC6A8 deficiency, as some patients with these conditions present with signs of myopathy and, at the biochemical level, mitochondrial dysfunction (e.g., lactic acidemia, or decreased ATP production in muscle biopsy samples; Anselm *et al.*, 2006; De Vries *et al.*, 2005; Stöckler *et al.*, 1996a).

Creatine has a protective effect on neuronal survival. This has been shown in neuronal cultures during exposure to high glutamate concentrations and in animal models of various neurodegenerative diseases (Brewer and Wallimann, 2000; Klivenyi *et al.*, 1999; Matthews *et al.*, 1998, 1999). The mechanisms by which intracellular creatine reduces excitotoxic neuronal death have been studied, including effects on the mitochondrial permeability transition (Dolder *et al.*, 2003) and on apoptotic signalling (Juravleva *et al.*, 2005). However, no definitive conclusions can yet be drawn. Regardless of the cellular mediators, if the presence of creatine confers advantages for neuronal survival, its absence may facilitate cellular apoptotic pathways. This could result in increased apoptotic rates during the regressive phase of brain development, or in an enhanced apoptotic response to excitotoxic or oxidative stress. The latter may contribute to the brain atrophy observed in some SLC6A8 patients.

6.4. Interstitial and Intracellular Accumulation of Guanidino Compounds and Other Metabolites

Accumulation of guanidino compounds in tissues and bodily fluids is a characteristic of GAMT deficiency. GAA is most prominent among these, but changes in other intermediates of guanidino compound metabolism have also been described and may contribute to the neurotoxic load. GAA displays epileptogenic potential and is held to be the main cause of intractable seizures which occur in about one third of GAMT patients (Mercimek-Mahmutoglu *et al.*, 2006). A clear correlation between seizure activity and GAA levels in plasma was demonstrated in one patient in whom intractable seizures were independent of cerebral creatine status, but responded promptly to a reduction in plasma GAA by dietary arginine restriction (which is the rate-limiting substrate for GAA synthesis; Schulze *et al.*, 1998). On a molecular level, GAA interacts as partial agonist with $GABA_A$ receptors at pathophysiologically relevant concentrations (Neu *et al.*, 2002). Based on these findings, it has been speculated that GAA may also be responsible in part for the development of the extra-pyramidal movement disorder, which manifests in about 50% of GAMT-deficient patients, primarily as athetosis, chorea, and/or ataxia (Mercimek-Mahmutoglu *et al.*, 2006).

Changes in guanidino compounds secondary to GAA accumulation may also play a role in the development of neurological manifestations in GAMT deficiency. Analyses of samples from a GAMT-deficient patient (Stöckler *et al.*, 1997) and of brain homogenates from a GAMT knock-out mouse model (Torremans *et al.*, 2005) revealed additional accumulation of β-guanidinosuccinate, β-guanidinopropionate, and γ-guanidinobutyrate in the brain. Guanidinosuccinate may cause seizures and activate NMDA receptors (De Deyn *et al.*, 2001), β-guanidinopropionate is a competitive inhibitor of the creatine transporter, SLC6A8 (Peral *et al.*, 2002), whereas convulsive properties have been ascribed to γ-guanidinobutyrate (Jinnai *et al.*, 1966). The relative contribution of these substances to the neurological symptoms in GAMT deficiency is unclear, and synergistic effects cannot be excluded at the present stage.

Creatine synthesis requires a high percentage of total body 'labile' methyl groups. Therefore, one might expect that in GAMT deficiency, *S*-adenosylmethionine accumulates, which acts as a methyl donor for the GAMT reaction. Analysis of this compound in CSF from GAMT-deficient patients and in brain homogenates of the GAMT knock-out mouse model will answer the question whether changes in the methylation/remethylation pathway contribute to the pathophysiology of GAMT deficiency.

6.5. A Role for Creatine in Neurotransmission?

In stress-induced anxiety studies in chickens, creatine was reported to mediate an anxiolytic (sedative-hypnotic) effect as partial agonist at the central $GABA_A$ receptors (Koga *et al.*, 2005). These findings suggest that, *in vivo*, creatine modulates GABAergic neurotransmission throughout the brain. However, it should be noted that conflicting data regarding this effect of creatine has been published in *in vitro* studies on cultured cortex neurons of neonatal mice (Neu *et al.*, 2002). Nevertheless, the ubiquitous presence of creatine as well as of its biosynthesis enzymes in brain, including central neurons (Braissant *et al.*, 2005), and the possible effects at the $GABA_A$ receptors suggest a putative role of creatine as a neuromodulator in the brain. In agreement with this putative role, it was shown that [^{3}H]creatine is taken up into rat brain (neocortex) slices in a Na^+-dependent manner, likely mediated by the SLC6A8 transporter. Additionally, both [^{3}H]creatine and endogenous creatine were shown to be released upon electrical field stimulation from superfused neocortex slices (Almeida *et al.*, 2006b). This release was Na^+- and Ca^{2+}-dependent, indicating the involvement of an exocytotic release mechanism. Moreover, this electrically evoked creatine release appeared to depend not only on the activation of voltage-gated Na^+-channels (i.e., electrically evoked creatine release blocked by tetrodotoxin), but also on that of K^+-channels (i.e., electrically evoked creatine release strongly increased by 4-aminopyridine), demonstrating that the release is action potential dependent. Similar findings have previously been reported for radiolabelled classical neurotransmitters (e.g., Limberger *et al.*, 1986; Schoffelmeer *et al.*, 1981). These novel findings in brain may very well contribute to an explanation of the clinical phenotype in CCDSs, and await further studies.

7. PROSPECTS

The knowledge of the natural history of CCDSs is limited to the clinical presentations of a few patients only. Most of the known patients are children, but also a few young adults as well as a few elderly have been diagnosed. As these patients have been followed for a few years only, no definite conclusions can be drawn in terms of disease progression and life expectancy. Furthermore, only patients with the most severe clinical phenotypes may have been diagnosed so far, while milder phenotypes might still be under-recognised.

CCDSs are potentially treatable disorders. However, progress in understanding efficacy of treatment and development of new treatment strategies has been delayed due to the rareness of the single disorders. As with other rare inborn errors of metabolism, worldwide networks and orphan disease registries are needed to facilitate progress in understanding the natural history and in the development of strategies for treatment and prevention. Creatine deficiency syndromes should be investigated using these tools in the future.

Development of treatment strategies depends on the understanding of pathophysiology. Our understanding of pathophysiology and pathobiochemistry in CCDSs is still incomplete. The pathobiochemical actions and pathophysiological consequences of accumulation of GAA, GAA's role in brain function, and pharmacological inhibition of GAA's action in the brain need to be understood fundamentally in order to develop more effective treatment strategies for GAMT deficiency. The lack of neurological symptoms in the GAMT knock-out mouse model makes the exploration of this issue even more difficult (Torremans *et al.*, 2005). New concepts are also needed for understanding the pathogenesis of brain dysfunction in creatine transporter or AGAT deficiency. What is the pathogenetic impact of creatine deficiency? Knowledge about the regulation of creatine transporter activity and its interaction with other genes might provide promising targets for alternative treatment strategies such as pharmacological gene therapy.

The development of SLC6A8 knock-out mice will yield further insights into these questions. Better understanding of creatine's neuroprotective role may be another source of inspiration for the development of new treatment strategies. An AGAT knock-out mouse model would represent a biological system of creatine depletion and would be fundamental for understanding creatine's effects beyond its energy-buffering function. This would also allow investigations into a possible neuroprotective role of creatine which still has not been proven convincingly in humans. The final goal of diagnosis and treatment of CCDSs is the prevention of the clinical phenotype. Therefore, it is mandatory that CCDSs are part of the routine diagnostic work-up in patients with mental retardation and/or epilepsy, and determination of marker substances (urinary GAA and the urinary creatine to creatinine ratio) needs to be introduced in metabolic laboratory panels.

ACKNOWLEDGEMENTS

G.S. Salomons is supported by the Dutch Society for Scientific Research (ZonMW/NWO), VIDI grant number 917.56.349.

REFERENCES

Almeida, L.S., Rosenberg, E.H., Verhoeven, N.M., Jakobs, C., and Salomons, G.S., 2006a, Are cerebral creatine deficiency syndromes on the radar screen? *Future Neurol.* **1:** 637–649.

Almeida, L.S., Salomons, G.S., Hogenboom, F., Jakobs, C., and Schoffelmeer, A.N., 2006b, Exocytotic release of creatine in rat brain. *Synapse* **60:** 118–123.

Almeida, L.S., Verhoeven, N.M., Roos, B., Valongo, C., Cardoso, M.L., Vilarinho, L., Salomons, G.S., and Jakobs, C., 2004, Creatine and guanidinoacetate: diagnostic markers for inborn errors in creatine biosynthesis and transport. *Mol. Genet. Metab.* **82:** 214–219.

Almeida, L.S., Vilarinho, L., Darmin, P.S., Rosenberg, E.H., Martinez-Munoz, C., Jakobs, C., and Salomons, G.S., 2007, A prevalent pathogenic GAMT mutation (c.59G>C) in Portugal. *Mol. Genet. Metab.* **91:** 1–6.

Anselm, I.A., Alkuraya, F.S., Salomons, G.S., Jakobs, C., Fulton, A.B., Mazumdar, M., Rivkin, M., Frye, R., Poussaint, T.Y., and Marsden, D., 2006, X-linked creatine transporter defect: a report on two unrelated boys with a severe clinical phenotype. *J. Inherit. Metab. Dis.* **29:** 214–219.

Battini, R., Alessandri, M.G., Leuzzi, V., Moro, F., Tosetti, M., Bianchi, M.C., and Cioni, G., 2006, Arginine:glycine amidinotransferase (AGAT) deficiency in a newborn: early treatment can prevent phenotypic expression of the disease. *J. Pediatr.* **148:** 828–830.

Battini, R., Leuzzi, V., Carducci, C., Tosetti, M., Bianchi, M.C., Item, C.B., Stöckler-Ipsiroglu, S., and Cioni, G., 2002, Creatine depletion in a new case with AGAT deficiency: clinical and genetic study in a large pedigree. *Mol. Genet. Metab.* **77:** 326–331.

Bianchi, M.C., Tosetti, M., Fornai, F., Alessandri', M.G., Cipriani, P., De, V.G., and Canapicchi, R., 2000, Reversible brain creatine deficiency in two sisters with normal blood creatine level. *Ann. Neurol.* **47:** 511–513.

Braissant, O., Henry, H., Loup, M., Eilers, B., and Bachmann, C., 2001, Endogenous synthesis and transport of creatine in the rat brain: an in situ hybridization study. *Brain Res. Mol. Brain Res.* **86:** 193–201.

Braissant, O., Henry, H., Villard, A.M., Zurich, M.G., Loup, M., Eilers, B., Parlascino, G., Matter, E., Boulat, O., Honegger, P., and Bachmann, C., 2002, Ammonium-induced impairment of axonal growth is prevented through glial creatine. *J. Neurosci.* **22:** 9810–9820.

Braissant, O., Henry, H., Villard, A.M., Speer, O., Wallimann, T., and Bachmann, C., 2005, Creatine synthesis and transport during rat embryogenesis: spatiotemporal expression of AGAT, GAMT and CT1. *BMC Dev. Biol.* **5:** 9.

Braissant, O., Bachmann, C., and Henry, H., 2007, Expression and function of AGAT, GAMT and CT1 in the mammalian brain. *Subcell. Biochem.* **46:** 67–81.

Brewer, G.J., and Wallimann, T.W., 2000, Protective effect of the energy precursor creatine against toxicity of glutamate and beta-amyloid in rat hippocampal neurons. *J. Neurochem.* **74:** 1968–1978.

Cecil, K.M., Salomons, G.S., Ball, W.S., Jr., Wong, B., Chuck, G., Verhoeven, N.M., Jakobs, C., and Degrauw, T.J., 2001, Irreversible brain creatine deficiency with elevated serum and urine creatine: a creatine transporter defect? *Ann. Neurol.* **49:** 401–404.

Christie, D.L., 2007, Functional insights into the creatine transporter. *Subcell. Biochem.* **46:** 99–118.

Clark, A.J., Rosenberg, E.H., Almeida, L.S., Wood, T.C., Jakobs, C., Stevenson, R.E., Schwartz, C.E., and Salomons, G.S., 2006, X-linked creatine transporter (SLC6A8) mutations in about 1% of males with mental retardation of unknown etiology. *Hum. Genet.* **119:** 604–610.

De Deyn, P.P., D'Hooge, R., Van Bogaert, P.P., and Marescau, B., 2001, Endogenous guanidino compounds as uremic neurotoxins. *Kidney Int. Suppl.* **78:** S77–S83.

De Vries, M., Morava, E., Rodenburg, R., Kluijtmans, L., Verhoeven, N.M., Salomons, G.S., Jakobs, C., and Smeitink, J., 2005, Mitochondrial dysfunction in creatine biosynthesis disorder. *J. Inherit. Metab. Dis.* **28:** 227.

DeGrauw, T.J., Cecil, K.M., Byars, A.W., Salomons, G.S., Ball, W.S., and Jakobs, C., 2003, The clinical syndrome of creatine transporter deficiency. *Mol. Cell. Biochem.* **244:** 45–48.

DeGrauw, T.J., Salomons, G.S., Cecil, K.M., Chuck, G., Newmeyer, A., Schapiro, M.B., and Jakobs, C., 2002, Congenital creatine transporter deficiency. *Neuropediatrics* **33:** 232–238.

Dolder, M., Walzel, B., Speer, O., Schlattner, U., and Wallimann, T., 2003, Inhibition of the mitochondrial permeability transition by creatine kinase substrates. Requirement for microcompartmentation. *J. Biol. Chem.* **278:** 17760–17766.

Dringen, R., Verleysdonk, S., Hamprecht, B., Willker, W., Leibfritz, D., and Brand, A., 1998, Metabolism of glycine in primary astroglial cells: synthesis of creatine, serine, and glutathione. *J. Neurochem.* **70:** 835–840.

Item, C.B., Stöckler-Ipsiroglu, S., Stromberger, C., Muhl, A., Alessandri, M.G., Bianchi, M.C., Tosetti, M., Fornai, F., and Cioni, G., 2001, Arginine:glycine amidinotransferase deficiency: the third inborn error of creatine metabolism in humans. *Am. J. Hum. Genet.* **69:** 1127–1133.

Jinnai, D., Sawai, A., and Mori, A., 1966, Gamma-guanidinobutyric acid as a convulsive substance. *Nature* **212:** 617.

Juravleva, E., Barbakadze, T., Mikeladze, D., and Kekelidze, T., 2005, Creatine enhances survival of glutamate-treated neuronal/glial cells, modulates Ras/NF-κB signaling, and increases the generation of reactive oxygen species. *J. Neurosci. Res.* **79:** 224–230.

Kleefstra, T., Rosenberg, E.H., Salomons, G.S., Stroink, H., van Bokhoven, H., Hamel, B.C., and de Vries, B.B., 2005, Progressive intestinal, neurological and psychiatric problems in two adult males with cerebral creatine deficiency caused by an SLC6A8 mutation. *Clin. Genet.* **68:** 379–381.

Klivenyi, P., Ferrante, R.J., Matthews, R.T., Bogdanov, M.B., Klein, A.M., Andreassen, O.A., Mueller, G., Wermer, M., Kaddurah-Daouk, R., and Beal, M.F., 1999, Neuroprotective effects of creatine in a transgenic animal model of amyotrophic lateral sclerosis. *Nat. Med.* **5:** 347–350.

Koga, Y., Takahashi, H., Oikawa, D., Tachibana, T., Denbow, D.M., and Furuse, M., 2005, Brain creatine functions to attenuate acute stress responses through GABAnergic system in chicks. *Neuroscience* **132:** 65–71.

Limberger, N., Spath, L., and Starke, K., 1986, Release of previously incorporated γ-[^{3}H]aminobutyric acid in rabbit caudate nucleus slices. *J. Neurochem.* **46:** 1102–1108.

Lion-Francois, L., Cheillan, D., Pitelet, G., Acquaviva-Bourdain, C., Bussy, G., Cotton, F., Guibaud, L., Gerard, D., Rivier, C., Vianey-Saban, C., Jakobs, C., Salomons, G.S., and des Portes, V., 2006, High frequency of creatine deficiency syndromes in patients with unexplained mental retardation. *Neurology* **67:** 1713–1714.

Loike, J.D., Zalutsky, D.L., Kaback, E., Miranda, A.F., and Silverstein, S.C., 1988, Extracellular creatine regulates creatine transport in rat and human muscle cells. *Proc. Natl. Acad. Sci. USA* **85:** 807–811.

Mancini, G.M., Catsman-Berrevoets, C.E., de Coo, I.F., Aarsen, F.K., Kamphoven, J.H., Huijmans, J.G., Duran, M., van der Knaap, M.S., Jakobs, C., and Salomons, G.S., 2005, Two novel mutations in SLC6A8 cause creatine transporter defect and distinctive X-linked mental retardation in two unrelated Dutch families. *Am. J. Med. Genet. A* **132:** 288–295.

Matthews, R.T., Ferrante, R.J., Klivenyi, P., Yang, L., Klein, A.M., Mueller, G., Kaddurah-Daouk, R., and Beal, M.F., 1999, Creatine and cyclocreatine attenuate MPTP neurotoxicity. *Exp. Neurol.* **157:** 142–149.

Matthews, R.T., Yang, L., Jenkins, B.G., Ferrante, R.J., Rosen, B.R., Kaddurah-Daouk, R., and Beal, M.F., 1998, Neuroprotective effects of creatine and cyclocreatine in animal models of Huntington's disease. *J. Neurosci.* **18:** 156–163.

Mercimek-Mahmutoglu, S., Muehl, A., Salomons, G.S., Waldhoer, T., Neophytou, B., Jakobs, C., and Stoeckler-Ipsiroglu, S., 2007, Simultaneous determination of urinary creatine and creatinine by electrospray tandem mass spectrometry for screening of X-linked creatine transporter (SLC6A8) deficiency. Submitted for publication.

Mercimek-Mahmutoglu, S., Stoeckler-Ipsiroglu, S., Adami, A., Appleton, R., Araujo, H.C., Duran, M., Ensenauer, R., Fernandez-Alvarez, E., Garcia, P., Grolik, C., Item, C.B., Leuzzi, V., Marquardt, I., Muhl, A., Saelke-Kellermann, R.A., Salomons, G.S., Schulze, A., Surtees, R., van der Knaap, M.S., Vasconcelos, R., Verhoeven, N.M., Vilarinho, L., Wilichowski, E., and Jakobs, C., 2006, GAMT deficiency: features, treatment, and outcome in an inborn error of creatine synthesis. *Neurology* **67:** 480–484.

Neu, A., Neuhoff, H., Trube, G., Fehr, S., Ullrich, K., Roeper, J., and Isbrandt, D., 2002, Activation of $GABA_A$ receptors by guanidinoacetate: a novel pathophysiological mechanism. *Neurobiol. Dis.* **11:** 298–307.

Newmeyer, A., Cecil, K.M., Schapiro, M., Clark, J.F., and Degrauw, T.J., 2005, Incidence of brain creatine transporter deficiency in males with developmental delay referred for brain magnetic resonance imaging. *J. Dev. Behav. Pediatr.* **26:** 276–282.

Peral, M.J., Garcia-Delgado, M., Calonge, M.L., Duran, J.M., De La Horra, M.C., Wallimann, T., Speer, O., and Ilundain, A., 2002, Human, rat and chicken small intestinal Na^+-Cl^--creatine transporter: functional, molecular characterization and localization. *J. Physiol.* **545:** 133–144.

Poo-Arguelles, P., Arias, A., Vilaseca, M.A., Ribes, A., Artuch, R., Sans-Fito, A., Moreno, A., Jakobs, C., and Salomons, G., 2006, X-Linked creatine transporter deficiency in two patients with severe mental retardation and autism. *J. Inherit. Metab. Dis.* **29:** 220–223.

Pyne-Geithman, G.J., Degrauw, T.J., Cecil, K.M., Chuck, G., Lyons, M.A., Ishida, Y., and Clark, J.F., 2004, Presence of normal creatine in the muscle of a patient with a mutation in the creatine transporter: a case study. *Mol. Cell. Biochem.* **262:** 35–39.

Renema, W.K., Schmidt, A., van Asten, J.J., Oerlemans, F., Ullrich, K., Wieringa, B., Isbrandt, D., and Heerschap, A., 2003, MR spectroscopy of muscle and brain in guanidinoacetate methyltransferase (GAMT)-deficient mice: validation of an animal model to study creatine deficiency. *Magn. Reson. Med.* **50:** 936–943.

Rosenberg, E.H., Martínez Muñoz, C., Betsalel, O.T., van Dooren, S.J., Fernandez, M., Jakobs, C., deGrauw, T.J., Kleefstra, T., Schwartz, C.E., and Salomons, G.S., 2007, Functional characterization of missense variants in creatine transporter/SLC6A8: improved diagnostic application. *Hum. Mutat.*, in press.

Rosenberg, E.H., Almeida, L.S., Kleefstra, T., deGrauw, R.S., Yntema, H.G., Bahi, N., Moraine, C., Ropers, H.H., Fryns, J.P., deGrauw, T.J., Jakobs, C., and Salomons, G.S., 2004, High prevalence of SLC6A8 deficiency in X-linked mental retardation. *Am. J. Hum. Genet.* **75:** 97–105.

Salomons, G.S., van Dooren, S.J., Verhoeven, N.M., Cecil, K.M., Ball, W.S., Degrauw, T.J., and Jakobs, C., 2001, X-linked creatine-transporter gene (SLC6A8) defect: a new creatine-deficiency syndrome. *Am. J. Hum. Genet.* **68:** 1497–1500.

Schmidt, A., Marescau, B., Boehm, E.A., Renema, W.K., Peco, R., Das, A., Steinfeld, R., Chan, S., Wallis, J., Davidoff, M., Ullrich, K., Waldschutz, R., Heerschap, A., De Deyn, P.P., Neubauer, S., and Isbrandt, D., 2004, Severely altered guanidino compound levels, disturbed body weight homeostasis and impaired fertility in a mouse model of guanidinoacetate N-methyltransferase (GAMT) deficiency. *Hum. Mol. Genet.* **13:** 905–921.

Schoffelmeer, A.N., Wemer, J., Mulder, A.H., 1981, Comparison between electrically evoked and potassium-induced ^{3}H-noradrenaline release from rat neocortex slices: role of calcium ions and transmitter pools. *Neurochem. Int.* **3:** 129–136.

Schulze, A., Mayatepek, E., Bachert, P., Marescau, B., De Deyn, P.P., and Rating, D., 1998, Therapeutic trial of arginine restriction in creatine deficiency syndrome. *Eur. J. Pediatr.* **157:** 606–607.

Schulze, A., Bachert, P., Schlemmer, H., Harting, I., Polster, T., Salomons, G.S., Verhoeven, N.M., Jakobs, C., Fowler, B., Hoffmann, G.F., and Mayatepek, E., 2003, Lack of creatine in muscle and brain in an adult with GAMT deficiency. *Ann. Neurol.* **53:** 248–251.

Schulze, A., Hoffmann, G.F., Bachert, P., Kirsch, S., Salomons, G.S., Verhoeven, N.M., and Mayatepek, E., 2006, Presymptomatic treatment of neonatal guanidinoacetate methyltransferase deficiency. *Neurology* **67:** 719–721.

Schulze, A., and Battini, R., 2007, Pre-symptomatic treatment of creatine biosynthesis defects. *Subcell. Biochem.* **46:** 167–181.

Speer, O., Neukomm, L.J., Murphy, R.M., Zanolla, E., Schlattner, U., Henry, H., Snow, R.J., and Wallimann, T., 2004, Creatine transporters: a reappraisal. *Mol. Cell. Biochem.* **256–257:** 407–424.

Stöckler, S., Holzbach, U., Hanefeld, F., Marquardt, I., Helms, G., Requart, M., Hanicke, W., and Frahm, J., 1994, Creatine deficiency in the brain: a new, treatable inborn error of metabolism. *Pediatr. Res.* **36:** 409–413.

Stöckler, S., Hanefeld, F., and Frahm, J., 1996a, Creatine replacement therapy in guanidinoacetate methyltransferase deficiency, a novel inborn error of metabolism. *Lancet* **348:** 789–790.

Stöckler, S., Isbrandt, D., Hanefeld, F., Schmidt, B., and von Figura, K., 1996b, Guanidinoacetate methyltransferase deficiency: the first inborn error of creatine metabolism in man. *Am. J. Hum. Genet.* **58:** 914–922.

Stöckler, S., Marescau, B., De Deyn, P.P., Trijbels, J.M., and Hanefeld, F., 1997, Guanidino compounds in guanidinoacetate methyltransferase deficiency, a new inborn error of creatine synthesis. *Metabolism* **46:** 1189–1193.

Stöckler-Ipsiroglu, S., and Salomons, G.S., 2006, Creatine deficiency syndromes. In: Fernandes, J., Saudubray, J.-M., van den Berghe, G., and Walter, J.H., eds., Inborn Metabolic Diseases, Diagnosis and Treatment, Springer, Heidelberg, Germany, pp. 211–217.

Tachikawa, M., Fukaya, M., Terasaki, T., Ohtsuki, S., and Watanabe, M., 2004, Distinct cellular expressions of creatine synthetic enzyme GAMT and creatine kinases uCK-Mi and CK-B suggest a novel neuron-glial relationship for brain energy homeostasis. *Eur. J. Neurosci.* **20:** 144–160.

Tachikawa, M., Hosoya, K.-i., Ohtsuki, S., and Terasaki, T., 2007, A novel relationship between creatine transport at the blood-brain and blood-retinal barriers, creatine biosynthesis, and its use for brain and retinal energy homeostasis. *Subcell. Biochem.* **46:** 83–98.

Tarnopolsky, M.A., Bourgeois, J.M., Snow, R., Keys, S., Roy, B.D., Kwiecien, J.M., and Turnbull, J., 2003, Histological assessment of intermediate- and long-term creatine monohydrate supplementation in mice and rats. *Am. J. Physiol. Regul. Integr. Comp. Physiol.* **285:** R762–R769.

Torremans, A., Marescau, B., Possemiers, I., Van Dam, D., D'Hooge, R., Isbrandt, D., and De Deyn, P.P., 2005, Biochemical and behavioural phenotyping of a mouse model for GAMT deficiency. *J. Neurol. Sci.* **231:** 49–55.

Volpe, J.J., 2001, Neurology of the Newborn, Saunders, Philadelphia.

Walker, J.B., 1979, Creatine: biosynthesis, regulation, and function. *Adv. Enzymol. Relat. Areas Mol. Biol.* **50:** 177–242.

Wang, Y.E., Esbensen, P., and Bentley, D., 1998, Arginine kinase expression and localization in growth cone migration. *J. Neurosci.* **18:** 987–998.

Wyss, M., and Kaddurah-Daouk, R., 2000, Creatine and creatinine metabolism. *Physiol. Rev.* **80:** 1107–1213.

CHAPTER 9

PRE-SYMPTOMATIC TREATMENT OF CREATINE BIOSYNTHESIS DEFECTS

ANDREAS SCHULZE[1] AND ROBERTA BATTINI[2]

[1] *University of Toronto, Department of Paediatrics, Division of Clinical and Metabolic Genetics, and Research Institute, The Hospital for Sick Children, 555 University Avenue, Toronto, ON. M5G 1X8, Canada.*

[2] *Department of Developmental Neuroscience, IRCCS Stella Maris, Calambrone, Pisa, Italy*

Abstract: Recent observations in two patients, one with AGAT deficiency (AGAT-D) and one with GAMT deficiency (GAMT-D), both diagnosed already at birth, provide first evidence for important therapeutic effects of pre-symptomatic treatment with creatine (Cr) supplementation in AGAT-D and Cr supplementation plus guanidinoacetate lowering strategies in GAMT-D. Although long-term data are lacking, the results suggest that complete prevention of neurological sequelae in early treated patients could be feasible (Battini *et al.*, 2006; Schulze *et al.*, 2006)

1. INTRODUCTION

Creatine (Cr) deficiency syndromes represent a group of recently discovered inborn errors of metabolism (Stöckler *et al.*, 1994; Schulze, 2003; Bianchi *et al.*, 2000; Salomons *et al.*, 2001). Cr deficiency syndromes are caused by defects in either the biosynthesis or the transport proteins of Cr, resulting in deficiency of Cr and phosphocreatine (PCr), mainly in the brain (Schulze, 2003). Within this group of disorders, there are two synthesis defects (arginine:glycine amidinotransferase deficiency, AGAT-D, and guanidinoacetate methyltransferase deficiency, GAMT-D) and one cellular Cr transport defect (CT1, SLC6A8); the first two defects respond to Cr supplementation.

2. CREATINE REPLACEMENT THERAPY IN AGAT AND GAMT DEFICIENCY

Cr is substituted as oral Cr monohydrate at a 15- to 20-fold dosage of the normal daily Cr requirement, corresponding to 350–400 mg/kg/d in children aged 4 to 12 years. A high blood Cr concentration is achieved by frequent administration of Cr

G.S. Salomons and M. Wyss (eds.), Creatine and Creatine Kinase in Health and Disease, 167–181.

monohydrate (6–8 times a day) which might be favourable with respect to transport of Cr across the blood-brain barrier (BBB) (Stöckler *et al.*, 1997). Administration of higher amounts of Cr (up to 1.5–2.0 g/kg/d), with larger time intervals in between individual servings (e.g. every 8–12 hours), or in an alternating manner (e.g. 4 days application followed by a break for 3 days) do not seem to be more effective in further enhancing transport of Cr across the BBB (Schulze, 2005).

The optimal dosage of Cr monohydrate for recovery and maintenance of the cerebral Cr pool still has to be established. Long-term observation of the three AGAT-D patients described so far has allowed confirmation of clinical and neuropsychological improvement even after a reduction of the dosage of daily Cr supplementation to 200 mg/kg after three years from the start of therapy (Bianchi *et al.*, 2007). Accordingly, when studied with proton magnetic resonance spectroscopy (^{1}H-MRS), the Cr pool did not change significantly over that time interval.

Besides only incomplete restoration of brain Cr pools by Cr replacement therapy in both AGAT-D and GAMT-D patients, the ^{1}H-MRS results indicate a faster slope and a more complete recovery of brain Cr concentrations in AGAT-D patients than in GAMT-D patients (Figure 1) (Schulze, 2005; Bianchi *et al.*, 2007). This

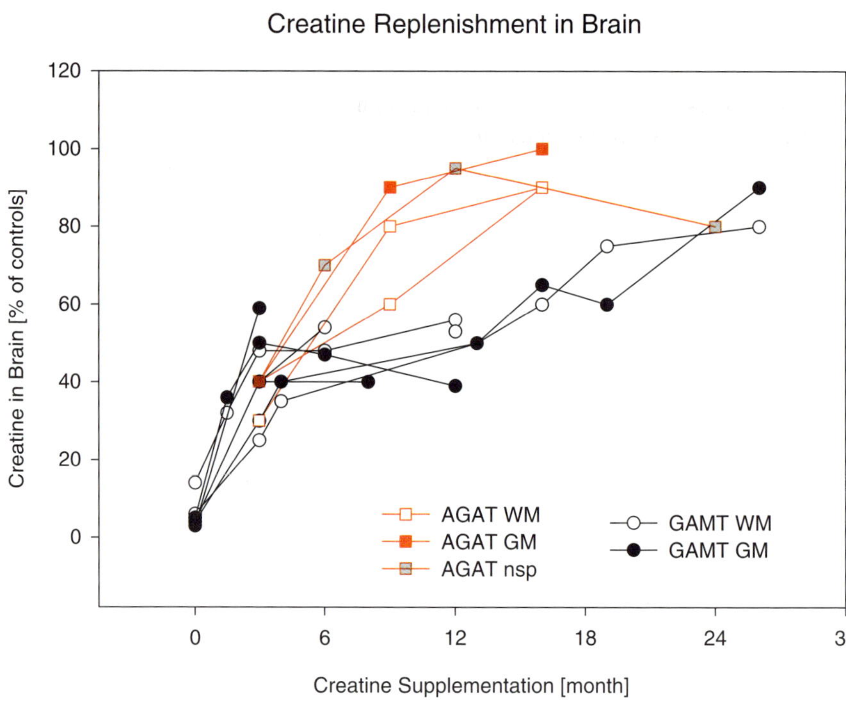

Figure 1. Creatine replenishment in the brain of patients with GAMT-D and AGAT-D in the course of creatine supplementation, estimated by *in vivo* proton magnetic resonance spectroscopy. WM, white matter; GM, grey matter; nsp, not specified. Reprinted from Schulze (2005) with kind permission from SPS Verlagsgesellschaft mbH, Heilbronn, Germany.

difference is likely correlated to guanidinoacetate (GAA) accumulation in GAMT-D. GAA, like some other guanidino compounds, is a competitive inhibitor of Cr transport (Ohtsuki *et al.*, 2002). In accordance with these findings, the clinical response to oral Cr supplementation is more pronounced for AGAT-D patients than for GAMT-D patients, including good language development and complete disappearance of autistic-like behaviour (Battini *et al.*, 2002; Bianchi *et al.*, 2000).

3. GUANIDINOACETATE LOWERING THERAPY IN GAMT DEFICIENCY

Accumulation of GAA, known for its neurotoxic and epileptogenic effects, contributes to the pathophysiology in GAMT-D. Cr replacement therapy causes an approximately 50% reduction of GAA in body fluids. Considering the 10- to 100-fold elevated GAA concentrations in affected individuals (Mercimek-Mahmutoglu *et al.*, 2006), this reduction is far from normal GAA levels. Two different approaches have proven effective in further lowering GAA in GAMT-D patients, namely (i) dietary arginine restriction by a protein restricted diet combined with low-dose ornithine supplementation (Schulze *et al.*, 2001) and (ii) high-dose ornithine treatment (Schulze, 2005).

The strategy for dietary treatment is substrate deprivation (arginine, glycine) of the AGAT reaction, and consists of protein restriction, supplementation of arginine-free amino acids, low-dose ornithine, and sodium benzoate. Intake of arginine by nutritional protein is restricted to 15–25 mg arginine/kg/d (corresponding to 0.4–0.7 g natural protein/kg/d). An arginine-free essential amino acid mixture is substituted with 0.2–0.7 g/kg/d to meet age-dependent physiological protein requirements. Ornithine in low dosage (50–100 mg/kg/d) seems necessary to keep arginine low. Sodium benzoate (100 mg/kg/d) is given for additional substrate deprivation. It removes glycine, the other substrate of the AGAT reaction, and lowers the flux through the arginine-forming urea cycle (Schulze *et al.*, 2001; Stoeckler-Ipsiroglu *et al.*, 2006).

More recently, treatment with high-dose ornithine has shown to be at least as effective in lowering GAA levels as dietary arginine restriction. Ornithine is a competitive inhibitor of the AGAT enzyme. AGAT inhibition is achieved by administration of 800 mg ornithine/kg/d given in 5–6 doses per day (dpd) (Schulze, 2005). Whether a combination of dietary treatment with high-dose ornithine therapy is best for decreasing GAA levels still has to be elucidated. Figure 2 illustrates our current knowledge about the effectiveness of different treatments on GAA reduction in blood and urine.

Cr replenishment and GAA lowering improve the clinical symptoms in GAMT-D patients. This applies mainly to the autistic behaviour, the movement disorder, and to some cognitive capabilities, whereas language skills and general cognitive development remain poor or improve only slightly. Control of severe seizures so far refractory to antiepileptic drugs and Cr treatment is obviously achievable by GAA lowering therapy (Schulze *et al.*, 2001). Even treatment only initiated in

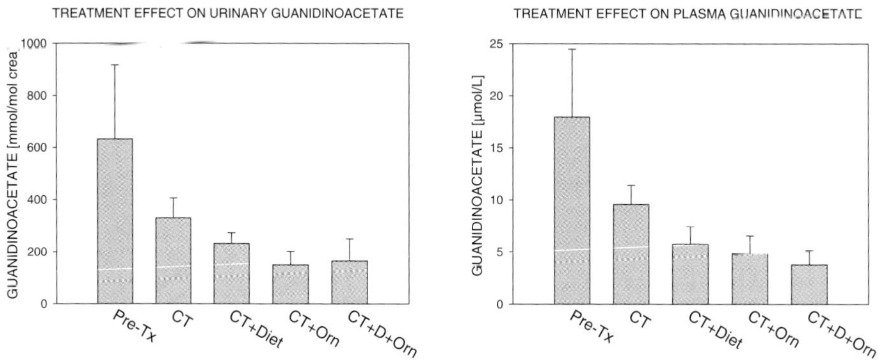

Figure 2. Effect of different treatments on guanidinoacetate levels in urine and plasma of patients with GAMT-D. Pre-Tx, pre-treatment (n = 6 patients), CT, Cr treatment (n = 3); CT+Diet, Cr treatment plus dietary arginine restriction (n = 1); CT+Orn, Cr treatment plus high-dose ornithine treatment (n = 5); CT+D+Orn, Cr treatment plus dietary arginine restriction plus high-dose ornithine treatment (n = 2).

adulthood led to an impressive improvement in epileptic seizures, mental capabilities, movement disorder and behaviour in one patient (Schulze, 2003).

4. PRE-SYMPTOMATIC TREATMENT IN AN AGAT-D PATIENT

AGAT-D (MIM 602360) is an autosomal recessive disease characterized by mental retardation, severe language impairment and behavioural disturbances. Supplementation of Cr has been shown to improve clinical symptoms in previously reported symptomatic cases (Battini *et al.*, 2002; Bianchi *et al.*, 2000). Pre-symptomatic treatment has been reported in only one subject (Battini *et al.*, 2006).

The boy is the third child of healthy non-consanguineous Italian parents. Both parents are carrying the c.446G>A mutation that results in the replacement of tryptophan by a stop codon at residue 149 (p.Trp149X) in the *AGAT* gene. His two older sisters were previously diagnosed with AGAT-D and are being treated for 6 years. Both sisters are homozygous for the above described pathogenic nonsense mutation (Bianchi *et al.*, 2000). The boy was born after unremarkable pregnancy and delivery (birth weight: 3 kg; length: 49 cm; head circumference: 35 cm). Neurological examination at the age of 3 days was normal. He was breast-fed. Serum panel chemistries were normal. Analysis of Cr deficiency metabolites by GC/MS showed the following: serum GAA (0.13 μM vs. normal range 0.22–3.14 μM) and Cr (16.2 μM vs. 18–141 μM); urine GAA (0.54 μM vs. 55–698 μM) and Cr (24.6 μM vs. 200–5500 μM). Sequencing of the proband's *AGAT* gene confirmed the presence of the same homozygous mutation as previously detected in DNA of the sisters. The diagnosis was also confirmed by means of a new AGAT activity assay in lymphocytes or lymphoblasts, using GC coupled with a quadrupole detector, without using labelled substrates. The lymphocytes or lymphoblasts were incubated in a reaction mixture containing glycine and arginine at pH 7.5. The reaction was

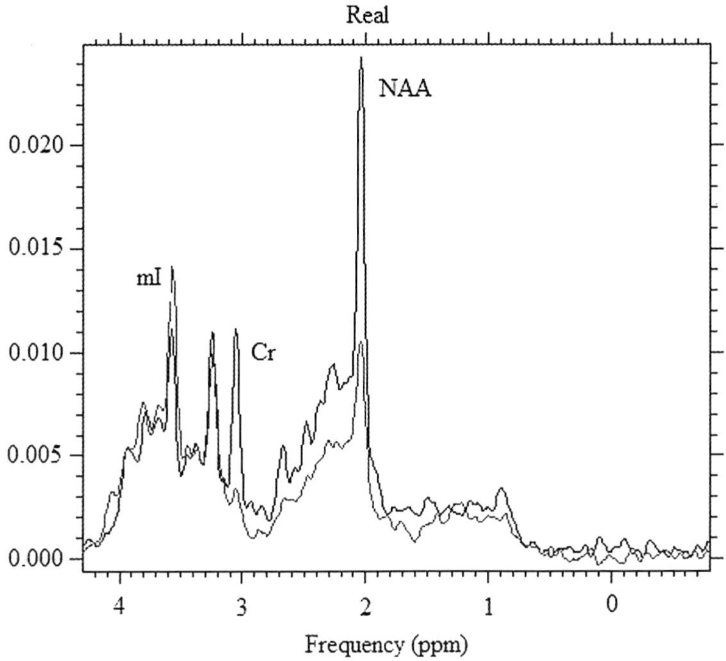

Figure 3. ^{1}H-MRS of the patient with AGAT deficiency at birth (grey line) and at 24 months, after 20 months of oral creatine supplementation (black line). Note the absence of a creatine (Cr) peak at 3.05 ppm at birth and its partial restoration after early treatment. Other differences in the spectrum represent the physiological changes with age and mainly reflect myelination during the first years of life. *Myo*-inositol (mI) is higher and N-acetylaspartate (NAA) is lower at birth.

stopped by addition of 1 N PCA after 4 hours of incubation at 37 °C. The end-product of the reaction (GAA) was separated and quantified with an Agilent GC/MS, set in Electron Ionization – SIM mode (Alessandri *et al.*, 2005). In control samples, AGAT activity was 0.243–0.425 and 0.950–1.470 nmol/mg/h for lymphocytes and lymphoblasts, respectively. Enzyme activity in AGAT-D patients was below the detection limit in both cell types.

During the first 15 days after birth, serum and urine Cr and GAA levels decreased even further (serum Cr 3.2 μM; serum GAA 0.04 μM; urine Cr 20.7 μM; urine GAA 0.34 μM). Brain MRI was normal but ^{1}H-MRS showed an almost complete absence of the cerebral Cr peak at 3.05 ppm, confirming the diagnosis of AGAT-D (Figure 3) (Battini *et al.*, 2006). In order to replenish the Cr depletion and considering that the child was breast-fed, we initially tried to supplement the maternal diet with Cr monohydrate (3 to 9 g/d). After one month of maternal supplementation, an increase in Cr concentration in the maternal milk (190 μM; normal range 82.3–128.9 μM) was detected; the serum and urine Cr concentrations were 4.04 μM and 6.3 μM, respectively, while GAA was undetectable in both serum and urine. Unfortunately, the Cr increase in blood, urine and brain of the child was unremarkable. At the age of 4 months, following weaning, dietary supplementation of the child with Cr monohydrate was initiated. We started with a low dosage of oral Cr (100 mg/kg/d) divided in five dpd. This dosage was selected based on (i) the lack of toxicological data in neonates and (ii) results obtained in his affected relatives for which optimization of Cr therapy led to the conclusion that 100 mg/kg/d might be the minimal optimal dose of Cr to be administered chronically (Bianchi *et al.*, 2007). Subsequent assessment of serum and urine Cr levels revealed a progressive replenishment of body Cr pools by the end of the first week of treatment: after 1 month of supplementation of the child, serum and urine Cr concentrations were 222 μM and 1.75 mM respectively; after 3 months 164 μM and 4.32 mM after 6 months 172 μM and 5.35 mM, and after 20 months 67.7 μM and 12.7 mM respectively. GAA levels in body fluids remained undetectable (Battini *et al.*, 2006). Serum concentration of Cr overlapped with those obtained in the affected relatives which were treated with a higher dosage of Cr (Battini *et al.*, 2002; Bianchi *et al.*, 2000). At the age of 12 months, after 8 months of treatment, we performed a control examination with brain ^{1}H-MRS and demonstrated the restoration of about 60% of normal brain Cr levels. Growth and psychomotor development of the child remained completely normal. At the age of 12 months, he walked unaided and uttered single words. At the age of 16 months he was able to ask by gestures and from the age of 18 months he produced some two-word combinations and understood simple verbal requests. His general developmental quotient was 105 (Griffiths Developmental Scales), and his growth was normal (Battini *et al.*, 2006). It is of interest that in the affected relatives, the first clinical symptoms appeared at an age of around 8–10 months, while at around 18 months, an important delay in somatic growth and psychomotor development, associated with hypotonia or autistic-like behaviour, was already apparent (Battini *et al.*, 2002; Bianchi *et al.*, 2000). At the age of 24 months, after 20 months of therapy at the same oral Cr dosage as from the start (100 mg/kg/d), a new ^{1}H-MRS examination showed similar Cr replenishment as at 12 months (Figure 3). The baby was healthy; his psychomotor development and social interaction were according to age. Cr supplementation was always well tolerated. No side effects were observed except for some episodes of diarrhoea at the beginning of treatment and when the dosage was increased according to weight gain.

This patient is the first subject with AGAT-D diagnosed in the neonatal period and treated when still asymptomatic. The observation in this patient has proven that blood GAA and Cr levels, which can be measured both in plasma and dried blood spots, are significantly low from the first days of life, supporting their use as early diagnostic markers for AGAT-D (Carducci *et al.*, 2002, 2006). Severe brain Cr depletion, as detected by ^{1}H-MRS, was already present since the first days of life and, even with Cr being a component of maternal milk, we found that this nutrient has a limited efficacy in the maintenance of Cr pools when a defect in endogenous Cr synthesis is present. The latency in clinical manifestation in Cr disorders may be related to a relatively low need for the Cr/PCr system during the early phases of brain development.

Despite the expression of the *GAMT*, *AGAT* and *CT1* genes in different stages of the developing rat embryo (Braissant *et al.*, 2005) and the disturbed Cr and

GAA levels in brain of neonates affected with AGAT-D or GAMT-D, no clinical abnormalities are found in newborns. Probably, brain Cr depletion causes a slow and cumulative effect on nervous development, in particular on higher cortical function, as shown by the presence of mental retardation and severe language deficits as specific hallmarks of all Cr disorders (Battini *et al.*, 2002; Schulze, 2003; Mancini *et al.*, 2005; Mercimek-Mahmutoglu *et al.*, 2006). Although congenital depletion of brain Cr would support very early treatment, it is of interest that a 4-month delay in starting Cr administration does not seem to have affected the development of the child so far. The Cr dosage we used was about a quarter of that administered initially in previously reported AGAT-D patients (Battini *et al.*, 2002; Bianchi *et al.*, 2000) and proved to be safe and effective in replenishing both the peripheral Cr pools and partially also the brain Cr pool.

5. PRE-SYMPTOMATIC TREATMENT IN A GAMT-D PATIENT

GAMT-D (MIM 601240) is an autosomal-recessive inherited Cr synthesis defect and has the most severe phenotype among the Cr deficiency syndromes. Patients are clinically affected by mental retardation, lack of speech, autistic behaviour, extra-pyramidal movement disorder, and epilepsy (sometimes refractory to therapeutic intervention) (Schulze, 2003). Cr deficiency and accumulation of GAA, the latter known for its neurotoxic and epileptogenic effects (da Silva *et al.*, 1999; D'Hooge *et al.*, 1992; Neu *et al.*, 2002), contribute to the pathophysiology of GAMT-D (Schulze *et al.*, 2001). Treatment in GAMT-D is directed towards Cr replenishment and decreasing of GAA levels (Schulze, 2005). Despite treatment, the clinical outcome of patients diagnosed in childhood or adulthood is still unfavourable (Mercimek-Mahmutoglu *et al.*, 2006). Pre-symptomatic disease detection and early initiation of treatment, which are potentially essential for a good outcome, have only been reported in one patient (Schulze *et al.*, 2006).

The girl is the second child of healthy, non-related Turkish parents. Her 5-year-old brother has GAMT-D which was diagnosed and treated at age 2 ¾ years. Because of the 25% probability of being affected, the girl was followed clinically and biochemically since birth. She was born at term by spontaneous delivery after an unremarkable pregnancy. Clinical examination up to age 3 weeks was normal. She was breast-fed with addition of preterm formula. Red blood cell count and serum chemistries (electrolytes, liver function tests, urea, glucose, protein) as well as clotting were normal. EEG and brain MRI revealed no abnormalities.

Blood was taken from the umbilical cord and subsequently every 12 hours until day 5 by heel prick, and was spotted onto filter paper. In the dried blood spot specimens, measurement of GAA, Cr, and creatinine was performed by means of electrospray-ionization tandem mass spectrometry (modified from Bodamer *et al.*, 2001). The results were informative for the diagnosis of GAMT-D in all specimens from birth to day 5 (Figure 4). GAA was already elevated in cord blood. A subsequent increase during the first 24 hours of life was followed by a decline thereafter. However, GAA remained permanently elevated until day 5 by exceeding the 99.5th percentile for healthy newborns. The levels of Cr and creatinine, both

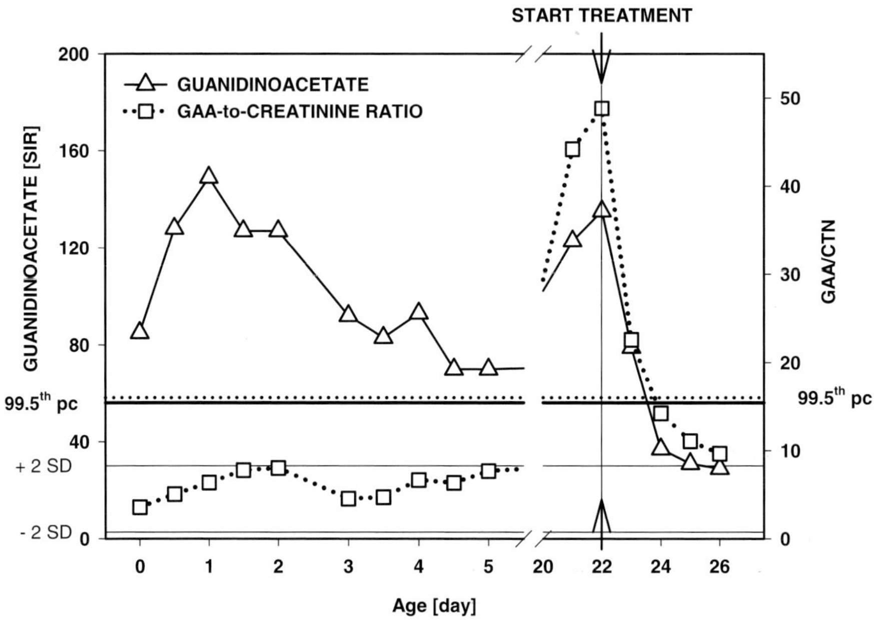

Figure 4. Course of guanidinoacetate (GAA) level and GAA-to-creatinine ratio (GAA/CTN) in dried blood spots of a patient with GAMT-D during the first 5 days of life and at the start of treatment. Guanidino compounds were analyzed by tandem-mass spectrometry. Normal range of GAA (± 2 standard deviations, SD) and cut-off limit for GAA and GAA/CTN in neonates (99.5th percentile, pc) were calculated from results in healthy neonates ($n = 3,407$). SIR, signal intensity ratio (Permitted reprint from: Schulze A., Hoffmann G.F., Bachert P., Kirsch S., Salomons G.S., Verhoeven N.M., and Mayatepek E., 2006, Presymptomatic treatment of neonatal guanidinoacetate methyltransferase deficiency. *Neurology* **67:** 719–721).

usually decreased in GAMT-D, were of no diagnostic relevance during the neonatal period. Creatinine was within the normal range at birth and decreased only gradually until day 5. The resulting GAA/creatinine ratio was unrevealing. Only in the third week of life did the GAA/creatinine ratio reach higher diagnostic sensitivity owing to a decrease in creatinine levels after the first week of life (Figure 4) (Schulze *et al.*, 2006).

Guanidino compounds in urine and plasma were measured by cation-exchange liquid chromatography (Schulze *et al.*, 1997). In urine, GAA was already slightly increased from the second day of life, whereas Cr was below the normal range. In the third week of life, the GAA increase in urine was even more pronounced, and in plasma, the GAA concentration was elevated, Cr decreased, and creatinine close to the lower limit of controls (Table 1). We investigated the total Cr content in brain by single-voxel ^{1}H-MRS (Schulze, 2003). Intensity ratios calculated from ^{1}H-MR spectra showed a largely decreased but still detectable Cr level in the brain at the age of three weeks ($I_{Cr}/I_{H2O} \sim 8 \times 10^{-5}$) (Figure 5) (Schulze *et al.*, 2006).

Table 1. Guanidino compounds in plasma and urine prior and during treatment of a pre-symptomatically diagnosed and treated patient with GAMT-D (Permitted reprint from: Schulze A., Hoffmann G.F., Bachert P., Kirsch S., Salomons G.S., Verhoeven N.M., and Mayatepek E., 2006, Presymptomatic treatment of neonatal guanidinoacetate methyltransferase deficiency. *Neurology* **67:** 719–721) .

Age, d	Plasma [μmol/L]			Urine [mmol/mol creatinine]	
	Guanidinoacetate	Creatine	Creatinine	Guanidinoacetate	Creatine
2				232	13
19	8.27	6	13	340	11
20				326	20
21	9.09	5	7	349	35
22					
	Start of Treatment				
23	3.60	238	54	456	6,795
26	6.75	258	39	454	8,026
28	2.88	512	14		
62	3.21	795	39	133	10,352
113	4.07	278	109		
216	4.15	388	16	174	8,297
419	4.10	498	25	179	13,195
421	3.23	251	28	173	8,665
Controls	0.20–1.46*	50–124*	5.2–35.2*	28–180†	28–1,700†

* Normal range, derived from healthy subjects at age 1 week to 2 years (n = 17)
† Normal range, derived from healthy subjects at age 1 month to 2 years (n = 16)

Mutation analysis of the *GAMT* gene of the girl (and her parents) using direct sequencing of genomic DNA (Caldeira Araujo *et al.*, 2005) confirmed compound heterozygosity for the two mutations known from her brother (Item *et al.*, 2004). c.152A>C in exon 1 results in replacement of a highly conserved histidine by proline at position 51 (p.His51Pro) and was not detected in 210 control chromosomes, suggesting that the mutation is pathogenic. The frameshift mutation c.526dupG in exon 5 (p.Glu176GlyfsX15) predicts a truncated protein. This indicates that the mutation is pathogenic. GAMT activity in lymphoblasts was analyzed by an enzyme assay using substrates labelled with stable isotopes (Verhoeven *et al.*, 2004). Activity was < 1 pmol/h/mg protein (control range: 63–443 pmol/h/mg protein) confirming the diagnosis of GAMT-D.

At the age of 22 days, treatment was initiated. Oral Cr monohydrate was given (400 mg/kg/d in 6 dpd). Sodium benzoate was given at 100 mg/kg/d in 3 dpd. We started oral ornithine hydrochloride supplementation with a dosage of 400 mg/kg/d in 6 dpd, increased the dosage to 600 mg/kg/d after two days, and to the final dosage of 800 mg/kg/d after another two days. The parents opted for an additional dietary treatment for their baby. Dietary arginine restriction was started with (per day) 50 ml breast milk, 8 g preterm formula, 35 g basic-p® (SHS, Germany), 20 g maltodextrin® (SHS, Germany), and 6 g E-AM1® (SHS, Germany). Thus, the daily intake of natural protein, of protein from arginine-free amino acid mixture, and of calories was 0.4 g/kg, 1.0 g/kg, and 102 kcal/kg, respectively. As a result of treatment, an immediate and distinct decrease of GAA levels in dried blood and plasma and, with some delay, in urine was observed (Figure 4 and Table 1) (Schulze *et al.*, 2006). At the age of 4 weeks after weaning, the diet was adapted to (per day) 27 g infant formula, 40 g basic-p®, 20 g maltodextrin®, and 7 g E-AM 1®. The resulting daily intake of natural protein, of protein from arginine-free amino acid mixture, and of calories was 0.6 g/kg, 1.0 g/kg, and 96 kcal/kg, respectively. At the age of 4 months, the parents felt comfortable with the treatment and the development of their child. Somatic growth, clinical status, and routine laboratory chemistries were normal, except for a slight metabolic acidosis (base excess [BE] ~ -5 mM). The diet and medication were adjusted further according to the child's weight. The Cr/PCr signal in ^{1}H-MRS increased significantly after 4 months of treatment ($I_{Cr}/I_{H2O} \sim 2.5 \times 10^{-4}$) and remained constant after 11 months of treatment. While I_{Cho}/I_{H2O} did not change, a continuous increase in the N-acetylaspartate level was seen during the examination period, indicating growth and formation of neuronal tissue (Figure 5) (Schulze *et al.*, 2006).

At the age of 7 months, the child's behaviour was reported as agile and friendly. She fixed and followed objects, and was babbling busily. She was able to freely sit without support already from the age of 6 months. Special neurodevelopmental investigation was normal except for slight dissociated motor development because she had not yet attained rotating skills. Medications and diet were tolerated well. The metabolic acidosis (BE -5.7 mM) persisted. The treatment was continued with adjustment for weight gain.

At 14 months of age, the girl is healthy and is developing normally. In the special neurodevelopmental investigation, her psychomotor and psychosocial development is according to age. She uses at least five meaning words correctly, understands simple verbal demands, and knows and shows several social gestures. Her gait is normal. Neurological status including muscle tone, tendon reflexes, and brain nerves are unaffected. She acquired the skills of crawling and free standing at 8 months, walking without support at 11 months, and stair climbing at 12 months. During treatment, the GAA concentration in plasma remained permanently reduced to about 50–60%, although still elevated compared to controls. GAA in urine was permanently close to normal. Creatinine in plasma and urine was normal. Cr levels in urine

Figure 5. Single-voxel 64-MHz ^{1}H-MR spectra of the brain of the patient with GAMT-D obtained at age 3 weeks (A), after 4 months (B) and after 11 months (C) of treatment. Measurement technique: double-spin echo sequence (PRESS), repetition time TR = 1500 ms, echo time TE = 135 ms, number of excitations nex = 256, water-signal suppression; B0 = 1.5 T, imaging head coil. The voxel [(1.5 cm)^3] was placed in the left occipital region. Peak assignments: N-acetyl-L-aspartate (NAA, chemical shift $\delta = 2.0$ ppm), creatines (Cr, $\delta = 3.0$ ppm), cholines (Cho, $\delta = 3.2$ ppm). These signals were normalized to the unsuppressed water signal obtained from the same voxel with the same measurement parameters except nex = 10. The intensity ratios I_{meta}/I_{H2O} are given in the boxes (Permitted reprint from: Schulze A., Hoffmann G.F., Bachert P., Kirsch S., Salomons G.S., Verhoeven N.M., and Mayatepek E., 2006, Presymptomatic treatment of neonatal guanidinoacetate methyltransferase deficiency. *Neurology* **67:** 719–721).

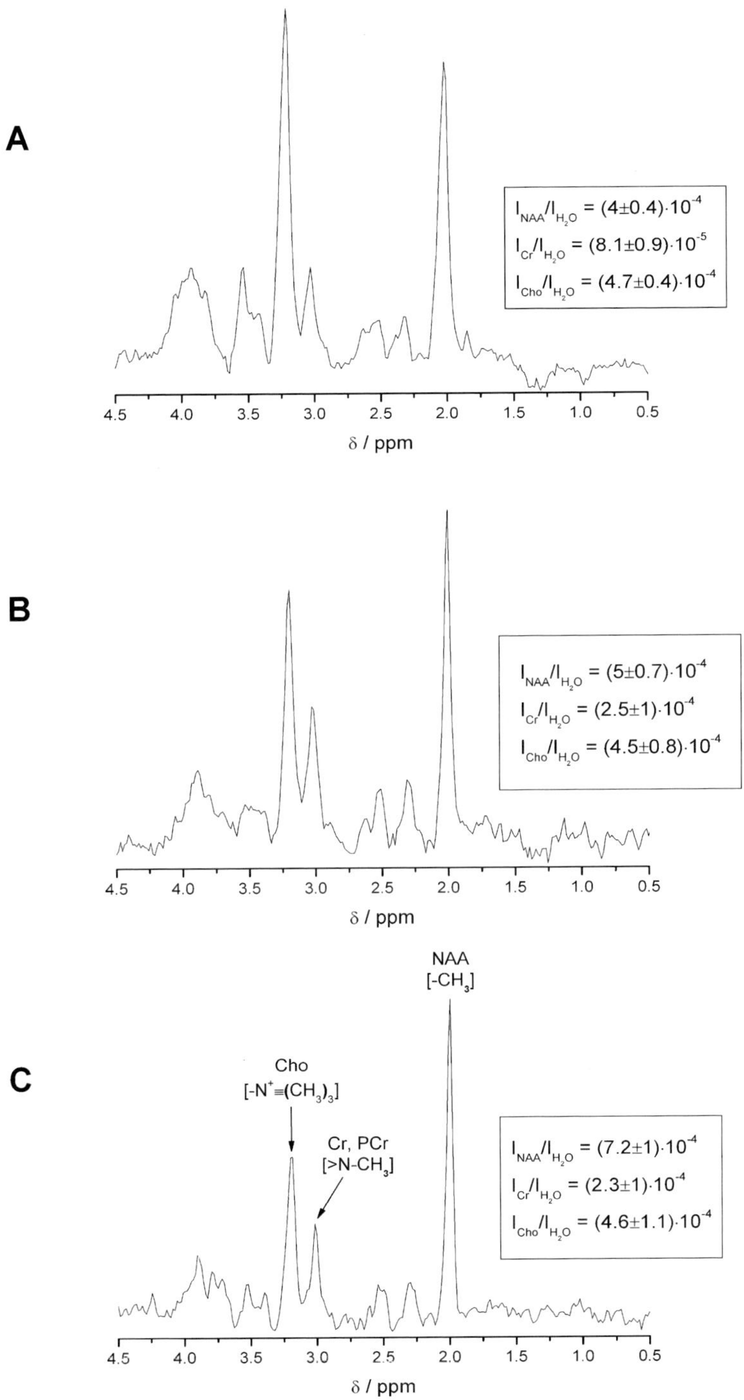
A
I_{NAA}/I_{H_2O} = (4±0.4)·10^{-4}
I_{Cr}/I_{H_2O} = (8.1±0.9)·10^{-5}
I_{Cho}/I_{H_2O} = (4.7±0.4)·10^{-4}
4.5 4.0 3.5 3.0 2.5 2.0 1.5 1.0 0.5
δ / ppm
B
I_{NAA}/I_{H_2O} = (5±0.7)·10^{-4}
I_{Cr}/I_{H_2O} = (2.5±1)·10^{-4}
I_{Cho}/I_{H_2O} = (4.5±0.8)·10^{-4}
4.5 4.0 3.5 3.0 2.5 2.0 1.5 1.0 0.5
δ / ppm
C
NAA
[-CH_3]
Cho
[-N^+≡(CH_3)$_3$]
Cr, PCr
[>N-CH_3]
I_{NAA}/I_{H_2O} = (7.2±1)·10^{-4}
I_{Cr}/I_{H_2O} = (2.3±1)·10^{-4}
I_{Cho}/I_{H_2O} = (4.6±1.1)·10^{-4}
4.5 4.0 3.5 3.0 2.5 2.0 1.5 1.0 0.5
δ / ppm

and plasma, due to supplementation, were elevated approximately 5-fold (Table 1). The mean plasma ornithine concentration was 160 μM, close to the treatment target concentration of ~ 200 μM. The persistent metabolic acidosis (BE −4.1 mM), also observed in other GAMT-D patients receiving high-dose ornithine hydrochloride, was assigned to the administration of ornithine as hydrochloride. We therefore changed to ornithine aspartate (800 mg ornithine/kg/d), which resolved the slight acidosis immediately and permanently. This approach resulted in a similar decrease of GAA levels, without having other adverse effects.

The first prospective observation and treatment in a neonate with GAMT-D thus provides evidence that diagnosis is possible at birth and that early treatment might be beneficial in the prevention of clinical symptoms. Biochemical findings at the age of 3 weeks were similar to those found in older patients, except for Cr in the brain which – in contrast to the findings in the AGAT patient above – did not seem to be lacking entirely. This suggests (near) normal prenatal Cr supply. This hypothesis is further supported by our finding of normal Cr and creatinine levels in dried blood during the early neonatal period. Slow postnatal release of Cr/PCr pools might explain the pre-symptomatic period of 3–6 months in all GAMT-D patients.

Treatment in GAMT-D is directed towards replenishment of Cr and reduction of GAA. GAA reduction is achieved by dietary arginine restriction combined with low-dose ornithine (100 mg/kg/d) (Schulze *et al.*, 2001). In addition, benzoate is given for substrate deprivation of the AGAT reaction. In our patient we have chosen a higher ornithine dosage because high-dose ornithine supplementation (800 mg/kg/d) is supposed to inhibit AGAT and to lower GAA formation more efficiently (Schulze, 2005). The treatment was well tolerated without adverse effects, except for slight metabolic acidosis which was caused by the ornithine hydrochloride formulation. Acidosis resolved after changing to ornithine aspartate. The biochemical effect of treatment comprised, besides partial Cr replenishment, the distinct and permanent reduction of GAA levels in plasma and urine.

Diagnostic work-up in patients is generally started only when deficits in cognitive function and absence of speech become clinically evident. If these symptoms are caused by irreversible brain impairment, it is self-evident why treatment usually fails. On the other hand, when treatment can be initiated before irreversible damage occurs, clinical symptoms may possibly be prevented entirely and permanently. These considerations may explain why the clinical course for the girl diagnosed and treated early is distinct from that of her brother who was treated late. His first clinical symptoms became obvious at age 6–9 months with few spontaneous movements, little interest in surrounding and playing, and delayed general development. At age 2 years he was able to use only few words correctly with no further speech development. At age 2 ½ years he developed epileptic seizures. The normal psychomotor and psychosocial development in the pre-symptomatically treated girl also differs from the common clinical course of all other GAMT-D patients reported so far.

6. CONCLUSION

Unspecific symptoms and normal results in standard metabolic testing may mean that a substantial number of AGAT-D and GAMT-D patients remain undiscovered and others will be diagnosed late. Independent of the age at diagnosis, all patients benefit from treatment. Different treatment approaches have shown to be efficient, even if neither Cr normalisation nor GAA normalisation is completely achieved. Perhaps, complete recovery is not required for normal functioning, and early treatment prior to irreversible brain impairment is much more crucial. Besides clinical and animal studies to further improve treatment, attempts at early diagnosis are mandatory. Both cases described above have proven not only the benefit of pre-symptomatic treatment but also that diagnosis is feasible already at birth. Classical criteria for the inclusion of inherited disorders in neonatal screening programmes are a pre-symptomatic phase of the disease, the availability of treatment options, the presence of simple diagnostic tools, and a relatively high frequency of the disease. Therefore, pilot studies for neonatal screening of AGAT-D and GAMT-D are warranted.

REFERENCES

Alessandri, M.G., Celati, L., Battini, R., Casarano, M., and Cioni, G., 2005, Gas chromatography/mass spectrometry assay for arginine:glycine-amidinotransferase deficiency. *Anal. Biochem.* **343:** 356–358.

Battini, R., Alessandri, M.G., Leuzzi, V., Moro, F., Tosetti, M., Bianchi, M.C., and Cioni, G., 2006, Arginine:glycine amidinotransferase (AGAT) deficiency in a newborn: early treatment can prevent phenotypic expression of the disease. *J. Pediatr.* **148:** 828–830.

Battini, R., Leuzzi, V., Carducci, C., Tosetti, M., Bianchi, M.C., Item, C.B., Stöckler-Ipsiroglu, S., and Cioni, G., 2002, Creatine depletion in a new case with AGAT deficiency: clinical and genetic study in a large pedigree. *Mol. Genet. Metab.* **77:** 326–331.

Bianchi, M.C., Tosetti, M., Battini, R., Leuzzi, V., Alessandri, M.G., Carducci, C., Antonozzi, I., and Cioni, G., 2007, Treatment monitoring of brain creatine deficiency syndromes: a 1H and 31P MR spectroscopy study. *Am. J. Neuroradiol.* **28:** 548–554.

Bianchi, M.C., Tosetti, M., Fornai, F., Alessandri, M.G., Cipriani, P., De Vito, G., and Canapicchi, R., 2000, Reversible brain creatine deficiency in two sisters with normal blood creatine level. *Ann. Neurol.* **47:** 511–513.

Bodamer, O.A., Bloesch, S.M., Gregg, A.R., Stöckler-Ipsiroglu, S., and O'Brien, W.E., 2001, Analysis of guanidinoacetate and creatine by isotope dilution electrospray tandem mass spectrometry. *Clin. Chim. Acta* **308:** 173–178.

Braissant, O., Henry, H., Villard, A.M., Speer, O., Wallimann, T., and Bachmann, C., 2005, Creatine synthesis and transport during rat embryogenesis: spatiotemporal expression of AGAT, GAMT and CT1. *BMC Dev. Biol.* **5:** 9.

Caldeira Araujo, H., Smit, W., Verhoeven, N.M., Salomons, G.S., Silva, S., Vasconcelos, R., Tomas, H., Tavares de Almeida, I., Jakobs, C., and Duran, M., 2005, Guanidinoacetate methyltransferase deficiency identified in adults and a child with mental retardation. *Am. J. Med. Genet. A* **133:** 122–127.

Carducci, C., Birarelli, M., Leuzzi, V., Carducci, C., Battini, R., Cioni, G., and Antonozzi, I., 2002, Guanidinoacetate and creatine plus creatinine assessment in physiologic fluids: an effective diagnostic tool for the biochemical diagnosis of arginine:glycine amidinotransferase and guanidinoacetate methyltransferase deficiencies. *Clin. Chem.* **48:** 1772–1778.

Carducci, C., Santagata, S., Leuzzi, V., Carducci, C., Artiola, C., Giovanniello, T., Battini, R., and Antonozzi, I., 2006, Quantitative determination of guanidinoacetate and creatine in dried blood spot by flow injection analysis-electrospray tandem mass spectrometry. *Clin. Chim. Acta* **364:** 180–187.

D'Hooge, R., Pei, Y.Q., Marescau, B., and De Deyn, P.P., 1992, Convulsive action and toxicity of uremic guanidino compounds: behavioral assessment and relation to brain concentration in adult mice. *J. Neurol. Sci.* **112:** 96–105.

da Silva, C.G., Parolo, E., Streck, E.L., Wajner, M., Wannmacher, C.M., and Wyse, A.T., 1999, In vitro inhibition of Na^+,K^+-ATPase activity from rat cerebral cortex by guanidino compounds accumulating in hyperargininemia. *Brain Res.* **838:** 78–84.

Item, C.B., Mercimek-Mahmutoglu, S., Battini, R., Edlinger-Horvat, C., Stromberger, C., Bodamer, O., Muhl, A., Vilaseca, M.A., Korall, H., and Stöckler-Ipsiroglu, S., 2004, Characterization of seven novel mutations in seven patients with GAMT deficiency. *Hum. Mutat.* **23:** 524.

Mancini, G.M., Catsman-Berrevoets, C.E., de Coo, I.F., Aarsen, F.K., Kamphoven, J.H., Huijmans, J.G., Duran, M., van der Knaap, M.S., Jakobs, C., and Salomons, G.S., 2005, Two novel mutations in SLC6A8 cause creatine transporter defect and distinctive X-linked mental retardation in two unrelated Dutch families. *Am. J. Med. Genet. A* **132:** 288–295.

Mercimek-Mahmutoglu, S., Stoeckler-Ipsiroglu, S., Adami, A., Appleton, R., Araujo, H.C., Duran, M., Ensenauer, R., Fernandez-Alvarez, E., Garcia, P., Grolik, C., Item, C.B., Leuzzi, V., Marquardt, I., Muhl, A., Saelke-Kellermann, R.A., Salomons, G.S., Schulze, A., Surtees, R., van der Knaap, M.S., Vasconcelos, R., Verhoeven, N.M., Vilarinho, L., Wilichowski, E., and Jakobs, C., 2006, GAMT deficiency: features, treatment, and outcome in an inborn error of creatine synthesis. *Neurology* **67:** 480–484.

Neu, A., Neuhoff, H., Trube, G., Fehr, S., Ullrich, K., Roeper, J., and Isbrandt, D., 2002, Activation of $GABA_A$ receptors by guanidinoacetate: a novel pathophysiological mechanism. *Neurobiol. Dis.* **11:** 298–307.

Ohtsuki, S., Tachikawa, M., Takanaga, H., Shimizu, H., Watanabe, M., Hosoya, K., and Terasaki, T., 2002, The blood-brain barrier creatine transporter is a major pathway for supplying creatine to the brain. *J. Cereb. Blood Flow Metab.* **22:** 1327–1335.

Salomons, G.S., van Dooren, S.J., Verhoeven, N.M., Cecil, K.M., Ball, W.S., Degrauw, T.J., and Jakobs, C., 2001, X-Linked creatine-transporter gene (SLC6A8) defect: a new creatine-deficiency syndrome. *Am. J. Hum. Genet.* **68:** 1497–1500.

Schulze, A., 2005, Strategies in the treatment of GAMT deficiency. In: Jakobs, C., and Stoeckler-Ipsiroglu, S., eds., Verhoeven, N.M., and Salomons, G.S., co-eds., Clinical and molecular aspects of defects in creatine & polyol metabolism, 1st ed., SPS Verlagsgesellschaft, Heilbronn, Germany, pp. 19–33.

Schulze, A., 2003, Creatine deficiency syndromes. *Mol. Cell. Biochem.* **244:** 143–150.

Schulze, A., Anninos, A., Hoffmann, G.F., Schwahn, B., Mayatepek, E., Waltz, S., and Rheingans, K., 2005, AGAT enzyme inhibition by high-dose ornithine: a new approach in treatment of GAMT deficiency (abstract). *J. Inherit. Metab. Dis.* **28:** 227.

Schulze, A., Bachert, P., Schlemmer, H., Harting, I., Polster, T., Salomons, G.S., Verhoeven, N.M., Jakobs, C., Fowler, B., Hoffmann, G.F., and Mayatepek, E., 2003, Lack of creatine in muscle and brain in an adult with GAMT deficiency. *Ann. Neurol.* **53:** 248–251.

Schulze, A., Ebinger, F., Rating, D., and Mayatepek, E., 2001, Improving treatment of guanidinoacetate methyltransferase deficiency: reduction of guanidinoacetic acid in body fluids by arginine restriction and ornithine supplementation. *Mol. Genet. Metab.* **74:** 413–419.

Schulze, A., Hess, T., Wevers, R., Mayatepek, E., Bachert, P., Marescau, B., Knopp, M.V., De Deyn, P.P., Bremer, H.J., and Rating, D., 1997, Creatine deficiency syndrome caused by guanidinoacetate methyltransferase deficiency: diagnostic tools for a new inborn error of metabolism. *J. Pediatr.* **131:** 626–631.

Schulze, A., Hoffmann, G.F., Bachert, P., Kirsch, S., Salomons, G.S., Verhoeven, N.M., and Mayatepek, E., 2006, Presymptomatic treatment of neonatal guanidinoacetate methyltransferase deficiency. *Neurology* **67:** 719–721.

Stöckler, S., Holzbach, U., Hanefeld, F., Marquardt, I., Helms, G., Requart, M., Hanicke, W., and Frahm, J., 1994, Creatine deficiency in the brain: a new, treatable inborn error of metabolism. *Pediatr. Res.* **36:** 409–413.

Stöckler, S., Marescau, B., DeDeyn, P.P., Trijbels, J.M.F., and Hanefeld, F., 1997, Guanidino compounds in guanidinoacetate methyltransferase deficiency, a new inborn error of creatine synthesis. *Metabolism* **46:** 1189–1193.

Stoeckler-Ipsiroglu, S., Battini, R., DeGrauw, T., and Schulze, A., 2006, Disorders of creatine metabolism. In: Blau, N., Hoffmann, G.F., Leonard, J., and Clarke, J.T.R., eds., Physician's Guide to the Treatment and Follow-Up of Metabolic Diseases, 1st ed., Springer-Verlag, Berlin, Heidelberg, New York, pp. 255–265.

Verhoeven, N.M., Roos, B., Struys, E.A., Salomons, G.S., van der Knaap, M.S., and Jakobs, C., 2004, Enzyme assay for diagnosis of guanidinoacetate methyltransferase deficiency. *Clin. Chem.* **50:** 441–443.

CHAPTER 10

CLINICAL USE OF CREATINE IN NEUROMUSCULAR AND NEUROMETABOLIC DISORDERS

MARK A. TARNOPOLSKY

Department of Pediatrics and Medicine (Neurology and Rehabilitation), Neuromuscular and Neurometabolic Clinic, Rm 2H26, McMaster University Medical Center, 1200 Main St. W., Hamilton, Ontario, Canada, L8N 3Z5

Abstract: Many of the neuromuscular (e.g., muscular dystrophy) and neurometabolic (e.g., mitochondrial cytopathies) disorders share similar final common pathways of cellular dysfunction that may be favorably influenced by creatine monohydrate (CrM) supplementation. Studies using the *mdx* model of Duchenne muscular dystrophy have found evidence of enhanced mitochondrial function, reduced intra-cellular calcium and improved performance with CrM supplementation. Clinical trials in patients with Duchenne and Becker's muscular dystrophy have shown improved function, fat-free mass, and some evidence of improved bone health with CrM supplementation. In contrast, the improvements in function in myotonic dystrophy and inherited neuropathies (e.g., Charcot-Marie-Tooth) have not been significant. Some studies in patients with mitochondrial cytopathies have shown improved muscle endurance and body composition, yet other studies did not find significant improvements in patients with mitochondrial cytopathy. Lower-dose CrM supplementation in patients with McArdle's disease (myophosphorylase deficiency) improved exercise capacity, yet higher doses actually showed some indication of worsened function. Based upon known cellular pathologies, there are potential benefits from CrM supplementation in patients with steroid myopathy, inflammatory myopathy, myoadenylate deaminase deficiency, and fatty acid oxidation defects. Larger randomized control trials (RCT) using homogeneous patient groups and objective and clinically relevant outcome variables are needed to determine whether creatine supplementation will be of therapeutic benefit to patients with neuromuscular or neurometabolic disorders. Given the relatively low prevalence of some of the neuromuscular and neurometabolic disorders, it will be necessary to use surrogate markers of potential clinical efficacy including markers of oxidative stress, cellular energy charge, and gene expression patterns

1. BACKGROUND OF THE CHAPTER

The focus of this chapter will be on the potential clinical utility of CrM as an adjunctive therapy for the treatment of patients with neuromuscular and neurometabolic disorders. By definition, the neuromuscular disorders represent genetic and acquired conditions

G.S. Salomons and M. Wyss (eds.), Creatine and Creatine Kinase in Health and Disease, 183–204.

that alter peripheral nerve and skeletal muscle function. Included in this category are also the neuromuscular junction defects such as myasthenia gravis which can be acquired (autoimmune) or congenital (due to defects in proteins involved in neuromuscular transition). The peripheral neuropathies affect motor and/or sensory function and often lead to distal atrophy and weakness with or without sensory symptoms. The resultant atrophy of skeletal muscle is due to the denervation process and could be a potential target for therapeutic intervention with CrM. Traditionally, amyotrophic lateral sclerosis (ALS) has been considered to be in the category of the neuromuscular disorders. ALS is a rapidly progressive neurological disorder affecting predominantly middle- and older-aged adults and leading to alpha motor neuron drop-out with resultant muscle weakness and death in, on average, three years. The topic of neurodegeneration and the potential role for creatine in the treatment of ALS have been well studied in the past and are the subject of chapter 11 (Klein and Ferrante, 2007). A list of some of the more common neuromuscular and neurometabolic disorders is presented in Table 1.

The most common peripheral neuropathy in North America and Europe is diabetic predominantly sensory polyneuropathy, with vitamin B12 deficiency being the next most common cause. The main peripheral neuropathy that will be discussed in this chapter will be Charcot-Marie-Tooth (CMT) disease which represents a group of hereditary motor and sensory neuropathies affecting either the myelin sheath or the axons. These patients have progressive distal atrophy with and without clinical sensory loss that initially affects gait but often impairs hand function. The estimated prevalence of CMT disease is 1:2,500 in the population.

Table 1. Common neuromuscular disorders .

Condition	Example
A. Myopathy	
Inherited	
Muscular dystrophy	Duchenne, limb-girdle, myotonic
Channelopathy	Thomsen's disease, malignant hyperthermia
Acquired	
Inflammatory	Dermatomyositis, inclusion body myositis
Toxin	Statins, corticosteroids
B. Neuropathy	
Inherited	
Charcot-Marie-Tooth disease	1A (*PMP22* duplication), 1B (MPZ deletion)
Spinal muscular atrophy	SM1 (*smn* deletion)
Acquired	
Nutritional deficiency	B_{12}, thiamine
Diabetes	Distal diabetic polyneuropathy

The myopathies represent a heterogenous group of disorders where skeletal muscle is the prime target for dysfunction. In general, the myopathies are classified as hereditary (muscular dystrophy, metabolic myopathies, channelopathies) or acquired (polymyositis, statin-induced myopathy). Most of the research evaluating the potential utility of CrM in neuromuscular disorders has focused on the dystrophinopathies, Duchenne muscular dystrophy (DMD) and Becker's muscular dystrophy (BMD). DMD is an X-linked progressive myopathy due to mutations within the dystrophin gene, with most boys becoming wheelchair-bound by age 12 y and dying in their early 20s of respiratory or cardiac failure (Brooke *et al.*, 1981a, 1989). The only therapy that has had any beneficial impact on the course of the disease are the corticosteroids (prednisone and deflazacort); however, they are associated with major side effects including osteopenia, dysglycemia, fluid retention, cataracts, and myopathy (Moxley *et al.*, 2005).

The neurometabolic disorders represent inborn errors of metabolism affecting intermediary metabolism. For purposes of this chapter, the focus will be on mitochondrial myopathies/cytopathies (e.g., mitochondrial encephalopathy, lactic acidosis, and stroke-like episodes, MELAS), glycolytic and glycogenolytic defects (e.g., McArdle's disease), and fat oxidation defects (e.g., carnitine palmitoyl transferase deficiency). The mitochondrial cytopathies are a group of disorders with a primary defect of the electron transport chain as the major feature. They can show maternal (MELAS), autosomal recessive (Leigh's disease) or autosomal dominant (chronic progressive external ophthalmoplegia, CPEO) inheritance patterns. They occur with a prevalence of about 1:8,000 and reveal extreme clinical and molecular heterogeneity, yet have reduced mitochondrial function/aerobic energy transduction as a common feature. The clinical symptoms include exercise intolerance, muscle weakness, peripheral neuropathy, strokes, migraines, seizures, cardiomyopathy, and ataxia. Defects in glycogenolysis are usually autosomal recessive and include myophosphorylase deficiency (glycogen storage disease [GSD] V, McArdle's disease) and acid maltase deficiency (GSD II, Pompe's disease); while an example of a glycolytic defect would be phosphofructokinase deficiency (GSD VII, Tarui's disease). Patients with these types of muscle GSDs have varying combinations of muscle cramping with exercise, myoglobinuria, proximal weakness (GSD II), and abdominal symptoms (GSD VII).

One of the confounding issues when evaluating the neuromuscular and neurometabolic disorders is the fact that these diseases *per se* lead to inactivity which can lead to disuse atrophy. It is known that disuse of skeletal muscle leads to atrophy, and some of the potential benefits of CrM supplementation could be merely functioning as a counter measure to the atrophy component. Nevertheless, there are a number of fundamental pathologic features that characterize the neuromuscular and neurometabolic disorders that could be amenable to creatine supplementation as outlined below (Table 2).

Table 2. Potential benefits of creatine monohydrate supplementation in neuromuscular and neurometabolic disorders. CHF = chronic heart failure; ↑ = increase; ↓ = decrease .

	Potential Benefit
A. General	
Weakness	↑ Strength
Fatigue	↑ Endurance
Muscle atrophy	↑ Muscle mass
↑ Oxidative stress	↓ Oxidative stress
B. Neuromuscular-specific	
↑ Intracellular Ca^{2+}	↓ Ca^{2+}
Side effects of corticosteroids	↑ Type II fibers
↓ Mitochondrial function	↑ Mitochondrial function
C. Neurometabolic-specific	
↑ Risk of strokes	CNS anoxia protection
Cardiomyopathy	Cardiac benefits, ↑ Strength in CHF patients
↓ Mitochondrial function	↑ Mitochondrial function
↓ Phosphocreatine	↑ Phosphocreatine

2. THEORY FOR THE CLINICAL USE OF CREATINE

2.1. Neuromuscular Disorders

There are a number of final common pathways that lead to skeletal muscle necrosis, apoptosis and autophagy in myopathies. These final pathways include cytoskeletal disruption, increased reactive oxygen species, electron transport chain dysfunction, increased protein degradation, secondary inflammation and elevated intracellular calcium levels (Mallouk *et al.*, 2000; Pulido *et al.*, 1998; Tarnopolsky and Beal, 2001). A number of studies have demonstrated that CrM supplementation can lead to a reduction in markers of oxidative stress in a variety of tissues, especially the central nervous system (Adcock *et al.*, 2002; Andreassen *et al.*, 2001; Klivenyi *et al.*, 1999, 2004; Lawler *et al.*, 2002; Malcon *et al.*, 2000; Sullivan *et al.*, 2000). Given the importance of the PCr system to calcium homeostasis (Duke and Steele, 1999) and the ability of creatine supplementation to enhance calcium ATPase function (SERCA channels) (Hespel *et al.*, 2002; Pulido *et al.*, 1998), this may be a mechanism whereby elevated intracellular calcium levels could be attenuated. Our group has shown a reduction in whole-body protein breakdown following creatine supplementation in young men (Parise *et al.*, 2001). Studies in *mdx* myoblasts (a murine model of DMD) have demonstrated that the pre-treatment with creatine (20 mM) attenuated the elevation of intracellular calcium and this appeared to be through a stimulation of calcium ATPase channels (Pulido *et al.*, 1998). Furthermore, myotube formation and survival were enhanced with the creatine supplementation strategy (Pulido *et al.*, 1998). In a second study by the same group, supplementation of newborn *mdx* mice with creatine caused an attenuation of the first wave of necrosis predominantly in white muscle, which was associated with enhancement of the mitochondrial respiratory capacity (Passaquin *et al.*, 2002).

Improved clinical outcomes such as increased strength have also been observed in trials with *mdx* mice (Granchelli *et al.*, 2000). Another group found that one month of creatine supplementation tended to increase tetanic tension (by 9.2%, $P = 0.08$), and significantly altered muscle relaxation time ($P < 0.001$) (Louis *et al.*, 2004). Our group found some benefits from creatine in slightly older *mdx* mice; however, the benefits were much more apparent when the animals were given a combination therapy consisting of conjugated linoleic acid, prednisolone (a corticosteroid), α-lipoic acid, CrM and β-OH-methylbutyrate (Payne *et al.*, 2006). These latter data suggest that while CrM may confer some therapeutic benefits on its own, it is likely more effective as part of a combination therapy with compounds that target overlapping and complementary pathways involved in muscle dysfunction in myopathies.

In the case of peripheral neuropathy, there is usually distal muscle atrophy and weakness due to the denervation process. Given that CrM may increase muscle mass and strength (Mihic *et al.*, 2000; Terjung *et al.*, 2000), and possibly through a long-term neuroprotective effect (Adcock *et al.*, 2002; Dedeoglu *et al.*, 2003; Malcon *et al.*, 2000; Sullivan *et al.*, 2000; Wilken *et al.*, 2000; Zhu *et al.*, 2004), there may be some role for creatine therapy in patients with CMT. Studies have demonstrated an increase in muscle strength with weight training in these patients (Chetlin *et al.*, 2004), and given that creatine may potentiate some of the strength-induced gains consequent to weight training in younger and older adults (Brose *et al.*, 2003; Burke *et al.*, 2003a; Chilibeck *et al.*, 2005; Tarnopolsky *et al.*, 2001; Terjung *et al.*, 2000), there could be a potential for an interactive role between creatine and strength training. To date, there have been no studies looking at potential therapeutic efficacy of CrM in diabetic neuropathy; however, it would be of interest to evaluate a potential role for creatine plus α-lipoic acid, given that α-lipoic acid can enhance creatine uptake (Burke *et al.*, 2003b), and that α-lipoic acid alone has demonstrated benefits in diabetic neuropathy (Ziegler *et al.*, 2004).

2.2. Neurometabolic Disorders

Mitochondrial disorders are associated with a reduction in aerobic ATP production and stores, increased utilization of anaerobic energy sources, and an increase in oxidative stress (Piccolo *et al.*, 1991; Raha and Robinson, 2000; Tarnopolsky and Beal, 2001; Tarnopolsky and Parise, 1999). Relevant to a potential for creatine as a therapy, patients with mitochondrial myopathy also show reductions in basal levels of phosphocreatine (Tarnopolsky and Parise, 1999), and impairment in phosphocreatine (PCr) re-synthesis (Argov *et al.*, 1996; Tartaglia *et al.*, 2000). Strategies to mitigate against these consequences have included: antioxidants (e.g., α-lipoic acid, vitamin E); bypass to overcome a specific defect (e.g., succinate to bypass complex I); direct cofactor provision (e.g., coenzyme Q_{10}); and CrM supplementation to provide an alternative energy route (Mahoney *et al.*, 2002; Tarnopolsky and Beal, 2001;

Tarnopolsky and Raha, 2005). Given that several of the mitochondrial cytopathies also predispose to cardiomyopathy and stroke-like episodes, there may be further neuro- and cardio-protective benefits from the administration of CrM in the long-term. Studies have demonstrated preservation of cerebral synaptic transmission during anoxia with creatine supplementation in hippocampal slices (Carter *et al.*, 1995; Whittingham and Lipton, 1981). This has also been demonstrated in the neonatal mouse brainstem where creatine supplementation maintained ATP and phosphocreatine concentrations higher than in control animals and demonstrated a neuroprotective effect (Wilken *et al.*, 2000). The neuroprotective effects of CrM are discussed in more detail in chapter 11 (Klein and Ferrante, 2007). Cardiomyopathy appears to be one of the more ominous signs in mitochondrial cytopathies, for its development is a negative predictive factor for survival (Scaglia *et al.*, 2004). It is unclear whether creatine would have any beneficial effects at the prevention of cardiomyopathy; however, in adult patients with congestive heart failure, CrM did enhance exercise performance (Gordon *et al.*, 1995; Kuethe *et al.*, 2006), and exercise is a potent and effective therapy for mitochondrial cytopathies (Taivassalo and Haller, 2005; Tartaglia *et al.*, 2000). Our group has reported a reversal of paracrystalline inclusions in skeletal muscle from a patient with a novel cytochrome b mutation (G15497A) following treatment with CrM (10 g/d × 5 wk) (Tarnopolsky *et al.*, 2004c). We generated neuroblastoma-based cybrids from platelets of the male patient and found enhanced production of reactive oxygen species, lower ATP content, and that these cybrids were much more sensitive to oxygen and glucose deprivation, as well as to peroxynitrite toxicity, as compared to control cybrids harboring wild-type mitochondrial DNA. Importantly, we found that the toxic effects of oxygen and glucose deprivation, as well as of peroxynitrite, were completely abolished with CrM administration (Tarnopolsky *et al.*, 2004c).

In disorders of glycogenolysis and glycolysis there is a "knock-out" of the major anaerobic energy yielding pathway in skeletal muscle. The pathways available for compensatory up-regulation are the PCr/Cr and myoadenylate deaminase pathways. The increased flux through myoadenylate deaminase in patients with McArdle's disease and phosphofructokinase (PFK) deficiency leads to hyperammonemia and hyperuricemia which can result in gout (myogenic hyperuricemia). The PCr/Cr pathway has a greater quantitative capacity for buffering energy charge during anaerobic exercise and was a major rationale for clinical studies of its use in McArdle's disease. Patients with McArdle's disease may develop fixed proximal weakness with aging, and patients with glycogen storage disease type 2 (acid maltase deficiency) usually have fixed and progressive proximal weakness with respiratory insufficiency. Given the potential for creatine to enhance muscle strength (Tarnopolsky *et al.*, 2004b; Tarnopolsky and Parise, 1999; Terjung *et al.*, 2000), there may be theoretical benefits in these groups of patients.

In patients with fatty acid oxidation defects there is often exercise impairment as the only manifestation during prolonged endurance activity and/or in the fasted state, and the quantitative contribution from the PCr/Cr pathway would be expected to be quite small. In contrast, in some of the β-oxidation defects including LCHAD, SCHAD and LCAD deficiency, there is progressive neuropathy and myopathy with severe atrophy. These latter individuals may derive benefit from creatine supplementation more from the strength and muscle mass perspective as opposed to an energy bypass strategy. There is only one published study which demonstrated "striking improvement of muscle strength..." and ataxia following creatine therapy in a girl with LCHAD deficiency (Korenke *et al.*, 2003).

3. STUDY DESIGN ISSUES IN CLINICAL TRIALS

The ultimate goal of a therapeutic intervention is to cure the disease. Given that a cure is not possible for essentially every neuromuscular and neurometabolic disorder, the goal is control of symptoms and functional improvement. Consequently, to determine the efficacy of a therapeutic intervention, the ultimate evidence of efficacy is an improvement in quality of life and functional capacity as well as a reduction in morbidity and mortality. An evaluation of the effect of an intervention on these clinical outcomes usually requires very large numbers of subjects followed for a long period of time to detect significant differences as is possible with common conditions such as cancer, heart disease and diabetes. Most of the neuromuscular and neurometabolic disorders are relatively rare and the recruitment of hundreds to thousands of patients is not feasible. Another issue is that of the cost of running such trials which can run into millions of dollars for clinical outcome evaluations. With the specific case of CrM, these types of resources are not likely to be invested by the large pharmaceutical companies given that creatine is a dietary supplement, and that many of the neuromuscular conditions are considered so rare that they only qualify for orphan drug trials with limited funding opportunities.

In order to complete trials in relatively rare diseases, study design issues and outcome variables become extremely important. From a design perspective, when subjects are limited, it is best to consider a repeated measures cross-over design where the individual becomes his or her own control (during the placebo arm). This design reduces biological variance between subjects and improves statistical power. An issue with this design is that there can be a carry-over effect and, in the case of CrM, the washout interval needs to be at least five to six weeks (Terjung *et al.*, 2000). Furthermore, certain disorders such as ALS or polymyositis show either a rapid decline or improvement in function, respectively, and a parallel design is more appropriate (however, sample size has to be larger). With smaller sample sizes it is also important to have outcome variables that have both validity and reliability. For example, it is clearly evident that manual muscle testing has a much higher variance than objective strength measurements in muscular dystrophy (Escolar *et al.*, 2001), which implies that either the effect size has to be much larger or the sample size increased to deal with the increased variance with manual muscle testing. Another important consideration is the use of surrogate markers. It is important that reproducible and valid surrogate markers are utilized that represent known pathophysiologic consequences of the disease. For example, the use of α-lipoic acid would be expected to reduce markers of oxidative stress (8-isoprostanes) in a two to four week trial; however, an improvement in strength or muscle function may not be measurable or apparent for many years.

The aforementioned design issues need to be considered when evaluating the literature and in the design of future studies. In the case of DMD, an evaluation of the use of prednisone as a therapy took many years, multiple centers, and millions of dollars to obtain the background information sufficient to initiate the early clinical trials (Brooke *et al.*, 1981a,b, 1989; Moxley *et al.*, 2005). In spite of these trials being completed almost 20 years ago, there still is uncertainty around the timing, type, dose and even need for corticosteroid therapy in DMD. In order to determine whether CrM has a therapeutic affect in a given disease state these design issues need to be considered, and it is likely that national and international collaborative efforts will be required to avoid type 2 errors (i.e. concluding that there is no effective therapy when in reality there is).

4. CLINICAL UTILITY OF CREATINE IN NEUROMUSCULAR DISORDERS

4.1. Muscular Dystrophy

Some of the earliest interest in the role of creatine and muscular dystrophy emerged in the 1950's (Benedict *et al.*, 1955; Milhorat, 1953). A review on biochemical aspects of muscular dystrophy summarized the available data showing that there was a marked increase in creatine excretion (Benedict *et al.*, 1955), to the extent that some investigators were suggesting that muscular dystrophy could be due to disturbance of creatine metabolism (which we now know is not correct). The hypothesis that the increased urinary creatine was due to an inability of the dystrophic muscle to transport creatine (Benedict *et al.*, 1955) is likely to be correct (Tarnopolsky *et al.*, 2003; Tarnopolsky and Parise, 1999). A re-emergence of the interest in the clinical utility of CrM in patients with neuromuscular disorders emerged following the increased interest in creatine metabolism in the early 1990's.

Our group reported the results of an open study (N = 21) followed by a single-blinded study looking at objective measures of strength and body composition in patients with a variety of neuromuscular disorders. On average the participants consumed 0.11 g creatine/kg/day for 11 days, and we found significant increases in dorsi-flexion, handgrip strength, and total body weight (we did not measure fat free mass) (Tarnopolsky and Martin, 1999). In the open study there was an equal distribution of patients with mitochondrial cytopathies (see below), inflammatory myopathies, muscular dystrophy, and peripheral neuropathic disorders (including CMT disease and post-polio muscle dysfunction). Most of the participants in the single-blinded component of the study had myoadenylate deaminase deficiency or muscular dystrophy (Tarnopolsky and Martin, 1999).

Further evidence for a role for creatine in muscle disorders came from a randomized double-blind cross-over study conducted by Walter and colleagues on 36 patients with muscular dystrophy (fascioscapulohumeral MD, N = 12; BMD, N = 10; DMD, N = 8;

and limb girdle muscular dystrophy, N = 6) (Klopstock *et al.*, 2000). The participants consumed 10 g creatine/d (children 5 g/d) or placebo for eight weeks with a three week washout and a cross-over. There was a statistically significant increase in the mean Medical Research Council score (MRC) and improvement in the mean neuromuscular symptoms score. In the proportion of patients who felt subjective improvement during the creatine phase, no change in vital capacity and serum creatine kinase activity, and no side effects were noted (Klopstock *et al.*, 2000).

In each of the previous two studies, the participants were not analyzed by specific subgroup. Subsequent studies have examined the clinical utility of creatine predominantly in the muscular dystrophy group, particularly in the dystrophinopathy (BMD and DMD) and myotonic (DM1 and DM2) muscular dystrophy groups.

Given the beneficial effects of creatine seen in heterogenous groups of patients with muscular dystrophy (Klopstock *et al.*, 2000; Tarnopolsky and Parise, 1999), several studies have evaluated its efficacy in boys with dystrophinopathy. The Cooperative International Neuromuscular Research Group (CINRG) completed a randomized parallel group double-blind placebo controlled study in 50 ambulant boys with DMD (Escolar *et al.*, 2005). Creatine was provided at 5 g/d for a six-month period with manual muscle testing and quantitative muscle testing completed as primary outcome variables and with time functional tests in pulmonary function as secondary outcome variables. A total of 16 boys were randomized to placebo and 15 to creatine (Escolar *et al.*, 2005). This study demonstrated a smaller decline in strength over the six month period for creatine versus placebo (P = 0.07). Secondary analysis showed significantly less deterioration in timed functional task scores in the creatine group compared to the placebo group in boys under 7 years of age, and lesser deterioration of strength in boys greater than 7 years of age. The functional task of timed climbing was significantly better at P = 0.015 for the creatine group (Escolar *et al.*, 2005). The conclusion of this group was that the effects seen with creatine were not as significant as those previously observed for prednisone or deflazacort (however, a parallel group randomized to corticosteroids was not included in this study). They concluded that a disease-modifying effect of treatment could not be excluded and suggested that larger age-stratified studies be completed with quantitative muscle strength testing (Escolar *et al.*, 2005).

Another study evaluated 12 boys with DMD and 3 boys with BMD using a randomized double-blind cross-over design with three months of CrM or placebo and a two-month washout period in between (Louis *et al.*, 2003). While on creatine supplementation, there was an improvement in maximal voluntary contraction strength and an increase in fatigue resistance with no change in total joint stiffness (whereas it was increased in the placebo group by 25%). In boys who were still ambulating there was an increase in bone mineral density and, interestingly, there was also a reduction in urinary N-telopeptide excretion (a marker of bone breakdown) (Louis *et al.*, 2003).

We studied 30 boys with Duchenne muscular dystrophy who completed a randomized double-blind crossover trial with two treatment periods (CrM = 0.1 g/kg/d, or placebo) of 4 months duration and a six-week washout period in between (Tarnopolsky *et al.*, 2004b). Fifty percent of the boys were taking corticosteroids throughout the duration of this study (Tarnopolsky *et al.*, 2004b). CrM treatment increased dominant handgrip strength and fat-free mass and reduced urinary N-telopeptide excretion (all at $P < 0.05$). We also found a strong trend towards an attenuated loss of global muscle strength ($P = 0.056$) for the creatine-treated arm of the trial. There was no effect on serum CK activity and liver function tests, but an expected slight increase in serum creatinine concentration (Tarnopolsky *et al.*, 2004b).

In summary, there have been three randomized double-blind trials looking at creatine supplementation in boys and men with dystrophinopathy (Escolar *et al.*, 2005; Louis *et al.*, 2003; Tarnopolsky *et al.*, 2004b). Each of these studies showed statistically significant improvements in strength and function. Given that most boys with dystrophinopathy are treated with corticosteroids, a very interesting finding was the reduction of N-telopeptide (a marker of bone breakdown) excretion with creatine supplementation in both studies that measured this outcome variable. This suggests that over a longer period of time there may be some bone protective effects of creatine which may provide a more inexpensive and safer treatment than with bisphosphonates which are currently the standard for treating osteoporosis.

The research group of Walter evaluated the potential efficacy of CrM (10.6 g/d for days 1 to 10, 5.3 g/d for days 11 to 56) in 34 patients with myotonic muscular dystrophy type 1 (DM1) using a double-blind cross-over design (Klopstock *et al.*, 2000). Treatment and placebo periods were eight weeks each, with a six-week washout period in between. There was a non-significant improvement in strength for the creatine versus the placebo group ($P = 0.11$) and during the second cross-over period, there was interaction where creatine-treated patients improved, while placebo patients worsened ($P = 0.005$) (Klopstock *et al.*, 2000). Similar trends were noted for neuromuscular symptom score with an overall improvement with creatine supplementation that failed to reach significance ($P = 0.15$); however, in the second cross-over period, the comparison between the groups showed a strong trend towards a significant effect for creatine supplementation ($P = 0.064$). A similar trend was noted for quantitative muscle testing of arm flexion with the creatine treatment arm showing improvement and the placebo showing a decline in function ($P = 0.067$). There were no effects seen for body composition or vital capacity, and the patients did not have a global perception of improvement in either phase. There were no adverse effects, and blood chemistries were not altered. Although several of the outcome trends consistently favored the creatine arm of the study, they were not statistically significant, which could indicate a type 2 error (Klopstock *et al.*, 2000).

We subsequently completed a randomized double-blind cross-over trial in 31 patients with DM1 with two four-month phases (creatine, 5 g/d and placebo) and a six-week washout period in between (Tarnopolsky *et al.*, 2004a). We found no effect of creatine supplementation on manual muscle strength testing, quantitative strength testing for hand-grip, dorsi-flexion and knee extension, high intensity exercise endurance, activities of daily living scales, urinary creatine kinase activity, body composition, liver function test and serum creatinine concentration. We also measured PCr/ATP ratios in forearm flexors using magnetic resonance spectroscopy (MRS) and did not find any significant increase in this ratio (Tarnopolsky *et al.*, 2004a). The conclusion from both studies was that there were no statistically significant changes; however, it will be of interest to combine the results from the two trials to increase sample size and determine whether some of the trends in the Walter study were due to a type 2 error or random chance. Given the lack of an increase in PCr content in the patients in response to creatine supplementation, strategies to enhance uptake into dystrophic muscle may be worth pursuing (for instance, a higher dose of creatine, or combination with α-lipoic acid supplementation).

Finally, a study on twenty patients with myotonic dystrophy type 2 (DM2) evaluated the efficacy of CrM supplementation (10 g/d) (Schneider-Gold *et al.*, 2003). This was a randomized parallel group design where half of the participants were randomized to creatine supplementation and half to placebo for a period of twelve weeks. Ten participants completed each arm of the trial. There was a significantly greater number of patients in the creatine group who reported subjective improvement in their overall function ($P < 0.01$). There was a non-significant trend towards an increase in MRC score (CrM = + 3.2 %; PL = + 1.8 %) and hand-held dynamometry strength of the proximal arms and legs (CrM = + 9.4%; PL = + 4.1%), with no effect on neuromuscular symptoms score or peak grip strength. Given that myalgias are a common symptom in DM2, it was interesting, and perhaps worthy of further investigation, that during the creatine administration period, leg pain resolved in two patients and chest pain resolved in another patient, with no improvements noted in those randomized to placebo (Schneider-Gold *et al.*, 2003).

Overall, there appears to be a consistent increase in several measurements of strength and an increase in fat-free mass in studies evaluating the short- (weeks) to intermediate- (months) term effects of CrM supplementation in patients with muscular dystrophy, with minimal changes seen in the myotonic muscular dystrophies. These conclusions are supported by a recent Cochrane review meta-analysis of all published literature on the subject (Kley *et al.*, 2007).

4.2. Peripheral Neuropathy

Our group evaluated a total of 39 participants with confirmed CMT disease type 1 (N = 34) and type 2 (N = 5) (Doherty *et al.*, 2001). This was a randomized double-blinded placebo-controlled cross-over study. Our participants took CrM (5 g/d) and a corresponding placebo for one month each, with a five-week washout period in between. We found no effect of creatine treatment on body composition, peak maximal voluntary contraction strength for grip, dorsi-flexion and knee extension, as well as on functional tasks (30 m walk and stair climbing) (Doherty *et al.*, 2001).

Another study evaluated 20 patients with CMT disease who completed a 12-week resistance exercise training program, with 50% of the patients taking creatine (5 g/d) and the balance taking placebo (Chetlin *et al.*, 2004). Although there were significant improvements in type 1 muscle fiber diameter, activities of daily living scores and objective measures of strength with training, there were no interactive affects with CrM supplementation. There were a few trends towards a greater response in the CrM supplemented group (Chetlin *et al.*, 2004). Consequently, the same group did further analysis on 18 of the subjects in the latter study, evaluating myosin heavy chain composition and relating this to functional outcomes (Smith *et al.*, 2006). There was a significant reduction in MHC type 1 in the participants who received creatine as compared to placebo, and the reduction in MHC type 1 content correlated with improved muscle performance (increase in chair rise). There was a non-significant increase in MHC type 2X and 2A content that was greater for the creatine group, and the training-induced increase in MHC type 2A content was inversely correlated with chair rise time for the creatine-supplemented group (i.e. an increase in MHC type 2A content improved performance) (Smith *et al.*, 2006).

Although further research needs to be conducted, it appears that CrM supplementation *per se* does not appear to improve muscle strength or body composition in patients with CMT disease in the short-term (one month). When combined with resistance exercise training, there is some suggestion that the patients receiving creatine showed greater improvement in function; however, this would be an expected finding given the consistency of studies in healthy and older adults showing evidence of enhancement of strength exercise induced gains in those supplemented with creatine versus placebo (Brose *et al.*, 2003; Burke *et al.*, 2003a; Chilibeck *et al.*, 2005; Tarnopolsky *et al.*, 2001). In the future, studies using techniques such as motor unit counting, nerve conduction studies and objective strength testing over long periods of time (one to two years) will be required to determine if there are long-lasting effects of creatine supplementation in patients with CMT disease, particularly when combined with resistance exercise training.

4.3. Other Disorders

Patients with chronic renal failure show evidence of mitochondrial dysfunction, muscular atrophy and weakness (Johansen *et al.*, 2005). Patients who are receiving dialysis for chronic renal failure often experience painful muscle cramps due to electrolyte imbalance. One study found that CrM given prior to dialysis over a period of four weeks reduced the frequency of muscle cramps by 60% in those randomized to CrM (Chang *et al.*, 2002). As expected there was a slight increase in serum creatinine due to an increased rate of appearance and decreased removal.

There are a large number of neuromuscular (Schneider-Gold *et al.*, 2005), and indeed many other rheumatologic (Rindfleisch and Muller, 2005), disorders where corticosteroids are a mainstay of therapy. As mentioned previously, corticosteroids are used also in the treatment of Duchenne muscular dystrophy (Moxley *et al.*, 2005). The side effects of corticosteroid use are well known and include type 2 muscle fiber atrophy with weakness, osteopenia and growth retardation in children. Studies using chondrocytes in culture have demonstrated a potential beneficial affect of creatine supplementation, enhancing their metabolic activity and mineralization (Gerber *et al.*, 2005). Furthermore, we have completed a study in growing Sprague-Dawley rats randomized to placebo, prednisolone injection, creatine, and prednisolone plus creatine. The animals taking creatine showed an increase in type 2 muscle fiber area even with concomitant corticosteroid treatment (Roy *et al.*, 2002). Importantly, from a pediatric perspective, the creatine treated rats also did not show the corticosteroid induced attenuation of growth in the latter study (Roy *et al.*, 2002). A similar paper in Golden hamsters showed that creatine administration prevented dexamethasone induced muscle fiber atrophy and impairment of spontaneous running (Campos *et al.*, 2006). This same group also showed that CrM prevented the dexamethasone induced loss of gastrocnemius and diaphragm muscle loss and type II fiber atrophy in the gastrocnemius (Menezes *et al.*, 2007). As mentioned above with our Duchenne muscular dystrophy trial, we found evidence that the clinical efficacy of creatine supplementation is equivalent for boys who are taking corticosteroids and those who are not (Tarnopolsky, 2004). This suggests that creatine can provide additive effects to the already well-known beneficial effects of prednisone in boys with DMD. The animal data suggests that there could be further benefit by attenuating some of the side effects of corticosteroids. Given the widespread use of corticosteroids and their severe and well-known side effects, further research to explore the potential for creatine to attenuate side effects is warranted.

Given that there is a reduction in PCr in patients with inflammatory myopathies (Park *et al.*, 1994, 1995, 2000), and that the mainstay of treatment for dermatomyositis and polymyositis is prednisone, there may well be a role for creatine supplementation in this group of patients. We had 14 patients with myositis in an initial open label study, and they showed improvements in function with creatine supplementation (Tarnopolsky and Parise, 1999); however, these findings clearly need to be repeated in a randomized double-blind crossover fashion. At the current time, inclusion body myositis is resistant to any immunosuppressive therapy, and the only intervention that has shown any benefit is resistance exercise training (Alexanderson, 2005; Arnardottir *et al.*, 2003; Isenberg *et al.*, 2004). Therefore, there is a need to complete a randomized double-blind trial to determine whether there are independent or interactive effects of CrM supplementation with exercise training in this group of individuals.

5. CLINICAL UTILITY OF CREATINE IN NEUROMETABOLIC DISORDERS

5.1. Mitochondrial Cytopathy

There have been a number of small case series and case reports on the effects of creatine supplementation in mitochondrial cytopathies. A case report of an 18 year old with MELAS syndrome taking creatine at 12 g/d for 12 days followed by 5g/d for 28 months showed a definite improvement in MRI and MR spectroscopy values with an increase in brain creatine as well as marked behavior and cognitive improvements. There appeared to be a deterioration of renal function with an impairment of renal creatinine clearance in this patient who had pre-existing nephropathy (Barisic *et al.*, 2002). A small open study in four patients with mitochondrial cytopathy (Kearns Sayre syndrome, NARP and MELAS) evaluated the effects of CrM supplementation (0.1–0.35 g/kg/day) (Komura *et al.*, 2003). This was actually a retrospective study for nine months to almost five years and found an improvement in cycle ergometry performance of about 12% during steady-state exercise with no significant effects on peak power

(Komura *et al.*, 2003). A similar study in four patients (Kearns Sayre syndrome, NARP and MELAS) evaluated the efficacy of CrM (0.1–0.2 g/kg/day for three months) on cycle ergometry performance (Borchert *et al.*, 1999). There was an increase in maximal power ranging from 8 to 17% with an increase in endurance performance ranging from 30 to 57%. The latter two studies, with a total of 9 patients, demonstrated improvements in cycle ergometry performance.

Our group used a randomized double-blind cross-over design in 7 patients with MELAS syndrome randomized to CrM (10 g/day for 2 wk → 3 g/d for 1 wk) and a placebo (3 wk) with a five week wash-out period (Tarnopolsky *et al.*, 1997). We found improvements in high-intensity repetitive grip strength and dorsi-flexion strength during a two-minute performance trial with no improvement in $VO_{2,max}$ in a cycle ergometry test, and no statistically significant improvements in body composition (Tarnopolsky *et al.*, 1997). Subsequent to this study, there were two randomized double-blind trials in patients with mitochondrial DNA deletions (predominantly CPEO) with a few patients with Kearns Sayre syndrome (Klopstock *et al.*, 2000; Kornblum *et al.*, 2005). One study evaluated the efficacy of CrM supplementation (20 g/d) in 16 patients with CPEO or isolated mitochondrial myopathy using a four-week randomized, double-blind, cross-over design. There were small trends towards an improvement in MRC score (P = 0.14) and non-ischemic isokinetic biceps flexion strength (P = 0.16); yet neuromuscular symptoms and Hammer-Smith score, functional tasks, ataxia and peak biceps strength were not different. The authors' conclusions were that the power of the study was limited and that in the future, larger multi-center trials are needed (Klopstock *et al.*, 2000). A subsequent study in patients with CPEO and Kearns Sayre syndrome evaluated the efficacy of CrM (150 mg/kg/d) or placebo using a cross-over design (Kornblum *et al.*, 2005). ^{31}P-MRS did not show any improvement with supplementation nor was there an enhancement in post-exercise PCr recovery (an indicator of mitochondrial function). There were no significant differences measured between the treatment arms; however, some of the outcomes such as the maximal voluntary contraction strength after aerobic exercise showed some trends towards an improvement on creatine (P = 0.14). The authors' conclusion was that the short study duration and limited number of patients may have led to a type 2 error (Kornblum *et al.*, 2005). Finally, our group has recently published the results of a randomized, double-blind trial showing that a combination of CrM + coenzyme Q10 + α-lipoic acid resulted in lower lactate and 8-isoprostanes (a marker of oxidative stress) and higher fat-free mass (only in the MELAS sub-group) as compared with a placebo (Rodriguez *et al.*, 2007). An issue with the latter study is that it was not possible to determine what proportion of the effects observed were due to the CrM component. Overall, the data with respect to the use of CrM supplementation in mitochondrial cytopathy is equivocal in spite of the strong theoretical evidence for benefit. Findings similar to those seen in young healthy individuals (e.g., an increase in repetitive high intensity activity) (Casey *et al.*, 1996; Mihic *et al.*, 2000) have been reported in these patients; however, whether this translates into functional improvement and long-term benefit remains to be seen. With a disease such as MELAS that has huge variability in clinical signs and symptoms (Tarnopolsky and Raha, 2005), the ability to detect differences in functional outcomes will be difficult and clearly will require large multi-center trials.

5.2. Glycogen Storage Disease

Given the increased reliance on alternative anaerobic energy pathways in response to severely attenuated glycogenolysis seen in McArdle's disease, it would appear to be an ideal candidate for a disease that should respond to CrM supplementation. One study evaluated 9 patients with McArdle's disease randomized to CrM (150 mg/kg/d for one week, followed by 60 mg/kg/d for 4 wk) versus placebo with a five-week washout period in between (Vorgerd *et al.*, 2000). There was a non-significant increase in muscle PCr content ($P = 0.15$) with no enhancement in PCr recovery following either aerobic or ischemic exercise. During ischemic exercise, the force-time integral ($P = 0.03$), PCr depletion and P_i accumulation ($P = 0.05$) were higher on creatine (Vorgerd *et al.*, 2000). During aerobic exercise there was a significant increase in PCr depletion ($P = 0.006$), and surface electromyography improved with respect to the mean frequency in the creatine group. There was no evidence of serological or clinical side effects during the trial (Vorgerd *et al.*, 2000). This same group completed a larger trial with 19 patients with McArdle's disease patients, using a similar design but a constant higher dose of CrM (150 mg/kg/d for 5 wk) (Vorgerd *et al.*, 2002). Interestingly, there was a worsening of muscle pain limitations in daily activity and negative affects on the surface EMG values for the creatine group (Vorgerd *et al.*, 2002). The overall conclusion from both studies was that high-dose creatine administration has a negative effect on function yet a dose of around 60 mg/kg/d appears to confer benefit with respect to muscle function. It may be that a limited ability to load muscle with creatine in these patients may play some role (Vorgerd *et al.*, 2002); however, that would not explain the worsening of symptoms with the higher dose. It is likely that some of the failure for creatine to be of clinical benefit in these patients is due to the inability of McArdle's disease patients to acidify muscle, and hence the flux in the direction of ATP generation by the PCr/Cr pathway would be inhibited. Currently, it is unclear why patients with McArdle's disease show an opposite response to different CrM doses, but more work is needed in this area.

6. FUTURE DIRECTIONS

The current data appear to support the concept that CrM administration over a period of several months can increase muscle strength and enhance body composition in patients with muscular dystrophy. It will be important to determine whether these objective markers of muscle strength translate into functional improvements and, ultimately, a reduction in clinical severity markers (such as the time to wheelchair dependence in boys with Duchenne muscular dystrophy). Clearly, these clinical outcome trials will require a multi-center approach. The potential for creatine to attenuate some of the side effects of corticosteroid treatment appears encouraging from preliminary data;

however, this needs to be carefully evaluated in a number of neuromuscular conditions where corticosteroid use is common (dystrophinopathy, polymyositis, myasthenia gravis). There does not appear to be a benefit from CrM supplementation in patients with inherited neuropathies at least over the short term; however, further work is needed to determine whether the favorable interactive effect between CrM supplementation and exercise training on MHC composition confers any functional long-term benefits to these patients. Given the high prevalence of diabetic neuropathy in the population and the encouraging results from preliminary trials with α-lipoic acid, a randomized double-blind trial of a combination of creatine plus α-lipoic in diabetic neuropathy would be warranted.

Statin-related muscle symptoms affect 3 to 5% of the population, and some of the pathology may be due to mitochondrial dysfunction (Baker and Tarnopolsky, 2001, 2003, 2005). Consequently, it may be of interest to determine whether CrM supplementation could have a beneficial affect with respect to myalgias and/or the development of myopathy. There is currently no proven therapy for inclusion body myositis except for strength training. Given that these patients are usually older men, and that creatine enhances the gains in strength in older adults during strength training, the potential for creatine as a therapy for inclusion body myositis does exist.

In patients with severe mitochondrial cytopathies such as MELAS syndrome, there appears to be an improvement in high-intensity exercise capacity with creatine supplementation, at least at the level seen for young healthy individuals. Given that a number of the pathophysiologic consequences of mitochondrial cytopathies could be favorably enhanced through CrM supplementation (Tarnopolsky and Beal, 2001), larger longer-term trials are warranted in this condition. Furthermore, given the increased propensity towards stroke and cardiomyopathy in patients with MELAS syndrome, the potential utility of CrM, particularly when used in combination with multiple cofactors targeting cellular energy dysfunction (Tarnopolsky and Beal, 2001), could be promising and is worthy of further investigation. The effect of CrM in patients with McArdle's disease appears to be somewhat minimal and this may be a function of poor muscle creatine uptake (Vorgerd *et al.*, 2002). There is a need for further investigations of creatine supplementation in patients with McArdle's disease with strategies that may enhance creatine uptake in skeletal muscle (post-contraction, α-lipoic acid, carbohydrate feeds). Unfortunately, however, even if creatine concentration does go up, the inability for McArdle's disease patients to acidify muscle may limit the creatine kinase reaction in the direction of ATP generation and mitigate against any potential therapeutic benefit from supplementation.

The traditional way to develop therapies for a disease is to identify the pathology of the disease and evaluate compounds that should mitigate against the pathological consequences. Preclinical studies usually focus on surrogate markers and eventually large long-term clinical trials are required. In rare disorders with typically a limited patient population, we feel that there is a need for a more rapid screening process, for even pre-clinical trials can be extremely expensive and time-consuming. To this end, we have started to evaluate transcription profiling (microarray-derived global mRNA profile for skeletal muscle) to determine a disease "signature" and to test the efficacy of some compounds. We have characterized the alterations in transcription profiles of skeletal muscle in response to MELAS syndrome and McArdle's disease. We have also evaluated the effect of short-term CrM supplementation in young healthy men. Following short-term CrM supplementation, there are a number of transcripts that were induced, including protein kinases that are sensitive to cell volume regulation and some proteins involved in apoptosis were also up-regulated. The concept that we are developing is the use of transcription profiling to evaluate the potential efficacy of therapeutic strategies on patients with well-characterized disorders. We feel that if a compound can return the pathogenic signature towards that of a healthy control, it has potential for further development as a therapeutic compound. This type of pre-clinical screening in humans and/or cell culture (e.g., cybrids from mitochondrial disease) could be the basis of high-throughput screening for therapies prior to the initiation of preclinical and large-scale clinical trials.

In summary, there does appear to be a biological rationale for the use of CrM supplementation in patients with several neuromuscular and neurometabolic disorders. Larger long-term careful clinical trials need to be employed to ensure that we do not commit a type II error (conclude that there is no affect when in reality there is) or a type I error (conclude that there is an effect when in reality there is not). These studies are particularly important as there are very few therapeutic options for the genetic muscular dystrophies, Charcot-Marie-Tooth disease, McArdle's disease, steroid myopathy, statin myopathy, or inclusion body myositis.

ACKNOWLEDGEMENTS

Creatine monohydrate for most of the studies was donated by Avicena. Funding for most of the basic research in the area has come from NSERC and for the clinical work has come from CIHR, Hamilton Health Science Corporation, and a donation from Warren Lammert and family.

REFERENCES

Adcock, K.H., Nedelcu, J., Loenneker, T., Martin, E., Wallimann, T., and Wagner, B.P., 2002, Neuroprotection of creatine supplementation in neonatal rats with transient cerebral hypoxia-ischemia. *Dev. Neurosci.* **24:** 382–388.

Alexanderson, H., 2005, Exercise: an important component of treatment in the idiopathic inflammatory myopathies. *Curr, Rheumatol. Rep.* **7:** 115–124.

Andreassen, O.A., Dedeoglu, A., Ferrante, R.J., Jenkins, B.G., Ferrante, K.L., Thomas, M., Friedlich, A., Browne, S.E., Schilling, G., Borchelt, D.R., Hersch, S.M., Ross, C.A., and Beal, M.F., 2001, Creatine increases survival and delays motor symptoms in a transgenic animal model of Huntington's disease. *Neurobiol. Dis.* **8:** 479–491.

Argov, Z., De Stefano, N., and Arnold, D.L., 1996, ADP recovery after a brief ischemic exercise in normal and diseased human muscle – a 31P MRS study. *NMR Biomed.* **9:** 165–172.

Arnardottir, S., Alexanderson, H., Lundberg, I.E., and Borg, K., 2003, Sporadic inclusion body myositis: pilot study on the effects of a home exercise program on muscle function, histopathology and inflammatory reaction. *J. Rehabil. Med.* **35:** 31–35.

Baker, S.K., and Tarnopolsky, M.A., 2001, Statin myopathies: pathophysiologic and clinical perspectives. *Clin. Invest. Med.* **24:** 258–272.

Baker, S.K., and Tarnopolsky, M.A., 2003, Targeting cellular energy production in neurological disorders. *Expert Opin. Investig. Drugs* **12:** 1655–1679.

Baker, S.K., and Tarnopolsky, M.A., 2005, Statin-associated neuromyotoxicity. *Drugs Today (Barc.)* **41:** 267–293.

Barisic, N., Bernert, G., Ipsiroglu, O., Stromberger, C., Muller, T., Gruber, S., Prayer, D., Moser, E., Bittner, R.E., and Stöckler-Ipsiroglu, S., 2002, Effects of oral creatine supplementation in a patient with MELAS phenotype and associated nephropathy. *Neuropediatrics* **33:** 157–161.

Benedict, J.D., Kalinsky, H.J., Scarrone, L.A., Wertbeim, A.R., and Stetten, D., Jr., 1955, The origin of urinary creatine in progressive muscular dystrophy. *J. Clin. Invest.* **34:** 141–145.

Borchert, A., Wilichowski, E., and Hanefeld, F., 1999, Supplementation with creatine monohydrate in children with mitochondrial encephalomyopathies. *Muscle Nerve* **22:** 1299–1300.

Brooke, M.H., Fenichel, G.M., Griggs, R.C., Mendell, J.R., Moxley, R., Florence, J., King, W.M., Pandya, S., Robison, J., Schierbecker, J., et al., 1989, Duchenne muscular dystrophy: patterns of clinical progression and effects of supportive therapy. *Neurology* **39:** 475–481.

Brooke, M.H., Griggs, R.C., Mendell, J.R., Fenichel, G.M., and Shumate, J.B., 1981a, The natural history of Duchenne muscular dystrophy: a caveat for therapeutic trials. *Trans. Am. Neurol. Assoc.* **106:** 195–199.

Brooke, M.H., Griggs, R.C., Mendell, J.R., Fenichel, G.M., Shumate, J.B., and Pellegrino, R.J., 1981b, Clinical trial in Duchenne dystrophy. I. The design of the protocol. *Muscle Nerve* **4:** 186–197.

Brose, A., Parise, G., and Tarnopolsky, M.A., 2003, Creatine supplementation enhances isometric strength and body composition improvements following strength exercise training in older adults. *J. Gerontol. A Biol. Sci. Med. Sci.* **58:** 11–19.

Burke, D.G., Chilibeck, P.D., Parise, G., Candow, D.G., Mahoney, D., and Tarnopolsky, M., 2003a, Effect of creatine and weight training on muscle creatine and performance in vegetarians. *Med. Sci. Sports Exerc.* **35:** 1946–1955.

Burke, D.G., Chilibeck, P.D., Parise, G., Tarnopolsky, M.A., and Candow, D.G., 2003b, Effect of alpha-lipoic acid combined with creatine monohydrate on human skeletal muscle creatine and phosphagen concentration. *Int. J. Sport Nutr. Exerc. Metab.* **13:** 294–302.

Campos, A.R., Serafini, L.N., Sobreira, C., Menezes, L.G., and Martinez, J.A., 2006, Creatine intake attenuates corticosteroid-induced impairment of voluntary running in hamsters. *Appl. Physiol. Nutr. Metab.* **31:** 490–494.

Carter, A.J., Muller, R.E., Pschorn, U., and Stransky, W., 1995, Preincubation with creatine enhances levels of creatine phosphate and prevents anoxic damage in rat hippocampal slices. *J. Neurochem.* **64:** 2691–2699.

Casey, A., Constantin-Teodosiu, D., Howell, S., Hultman, E., and Greenhaff, P.L., 1996, Creatine ingestion favorably affects performance and muscle metabolism during maximal exercise in humans. *Am. J. Physiol.* **271:** E31–E37.

Chang, C.T., Wu, C.H., Yang, C.W., Huang, J.Y., and Wu, M.S., 2002, Creatine monohydrate treatment alleviates muscle cramps associated with haemodialysis. *Nephrol. Dial. Transplant.* **17:** 1978–1981.

Chetlin, R.D., Gutmann, L., Tarnopolsky, M.A., Ullrich, I.H., and Yeater, R.A., 2004, Resistance training exercise and creatine in patients with Charcot-Marie-Tooth disease. *Muscle Nerve* **30:** 69–76.

Chilibeck, P.D., Chrusch, M.J., Chad, K.E., Shawn Davison, K., and Burke, D.G., 2005, Creatine monohydrate and resistance training increase bone mineral content and density in older men. *J. Nutr. Health Aging* **9:** 352–353.

Dedeoglu, A., Kubilus, J.K., Yang, L., Ferrante, K.L., Hersch, S.M., Beal, M.F., and Ferrante, R.J., 2003, Creatine therapy provides neuroprotection after onset of clinical symptoms in Huntington's disease transgenic mice. *J. Neurochem.* **85:** 1359–1367.

Doherty, T.J., Lougheed, K., Markez, J., and Tarnopolsky, M.A., 2001, Creatine monohydrate does not increase strength in patients with hereditary neuropathy. *Neurology* **57:** 559–560.

Duke, A.M., and Steele, D.S., 1999, Effects of creatine phosphate on Ca^{2+} regulation by the sarcoplasmic reticulum in mechanically skinned rat skeletal muscle fibres. *J. Physiol.* **517:** 447–458.

Escolar, D.M., Buyse, G., Henricson, E., Leshner, R., Florence, J., Mayhew, J., Tesi-Rocha, C., Gorni, K., Pasquali, L., Patel, K.M., McCarter, R., Huang, J., Mayhew, T., Bertorini, T., Carlo, J., Connolly, A.M., Clemens, P.R., Goemans, N., Iannaccone, S.T., Igarashi, M., Nevo, Y., Pestronk, A., Subramony, S.H.,

Vedanarayanan, V.V., and Wessel, H., 2005, CINRG randomized controlled trial of creatine and glutamine in Duchenne muscular dystrophy. *Ann. Neurol.* **58:** 151–155.

Escolar, D.M., Henricson, E.K., Mayhew, J., Florence, J., Leshner, R., Patel, K.M., and Clemens, P.R., 2001, Clinical evaluator reliability for quantitative and manual muscle testing measures of strength in children. *Muscle Nerve* **24:** 787–793.

Gerber, I., ap Gwynn, I., Alini, M., and Wallimann, T., 2005, Stimulatory effects of creatine on metabolic activity, differentiation and mineralization of primary osteoblast-like cells in monolayer and micromass cell cultures. *Eur. Cells Mater.* **10:** 8–22.

Gordon, A., Hultman, E., Kaijser, L., Kristjansson, S., Rolf, C.J., Nyquist, O., and Sylven, C., 1995, Creatine supplementation in chronic heart failure increases skeletal muscle creatine phosphate and muscle performance. *Cardiovasc. Res.* **30:** 413–418.

Granchelli, J.A., Pollina, C., and Hudecki, M.S., 2000, Pre-clinical screening of drugs using the *mdx* mouse. *Neuromusc. Disord.* **10:** 235–239.

Hespel, P., Op't Eijnde, B., and Van Leemputte, M., 2002, Opposite actions of caffeine and creatine on muscle relaxation time in humans. *J. Appl. Physiol.* **92:** 513–518.

Isenberg, D.A., Allen, E., Farewell, V., Ehrenstein, M.R., Hanna, M.G., Lundberg, I.E., Oddis, C., Pilkington, C., Plotz, P., Scott, D., Vencovsky, J., Cooper, R., Rider, L., and Miller, F., 2004, International consensus outcome measures for patients with idiopathic inflammatory myopathies. Development and initial validation of myositis activity and damage indices in patients with adult onset disease. *Rheumatology (Oxford)* **43:** 49–54.

Johansen, K.L., Doyle, J., Sakkas, G.K., and Kent-Braun, J.A., 2005, Neural and metabolic mechanisms of excessive muscle fatigue in maintenance hemodialysis patients. *Am. J. Physiol. Regul. Integr. Comp. Physiol.* **289:** R805–R813.

Klein, A.M., and Ferrante, R.J., 2007, The neuroprotective role of creatine. *Subcell. Biochem.* **46:** 205–243.

Kley, R., Vorgerd, M., and Tarnopolsky, M., 2007, Creatine for treating muscle disorders. *Cochrane Database Syst. Rev.* CD004760.

Klivenyi, P., Calingasan, N.Y., Starkov, A., Stavrovskaya, I.G., Kristal, B.S., Yang, L., Wieringa, B., and Beal, M.F., 2004, Neuroprotective mechanisms of creatine occur in the absence of mitochondrial creatine kinase. *Neurobiol. Dis.* **15:** 610–617.

Klivenyi, P., Ferrante, R.J., Matthews, R.T., Bogdanov, M.B., Klein, A.M., Andreassen, O.A., Mueller, G., Wermer, M., Kaddurah-Daouk, R., and Beal, M.F., 1999, Neuroprotective effects of creatine in a transgenic animal model of amyotrophic lateral sclerosis. *Nat. Med.* **5:** 347–350.

Klopstock, T., Querner, V., Schmidt, F., Gekeler, F., Walter, M., Hartard, M., Henning, M., Gasser, T., Pongratz, D., Straube, A., Dieterich, M., and Muller-Felber, W., 2000, A placebo-controlled crossover trial of creatine in mitochondrial diseases. *Neurology* **55:** 1748–1751.

Komura, K., Hobbiebrunken, E., Wilichowski, E.K., and Hanefeld, F.A., 2003, Effectiveness of creatine monohydrate in mitochondrial encephalomyopathies. *Pediatr. Neurol.* **28:** 53–58.

Korenke, G.C., Wanders, R.J., and Hanefeld, F., 2003, Striking improvement of muscle strength under creatine therapy in a patient with long-chain 3-hydroxyacyl-CoA dehydrogenase deficiency. *J. Inherit. Metab. Dis.* **26:** 67–68.

Kornblum, C., Schroder, R., Muller, K., Vorgerd, M., Eggers, J., Bogdanow, M., Papassotiropoulos, A., Fabian, K., Klockgether, T., and Zange, J., 2005, Creatine has no beneficial effect on skeletal muscle energy metabolism in patients with single mitochondrial DNA deletions: a placebo-controlled, double-blind 31P-MRS crossover study. *Eur. J. Neurol.* **12:** 300–309.

Kuethe, F., Krack, A., Richartz, B.M., and Figulla, H.R., 2006, Creatine supplementation improves muscle strength in patients with congestive heart failure. *Pharmazie* **61:** 218–222.

Lawler, J.M., Barnes, W.S., Wu, G., Song, W., and Demaree, S., 2002, Direct antioxidant properties of creatine. *Biochem. Biophys. Res. Commun.* **290:** 47–52.

Louis, M., Lebacq, J., Poortmans, J.R., Belpaire-Dethiou, M.C., Devogelaer, J.P., Van Hecke, P., Goubel, F., and Francaux, M., 2003, Beneficial effects of creatine supplementation in dystrophic patients. *Muscle Nerve* **27:** 604–610.

Louis, M., Raymackers, J.M., Debaix, H., Lebacq, J., and Francaux, M., 2004, Effect of creatine supplementation on skeletal muscle of *mdx* mice. *Muscle Nerve* **29:** 687–692.

Mahoney, D.J., Parise, G., and Tarnopolsky, M.A., 2002, Nutritional and exercise-based therapies in the treatment of mitochondrial disease. *Curr. Opin. Clin. Nutr. Metab. Care* **5:** 619–629.

Malcon, C., Kaddurah-Daouk, R., and Beal, M.F., 2000, Neuroprotective effects of creatine administration against NMDA and malonate toxicity. *Brain Res.* **860:** 195–198.

Mallouk, N., Jacquemond, V., and Allard, B., 2000, Elevated subsarcolemmal Ca^{2+} in *mdx* mouse skeletal muscle fibers detected with Ca^{2+}-activated K^{+} channels. *Proc. Natl. Acad. Sci. USA* **97:** 4950–4955.

Menezes, L.G., Sobreira, C., Neder, L., Rodrigues-Junior, A.L., and Martinez, J.A., 2007, Creatine supplementation attenuates corticosteroid-induced muscle wasting and impairment of exercise performance in rats. *J. Appl. Physiol.* **102:** 698–703.

Mihic, S., MacDonald, J.R., McKenzie, S., and Tarnopolsky, M.A., 2000, Acute creatine loading increases fat-free mass, but does not affect blood pressure, plasma creatinine, or CK activity in men and women. *Med. Sci. Sports Exerc.* **32:** 291–296.

Milhorat, A.T., 1953, Creatine and creatinine metabolism and diseases of the neuromuscular system. *Res. Publ. Assoc. Res. Nerv. Ment. Dis.* **32:** 400–421.

Moxley, R.T., 3rd, Ashwal, S., Pandya, S., Connolly, A., Florence, J., Mathews, K., Baumbach, L., McDonald, C., Sussman, M., and Wade, C., 2005, Practice parameter: corticosteroid treatment of Duchenne dystrophy: report of the Quality Standards Subcommittee of the American Academy of Neurology and the Practice Committee of the Child Neurology Society. *Neurology* **64:** 13–20.

Parise, G., Mihic, S., MacLennan, D., Yarasheski, K.E., and Tarnopolsky, M.A., 2001, Effects of acute creatine monohydrate supplementation on leucine kinetics and mixed-muscle protein synthesis. *J. Appl. Physiol.* **91:** 1041–1047.

Park, J.H., Niermann, K.J., Ryder, N.M., Nelson, A.E., Das, A., Lawton, A.R., Hernanz-Schulman, M., and Olsen, N.J., 2000, Muscle abnormalities in juvenile dermatomyositis patients: P-31 magnetic resonance spectroscopy studies. *Arthritis Rheum.* **43:** 2359–2367.

Park, J.H., Olsen, N.J., King, L., Jr., Vital, T., Buse, R., Kari, S., Hernanz-Schulman, M., and Price, R.R., 1995, Use of magnetic resonance imaging and P-31 magnetic resonance spectroscopy to detect and quantify muscle dysfunction in the amyopathic and myopathic variants of dermatomyositis. *Arthritis Rheum.* **38:** 68–77.

Park, J.H., Vital, T.L., Ryder, N.M., Hernanz-Schulman, M., Partain, C.L., Price, R.R., and Olsen, N.J., 1994, Magnetic resonance imaging and P-31 magnetic resonance spectroscopy provide unique quantitative data useful in the longitudinal management of patients with dermatomyositis. *Arthritis Rheum.* **37:** 736–746.

Passaquin, A.C., Renard, M., Kay, L., Challet, C., Mokhtarian, A., Wallimann, T., and Ruegg, U.T., 2002, Creatine supplementation reduces skeletal muscle degeneration and enhances mitochondrial function in *mdx* mice. *Neuromusc. Disord.* **12:** 174–182.

Payne, E.T., Yasuda, N., Bourgeois, J.M., Devries, M.C., Rodriguez, M.C., Yousuf, J., and Tarnopolsky, M.A., 2006, Nutritional therapy improves function and complements corticosteroid intervention in *mdx* mice. *Muscle Nerve* **33:** 66–77.

Piccolo, G., Banfi, P., Azan, G., Rizzuto, R., Bisson, R., Sandona, D., and Bellomo, G., 1991, Biological markers of oxidative stress in mitochondrial myopathies with progressive external ophthalmoplegia. *J. Neurol. Sci.* **105:** 57–60.

Pulido, S.M., Passaquin, A.C., Leijendekker, W.J., Challet, C., Wallimann, T., and Ruegg, U.T., 1998, Creatine supplementation improves intracellular Ca^{2+} handling and survival in *mdx* skeletal muscle cells. *FEBS Lett.* **439:** 357–362.

Raha, S., and Robinson, B.H., 2000, Mitochondria, oxygen free radicals, disease and ageing. *Trends Biochem. Sci.* **25:** 502–508.

Rindfleisch, J.A., and Muller, D., 2005, Diagnosis and management of rheumatoid arthritis. *Am. Fam. Physician* **72:** 1037–1047.

Rodriguez, M.C., MacDonald, J.R., Mahoney, D.J., Parise, G., Beal, M.F., and Tarnopolsky, M.A., 2007, Beneficial effects of creatine, CoQ10, and lipoic acid in mitochondrial disorders. *Muscle Nerve* **35:** 235–242.

Roy, B.D., Bourgeois, J.M., Mahoney, D.J., and Tarnopolsky, M.A., 2002, Dietary supplementation with creatine monohydrate prevents corticosteroid-induced attenuation of growth in young rats. *Can. J. Physiol. Pharmacol.* **80:** 1008–1014.

Scaglia, F., Towbin, J.A., Craigen, W.J., Belmont, J.W., Smith, E.O., Neish, S.R., Ware, S.M., Hunter, J.V., Fernbach, S.D., Vladutiu, G.D., Wong, L.J., and Vogel, H., 2004, Clinical spectrum, morbidity, and mortality in 113 pediatric patients with mitochondrial disease. *Pediatrics* **114:** 925–931.

Schneider-Gold, C., Beck, M., Wessig, C., George, A., Kele, H., Reiners, K., and Toyka, K.V., 2003, Creatine monohydrate in DM2/PROMM: a double-blind placebo-controlled clinical study. Proximal myotonic myopathy. *Neurology* **60:** 500–502.

Schneider-Gold, C., Gajdos, P., Toyka, K.V., and Hohlfeld, R.R., 2005, Corticosteroids for myasthenia gravis. *Cochrane Database Syst. Rev.* CD002828.

Smith, C.A., Chetlin, R.D., Gutmann, L., Yeater, R.A., and Alway, S.E., 2006, Effects of exercise and creatine on myosin heavy chain isoform composition in patients with Charcot-Marie-Tooth disease. *Muscle Nerve* **34:** 586–594.

Sullivan, P.G., Geiger, J.D., Mattson, M.P., and Scheff, S.W., 2000, Dietary supplement creatine protects against traumatic brain injury. *Ann. Neurol.* **48:** 723–729.

Taivassalo, T., and Haller, R.G., 2005, Exercise and training in mitochondrial myopathies. *Med. Sci. Sports Exerc.* **37:** 2094–2101.

Tarnopolsky, M.A., Roy, B.D., and MacDonald, J.R., 1997, A randomized, controlled trial of creatine monohydrate in patients with mitochondrial cytopathies. *Muscle Nerve* **20:** 1502–1509.

Tarnopolsky, M., and Martin, J., 1999, Creatine monohydrate increases strength in patients with neuromuscular disease. *Neurology* **52:** 854–857.

Tarnopolsky, M.A., and Parise, G., 1999, Direct measurement of high-energy phosphate compounds in patients with neuromuscular disease. *Muscle Nerve* **22:** 1228–1233.

Tarnopolsky, M.A., Parise, G., Yardley, N.J., Ballantyne, C.S., Olatinji, S., and Phillips, S.M., 2001, Creatine-dextrose and protein-dextrose induce similar strength gains during training. *Med. Sci. Sports Exerc.* **33:** 2044–2052.

Tarnopolsky, M.A., and Beal, M.F., 2001, Potential for creatine and other therapies targeting cellular energy dysfunction in neurological disorders. *Ann. Neurol.* **49:** 561–574.

Tarnopolsky, M., Parise, G., Fu, M.H., Brose, A., Parshad, A., Speer, O., and Wallimann, T., 2003, Acute and moderate-term creatine monohydrate supplementation does not affect creatine transporter mRNA or protein content in either young or elderly humans. *Mol. Cell. Biochem.* **244:** 159–166.

Tarnopolsky, M., 2004, Exercise testing as a diagnostic entity in mitochondrial myopathies. *Mitochondrion* **4:** 529–542.

Tarnopolsky, M., Mahoney, D., Thompson, T., Naylor, H., and Doherty, T.J., 2004a, Creatine monohydrate supplementation does not increase muscle strength, lean body mass, or muscle phosphocreatine in patients with myotonic dystrophy type 1. *Muscle Nerve* **29:** 51–58.

Tarnopolsky, M.A., Mahoney, D.J., Vajsar, J., Rodriguez, C., Doherty, T.J., Roy, B.D., and Biggar, D., 2004b, Creatine monohydrate enhances strength and body composition in Duchenne muscular dystrophy. *Neurology* **62:** 1771–1777.

Tarnopolsky, M.A., Simon, D.K., Roy, B.D., Chorneyko, K., Lowther, S.A., Johns, D.R., Sandhu, J.K., Li, Y., and Sikorska, M., 2004c, Attenuation of free radical production and paracrystalline inclusions by creatine supplementation in a patient with a novel cytochrome b mutation. *Muscle Nerve* **29:** 537–547.

Tarnopolsky, M.A., and Raha, S., 2005, Mitochondrial myopathies: diagnosis, exercise intolerance, and treatment options. *Med. Sci. Sports Exerc.* **37:** 2086–2093.

Tartaglia, M.C., Chen, J.T., Caramanos, Z., Taivassalo, T., Arnold, D.L., and Argov, Z., 2000, Muscle phosphorus magnetic resonance spectroscopy oxidative indices correlate with physical activity. *Muscle Nerve* **23:** 175–181.

Terjung, R.L., Clarkson, P., Eichner, E.R., Greenhaff, P.L., Hespel, P.J., Israel, R.G., Kraemer, W.J., Meyer, R.A., Spriet, L.L., Tarnopolsky, M.A., Wagenmakers, A.J., and Williams, M.H., 2000, American College of Sports Medicine roundtable. The physiological and health effects of oral creatine supplementation. *Med. Sci. Sports Exerc.* **32:** 706–717.

Vorgerd, M., Grehl, T., Jager, M., Muller, K., Freitag, G., Patzold, T., Bruns, N., Fabian, K., Tegenthoff, M., Mortier, W., Luttmann, A., Zange, J., and Malin, J.P., 2000, Creatine therapy in myophosphorylase deficiency (McArdle disease): a placebo-controlled crossover trial. *Arch. Neurol.* **57:** 956–963.

Vorgerd, M., Zange, J., Kley, R., Grehl, T., Husing, A., Jager, M., Muller, K., Schroder, R., Mortier, W., Fabian, K., Malin, J.P., and Luttmann, A., 2002, Effect of high-dose creatine therapy on symptoms of exercise intolerance in McArdle disease: double-blind, placebo-controlled crossover study. *Arch. Neurol.* **59:** 97–101.

Whittingham, T.S., and Lipton, P., 1981, Cerebral synaptic transmission during anoxia is protected by creatine. *J. Neurochem.* **37:** 1618–1621.

Wilken, B., Ramirez, J.M., Probst, I., Richter, D.W., and Hanefeld, F., 2000, Anoxic ATP depletion in neonatal mice brainstem is prevented by creatine supplementation. *Arch. Dis. Child. Fetal Neonatal Ed.* **82:** F224–F227.

Zhu, S., Li, M., Figueroa, B.E., Liu, A., Stavrovskaya, I.G., Pasinelli, P., Beal, M.F., Brown, R.H., Jr., Kristal, B.S., Ferrante, R.J., and Friedlander, R.M., 2004, Prophylactic creatine administration mediates neuroprotection in cerebral ischemia in mice. *J. Neurosci.* **24:** 5909–5912.

Ziegler, D., 2004, Thioctic acid for patients with symptomatic diabetic polyneuropathy: a critical review. *Treat. Endocrinol.* **3:** 173–189.

CHAPTER 11

THE NEUROPROTECTIVE ROLE OF CREATINE

AUTUMN M. KLEIN[1] AND ROBERT J. FERRANTE[2]

[1]*Neurology Department, Brigham and Womens Hospital, Boston, MA 02115, USA; and Harvard Medical School, Boston, MA 02115, USA*

[2]*Geriatric Research Education and Clinical Center, Bedford VA Medical Center, Bedford, MA 01730, USA; and Neurology, Laboratory Medicine and Pathology, and Psychiatry Departments, Boston University School of Medicine, Boston, MA 02180, USA*

Abstract: Significant progress has been made in identifying neuroprotective agents and their translation to patients with neurological disorders. While the direct causative pathways of neurodegeneration remain unclear, they are under great clinical and experimental investigation. There are a number of interrelated pathogenic mechanisms triggering molecular events that lead to neuronal death. One putative mechanism reported to play a prominent role in the pathogenesis of neurological diseases is impaired energy metabolism. If reduced energy stores play a role in neuronal loss, then therapeutic strategies that buffer intracellular energy levels may prevent or impede the neurodegenerative process. Recent studies suggest that impaired energy production promotes neurological disease onset and progression. Sustained ATP levels are critical to cellular homeostasis and may have both direct and indirect influence on pathogenic mechanisms associated with neurological disorders. Creatine is a critical component in maintaining cellular energy homeostasis, and its administration has been reported to be neuroprotective in a wide number of both acute and chronic experimental models of neurological disease. In the context of this chapter, we will review the experimental evidence for creatine supplementation as a neurotherapeutic strategy in patients with neurological disorders, including Huntington's disease, Parkinson's disease, amyotrophic lateral sclerosis, and Alzheimer's disease, as well as in ischemic stroke, brain and spinal cord trauma, and epilepsy

1. INTRODUCTION

There is substantial evidence to suggest that impaired energy metabolism, in concert with mitochondrial dysfunction, plays a critical role in the pathogenesis and progression of neurological diseases as a primary and/or secondary mechanism in the neuronal death cascade (Beal, 2005; Tarnopolsky and Beal, 2001). This not only includes neurodegenerative disorders, but also acute and chronic conditions

G.S. Salomons and M. Wyss (eds.), Creatine and Creatine Kinase in Health and Disease, 205–243.

involving the central and peripheral nervous systems. While other mechanisms may exacerbate bioenergetic dysfunction, such as protein aggregation and altered transcription, impaired energy metabolism may trigger pro-apoptotic signaling, oxidative damage, excitotoxicity, and impede nuclear and mitochondrial DNA repair (Figure 1). These pathologic messages can interact and potentiate one another, resulting in a continued cycle of energy depletion. Energy is critical to the biological and molecular regulation of multiple cellular functions and, as such, reduced energy levels threaten cellular homeostasis and integrity.

While the concept of neuroprotection was first considered by ancient Greek physicians using hypothermia to treat stroke, a modern-day therapeutic strategy

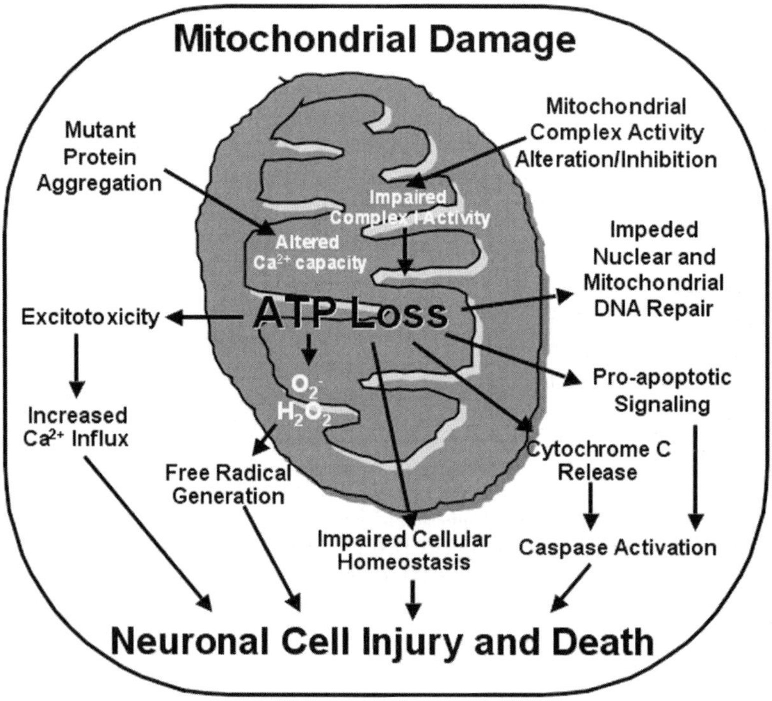

Figure 1. Mitochondria-mediated neuronal dysfunction in neurological disease. Protein aggregation may disrupt the mitochondrial membrane potential and alter excitotoxin-induced Ca^{2+} influx. In addition, impaired respiratory enzyme activities may result in reduced ATP generation and increased levels of reactive oxygen species (O_2^- and H_2O_2). Free radical generation may then result in further damage to cellular macromolecules through nitrosylation, oxidation, and peroxidation, which can directly contribute to neuronal injury. In addition, the release of cytochrome c from damaged/dysfunctional mitochondria triggers the activation of the apoptotic cascade and the release of initiator and executioner caspases, resulting in neuronal cell death. Impaired energy metabolism may trigger pro-apoptotic signaling, impair cellular homeostasis, and impede nuclear and mitochondrial DNA repair. Each of these pathological mechanisms may in turn further exacerbate mitochondrial dysfunction and energy loss. Creatine supplementation improves intracellular high-energy phosphate levels, improving multiple cellular functions and neuronal survival.

inducing sustained ATP levels may have both direct and indirect importance in ameliorating the severity of many of the pathogenic mechanisms associated with neurological disorders. Indeed, if neuronal dysfunction and loss are caused by reduced energy stores, then therapeutic regimens that buffer intracellular energy levels may impede progression or prevent the neurodegenerative process. Creatine, a naturally occurring compound produced endogenously and acquired exogenously through the diet, is important in maintaining cellular energy homeostasis. As such, creatine has been shown to be one of a number of ergogens that may provide a relatively safe and immediately available therapeutic strategy to patients with neurological disease. Creatine, through its metabolite phosphocreatine, provides a cellular reserve of high-energy phosphates. It plays an important role in cellular high-energy phosphate transport. In addition to ameliorating bioenergetic defects, creatine may also indirectly benefit other pathophysiological mechanisms by improving cellular homeostasis. While the investigational use of creatine as a supplement began in the early 20th century, interest in the use of oral creatine supplementation began in earnest in the 1990s, as it became a widely used ergogenic supplement for performance enhancement by professional and amateur athletes (Greenhaff *et al.*, 1993; Harris *et al.*, 1992; Persky and Brazeau, 2001; Tarnopolsky *et al.*, 1992; Wyss and Kaddurah-Daouk, 2000). Creatine (methylguanidino-acetic acid) is a guanidino compound synthesized endogenously from arginine, methionine, and glycine, predominantly in the liver, as well as in the kidneys, pancreas, testes (Bloch and Schoenheimer, 1941; Persky and Brazeau, 2001; Walker, 1979), and the brain (Braissant *et al.*, 2001). Creatine is also derived exogenously through the diet in the consumption of meat and fish (Balsom *et al.*, 1994). In order to maintain sufficient body stores of creatine, approximately 2 g are required daily through both diet and endogenous synthesis (Casey and Greenhaff, 2000). Over 90% of creatine is found in skeletal muscle, mostly as phosphocreatine (PCr), with the remaining stores in the brain and other organs (Walker, 1979; Wyss and Kaddurah-Daouk, 2000). Significant amounts of creatine are present in tissues with high and fluctuating energy demands. Creatine is non-enzymatically converted to creatinine and excreted through the kidneys (Casey and Greenhaff, 2000). Creatine is also an excellent stimulant of mitochondrial respiration, resulting in the generation of PCr (Kernec *et al.*, 1996; O'Gorman *et al.*, 1996). It is a critical component of the creatine kinase system in maintaining cellular energy needs.

The major source of energy in the brain is ATP, which is tightly coupled to creatine and PCr levels within the cell. Creatine is shuttled across membranes via a sodium-dependent creatine transporter protein, CreaT (Schloss *et al.*, 1994; Snow and Murphy, 2001; Willott *et al.*, 1999). CreaT regulates tissue levels in response to low dietary intake or high endogenous creatine levels (Guerrero-Ontiveros and Wallimann, 1998; Loike *et al.*, 1988). Creatine kinase catalyzes the reversible transfer of a phosphoryl group from PCr to adenosine diphosphate (ADP), forming adenosine triphosphate (ATP). As such, creatine offsets energy depletion by forming PCr, providing a spatial energy buffer to re-phosphorylate ADP to ATP at cellular sites of energy consumption and, in the reversible reaction, forming PCr and ADP from creatine and ATP at cellular sites of high-energy phosphate production (PCr shuttle hypothesis) (Bessman and Carpenter, 1985; Meyer *et al.*, 1984; Tombes and Shapiro, 1985; Van Brussel *et al.*, 1983). Increasing creatine levels may, therefore, help to prevent reduced energy stores and improve neuronal function. Creatine is also involved in regulating glycolysis, stabilizing the mitochondrial, octameric form of creatine kinase, and inhibiting the mitochondrial permeability transition pore (O'Gorman *et al.*, 1996, 1997). The opening of the mitochondrial permeability transition pore is associated with both apoptotic and necrotic cell death mechanisms (Bernardi *et al.*, 1998). Another potential neuroprotective mechanism of creatine supplementation is the ability of PCr to stimulate synaptic glutamate uptake and thereby reduce the accumulation of extracellular glutamate and the potential for excitotoxicity (Xu *et al.*, 1996). Creatine has also been reported to act as a direct anti-oxidant, scavenging reactive oxygen species that may further potentiate mitochondrial dysfunction if left unchecked (Lawler *et al.*, 2002). While creatine did not significantly reduce levels of hydrogen peroxide or lipid peroxidation in these studies, creatine was effective in reducing superoxide anions and peroxinitrite. Interestingly, the neuroprotective effects of creatine may be independent of mitochondrial creatine kinase (Brustovetsky *et al.*, 2001; Klivenyi *et al.*, 2004). Brustovetsky and colleagues reported that creatine had no effect on the mitochondrial permeability transition pore in isolated brain mitochondria (Brustovetsky *et al.*, 2001). In addition, Beal and colleagues showed that creatine administration in mice deficient in ubiquitous mitochondrial creatine kinase increased brain levels of creatine and PCr, suggesting that the neuroprotective effects of creatine are the result of maintaining PCr and ATP levels and are not due to inhibition of the mitochondrial permeability transition pore (Klivenyi *et al.*, 2004). Of great interest is the fact that oral supplementation of creatine monohydrate (20 g/d over 4 wk) in healthy human volunteers significantly increases the levels of total creatine in the brain (Dechent *et al.*, 1999). Quantitative localized proton magnetic resonance spectroscopy *in vivo* yielded a statistically significant increase (8.7%) of the mean concentration of total creatine in the brain with region-dependent increases, least in white matter and greatest in the thalamus.

2. CREATINE SUPPLEMENTATION IN NEURODEGENERATIVE DISORDERS

2.1. Huntington's Disease

Huntington's disease (HD) is an autosomal dominant inherited neurodegenerative disorder that is characterized by progressive motor dysfunction, emotional disturbances, dementia, and weight loss. HD occurs worldwide, in all races and ethnic groups (Kremer *et al.*, 1994). Its prevalence is 5–10 per 100,000, with a new mutation rate as high as 1–3% (Myers *et al.*, 1993). There are about 30,000 affected individuals in the United States. Another 150,000 Americans have a genetic risk for developing the disease. The average age of onset is approximately 37 years, however the range is from infancy into the 9th decade. There is increasing reason to believe that pathologic alterations occur in the brain for years before symptoms manifest themselves (Paulsen *et al.*, 2001). Once symptomatic, affected individuals are rapidly disabled by early functional decline, and require increasing care and supervision for another 15–25 years before succumbing to the effects of severe physical and mental deterioration. Because of the chronic and increasingly intensive multidisciplinary care it requires, and its genetic nature, HD disproportionately consumes medical, social, and family resources (Helder *et al.*, 2001). The neuropathological hallmark of HD is selective neuronal degeneration, particularly within the neostriatum, in which medium-sized spiny striatal projection neurons are disproportionately affected early and most severely, while large and medium-sized aspiny interneurons are relatively spared (Ferrante *et al.*, 1985; Hersch *et al.*, 2004). While the earliest and most striking neuropathological changes are found in the neostriatum, neuronal loss has also been identified in other regions of the brain (Hersch *et al.*, 2004). Proliferative and degenerative changes in vulnerable neurons suggest that the presence of mutant huntingtin leads to both compensatory and degenerative genetic programs in a prolonged process leading to neuronal dysfunction and death (Ferrante *et al.*, 1991; Graveland *et al.*, 1985). HD is caused by an expanded trinucleotide CAG repeat in the gene coding for the large, highly conserved protein, huntingtin. The huntingtin gene was first cloned in 1993 (Huntington's Disease Collaborative Research Group, 1993). Emerging evidence suggests it is involved in fast axonal transport (Szebenyi *et al.*, 2003; Trushina *et al.*, 2003), specifically enhancing vesicular transport of brain-derived neurotrophic factor along microtubules (Gauthier *et al.*, 2004; Ross, 2004). In individuals with HD, the polymorphic CAG repeat region near the 5' end of the gene is expanded beyond the normal range, leading to translation of an expanded polyglutamine stretch in the protein (Huntington's Disease Collaborative Research Group, 1993). In the normal population the number of CAG repeats varies from 17 to 29. In individuals with HD there are more than 38 repeats. Once expanded into the pathogenic range, there is an inverse relationship between the CAG repeat number and the age of disease onset, with higher repeat numbers associated with younger age. HD is one member of the family of neurodegenerative triplet repeat disorders with anticipation and a gain of function mutation (Robitaille *et al.*, 1997). These include spinocerebellar ataxias, dentato-rubro-pallido-luysian atrophy, Machado-Joseph disease, and spinal bulbar muscular atrophy. As in HD, selective loss of neurons underlies these diseases, and misfolding and abnormal aggregation of the mutant protein occurs. It has thus been hypothesized that neurodegeneration in these disorders may have similar molecular bases (Ross, 1997). As such, therapeutic strategies that are effective in HD will likely be highly relevant for these other disorders as well.

While mitochondrial dysfunction and impaired energy metabolism have not been definitively determined in neurodegenerative disorders, there is strong evidence, particularly in HD, that an energy disturbance plays a prominent role in the pathogenesis of these diseases (Beal, 2005; Ryu *et al.*, 2005; Ryu and Ferrante, 2005). The concept of defective cellular energy metabolism in neurological diseases, particularly HD, was suggested by Beal and Albin in separate publications as an alternative excitotoxic hypothesis and has been coined 'slow onset excitotoxicity' (Albin and Greenamyer, 1992; Beal, 1992). The relevant observations were first made by Olney, showing that partial membrane depolarization produced NMDA receptor-mediated excitotoxicity in which the voltage-dependent magnesium block was released from the NMDA calcium channel (Olney and de Gubareff, 1978a,b; Novelli *et al.*, 1998; Zeevalk and Nicklas, 1991). Strong evidence exists for early metabolic dysfunction, energy depletion, and a role of increased oxidative signaling in patients with HD. Weight loss occurs early, frequently prior to the onset of the movement disorder (Djousse *et al.*, 2002). Positron emission tomography studies have demonstrated reduced glucose utilization in both presymptomatic and symptomatic HD patients (Kuhl *et al.*, 1985; Kuwert *et al.*, 1990; Mazziotta *et al.*, 1987). Glucose hypometabolism appears early, prior to striatal atrophy (Kuhl *et al.*, 1982, 1985). Magnetic resonance spectroscopy (MRS) has demonstrated a significant decrease in the phosphocreatine to inorganic phosphate ratio in resting muscle in patients with HD and increased lactate concentrations in the cerebral cortex (Koroshetz *et al.*, 1997). In another study, increased lactate levels were reported in both the basal ganglia and occipital cortex in symptomatic HD patients; however, they were not found in HD patients asymptomatic or at risk for the disease (Jenkins *et al.*, 1993, 1998). These findings suggest that there is a progression of impaired energy metabolism.

Energy defects may also result from mitochondrial damage caused by oxidative stress as a consequence of free radical generation. Ultrastructural studies on brain biopsies of HD patients have provided evidence of mitochondrial abnormalities and increased levels of lipofuscin, a pigment that accumulates as a consequence of free radical mediated membrane damage (Tellez-Nagel *et al.*, 1974). Evidence for oxidative damage in HD also includes DNA fragmentation, increased oxidative damage products of protein nitration, lipid peroxidation, and DNA oxidation, and inducible markers for oxidative stress (Browne *et al.*, 1999). Consistent with the findings of mitochondrial damage, we have preliminary data that shows a significant reduction in mitochondrial size and number in striatal caudate neurons in presymptomatic HD patients, with greater loss of mitochondria and further reduced mitochondrial size in moderate to severe grades of HD (Ryu *et al.*, 2005). Impaired energy metabolism may be the result of altered electron transport activities. Indeed, while significant reductions in complex I, II-III, and IV activities are present in the neostriatum from HD patients, they were not found in other brain regions (Brennan *et al.*, 1985; Browne and Beal, 1994; Gu *et al.*, 1996; Mann *et al.*, 1990; Parker *et al.*, 1990). Electron transport chain complex subunits have been reported to be involved in selective degeneration of the basal ganglia in Leber's optic neuropathy (Howell *et al.*, 1991; Jun *et al.*, 1994) and, as such, it is not unreasonable to suggest that the genetic HD mutation may alter nuclear encoded components of electron transport complexes, resulting in a primary bioenergetic defect. It is of interest to note that in patients with other trinucleotide repeat diseases, such as spinocerebellar ataxias, mitochondrial abnormalities and metabolic defects are present, linking a common mechanism of energy deficiency to the polyglutamine gene mutation (Mastrogiacomo *et al.*, 1996; Matsuishi *et al.*, 1996).

While the role of huntingtin aggregates continues to be debated, the evidence points to a proximal toxicity residing in mutant huntingtin, its proteolytic fragments, and their interactions with other proteins (Ryu and Ferrante, 2005). A secondary consequence of the gene defect may be impaired energy metabolism via mitochondrial dysfunction (Beal, 2005). N-terminal huntingtin fragments may directly impair mitochondrial function, leading to increased oxidative damage, as mitochondria are a major source of free radicals in the cell. Indeed, it appears clear that such factors as the mass effect of cytosolic and nuclear huntingtin aggregate burden, the sequestration by huntingtin aggregates of transcription factors and neuronal proteins that are essential for neuronal survival and their subsequent reduced activity (Cha, 2000; Sugars and Rubinsztein, 2003), altered proteasomal function (Bence *et al.*, 2001), and the localization of mutant huntingtin aggregates to cellular organelles such as mitochondria (Panov *et al.*, 2002) have a deleterious effect upon neuronal function and survival. Therefore, therapeutic strategies that buffer intracellular energy levels may play an important role in the treatment of HD, and polyglutamine diseases in general.

There is also substantial evidence from experimental models of HD that suggests an important interplay between energy metabolism defects, aberrant mitochondrial function, and excitotoxicity in the pathogenesis of HD (Beal, 1996, 2000; Beal *et al.*, 1993; Browne and Beal, 2004; Browne *et al.*, 1999; Grunewald and Beal, 1999; Palfi *et al.*, 1996; Schulz *et al.*, 1995; Tarnopolsky and Beal, 2001). Both necrotic and apoptotic cell death may be triggered by reduced cellular energy states (Desagher and Martinou, 2000; Green and Reed, 1998; Roy and Nicholson, 2000). Other possible sequelae of energetic deficiency and mitochondrial dysfunction include reduced redox potentials of cellular membranes, dysfunction of the mitochondrial permeability transition pore, and activation of initiator and executioner caspases, each one of which may further contribute to the cell death cascade (Beal, 2000; Green and Reed, 1998; Kiechle *et al.*, 2002). *In vitro* studies have shown that N-terminal huntingtin fragments may directly impair mitochondrial function, resulting in calcium abnormalities and subsequent energy deficiency (Panov *et al.*, 2002). Consistent with a link between energy depletion and the pathological phenotype of HD, there are a number of mitochondrial inhibitors that act at complexes of the electron transport chain, resulting in high-energy phosphate deficiency and reduced cellular levels of ATP, and mimicking the behavioral and neuropathological phenotype of HD in both primates and rodents (Alston *et al.*, 1977; Beal *et al.*, 1993; Brouillet *et al.*, 1993; Brouillet *et al.*, 1995; Henshaw *et al.*, 1994; Palfi *et al.*, 1996; Schulz *et al.*, 1995). One such naturally occurring plant toxin, 3-nitropropionic acid (3-NP), is an irreversible inhibitor of succinate dehydrogenase and, thus, both the Krebs cycle and complex II activity of the electron transport chain (Candlish *et al.*, 1969; Ludolph *et al.*, 1991). 3-NP exposure is associated with HD-like symptoms in both humans and animals and, as such, has been used as an experimental model for HD (Alston *et al.*, 1977; Beal *et al.*, 1993; Brouillet *et al.*, 1995; Ludolph *et al.*, 1991, 1992; Palfi *et al.*, 1996). Accidental ingestion of 3-NP in humans results in dystonia with jerk-like movements and bilateral damage to the basal ganglia, as determined by brain imaging (Ludolph *et al.*, 1991). In animals, 3-NP-induced experimental striatal lesions are associated with energetic deficiency, showing marked reductions in cellular levels of ATP (Hamilton and Gould, 1987; Ludolph *et al.*, 1992). Sodium azide, a complex IV (cytochrome oxidase) inhibitor, produces striatal damage and a hyperkinetic movement disorder in primates (Mettler, 1972).

As noted above, creatine administration has several potential neuroprotective effects including buffering of intracellular energy reserves, stabilizing intracellular calcium, reducing extracellular glutamate, inhibiting activation of the mitochondrial permeability transition, and acting as an anti-oxidant. The salubrious neuroprotective effects of creatine have been widely reported in experimental models of neurological diseases, particularly in neurotoxin and transgenic models of HD (Andreassen *et al.*, 2001a; Balestrino *et al.*, 1999; Brewer and Wallimann, 2000; Brustovetsky *et al.*, 2001; Carter *et al.*, 1995; Dedeoglu *et al.*, 2003; Ferrante *et al.*, 2000; Hausmann *et al.*, 2002; Ikeda *et al.*, 2000; Klivenyi *et al.* 2003, 2004a,b; Malcon *et al.*, 2000; Matthews *et al.*, 1998, 1999; Royes *et al.*, 2003; Ryu *et al.*, 2005; Shear *et al.*, 2000; Sullivan *et al.*, 2000; Zhu *et al.*, 2004). Both creatine and phosphocreatine prevented death of cultured striatal and hippocampal neurons when exposed to 3-NP (Brustovetsky *et al.*, 2001). These studies also showed that creatine decreased mitochondrial swelling induced by inhibitors of creatine kinase octamer-dimer transition. Creatine administration in rodents provided significant protection against 3-NP-induced behavioral and neuropathological phenotype (Shear *et al.*, 2000). Using the mitochondrial toxins 3-NP and malonate to mimic the energy deficiency found in HD, creatine supplementation significantly reduced striatal lesion volumes (Matthews *et al.*, 1998). The neuroprotection by creatine was associated with higher levels of PCr and creatine and reduced lactate levels in the brain, consistent with improved energy homeostasis.

Transgenic animal models have greatly advanced the study of human neurological diseases, providing experimental systems to study molecular pathogenesis and to test potential therapeutic strategies for translation to humans experiencing these diseases. Transgenic mouse models of HD, which have been used extensively to assess potential neuroprotective therapies, have energetic deficits in the brain and are ideal for examining the therapeutic potential of creatine (Beal and Ferrante, 2004; Hersch and Ferrante, 2004). While there are a number of bioenergetic therapeutic agents, including creatine, coenzyme Q_{10}, and lipoic acid, that augment energy levels and improve the behavioral and neuropathological phenotype in transgenic R6/2 HD mice (Andreassen *et al.*, 2001b,c; Dedeoglu *et al.*, 2003; Ferrante *et al.*, 2000, 2002), dietary creatine supplementation has had the greatest efficacy (Ferrante *et al.*, 2000). Creatine supplementation formulated at 1%, 2% and 3% in the diet (chow), starting at three weeks of age, significantly improved survival, reduced gross brain atrophy, delayed atrophy of striatal neurons, and reduced the formation of mutant huntingtin protein aggregates in both the striatum and neocortex. In addition, motor performance was improved and body weight loss was reduced in the creatine-treated R6/2 mice. The brain levels of creatine were significantly increased in the treated mice, as determined using nuclear magnetic resonance spectroscopy, while decreases in *N*-acetylaspartate (NAA) concentrations were delayed (Ferrante *et al.*, 2000). There was an inverted 'U' shaped efficacy curve according to dose, such that 2% creatine in the diet resulted in an approximately 18% extension of survival in the R6/2 mice, while both 1% and 3% creatine chow formulations resulted in approximately 9% and 5% in survival extension, respectively. The reduced efficacy with 3% creatine may reflect down-regulation of the creatine transporter, leading to a relative reduction in brain creatine levels. Creatine treatment in another transgenic model of HD, the N171-82Q HD mouse model, confirmed the initial results (Andreassen *et al.*, 2001a).

The effectiveness of creatine at different stages of the R6/2 HD phenotype have also been examined by initiating creatine administration (2% in the chow) after clinical symptoms appear in the R6/2 HD mice at 6, 8, and 10 weeks of age (Dedeoglu *et al.*, 2003). These time points are analogous to early, middle, and late stage disease in human HD. There was a significant extension in survival in the 6- and 8-week start groups, as well as improved motor performance, body weight, and neuropathology. There is marked bioenergetic impairment in the R6/2 HD mice (Dedeoglu *et al.*, 2003; Smith *et al.*, 2006). While there was a significant reduction in creatine and ATP in the striatum of untreated R6/2 HD mice, creatine and ATP levels were markedly improved by 39% and 65%, respectively, in the creatine-treated R6/2 mice. We recently performed a dose-ranging and efficacy study of high-dose creatine (Foran *et al.*, 2006). The preliminary data shows a dose response effect using 2%, 4%, 6%, and 10% creatine, with greatest survival (22% extended survival) from 6% creatine in the diet. Creatine at the optimal dose (6% in the diet) improved gross brain atrophy, and normalized brain weight and striatal neuronal size to wild-type levels in littermate control mice at 91 days, in comparison to untreated R6/2 mice. These improvements are significantly greater than those previously published (Ferrante *et al.*, 2000). In addition, administration of 6% creatine significantly improved brain ATP and creatine levels, as well as urine 8-hydroxy-2'-deoxyguanosine levels. This is the first instance of parallel efficacy with HD patients (Hersch *et al.*, 2006). These findings are consistent with a role for energy deficiency in HD pathogenesis and suggest that creatine therapy may benefit HD patients if started before or after clinical symptoms are present. Because there have been few phase III studies in HD patients at this time, it has not yet been confirmed that experiments demonstrating improved phenotypes in transgenic mice are predictive of benefits in humans. Similarly, it is unknown whether the magnitude of benefit in mice predicts the magnitude of benefit in humans. Nevertheless, preclinical studies using creatine in HD mouse models have provided the therapeutic rationale for the use of creatine in HD patients. Results from human clinical trials will illuminate the value of mouse clinical trials.

There is strong evidence to suggest that a bioenergetic defect exists in HD, as discussed above. Creatine supplementation is intended to augment cerebral energy reserves and thereby reduce neuronal metabolic and oxidative stress, and slow neurodegeneration. While there have been several clinical trials of creatine supplementation in HD, none of these trials have been powered to detect significant slowing of progression, although some have revealed improvement in clinical outcome measures. These trials have largely demonstrated safety and tolerability of creatine in HD patients. In addition, the biomarkers used in these studies have provided information that supports further trials. Creatine, 3–5 g/d, has been shown to be safe and well tolerated by early-stage HD patients, with blood serum creatine levels increasing over two-fold (Kieburtz, 2001). Verbessem *et al.* (2003) treated 26 HD patients with 5 g/day creatine and 15 patients with placebo for one year and found no differences in measures of strength, neurological status, or cognitive status. In a one-year open-label pilot study, Tabrizi *et al.* (2003) treated 13 individuals with HD (3 were presymptomatic, 10 were symptomatic) with creatine (10 g/d) for 12 months. Creatine administration in this study was safe and well tolerated and resulted in increased brain concentrations of creatine as demonstrated by MRS. The United Huntington Disease Rating Scale (UHDRS) scores were unchanged after 12 months. For this reason, the authors suggest that creatine supplementation at 10 g/d may be effective in stabilizing disease progression. In a multi-center double-blind placebo-controlled study of 8 g/day of creatine in 32 HD patients compared to 32 patients on placebo for four months, creatine was safe and well tolerated; however, no effects on the UHDRS were observed due to the short study period. Serum levels of creatine were increased up to 15-fold. Brain levels of creatine were significantly increased by 7.2% and NAA levels (a biomarker of neuroprotection) were increased by 16% (NAA/tCr) as measured by MRS (Hersch *et al.*, 2006). Bender and colleagues used MRS to examine another biomarker of creatine's activity in HD patients treated with 20 g/day for 5 days, followed by 6 g/day for 8–10 weeks (Bender *et al.*, 2005). They demonstrated a significant reduction in glutamate levels in the parieto-occipital cortex. This is very interesting because glutamate release and excitotoxicity are enhanced by energetic deficiency and are considered to play a significant role in the pathogenesis of HD. While none of these studies were sufficiently powered to be informative about whether or not creatine slows the clinical progression of HD, they do attest to its safety and tolerability, and to its favorable effects on serum and brain levels of creatine and on biomarkers of HD pathology. It remains unclear whether optimal creatine dosing has been determined. It may well be that higher doses of creatine are necessary to significantly slow the disease process. The most efficacious neuroprotective dose of creatine in transgenic mouse studies was 2% of the diet, corresponding to 30–35 g/d in HD patients weighing 70 kg, suggesting that the dose of creatine supplementation in HD patients may have been underestimated. While mouse and human bioavailability may not correspond well, such a dose is at least feasible for humans. In addition, creatine is taken up into the brain by a sodium- and chloride-dependent transporter (Schloss *et al.*, 1994; Snow and Murphy, 2001; Willott *et al.*, 1999). Down-regulation of the creatine transporter may occur with extended constant consumption at high doses and may alter the efficacy of creatine supplementation. Improved efficacy may be accomplished using an intermittent dosing regimen of creatine. Consequently, a dose escalation study has been initiated to determine whether there is a maximally tolerated dose in HD, as well as whether there are doses at which serum and brain levels of creatine are maximized.

2.2. Parkinson's Disease

Parkinson's disease (PD) is characterized by chronic progressive neurodegeneration of brainstem neurons, particularly dopaminergic neurons in the substantia nigra pars compacta, and the loss of their projection axon terminals in the neostriatum. This loss of dopaminergic neurons is associated with a slow onset of clinical symptoms associated with motor disability. There is a direct correlation between disease progression, neuronal loss, and motor dysfunction (Bernheimer *et al.*, 1973; Damier *et al.*, 1999). Abnormal movements are characterized by both akinesia and bradykinesia, resting tremor, rigidity, and gait abnormalities with postural instability and difficulty in initiating movement. The presence of symptoms in PD can be correlated with the loss of dopaminergic neurons when depletion reaches 60%, and with dopamine depletion in the striatum by 60% to 80% (Bernheimer *et al.*, 1973). Although, recent imaging studies show that much less dopamine depletion may be required for detection of early symptoms (Hilker *et al.*, 2005). PD affects approximately 1% of the population over 65 years of age. It is second to Alzheimer's disease as the most common neurodegenerative disorder. The neuropathological phenotype has been replicated in rodents, using a number of selective toxins, or through genetic manipulation. The toxin-induced models, however, result in acute loss of dopamine, while genetic models have produced more exact parkinsonian pathology and may be more relevant to human PD (Hwang *et al.*, 2003; Jiang *et al.*, 2005; Nunes *et al.*, 2003; Simon *et al.*, 2001; van den Munckhof *et al.*, 2003). Nevertheless, both toxin-induced and genetic models have been important in translational studies for therapeutic clinical trials in PD patients.

While the exact cause of PD is unknown, there is significant evidence to suggest that impaired energy metabolism, as a result of mitochondrial dysfunction, may play a role in the pathogenesis of Parkinson's disease (Beal, 2005). There is evidence that mitochondrial complex I defects occur in parkinsonism. In idiopathic PD, complex I activity in the electron transport chain is reduced by 30–40% in the substantia nigra (Bindoff *et al.*, 1989; Nicklas *et al.*, 1985). In addition, studies show reduced complex I activity in PD muscle and platelets (Hass *et al.*, 1995; Parker *et al.*, 1989). These findings have been strengthened by recent studies in animal models using selective inhibitors of complex I of the electron transport chain. Administration of 1-methyl-4-phenyl-1,2,3,6-tetrahydropyridine (MPTP) causes a Parkinson syndrome in both humans and in experimental animals that produces motor abnormalities consisting of bradykinesia, increased muscle tone and a characteristic resting tremor. Furthermore it induces marked depletion of dopamine stores and results in progressive degeneration of dopaminergic nigrostriatal neurons by a mechanism involving impaired high-energy phosphate production (Beal, 2001). The pathogenesis of neuronal degeneration following MPTP administration has been intensively investigated (Bove *et al.*, 2005). The neurotoxic effects of MPTP are thought to be mediated by its metabolite 1-methyl-4-phenylpyridinium (MPP^+). MPTP is converted by monoamine oxidase B to MPP^+, and MPP^+ is selectively taken up by the high-affinity dopamine and noradrenaline uptake systems, and is subsequently accumulated within mitochondria of dopaminergic neurons. There MPP^+ disrupts oxidative phosphorylation by inhibiting complex I of the electron transport chain. This can lead to a number of deleterious effects on cellular function. These include impaired intracellular calcium buffering, generation of free radicals in mitochondria, and activation of neuronal nitric oxide synthase, a calmodulin-dependent enzyme.

The effects of creatine supplementation on MPTP-induced parkinsonism have been examined. It has been reported that there was a dose-dependent neuroprotection against dopaminergic neuronal loss in the substantia nigra of mice treated with MPTP (Matthews *et al.*, 1999). The effects of creatine administration on the dopamine metabolites homovanillic acid (HVA) and 3,4-dihydroxyphenylacetic acid (DOPAC) paralleled those seen with dopamine. Creatine produced marked protection against MPTP-induced depletions of dopamine, DOPAC, and HVA. There was no significant loss of nigral neurons in creatine/MPTP-treated animals, as compared to control non-MPTP lesioned mice. Creatine supplementation has also been shown to be protective in organotypic cell cultures of mesencephalic neurons. Creatine provided neuroprotection against MPP^+-induced tyrosine hydroxylase immunoreactive cell loss. The decrease of TH-immunoreactive neurons in the MPP^+-treated cells correlated with an increase in immunoreactivity for active caspase-3, which was not seen in cells receiving creatine supplementation (Andres *et al.*, 2005). Recent combination therapeutic studies using creatine and rofecoxib resulted in an additive neuroprotective effect of the two compounds in MPTP-treated mice (Klivenyi *et al.*, 2003).

These findings set the stage for a recent randomized, double-blind, futility clinical trial in PD patients using creatine (NINDS NET-PD Investigators, 2006). Creatine supplementation (10 g/d creatine monohydrate) in PD patients was not rejected as futile and met the criteria for further clinical testing. Creatine was considered well tolerated in this patient population, and 91% of the test population continued throughout the full course of the study. In the 12-month evaluation of the clinical progression of PD, there were no significant differences in the Unified Parkinson's Disease Rating Scale (UPDRS) scores in comparison to the futility threshold values. While the study was not designed to determine whether creatine was effective in slowing clinical disease progression, the magnitude of the effects of creatine on the UPDRS score progression was comparable to that observed in a recent PD clinical trial with 1,200 mg/d coenzyme Q_{10} in which the rate of deterioration in the UPDRS was markedly slowed. These findings suggest that interventions, such as creatine, that improve bioenergetic dysfunction may hold therapeutic promise in PD. A randomized, placebo-controlled Phase III clinical trial is being planned.

2.3. Amyotrophic Lateral Sclerosis

Amyotrophic lateral sclerosis (ALS) is a clinically severe and fatal neurological disorder characterized by a range of progressive motor disabilities. Degeneration of motor neurons throughout the central nervous system results in progressive muscle wasting and paralysis that leads inexorably to death (Rowland and Shneider, 2001). The cause of selective motor neuron death in ALS is not known. Generally fatal within 2–5 years of onset, ALS has a prevalence of 2–3 per 100,000 people. The incidence of ALS is 1–2/100,000/year and may be rising. It has recently been reported that there is a significantly increased risk of developing ALS in Gulf War veterans and within the military service outside of the Gulf War (Haley, 2003; Horner *et al.*, 2003; Weisskopf *et al.*, 2005). The vast majority of ALS cases occur sporadically, but about 5–10% of ALS cases are familial. The genetic linkage of several mutations in the gene for Cu/Zn superoxide dismutase (SOD1) with some cases of familial ALS provided the first indication of a potential causal factor in the disease process (Rosen *et al.*, 1993). The similarity in the clinical course and pathological features of familial (FALS) and sporadic ALS (SALS) has led a number of investigators to search for genetic mutations associated with FALS as a strategy for elucidating disease pathogenesis and defining novel treatments in both sporadic and inherited forms of the disease. Current medical care focuses on symptom management. Supportive care ameliorates symptoms and makes ALS more manageable for patients and their families, but does not significantly affect the primary disease process. Riluzole, the only FDA-approved ALS therapy, is associated with a 2–3 month prolongation of survival (Bensimon *et al.*, 1994; Miller *et al.*, 2003). To date, no other drug therapies slow or abrogate the disease process in ALS.

Mitochondrial abnormalities have been reported in sporadic ALS (Hirano *et al.*, 1984; Okamoto *et al.*, 1990; Sasaki and Iwata, 1996). Decreased mitochondrial respiratory chain activities have also been found in patients with ALS (Borthwick *et al.*, 1999; Wiedemann *et al.*, 2002). Missense mutations in the enzyme copper/zinc superoxide dismutase (SOD1) cause about 25% of FALS cases (Rosen *et al.*, 1993), whose clinical and pathological features are indistinguishable from those in sporadic ALS. This has prompted the view that all forms of the disease may be better understood and ultimately treated by studying pathogenesis and therapy in rodent models of ALS transgenic mice and rats expressing mutant forms of SOD1 (Andersen *et al.*, 2003; Brown and Robberecht, 2001). Transgenic mice with high expression of FALS mutant SOD genes develop clinical and pathological features similar to those seen in the human disease, including hindlimb weakness and loss of motor neurons (Bruijn *et al.*, 1997; Gurney *et al.*, 1994; Newbery and Abbott, 2002). Beneficial therapeutic trials in transgenic ALS mice have generated clinical trials in humans with ALS (Choudry and Cudkowicz, 2005; Festoff *et al.*, 2003; McGeer and McGeer, 2005; Rothstein, 2003).

While a number of pathogenic mechanisms of neuronal damage have been implicated (Bruijn *et al.*, 2004; Cleveland and Rothstein, 2001), there is experimental evidence to suggest that they are interdependent and may converge with mitochondrial physiology. One putative mechanism reported to play a role in the pathogenesis of ALS is mitochondrial dysfunction and subsequent impairment of cellular energy metabolism. Mitochondria play a critical role in associated metabolic and apoptotic pathways that regulate neuronal survival. Their altered function may be relevant to motor neuron disease. Evidence that impaired energy metabolism may contribute to the death of motor neurons in ALS comes primarily from model systems. Indeed, both reduced cellular energy and mitochondrial dysfunction have been reported to be early and significant features of motor neuron disease in ALS mice (Kong and Xu, 1998). Cytochrome c release is a marker for mitochondrial dysfunction in ALS models (Zhu *et al.*, 2002). Neuropathological studies of motor neurons in SOD1 transgenic mice revealed mitochondrial vacuolation prior to cell loss (Dal Canto and Gurney, 1994; Wong *et al.*, 1995). SOD1 accumulates in vacuolated mitochondria (Jaarsma, 2006) and mutant SOD1 aberrantly binds to mitochondria where it forms high-molecular weight, SDS-insoluble species (Pasinelli *et al.*, 2004). Mutated SOD1 in mitochondria may cause mitochondrial defects, which contribute to precipitating the neurodegenerative process in motor neurons (Mattiazzi *et al.*, 2002). It has been reported that chronic mitochondrial inhibition leads to selective motor neuron death *in vitro* (Kaal *et al.*, 2000). Observations in sporadic ALS have found altered mitochondrial morphology in liver, suggesting a role for mitochondrial dysfunction (Masui *et al.*, 1985; Nakano *et al.*, 1987). There is also a high frequency of mitochondrial DNA mutations in cortical ALS motor neurons (Dhaliwal and Grewal, 2000); however, other studies question whether these mutations in fact impair mitochondrial respiration (Gajewski *et al.*, 2003; Swerdlow *et al.*, 1998). Finally, PCr has been shown to be a direct energy source for glutamate uptake into synaptic vesicles. Impaired glutamate uptake has been demonstrated in a transgenic animal model of ALS and is implicated in the pathogenesis of human ALS, linked to loss of the EAAT2 glutamate transporter (Howland *et al.*, 2002; Rothstein *et al.*, 1992, 1995). As such, compounds with beneficial effects on bioenergetics are therefore rational therapeutic candidates for ALS.

Several preclinical studies have found that creatine is beneficial in the SOD1 G93A ALS mouse model (Dupuis *et al.*, 2004; Klivyeni *et al.*, 1999, 2004). In addition, creatine in combination with minocycline demonstrated additive neuroprotective effects in ALS mice (Zhang *et al.*, 2003). Oral administration of creatine has also been shown to reduce the progression of motor neuron disease in the wobbler mouse (Ikeda *et al.*, 2000). It has been reported that in the G93A transgenic SOD1 mouse model of ALS, creatine supplementation delays onset and slows progression of the clinical and pathological phenotypes at doses of 1–2% in the diet (Klivenyi *et al.*, 1999, 2004). Creatine treatment improved the mean survival for animals by 18% and showed a significant increase in motor performance. There was a marked amelioration of ventral horn neuron loss in comparison to untreated G93A mice. Cortical concentrations of glutamate as measured by *in vivo* microdialysis and nuclear magnetic resonance (NMR) spectroscopy were significantly higher in G93A mice compared to littermate wild-type mice at 115 days of age, with attenuated increases in glutamate measured in creatine-treated G93A mice. These results are consistent with impaired glutamate transport in G93A transgenic SOD1 mice. In addition, significant depletions in cortical ATP content in presymptomatic G93A mice have been reported. Reduced ATP was partially ameliorated by creatine administration in these studies, suggesting that bioenergetic defects are involved in the initial stages of mutant SOD1-induced toxicity in G93A mice (Browne *et al.*, 2006).

These translational therapeutic trials in ALS mice suggest that creatine supplementation might be neuroprotective in ALS patients. On this basis, there have been several clinical trials. A double-blind, placebo-controlled, sequential clinical trial was performed to assess the efficacy of 10 g/d creatine monohydrate in 175 ALS patients (Groeneveld *et al.*, 2003). Creatine supplementation, while being well tolerated, showed no benefit on survival or disease progression. In another randomized double-blind, placebo-controlled trial in 104 patients with ALS treated with 5 g/d creatine over six months, there was also no clinical benefit observed (Shefner *et al.*, 2004). It has been suggested, however, that the target dose using bioenergetic nutraceutical compounds may need to be much greater in treating neurodegenerative disorders, particularly ALS, given the fulminant progression of disease (Ferrante *et al.*, 2005). To that end, an open-label dose-escalation trial was recently performed to assess the safety and tolerability of high doses of coenzyme Q_{10} in ALS. In this study, coenzyme Q_{10} was considered safe and well tolerated in 31 subjects treated with doses as high as 3 g/d for 8 months. Based upon these findings and the proposed high-dose (30–35 g/d) clinical trial using creatine in HD, a safety and tolerability trial using high-dose creatine (30 g/d) is being considered for further clinical testing in ALS patients.

2.4. Alzheimer's Disease

Alzheimer's disease (AD) is a devastating neurodegenerative disease with associated manifest dementia that affects roughly 4.5 million aged adults in the United States today (Bossy-Wetzel *et al.*, 2004). Shifts that will occur in population demographics anticipate the number of affected individuals to exceed 13 million by 2050 (Hebert *et al.*, 2003). With age being a major risk factor, most AD patients become symptomatic in the seventh decade (Mesulam, 2000). AD is the fourth leading cause of death in the United States. The most common feature of the early stages of AD, and perhaps the hallmark symptom of the disease, is the impairment of episodic memory. However, there has been some debate as to whether alterations in sensory and information processing precede these episodic memory impairments (Dudkin *et al.*, 2005). Once manifest, the disease progresses relentlessly. Currently there is no effective treatment to significantly alter the course of AD. The etiology of AD has been well documented in pathological specimens as concentrated collections of neuritic plaques and neurofibrillary tangles that accumulate in the brain. AD is a complex, multifactorial disease in which several genes act independently or in concert with each other or with environmental agents, resulting in beta-amyloid (Aβ) deposition in the brain, neurofibrillary tangle formation and cell death.

AD has a number of fundamental etiologies that lead to the neuropathological manifestations of the disease. AD patients that are at high risk for developing the disease have impaired cerebral metabolic rate *in vivo* before clinical evidence of disease is present, suggesting that impaired energy metabolism is not solely attributed to the loss of brain substance, and may be a harbinger of disease onset (Blass *et al.*, 2000). Previous data has demonstrated that mitochondrial dysfunction and subsequent impaired energy metabolism are central features of the neuropathological findings that occur in AD brain (Beal, 2005), although this has been somewhat controversial. While alterations in ATP production and concentration have been reported in human AD (Aksenov *et al.*, 2000; Hoyer, 1993; Shoffner, 1997), a spectrophotometric analysis in AD brain homogenates did not find a primary dysfunction in mitochondrial respiratory chain complexes (Casademont *et al.*, 2005). The cytoplasmic BB-CK isoenzyme has been reported to be reduced by 80% in AD brain homogenates (Aksenov *et al.*, 1997). Impaired energy metabolism in AD is also supported by decreases in PCr levels in AD brain as revealed by magnetic resonance spectroscopy (Pettegrew *et al.*, 1994). The loss of ATP is an early event in Aβ neurotoxicity (Mark *et al.*, 1997). Aβ has been reported to cause mitochondrial dysfunction in neurons and, conversely, impaired energy metabolism increases amyloidogenic amyloid precursor protein (APP) processing (Gabuzda *et al.*, 1994; Gasparini *et al.*, 1997, 1999; Webster *et al.*, 1998). Increased mitochondrial dysfunction is demonstrated by a reduction in cytochrome oxidase activity in AD cortex (Mutisya *et al.*, 1994; Parker and Parks, 1995) and an increase in mitochondrial DNA damage in postmortem human AD brain (Beal, 1995; Mecocci *et al.*, 1994). Similarly, mitochondrial mass is reduced in AD cortex. Reductions in ATP formation and an increase in reactive oxygen species, as a consequence of Aβ exposure, have also been observed in cytoplasmic hybrid NT2 cells containing mitochondrial DNA from AD and control platelets (Cardoso *et al.*, 2004; Swerdlow *et al.*, 1997).

Creatine administration has been shown to be neuroprotective against Aβ toxicity in embryonic hippocampal neurons isolated from rat brains (Brewer and Wallimann, 2000). Treatment with creatine increased the molar ratio of PCr/ATP in such neurons by 61%, with the reserve energy ratio rising 4-fold. Neurons were also partially protected from exposure to Aβ using creatine in adult rats, but not in older rats. Many of the mitochondrial abnormalities, such as an age-dependent increase in anti-oxidant enzymes and oxidative protein modifications (Pappolla *et al.*, 1998; Praticò *et al.*, 2001; Rodrigo *et al.*, 2004; Smith *et al.*, 1998), are also observed in transgenic murine models of AD. As such, we have performed preclinical trials in APP Tg2576 mice. By 6 months of age, ATP levels are significantly reduced in Tg2576 mice relative to wild-type littermate control mice. Creatine supplementation in chow (2% in the diet) starting at weaning resulted in significant neuroprotection in the Tg2576 mice. Creatine treatment protected against reductions in brain weight in APP Tg2576 mice and reduced Aβ plaque burden by 6 months. The fact that there is convincing evidence of impaired bioenergetics in AD and that creatine supplementation ameliorates the biochemical and neuropathological phenotype in APP Tg2576 mice may well be sufficient rationale for supporting clinical trials using creatine supplementation in AD patients.

3. MITOCHONDRIAL ENCEPHALOPATHIES

Metabolic activity within the central nervous system is relatively high, making brain tissue especially vulnerable to mitochondrial respiratory chain abnormalities. Mitochondrial encephalopathies are multi-system diseases characterized by biochemical and genetic mitochondrial defects with differing clinical symptomatologies and brain area involvements. Mitochondrial encephalopathies are age-related with respect to disease onset, occurring mostly in infants and young children, but also in adulthood in some patients (Barisic *et al.*, 2002; Komura *et al.*, 2003; Kremer *et al.*, 1993; Lofberg *et al.*, 2001; Scaglia and Northrop, 2006; Tarnopolsky *et al.*, 1998; Walter *et al.*, 2002). Included in this group of mitochondrial dysfunction disorders are lactic acidosis with stroke-like episodes (MELAS); Leigh syndrome; myoclonus epilepsy and ragged-red-fibers (MERRF); Kearns-Sayre syndrome (KSS); neuropathy, ataxia, and retinitis pigmentosa (NARP); mitochondrial neurogastrointestinal encephalopathy (MNGIE); Alpers syndrome; and coenzyme Q_{10} deficiency. In these diseases, specific and/or multiple mitochondrial DNA mutations have been detected (Hirano *et al.*, 2006). The exact pathophysiology of these diseases is not completely understood. While there has been little or no consensus in treatment, therapeutic strategies have been adopted as the result of single case studies and are based on the use of anti-oxidants and respiratory chain substrates and cofactors such as creatine and coenzyme Q_{10}. Adult onset encephalopathy involving mitochondrial dysfunction can also occur, as in Wernicke's encephalopathy (Desjardins and Butterworth, 2005). The latter disorder is more commonly encountered in chronic alcoholism, as well as in AIDS. Chronic alcoholism results in thiamine deficiency with a corresponding decrease in activity of the rate-limiting enzyme of the tricarboxylic acid cycle, which is thiamine diphosphate-dependent. Thiamine deficiency results in selective neurodegeneration, particularly in the thalamus, pons, and cerebellum. While the exact cause of specific neuronal death is unknown, mechanisms associated with neuronal energetic deficiency and subsequent lactic acidosis, and NMDA receptor-mediated excitotoxicity have been implicated.

As such, it has been suggested that creatine supplementation may improve the clinical symptoms of mitochondrial encephalopathies associated with energetic dysfunction (Baker and Tarnopolsky, 2003; Barisic *et al.*, 2002; Hagenfeldt *et al.*, 1994; Komura *et al.*, 2003, 2006; Kremer *et al.*, 1993). In a single case study of a patient with MELAS treated with creatine monohydrate, clinical symptoms were ameliorated after four weeks of treatment (Barisic *et al.*, 2002). Proton magnetic resonance spectroscopy performed at 6 and 12 months of treatment showed high creatine levels in the brain. Urinary creatinine excretion increased upon short-term (12 days) high-dosage creatine supplementation (20 g per day), most likely because creatinine is a metabolic product of creatine. In a small study of 5 patients with either KSS or MELAS, creatine administration (0.08–0.35 g/kg body weight/day) resulted in improved measures of skeletal muscle function, with an increase in maximum performance by 12% on average (Komura *et al.*, 2003). In addition, these investigators have shown that in a single study of Leigh syndrome, creatine monohydrate supplementation improved gross and fine motor skills and respiratory and cardiac functions in their patients (Komura *et al.*, 2006). Creatine has also been shown to be effective in patients with mitochondrial cytopathies (Tarnopolsky *et al.*, 1997). A randomized, controlled trial of creatine monohydrate administration in seven patients with mitochondrial cytopathies, primarily MELAS, showed an increased high-intensity anaerobic and aerobic exercise performance. Creatine supplementation resulted in a 19% increase in ischemic, isometric grip strength ($P < 0.01$) and an overall 11% increase ($P < 0.01$) in non-ischemic dorsiflexion torque. There is also strong evidence that coenzyme Q_{10} is effective in mitochondrial encephalopathies, especially in MELAS (Abe *et al.*, 1999; Berbel-Garcia *et al.*, 2004; Chen *et al.*, 1997; Ihara *et al.*, 1989; Shinkai *et al.*, 2000). As the therapeutic mechanisms of action of coenzyme Q_{10} and creatine are similar, these studies may be further evidence of the potential efficacy of creatine in mitochondrial encephalopathies.

4. GENETIC AND SECONDARY DISORDERS OF CREATINE METABOLISM

Inborn errors of creatine metabolism in humans may provide better insight into the pathogenic mechanisms associated with these diseases. Mutations in two enzymes responsible for creatine biosynthesis, L-arginine:glycine amidinotransferase (AGAT) and guanidinoacetate methyltransferase (GAMT), as well as mutations in a sodium- and chloride-dependent creatine transporter (SLC6A8), have been identified. In patients with GAMT and AGAT mutations, clinical symptoms include mental and motor retardation, extra-pyramidal symptoms, and seizures (see chapter 8; Stockler *et al.*, 2007; Item *et al.*, 2001; Stromberger *et al.*, 2003). The gene encoding SLC6A8 is located on the X-chromosome, thereby usually affecting males, who show more severe mental retardation and absence of speech than affected females (Mancini *et al.*, 2004). While not always consistent, plasma and CSF levels of creatine are usually low in these patients with variable, but usually elevated creatinine levels in urine (Verhoeven *et al.*, 2005). There have been some successes in the use of creatine supplementation in these disorders. Both AGAT and GAMT disorders, which are associated with a lack of creatine and PCr in the brain and other bodily tissues, are improved clinically by creatine supplementation (Mercimek-Mahmutoglu *et al.*, 2006; Stöckler *et al.*, 1996a,b; Sykut-Cegielska *et al.*, 2004;

Wyss and Schulze, 2002). On the other hand, creatine supplementation has shown no benefit in patients with mutations in SLC6A8 (Anselm *et al.*, 2006). Secondary disorders of creatine metabolism also respond to creatine supplementation. In gyrate atrophy, an autosomal recessive disorder causing hyperornithinemia and resulting in chorioretinal degeneration and type 2 muscle fiber atrophy, creatine therapy increases muscle area, augments PCr/P_i ratios by 1.5-fold, normalizes PCr/ATP ratios, and reduces tubular aggregate abnormalities in skeletal muscle (Heinanan *et al.*, 1999a,b). These patients were treated with creatine for 8–15 years, using 1.5–2.0 g/d, thereby providing evidence that prolonged creatine treatment may be possible in other disorders of creatine and PCr deficiency or dysregulation.

5. ISCHEMIC STROKE

Stroke remains one of the major causes of death and a leading cause of functional impairment, resulting in long-term disability. The latter is manifest by neurological dysfunction and significant reduction in the ability to perform activities of daily living. There are greater than 20 million incident strokes worldwide each year, resulting in more than 5.5 million annual deaths (World Health Report, 2002). Ischemic stroke is by far the most prevalent, accounting for about 88% of all strokes. Despite the advent of treatment using intravenous tissue-type plasminogen activator and the promise of additional acute therapies, effective pre- and post-stroke prevention are paramount for reducing the burden of stroke (Adams *et al.*, 2005; Sacco *et al.*, 1997).

The maintenance of ion gradients across the neuronal cell membrane involves a significant degree of metabolic energy provided by ATP. Within minutes of reduced or lack of blood flow, a cascade of events occur resulting in failure of sodium/potassium pumps, influx of extracellular calcium ions, and subsequent excitotoxicity that ultimately results in neuronal death. Neurons most severely affected by hypoxic injury die rapidly by necrosis, while those neurons that are exposed to a lesser degree of hypoxia in the penumbral zone succumb via apoptosis (Tatsumi *et al.*, 2003). Creatine supplementation may result in improved ability of the neuron to withstand ischemia-mediated energetic deficiency.

Creatine supplementation has been reported to be neuroprotective in an experimental model of anoxia in neonatal mice (Wilken *et al.*, 2000). After 30 minutes of anoxia, both ATP and PCr concentrations were significantly higher in creatine-treated pups than unsupplemented controls, suggesting that hypoxic energy failure in neonatal mice can be prevented by creatine applied before hypoxic events. In a model of transient hypoxic ischemia, six-day-old rats received creatine (3 g/kg/d) for 3 days prior to unilateral common carotid artery ligation followed by hypoxia. The creatine treated rats showed a significant reduction in volume of brain edema and an increased 'energy potential' as reflected by the ratio of PCr to inorganic phosphate that was measured by ^{31}P-magnetic resonance spectroscopy (Adcock *et al.*, 2002). Neuronal cell injury was significantly lower in the cortex of the animals that had received creatine. In a separate report, oral creatine administration resulted in a marked reduction in ischemic brain infarction and neuroprotection after cerebral ischemia in mice, with a direct correlation between the preservation of bioenergetic cellular status and the inhibition of activation of caspase cell-death pathways *in vivo* (Zhu *et al.*, 2004). Post-ischemic caspase-3 activation and cytochrome c release were significantly reduced in creatine-treated mice. In addition, creatine administration buffered ischemia-mediated cerebral ATP depletion, suggesting that creatine may be neuroprotective in this experimental paradigm through mechanisms independent of mitochondrial cell-death pathways. With respect to the latter, it has been shown that creatine-mediated neuroprotection may involve improved cerebrovascular function (Prass *et al.*, 2007). The authors found a 40% reduction in infarct volume from transient focal cerebral ischemia after 3 weeks of dietary creatine administration without any changes in bioenergetic status as reflected by brain creatine, PCr, and ATP concentrations. There were, however, increased cerebral blood flow and vasodilatory responses after stroke in creatine-treated mice, suggesting that creatine-mediated neuroprotection may be associated with improved cerebrovascular function. This experimental data demonstrates that there is increased ability to resist ischemic injury by creatine supplementation and that these findings correlate with improved maintenance of energy metabolism and cellular homeostasis. In ischemic stroke, creatine may be multimodal by also inhibiting hypoxia-mediated release of cytochrome c and downstream activation of caspase-3, and may improve cerebrovascular flow. Prophylactic creatine supplementation may be beneficial in patients at high risk for stroke in preventing neuronal damage and loss.

6. TRAUMATIC BRAIN AND SPINAL CORD INJURY

Traumatic brain (TBI) and spinal cord injuries (SCI) remain a major health and social problem. These types of injuries often occur in early adulthood and have a major impact for society. Traumatic brain and spinal cord insults result in both immediate mechanical damage and subsequent tissue damage and loss. The frequency, complexity, severity, and diversity of head and spinal cord injury are myriad with extensive long-term disabilities. Approximately 15 million people sustain a TBI or SCI each year in North America. According to the Centers for Disease Control and Prevention, TBI and SCI cost the nation an estimated tens of billions of dollars each year, and treatment of secondary conditions comes at still much higher costs. The leading causes of TBI and SCI are motor vehicle accidents, falls, and being struck by a foreign object. TBI can be quite diverse, depending upon the areas of the brain that are involved. Trauma to the vertebral column alters the spinal cord's ability to send and receive messages from the brain, resulting in loss of sensory, autonomic, and motor function below the level of injury. The incidence of SCI is highest among young individuals, age 16–30, with males representing approximately 80%. While there is no current cure for TBI and SCI, ongoing studies in drug testing, transplantation, and surgical therapies continue to make progress. Education and prevention may have the greatest impact in reducing these injuries. Injury can arise immediately after trauma or within days of injury. While the primary cause of neuronal death is the physical damage of impact, the secondary effects of trauma are less apparent and are associated with edema, ischemia, inflammation, altered calcium homeostasis, and impaired energy metabolism related to mitochondrial dysfunction and ATP loss (Sullivan *et al.*, 1998, 1999; Taoka and Okajima, 1998; Xiong *et al.*, 1997, 1998). ATP loss may also be the result of vascular damage and consequent hypoxia. As such, enhanced neuronal survival in traumatic injuries may be dependent upon adequate ATP supplies. As discussed above, creatine may provide enhanced neuroprotection after these cellular events.

There have been efforts to demonstrate the efficacy of creatine in experimental models of TBI and SCI (Hausmann *et al.*, 2002; Rabchevsky *et al.*, 2003; Scheff and Dhillon, 2004). Surgically-induced cortical contusions representing the sequelae of human closed-head injury were performed in mice. These resulted in severe behavioral deficits, cortical tissue loss, neuronal loss in the hippocampus, and blood brain barrier damage. Intraperitoneal injection of creatine, 3 mg/kg/day up to five days before injury, significantly reduced the TBI tissue damage. Neuroprotection was also observed in TBI rats fed creatine. Mitochondrial membrane potential was significantly greater in creatine-treated rats than untreated animals, with reduced levels of reactive oxygen species in treated rats. Induction of mitochondrial permeability transition was significantly inhibited in creatine-treated TBI animals. In a subsequent study, creatine administration significantly lowered free fatty acid and lactate levels in comparison to untreated rats. Both are markers of cellular injury following TBI (Scheff and Dhillon, 2004). In addition, the neuroprotective effects of creatine have been reported in SCI (Hausmann *et al.*, 2002). Following hindlimb paralysis by surgical laminectomy in adult rats, creatine-treated animals recovered better with greater scores on an open field locomotor test. Creatine-treated rats had significantly smaller volumes of scar tissue. These experimental studies support the concept that creatine supplementation may provide neuroprotection after CNS injury and that a prophylactic dose may be efficacious when given prior to elective brain and spinal cord surgery.

7. EPILEPSY

It is estimated that epilepsy affects 0.5–1% of the population worldwide, with higher rates in developing countries (Hauser and Kurland, 1975). In the United States, there are approximately 2.5–3 million people with epilepsy, with a slightly higher prevalence in men. There are numerous types of epilepsy, primarily because the etiology of epilepsy is related to various risk factors at different ages. In Western developed countries, the risk of epilepsy increases with age, due to tumors, strokes, and neurodegenerative diseases. Epilepsy in children and young adults is usually due to congenital brain disease or one of the idiopathic generalized epilepsies, such as absence epilepsy or juvenile myoclonic epilepsy. Overall, most epilepsies are localization related, with a small percentage being due to idiopathic generalized epilepsies (Loiseau *et al.*, 1990). Localization-related, or focal, epilepsies begin in a small area of dysfunctional cortex and may spread to adjacent areas of the brain or, eventually, throughout the entire brain. These are usually due to trauma, infection, tumor, stroke, or rare cortical malformations. Idiopathic generalized epilepsies are thought to originate from the thalamus and cause the entire cortex to seize nearly simultaneously.

A seizure is an abnormal, synchronized discharge of a large population of neurons, while epilepsy is defined as recurrent unprovoked seizures. The pathophysiology of seizures is simply an abnormal excitation of neurons with a lack of normal neuronal inhibition. The basic principles of seizures arise from dysfunction of the neuronal membrane resting potential. Intracellularly, sodium is low and potassium is high, while the opposite is true in the extracellular space. Energy, in the form of ATP, is required to maintain this gradient, thereby creating a negative neuronal membrane potential. When there are sufficient excitatory action potentials, voltage-gated sodium channels open and sodium rapidly enters the cell, thereby depolarizing the membrane and creating an action potential. This action potential is propagated along the axon, resulting in neurotransmitter release, which in turn excites an adjacent neuron. When the resting membrane of the cell reaches a certain level, potassium channels are activated, thereby hyperpolarizing the cell and creating a refractory period where the neuron cannot be excited. Sodium-calcium ion channels are believed to be involved in seizure termination. Ion channel mutations have been identified in several nervous system diseases, some of which are inherited epilepsies (Dichter and Wilcox, 1997).

There is substantial evidence for increased metabolism, oxidative damage, and cell death in epilepsy. During a seizure, there is a sudden increase in metabolic demand as evidenced by increased glucose uptake and blood flow to the involved tissue with a subsequent enhanced production of lactate (Theodore, 1999). In animal models, repeated seizures have also been shown to be associated with oxidative damage to cellular components (Liang and Patel, 2004). Neuronal loss and cell death in epilepsy occur through excitatory amino acids, such as glutamate, leading to a sustained depolarization and activation of voltage-dependent sodium channels, influx of chloride and water, and necrotic cell death. In parallel, activation of NMDA receptors by glutamate causes calcium influx into neurons, leading to mitochondrial dysfunction, inhibition of ATP production, increased production of free radicals, and eventually apoptotic cell death.

Mitochondria have been shown to be the major site of oxidative damage in animal models of chemically-induced seizures. Kainate, an excitotoxin resembling glutamate, induces status epilepticus and has been shown to lead to free radical production in mitochondria by inhibition of aconitase with eventual death of hippocampal neurons (Liang *et al.*, 2000). Oxidative damage to mitochondrial lipids (Baran *et al.*, 1987) and DNA (Liang *et al.*, 2000) has also been shown in kainate models of epilepsy. Oxidative damage is found in areas more susceptible to kainate injury, such as in hippocampal CA1 and CA3 fields. There is also a strong correlation between seizure-induced mitochondrial oxidative stress and age (Patel, 2004). Conversely, known mitochondrial mutations are associated with seizures. In MERRF, as discussed above, there are defects in mitochondrial complexes I and IV that lead to increased oxidative changes and disease pathology (Shoffner *et al.*, 1990). Finally, increased oxidative damage, whether through neurodegenerative conditions or normal aging, may also lead to increased seizure susceptibility. Mice heterozygous for superoxide dismutase 2 (SOD2) have an increased risk of kainate-induced seizures and develop more seizures with age, correlating with mitochondrial dysfunction (Liang and Patel, 2004).

Therapeutic antioxidant compounds have been demonstrated to decrease free radical generation during seizures. In trimethyltin-induced seizures in rats, hippocampi showed an increase in oxidative damage that was ameliorated, as were seizures, with ascorbic acid administration (Shin *et al.*, 2005). Melatonin, an oxygen free radical scavenger and a GABA receptor regulator, when administered to kainate-injected mice, significantly reduced clinical seizure activity and almost completely abolished lipoperoxidation products in the brain (Tan *et al.*, 1998). A ketogenic diet, which decreases blood glucose, has been shown to decrease seizure frequency. In mice fed a ketogenic diet, it has been shown that there is an increase in mitochondrial uncoupling proteins with an associated significant reduction in reactive oxygen species (Sullivan *et al.*, 2004). Finally, some anti-epileptic drugs have been shown to have anti-oxidant properties. Zonisamide has been shown to decrease free radical production associated with seizures in kainate animal models (Ueda *et al.*, 2005; Masumizu *et al.*, 2005).

Creatine, as an intracellular facilitator of high-energy phosphate transport and as an anti-oxidant, makes it a likely therapeutic candidate in epilepsy. There have been many recent spectroscopy studies examining alterations in creatine levels *in vitro* as well as *in vivo* in animal models of epilepsy and in humans. *In vitro* studies showed that stimulated hippocampal neurons fired faster and had faster recovery rates when phosphocreatine (PCr)/NAA levels were higher (Williamson *et al.*, 2005). In *in vivo* trials, creatine administration has been shown to ameliorate hypoxic seizures in both rabbit pups and rats (Holtzman *et al.*, 1998, 1999). Rabbit pups were given subcutaneous creatine injections three days before inducing hypoxia. Early seizures were prevented and late seizures were significantly reduced by 60%. Additionally, brain levels of creatine, as detected by NMR spectroscopy, initially doubled as compared to controls and eventually reached adult levels (Holtzman *et al.*, 1999). More recently, it has been shown that seizures and increased lactate production after intrastriatal methylmalonate (MMA) injections are significantly decreased by creatine pre-administration (Royes *et al.*, 2003). Brain levels of phosphocreatine, as assayed by HPLC of tissue homogenates, were higher in animals supplemented with creatine, suggesting that creatine may protect against seizure-induced metabolic changes by inhibition of secondary glutamate release. In a subsequent study using the same MMA model, creatine offered protection against protein carbonylation in animals pre-fed creatine (Royes *et al.*, 2006). There has been only one report documenting no protective effect of creatine against seizures (Mikati *et al.*, 2004). Pre-pubescent rats were injected with kainate, leading to status epilepticus, and they then received creatine chow for the subsequent 50 days. Creatine-fed rats showed selective learning impairments, but no difference in spontaneous recurrent seizures or hippocampal pathology. These findings might, in fact, be attributable to creatine supplementation occurring after instead of before kainate-induced status epilepticus. R6/2 HD mice also have increased frequency of seizures early on in the disease pathogenesis. As stated above, there is significantly impaired energy metabolism in these mice, consistent with reduced creatine, PCr, and ATP levels (Dedeoglu *et al.*, 2003; Smith *et al.*, 2006). Creatine treatment blocks seizure activity in these mice.

While there are no documented therapeutic trials of creatine supplementation in patients with epilepsy, there is strong evidence that this patient population has decreased intracellular creatine concentrations in the epileptogenic zones. In patients who were candidates for epilepsy surgery, Pan and colleagues found that there was a significant decrease in the PCr/ATP ratio in the ipsilateral amygdala (Pan *et al.*, 2005). It has also been shown that the periphery of cortical malformations is associated with a decrease in NAA/Cr in about 15% of the patients (Mueller *et al.*, 2005). Interestingly, Cohen-Gadol *et al.* (2004) showed that in mesial temporal lobectomy patients who had undergone pre-operative NMR spectroscopy and post-operative hippocampal pathological analysis, NAA/Cr ratios correlated with neuronal cell loss in most, but not all, hippocampal regions. Surprisingly, the NAA/Cr ratio did not correlate with the duration of seizures. Similar findings have been reported by other investigators (Burneo *et al.*, 2004), suggesting that these changes may be primary and not secondary to seizure activity.

8. CONCLUSION

Neuroprotective compounds targeting identified pathologic mechanisms of disease have the potential to delay the onset and slow the progression of neurological diseases. A large number of studies in both patients and experimental model systems have validated high-energy phosphate deficiency as a promising therapeutic target in neurological diseases. Compounds such as creatine buffer neuronal energy demands and are attractive candidates for targeting this important disease mechanism. Creatine has some advantages over other similar-acting compounds, which include more straightforward bioavailability and the potential to serve as an *in vivo* biomarker of premanifest and manifest disease. Creatine is available for human use and represents an immediate candidate as a neuroprotective agent for large-scale clinical trials in neurological diseases. Given the safety and tolerability of creatine, it may be especially well suited for long-term use in neurological disorders, providing prophylactic treatment in patients at high risk for recurrent events. Ongoing early phase studies will soon determine an optimal dose range for patients and provide useful biomarkers. An important implication of the multiple levels of molecular pathology and treatment is that it will most likely be necessary to combine neuroprotective therapies to maximize neuroprotection in order to reach the greatest efficacy. As such, creatine may well be suited for use in combination with neuroprotective agents targeting other pathologic mechanisms of disease. Preclinical studies in animal models of neurological diseases testing creatine in combination with other neuroprotective agents have reported additive efficacy in both clinical and neuropathological outcome measures. It has yet to be determined which combinations might have the most promise for translation to human clinical trials.

ACKNOWLEDGEMENTS

This work was supported by NIH grants AT00613, NS045806, NS045242 (RJF), the Veterans Administration (RJF), and the Scholars in Clinical Science Program, Harvard Medical School (AMK). We would like to give thanks to Steve Del Signore for his help in editing this manuscript.

REFERENCES

Abe, K., Matsuo, Y., Kadekawa, J., Inoue, S., and Yanagihara, T., 1999, Effect of coenzyme Q10 in patients with mitochondrial myopathy, encephalopathy, lactic acidosis, and stroke-like episodes (MELAS): evaluation by noninvasive tissue oximetry. *J. Neurol. Sci.* **162:** 65–68.

Adams, H., Adams, R., Del Zoppo, G., Goldstein, L.B., and Stroke Council of the American Heart Association, American Stroke Association, 2005, Guidelines for the early management of patients with ischemic stroke: 2005 guidelines update: a scientific statement from the Stroke Council of the American Heart Association/American Stroke Association. *Stroke* **36:** 916–923 (published corrections appear in *Stroke* **36:** 1352 and *Stroke* **36:** 1626).

Adcock, K.H., Nedelcu, J., Loenneker, T., Martin, E., Wallimann, T., and Wagner, B.P., 2002, Neuroprotection of creatine supplementation in neonatal rats with transient cerebral hypoxia-ischemia. *Dev. Neurosci.* **24:** 382–388.

Aksenov, M., Aksenova, M., Butterfield, D.A., and Markesbery, W.R., 2000, Oxidative modification of creatine kinase BB in Alzheimer's disease brain. *J. Neurochem.* **74:** 2520–2527.

Albin, R.L., and Greenamyer, J.T., 1992, Alternative excitotoxic hypotheses. *Neurology* **42:** 733–738.

Alston, T.A., Mela, L., and Bright, H.J., 1977, 3-Nitropropionate, the toxic substance of Indigofera, is a suicide inactivator of succinate dehydrogenase. *Proc. Natl. Acad. Sci. USA* **74:** 3767–3771.

Andersen, P.M., Sims, K.B., Xin, W.W., Kiely, R., O'Neill, G., Ravits, J., Pioro, E., Harati, Y., Brower, R.D., Levine, J.S., Heinicke, H.U., Seltzer, W., Boss, M., and Brown, Jr. R.H., 2003, Sixteen novel mutations in the Cu/Zn superoxide dismutase gene in amyotrophic lateral sclerosis: a decade of discoveries, defects and disputes. *Amyotroph. Lateral Scler. Other Motor Neuron Disord.* **4:** 62–73.

Andreassen, O.A., Dedeoglu, A., Ferrante, R.J., Jenkins, B.G., Ferrante, K.L., Thomas, M., Friedlich, A., Browne, S.E., Schilling, G., Borchelt, D.R., Hersch, S.M., Ross, C.A., and Beal, M.F., 2001a, Creatine increase survival and delays motor symptoms in a transgenic animal model of Huntington's disease. *Neurobiol. Dis.* **8:** 479–491.

Andreassen, O.A., Ferrante, R.J., Dedeoglu, A., and Beal, M.F., 2001b, Lipoic acid improves survival in transgenic mouse models of Huntington's disease. *NeuroReport* **12:** 3371–3373.

Andreassen, O.A., Ferrante, R.J., Huang, H.M., Dedeoglu, A., Park, L., Ferrante, K.L., Kwon, J., Borchelt. D.R., Ross, C.A., Gibson, G.E., and Beal, M.F., 2001c, Dichloroacetate exerts therapeutic effects in transgenic mouse models of Huntington's disease. *Ann. Neurol.* **50:** 112–117.

Andres, R.H., Ducray, A.D., Perez-Bouza, A., Schlattner, U., Huber, A.W., Krebs, S.H., Seiler, R.W., Wallimann, T., and Widmer, H.R., 2005, Creatine supplementation improves dopaminergic cell survival and protects against MPP$^+$ toxicity in an organotypic tissue culture system. *Cell Transplant.* **14:** 537–550.

Anselm, I.M., Alkuraya, F.S., Salomons, G.S., Jakobs, C., Fulton, A.B., Mazumdar, M., Rivkin, M., Frye, R., Poussaint, T.Y., and Marsden, D., 2006, AX-linked creatine transporter defect: A report on two unrelated boys with a severe clinical phenotype. *J. Inherit. Met. Dis.* **29:** 214–219.

Baker, S.K., and Tarnopolsky, M.A., 2003, Targeting cellular energy production in neurological disorders. *Expert Opin. Investig. Drugs* **12:** 1655–1679.

Balestrino, M., Rebaudo, R., and Lunardi, G., 1999, Exogenous creatine delays anoxic depolarization and protects from hypoxic damage: dose-effect relationship. *Brain Res.* **816:** 124–130.

Balsom, P.D., Soderlund, K., and Ekblom, B., 1994, Creatine in humans with special reference to creatine supplementation. *Sports Med.* **18:** 268–280.

Baran, H., Heldt, R., and Hertting, G., 1987, Increased prostaglandin formation in rat brain following systemic application of kainic acid. *Brain Res.* **404:** 107–112.

Barisic, N., Bernert, G., Ipsiroglu, O., Stromberger, C., Muller, T., Gruber, S., Prayer, D., Moser, E., Bittner, R.E., and Stöckler-Ipsiroglu, S., 2002, Effects of oral creatine supplementation in a patient with MELAS phenotype and associated nephropathy. *Neuropediatrics* **33:** 157–161.

Beal, M.F., 1992, Does impairment of energy metabolism result in excitotoxic neuronal death in neurodegenerative illnesses? *Ann. Neurol.* **31:** 119–130.

Beal, M.F., 1995, Aging, energy, and oxidative stress in neurodegenerative diseases. *Ann. Neurol.* **38:** 357–366.

Beal, M.F., 1996, Mitochondria, free radicals, and neurodegeneration. *Curr. Opin. Neurobiol.* **6:** 661–666.

Beal, M.F., 2000, Energetics in the pathogenesis of neurodegenerative diseases. *Trends Neurosci.* **23:** 298–304.

Beal, M.F., 2001, Mitochondria and oxidative damage in amyotrophic lateral sclerosis. *Funct. Neurol.* **16:** 161–169.

Beal, M.F., 2003, Bioenergetic approaches for neuroprotection in Parkinson's disease. *Ann. Neurol.* **53 (Suppl. 3):** S39–S47.

Beal, M.F., 2005, Mitochondria take center stage in aging and neurodegeneration. *Ann. Neurol.* **58:** 495–505.

Beal, M.F., and Ferrante, R.J., 2004, Experimental therapeutics in transgenic mouse models of Huntington's disease. *Nat. Rev. Neurosci.* **5:** 373–384.

Beal, M.F., Brouillet, E., Jenkins, B.G., Ferrante, R.J., Kowall, N.W., Miller, J.M., Storey, E., Srivastava, R., Rosen, B.R., and Hyman, B.T., 1993, Neurochemical and histologic characterization of striatal excitotoxic lesions produced by the mitochondrial toxin 3-nitropropionic acid. *J. Neurosci.* **13:** 4181–4192.

Bence, N.F., Sampat, R.M., and Kopito, R.R., 2001, Impairment of the ubiquitin-proteasome system by protein aggregation. *Science* **292:** 1552–1555.

Bender, A., Auer, D.P., Merl, T., Reilmann, R., Saemann, P., Yassouridis, A., Bender, J., Weindl, A., Dose, M., Gasser, T., and Klopstock, T., 2005, Creatine supplementation lowers brain glutamate levels in Huntington's disease. *J. Neurol.* **252:** 36–41.

Bensimon, G., Lacomblez, L., Meininger, V., and The ALS/Riluzole Study Group, 1994, A controlled trial of riluzole in amyotrophic lateral sclerosis. *New Engl. J. Med.* **330:** 585–591.

Berbel-Garcia, A., Barbera-Farre, J.R., Etessam, J.P., Salio, A.M., Cabello, A., Gutierrez-Rivas, E., and Campos, Y., 2004, Coenzyme Q10 improves lactic acidosis, strokelike episodes, and epilepsy in a patient with MELAS (mitochondrial myopathy, encephalopathy, lactic acidosis, and strokelike episodes). *Clin. Neuropharmacol.* **27:** 187–191.

Bernardi, P., Colonna, R., Costantini, P., Eriksson, O., Fontaine, E., Ichas, F., Massari, S., Nicolli, A., Petronilli, V., and Scorrano, L., 1998, The mitochondrial permeability transition. *Biofactors* **8:** 273–281.

Bernheimer, H., Birkmayer, W., Hornykiewicz, O., Jellinger, K., and Seitelberger, F., 1973, Brain dopamine and the syndromes of Parkinson and Huntington. Clinical, morphological and neurochemical correlations. *J. Neurol. Sci.* **20:** 415–455.

Bessman, S.P., and Carpenter, C.L., 1985, The creatine-creatine phosphate energy shuttle. *Annu. Rev. Biochem.* **54:** 831–862.

Bindoff, L.A., Birch-Machin, M.A., Farnsworthm L., Gardner-Medwin, D., Lindsay, J.G., and Turnbull, D.M., 1989, Familial intermittent ataxia due to a defect of the E1 component of pyruvate dehydrogenase complex. *J. Neurol. Sci.* **93:** 311–318.

Blass, J.P., Sheu, R.F., and Gibson, G.E., 2000, Inherent abnormalities in energy metabolism in Alzheimer disease: Interaction with cerebrovascular compromise. *Ann. N.Y. Acad. Sci.* **903:** 204–221.

Bloch, K., and Schoenheimer, R., 1941, The biological precursors of creatine. *J. Biol. Chem.* **138:** 167–194.

Borthwick, G.M., Johnson, M.A., Ince, P.G., Shaw, P.J., and Turnbull, D.M., 1999, Mitochondrial enzyme activity in amyotrophic lateral sclerosis: implications for the role of mitochondria in neuronal cell death. *Ann. Neurol.* **46:** 787–790.

Bossy-Wetzel, E., Schwarzenbacher, R., and Lipton, S.A., 2004, Molecular pathways to neurodegeneration. *Nat. Med.* **10 (Suppl.):** S2–S9.

Bove, J., Prou, D., Perier, C., and Przedborski, S., 2005, Toxin-induced models of Parkinson's disease. *NeuroRx* **2:** 484–494.

Braissant, O., Henry, H., Loup, M., Eilers, B., and Bachmann, C., 2001, Endogenous synthesis and transport of creatine in the rat brain: an in situ hybridization study. *Brain Res. Mol. Brain Res.* **86:** 193–201.

Brennan, Jr. W.A., Bird, E.D., and Aprille, J.R., 1985, Regional mitochondrial respiratory activity in Huntington's disease brain. *J. Neurochem.* **44:** 1948–1950.

Brewer, G.J., and Wallimann, T.W., 2000, Protective effect of the energy precursor creatine against toxicity of glutamate and beta-amyloid in rat hippocampal neurons. *J. Neurochem.* **74:** 1968–1978.

Brouillet, E., Jenkins, B.G., Hyman, B.T., Ferrante, R.J., Kowall, N.W., Srivastava, R., Roy, D.S., Rosen, B.R., and Beal, M.F., 1993, Age-dependent vulnerability of the striatum to the mitochondrial toxin 3-nitropropionic acid. *J. Neurochem.* **60:** 356–359.

Brown, R.H., Jr, and Robberecht, W., 2001, Amyotrophic lateral sclerosis: Pathogenesis. *Semin. Neurol.* **21:** 131–140.

Browne, S.E., and Beal, M.F., 1994, Oxidative damage and mitochondrial dysfunction in neurodegenerative diseases. *Biochem. Soc. Trans.* **22:** 1002–1006.

Browne, S.E., and Beal, M.F., 2004, The energetics of Huntington's disease. *Neurochem. Res.* **29:** 531–546.

Browne, S.E., Ferrante, R.J., Beal, M.F., 1999, Oxidative stress in Huntington's disease. *Brain Pathol.* **9:** 147–163.

Browne, S.E., Yang, L., DiMauro, J.P., Fuller, S.W., Licata, S.C., and Beal, M.F., 2006, Bioenergetic abnormalities in discrete cerebral motor pathways presage spinal cord pathology in the G93A SOD1 mouse model of ALS. *Neurobiol. Dis.* **22:** 599–610.

Bruijn, L.I., Becher, M.W,. Lee, M.K., Anderson, K.L., Jenkins, N.A., Copeland, N.G., Sisodia, S.S., Rothstein, J.D., Borchelt, D.R., Price, D.L., and Cleveland, D.W., 1997, ALS-linked SOD1 mutant G85R mediates damage to astrocytes and promotes rapidly progressive disease with SOD1-containing inclusions. *Neuron* **18:** 327–338.

Bruijn, L.I., Miller, T.M., and Cleveland, D.W., 2004, Unraveling the mechanisms involved in motor neuron degeneration in ALS. *Annu. Rev. Neurosc.* **27:** 723–749.

Brustovetsky, N., Brustovetsky, T., and Dubinsky, J.M., 2001, On the mechanisms of neuroprotection by creatine and phosphocreatine. *J. Neurochem.* **76:** 425–434.

Burneo, J.G., Knowlton, R.C., Faught, E., Martin, R., Sawrie, S., and Kunzniecky, R.I., 2004, Chronic temporal lobe epilepsy; spatial extent and degree of metabolic dysfunction studied with magnetic resonance spectroscopy (MRS). *Epilepsy Res.* **62:** 119–124.

Candlish, E., La Croix, J., and Unrau, A.M., 1969, The biosynthesis of 3-nitropropionic acid in creeping indigo (Indigofera spicata). *Biochemistry* **8:** 182–186.

Cardoso, S.M., Santana, I., Swerdlow, R.H., and Oliveira, C.R., 2004, Mitochondria dysfunction of Alzheimer's disease cybrids enhances Abeta toxicity. *J. Neurochem.* **89:** 1417–1426.

Carter, A.J., Muller, R.E., Pschorn, U., and Stransky, W., 1995, Preincubation with creatine enhances levels of creatine phosphate and prevents anoxic damage in rat hippocampal slices. *J. Neurochem.* **64:** 2691–2699.

Casademont, J., Rodriguez-Santiago, B., Miro, O., Beato, A., Lopez, S., Nunes, V., and Cardellach, F., 2005, Mitochondrial respiratory chain in brain homogenates: activities in different brain areas in patients with Alzheimer's disease. *Aging Clin. Exp. Res.* **17:** 1–7.

Casey, A., and Greenhaff, P.L., 2000, Does dietary creatine supplementation play a role in skeletal muscle metabolism and performance? *Am. J. Clin. Nutr.* **72 (2 Suppl.):** 607S–617S.

Cha, J.H., 2000, Transcriptional dysregulation in Huntington's disease. *Trends Neurosci.* **23:** 387–392.

Chen, R.S., Huang, C.C., and Chu, N.S., 1997, Coenzyme Q10 treatment in mitochondrial encephalomyopathies. Short-term double-blind, crossover study. *Eur. Neurol.* **37:** 212–218.

Choudry, R.B., and Cudkowicz, M.E., 2005, Clinical trials in amyotrophic lateral sclerosis: The tenuous past and the promising future. *J. Clin. Pharmacol.* **45:** 1334–1344.

Cleveland, D.W., and Rothstein, J.D., 2001, From Charcot to Lou Gehrig: deciphering selective motor neuron death in ALS. *Nat. Rev. Neurosci.* **2:** 806–819.

Cohen-Gadol, A.A., Pan, J.W., Kim, J.H., Spencer, D.D., and Hetherington, H.H., 2004, Mesial temporal lobe epilepsy; a proton magnetic resonance spectroscopy study and a histopathological analysis. *J. Neurosurg.* **1010:** 613–620.

Dal Canto, M.C., and Gurney, M.E., 1994, Neuropathological changes in two lines of mice carrying a transgene for mutant human Cu,Zn SOD, and in mice overexpressing wild type human SOD: a model of familial amyotrophic lateral sclerosis (FALS). *Brain Res.* **676:** 25–40.

Damier, P., Hirsch, E.C., Agid, Y., and Graybiel, A.M., 1999, The substantia nigra of the human brain. II. Patterns of loss of dopamine-containing neurons in Parkinson's disease. *Brain* **122:** 1437–1448.

Dedeoglu, A., Kubilus, J.K., Yang, L., Ferrante, K.L., Hersch, S.M., Beal, M.F., and Ferrante, R.J., 2003, Creatine therapy provides neuroprotection after onset of clinical symptoms in Huntington's disease transgenic mice. *J. Neurochem.* **85:** 1359–1367.

Dechent, P., Pouwels, P.J., Wilken, B., Hanefeld, F., and Frahm, J., 1999, Increase of total creatine in human brain after oral supplementation of creatine-monohydrate. *Am. J. Physiol.* **277:** R698–R704.

Desagher, S., and Martinou, J.C., 2000, Mitochondria as the central control point of apoptosis. *Trends Cell Biol.* **10:** 369–377.

Desjardins, P., and Butterworth, R.F., 2005, Role of mitochondrial dysfunction and oxidative stress in the pathogenesis of selective neuronal loss in Wernicke's encephalopathy. *Mol. Neurobiol.* **31:** 17–25.

Dhaliwal, G.K., and Grewal, R.P., 2000, Mitochondrial DNA deletion mutation levels are elevated in ALS brains. *Neuroreport* **11:** 2507–2509.

Dichter, M., and Wilcox, K.S., 1997, Excitatory synaptic transmission. In: Engel, J., and Pedley, T.A. (eds.), Epilepsy: A Comprehensive Textbook. Lippincott Raven, Philadelphia, pp. 251–263.

Djousse, L., Knowlton, B., Cupples, L.A., Marder, K., Shoulson, I., and Myers, R.H., 2002, Weight loss in early stage of Huntington's disease. *Neurology* **59:** 1325–1330.

Dudkin, K.N., Chueva, I.V., and Makarov, F.N., 2005, Interaction of sensory and cognitive processes during visual recognition: the role of the associative areas of the cerebral cortex. *Neurosci. Behavioral Physiol.* **35:** 407–416.

Dupuis, L., Oudart, H,. René, F., Gonzalez de Aguilar, J-L., and Loeffler, J-P., 2004, Evidence for defective energy homeostasis in amyotrophic lateral sclerosis: Benefit of a high-energy diet in a transgenic mouse model. *Proc. Natl. Acad. Sci. USA* **101:** 11159–11164.

Ferrante, K.L., Shefner, J., Zhang, H., Betensky, R., O'Brien, M., Yu, H., Fantasia, M., Taft, J., Beal, M.F., Traynor, B., Newhall, K., Donofrio, P., Caress, J., Ashburn, C., Freiberg, B., O'Neill, C., Paladenech, C., Walker, T., Pestronk, A., Abrams, B., Florence, J., Renna, R., Schierbecker, J., Malkus, B., and Cudkowicz, M., 2005, Tolerance of high-dose (3,000 mg/day) coenzyme Q10 in ALS. *Neurology* **65:** 1834–1836.

Ferrante, R.J., Kowall, N.W., Beal, M.F., Richardson, Jr. E.P., Bird, E.D., and Martin, J.B., 1985, Selective sparing of a class of striatal neurons in Huntington's disease. *Science* **230:** 561–563.

Ferrante, R.J., Kowall, N.W., and Richardson, Jr. E.P., 1991, Proliferative and degenerative changes in striatal spiny neurons in Huntington's disease: a combined study using the section-Golgi method and calbindin D28k immunocytochemistry. *J. Neurosci.* **11:** 3877–3887.

Ferrante, R.J., Andreassen, O.A., Jenkins, B.G., Dedeoglu, A., Kuemmerle, S., Kubilus, J.K., Kaddurah-Daouk, R., Hersch, S.M., and Beal, M.F., 1999, Neuroprotective effects of creatine in a transgenic animal model of ALS. *Nat. Med.* **5:** 347–350.

Ferrante, R.J., Andreassen, O.A., Dedeoglu, A., Kuemmerle, S., Kubilus, J.K., Kaddurah-Daouk, R., Hersch, S.M., and Beal, M.F., 2000, Neuroprotective effects of creatine in a transgenic mouse model of Huntington's disease. *J. Neurosci.* **20:** 4389–4397.

Ferrante, R.J., Andreassen, O.A., Dedeoglu, A., Ferrante, K.L., Jenkins, B.G., Hersch, S.M., and Beal, M.F., 2002, Therapeutic effects of coenzyme Q10 and remacemide in transgenic mouse models of Huntington's disease. *J. Neurosci.* **22:** 1592–1599.

Festoff, B.W., Suo, Z., and Citron, B.A., 2003, Prospects for the pharmacotherapy of amyotrophic lateral sclerosis: old strategies and new paradigms for the third millennium. *CNS Drugs* **17:** 699.

Foran, E., Del Signore, S.J., Markey, A., Matson, S., Smith, K.M., Cormier, K., Stack, E.C., Hersch, S.M., Ryu, H., and Ferrante, R.J., 2006, Dose ranging and efficacy study of high-dose creatine in Huntington's disease mouse models. Program No. 758.7/EE20, 2006 Neuroscience Meeting Planner, Atlanta, GA. Society for Neuroscience, Washington, DC, online (http://www.sfn.org).

Gabuzda, D., Busciglio, J., Chen, L.B., Matsudaira, P., and Yankner, B.A., 1994, Inhibition of energy metabolism alters the processing of amyloid precursor protein and induces a potentially amyloidogenic derivative. *J. Biol. Chem.* **269:** 13623–13628.

Gajewski, C.D., Lin, M.T., Cudkowicz, M.E., Beal, M.F., and Manfredi, G., 2003, Mitochondrial DNA from platelets of sporadic ALS patients restores normal respiratory functions in rho(0) cells. *Exp. Neurol.* **179:** 229–235.

Gasparini, L., Benussi, L., Bianchetti, A., Binetti, G., Curti, D., Govoni, S., Moraschi, S., Racchi, M., and Trabucchi, M., 1999, Energy metabolism inhibition impairs amyloid precursor protein secretion from Alzheimer's fibroblasts. *Neurosci. Lett.* **263:** 197–200.

Gasparini, L., Racchi, M., Benussi, L., Curti, D., Binetti, G., Bianchetti, A., Trabucchi, M., and Govoni, S., 1997, Effect of energy shortage and oxidative stress on amyloid precursor protein metabolism in COS cells. *Neurosci. Lett.* **231:** 113–117.

Gauthier, L.R., Charrin, B.C., Borrell-Pages, M., Dompierre, J.P., Rangone, H., Cordelieres, F.P., De Mey, J., MacDonald, M.E., Lessmann, V., Humbert, S., and Saudou, F., 2004, Huntingtin controls neurotrophic support and survival of neurons by enhancing BDNF vesicular transport along microtubules. *Cell* **118:** 127–138.

Graveland, G.A., Williams, R.S., and Difiglia, M.A., 1985, Evidence for degenerative and regenerative changes in neostriatal spiny neurons in Huntington's disease. *Science* **227:** 770–773.

Green, D.R., and Reed, J.C., 1998, Mitochondria and apoptosis. *Science* **281:** 1309–1312.

Greenhaff, P.L., Casey, A., Short, A.H., Harris, R., Soderlund, K., and Hultman, E., 1993, Influence of oral creatine supplementation of muscle torque during repeated bouts of maximal voluntary exercise in man. *Clin. Sci. (London)* **84:** 565–571.

Groeneveld, G.J., Veldink, J.H., van der Tweel, I., Kalmijn, S., Beijer, C., de Visser, M., Wokke, J.H.J., Franssen, H., and van den Berg, L.H., 2003, A randomized sequential trial of creatine in amyotrophic lateral sclerosis. *Ann. Neurol.* **53:** 437–445.

Grunewald, T., and Beal, M.F., 1999, Bioenergetics in Huntington's disease. *Ann. N.Y. Acad. Sci.* **893:** 203–213.

Gu, M., Gash, M.T., Mann, V.M., Javoy-Agid, F., Cooper, J.M., and Schapira, A.H., 1996, Mitochondrial defect in Huntington's disease caudate nucleus. *Ann. Neurol.* **39:** 385–389.

Guerrero-Ontiveros, M.L., and Wallimann, T., 1998, Creatine supplementation in health and disease. Effects of chronic creatine ingestion in vivo: down-regulation of the expression of creatine transporter isoforms in skeletal muscle. *Mol. Cell. Biochem.* **184:** 427–437.

Gurney, M.F., Pu, H., Chiu, A.Y., Dal Canto, M.C., Poichow, C.Y., Alexander, D.D., Caliendo, J., Hentati, A., Kwon, Y.W., Deng, H-X., Chen, W., Zhai, P., Sufit, R.L., and Siddique, T., 1994, Motor neuron degeneration in mice that express a human Cu,Zn superoxide dismutase mutation. *Science* **264:** 1772–1775.

Hagenfeldt, L., von Dobeln, U., Solders, G., and Kaijser, L., 1994, Creatine treatment in MELAS. *Muscle Nerve* **17:** 1236–1237.

Haley, R.W., 2003, Excess incidence of ALS in young Gulf War veterans. *Neurology* **61:** 750.

Hamilton, B.F., and Gould, D.H., 1987, Nature and distribution of brain lesions in rats intoxicated with 3-nitropropionic acid: a type of hypoxic (energy deficient) brain damage. *Acta Neuropathol. (Berlin)* **72:** 286–297.

Harris, R.C., Soderlund, K., and Hultman, E., 1992, Elevation of creatine in resting and exercised muscle of normal subjects by creatine supplementation. *Clin. Sci. (London)* **83:** 367–374.

Hass, R.H., Nasirian, F., Nakano, K., Ward, D., Pay, M., Hill, R., and Shults, C.W., 1995, Low platelet mitochondrial complex I and complex II/III activity in early untreated Parkinson's disease. *Ann. Neurol.* **37:** 714–722.

Hauser, W.A., and Kurland, L.T., 1975, The epidemiology of epilepsy in Rochester, Minnesota, 1935 through 1967. *Epilepsia* **16:** 1–66.

Hausmann, O.N., Fouad, K., Wallimann, T., and Schwab, M.E., 2002, Protective effects of oral creatine supplementation on spinal cord injury in rats. *Spinal Cord* **40:** 449–456.

Hebert, L.E., Scherr, P.A., Bienias, J.L., Bennett, D.A., and Evans, D.A., 2003, Alzheimer disease in the US population: prevalence estimates using the 2000 census. *Arch. Neurol.* **60:** 1119–1122.

Heinanen, K., Nanto-Salonen, K., Komu, M., Erkintalo, M., Alanen, A., Heinonen, O.J., Pulkki, K., Nikoskelainen, E., Sipila, I., and Simell, O., 1999a, Creatine corrects muscle 31P spectrum in gyrate atrophy with hyperornithinaemia. *Eur. J. Clin. Invest.* **29:** 1060–1065.

Heinanen, K., Nanto-Salonen, K., Komu, M., Erkintalo, M., Heinonen, O.J., Pulkki, K., Valtonen, M., Nikoskelainen, E., Alanen, A., and Simell, O., 1999b, Muscle creatine phosphate in gyrate atrophy of the choroid and retina with hyperornithinaemia—clues to pathogenesis. *Eur. J. Clin. Invest.* **29:** 426–431.

Helder, D.I., Kaptein, A.A., Van Kempen, G.M., Van Houwelingen, J.C., and Roos, R.A., 2001, Impact of Huntington's disease on quality of life. *Mov. Disord.* **16:** 325–330.

Henshaw, R., Jenkins, B.G., Schulz, J.B., Ferrante, R.J., Kowall, N.W., Rosen, B.R., and Beal, M.F., 1994, Malonate produces striatal lesions by indirect NMDA receptor activation. *Brain Res.* **647:** 161–166.

Hersch, S.M., and Ferrante, R.J., 2004, Translating therapies for Huntington's disease from genetic animal models to clinical trials. *NeuroRx* **1:** 298–306.

Hersch, S.M., Rosas, H.D., and Ferrante, R.J., 2004, Neuropathology and pathophysiology of Huntington's disease, in movement disorders. In: Koller, W. (ed.), Neurologic Principles and Practice. McGraw-Hill, New York, pp. 503–526.

Hersch, S.M., Gevorkian, S., Marder, K., Moskowitz, C., Feigin, A., Cox, M., Como, P., Zimmerman, C., Lin, M., Zhang, L., Ulug, A.M., Beal, M.F., Matson, W., Bogdanov, M., Ebbel, E., Zaleta, A., Kaneko, Y., Jenkins, B., Hevelone, N., Zhang, H., Yu, H., Schoenfeld, D., Ferrante, R.J., and Rosas, H.D., 2006, Creatine in Huntington's disease is safe, tolerable, and bioavailable in brain and reduces serum 8-OHDG. *Neurology* **66:** 250–252.

Hilker, R., Schweitzer, K., Coburger, S., Ghaemi, M., Weisenbach, S., Jacobs, A.H., Rudolf, J., Herholz, K., and Heiss, W.D., 2005, Nonlinear progression of Parkinson disease as determined by serial positron emission tomographic imaging of striatal fluorodopa F 18 activity. *Arch. Neurol.* **62:** 378–382.

Hirano, A., Nakano, I., Kurland, L.T., Mulder, D.W., Holley, P.W., and Saccomanno, G., 1984, Fine structural study of neurofibrillary changes in a family with amyotrophic lateral sclerosis. *J. Neuropathol. Exp. Neurol.* **43:** 471–480.

Hirano, M., Kaufmann, P., DeVivo, D., and Kurenarai, T., 2006, Mitochondrial Neurology I: Encephalopathies. In: DiMauro, S., Hirano, S., and Schon, E. (eds.), Mitochondrial Medicine. Informa Healthcare, Oxon, pp. 27–44.

Holtzman, D., Khait, I., Mulkern, R., Allred, E., Rand, T., Jensen, F., and Kraft, R., 1999, In vivo development of brain phosphocreatine in normal and creatine-treated rabbit pups. *J. Neurochem.* **73:** 2477–2484.

Holtzman, D., Togliatti, A., Khait, I., and Jensen, F., 1998, Creatine increases survival and suppresses seizures in the hypoxic immature rat. *Pediatr. Res.* **44:** 410–414.

Horner, R.D., Kamins, K.G., Feussner, J.R., Grambow, S.C., Hoff-Lindquist, J., Harati, Y., Mitsumoto, H., Pascuzzi, R., Spencer, P.S., Tim. R., Howard, D., Smith, T.C., Ryan, M.A.K., Coffman, C.J., and Kasarskis, E.J., 2003, Occurrence of amyotrophic lateral sclerosis among Gulf War veterans. *Neurology***61:** 742.

Howell, N., Bindoff, L.A., McCullough, D.A., Kubacka, I., Poulton, J., Mackey, D., Taylor, L., and Turnbull, D.M., 1991, Leber hereditary optic neuropathy: identification of the same mitochondrial ND1 mutation in six pedigrees. *Am. J. Hum. Genet.* **49:** 939–950.

Howland, D., Liu, J., She, Y., Goad, B., Maragakis, N., Kim, B., Erickson, J., Kulik, J., DeVito, L., Psaltis, G., DeGennaro, L., Cleveland, D., and Rothstein, J., 2002, Focal loss of the glutamate transporter EAAT2 in a transgenic rat model of SOD1 mutant-mediated amyotrophic lateral sclerosis (ALS). *Proc. Natl. Acad. Sci. USA* **99:** 1604–1609.

Hoyer, S., 1993, Brain oxidative energy and related metabolism, neuronal stress, and Alzheimer's disease: a speculative synthesis. *J. Geriatr. Psychiatry Neurol.* **6:** 3–13.

Huntington's Disease Collaborative Research Group, 1993, A novel gene containing a trinucleotide repeat that is expanded and unstable on Huntington's disease chromosomes. *Cell* **72:** 971–983.

Hwang, D.Y., Ardayfio, P., Kang, U.J., Semina, E.V., and Kim, K.S., 2003, Selective loss of dopaminergic neurons in the substantia nigra of Pitx3-deficient aphakia mice. *Brain Res. Mol. Brain Res.* **114:** 123–131.

Ihara, Y., Namba, R., Kuroda, S., Sato, T., and Shirabe, T., 1989, Mitochondrial encephalomyopathy (MELAS): pathological study and successful therapy with coenzyme Q10 and idebenone. *J. Neurol. Sci.* **90:** 263–271.

Ikeda, K., Iwasaki, Y., and Kinoshita, M., 2000, Oral administration of creatine monohydrate retards progression of motor neuron disease in the wobbler mouse. *Amyotroph. Lateral Scler. Other Motor Neuron Disord.* **1:** 207–12.

Item, C.B., Stöckler-Ipsiroglu, S., and Stromberger, C., 2001, Arginine:glycine amidinotransferase deficiency; the third inborn error of creatine metabolism in humans. *Am. J. Hum. Genet.* **69:** 1127–1133.

Jaarsma, D., 2006, Swelling and vacuolisation of mitochondria in transgenic SOD1-ALS mice: a consequence of supranormal SOD1 expression? *Mitochondrion* **5:** 77–87.

Jenkins, B.G., Koroshetz, W., Beal, M.F., and Rosen, B., 1993, Evidence for an energy metabolism defect in Huntington's disease using localized proton spectroscopy. *Neurology* **43:** 2689–2695.

Jenkins, B.G., Rosas, H.D., Chen, Y.C., Makabe, T., Myers, R., MacDonald, M., Rosen, B.R., Beal, M.F., and Koroshetz, W.J., 1998, 1H-NMR spectroscopy studies of Huntington's disease. *Neurology* **50:** 1357–1365.

Jiang, C., Wan, X., He, Y., Pan, T., Jankovic, J., and Le, W., 2005, Age-dependent dopaminergic dysfunction in Nurr1 knockout mice. *Exp. Neurol.* **191:** 154–162.

Jun, A.S., Brown, M.D., and Wallace, D.C., 1994, A mitochondrial DNA mutation at nucleotide pair 14459 of the NADH dehydrogenase subunit 6 gene associated with maternally inherited Leber hereditary optic neuropathy and dystonia. *Proc. Natl. Acad. Sci. USA* **91:** 6206–6210.

Kaal, E., Vlug, A., Versleijen, M., Kuilman, M., Joosten, E., and Dop Bar, P., 2000, Chronic mitochondrial inhibition induces selective motoneuron death in vitro: a new model for amyotrophic lateral sclerosis. *J. Neurochem.* **74:** 1158–1165.

Kernec, F., Le Tallec, N., Nadal, L., Begue, J.M., and Le Rumeur, E., 1996, Phosphocreatine synthesis by isolated rat skeletal muscle mitochondria is not dependent upon external ADP: a 31P NMR study. *Biochem. Biophys. Res. Commun.* **225:** 819–825.

Kieburtz, K., 2001, Placebo-controlled trial of creatine in Huntington's disease. *Neurology* **56 (Suppl. 3):** A334.

Kiechle, T., Dedeoglu, A., Kubilus, J., Kowall, N.W., Beal, M.F., Friedlander, R.M., Hersch, S.M., and Ferrante, R.J., 2002, Cytochrome C and caspase-9 expression in Huntington's disease. *Neuromol. Med.* **1:** 183–195.

Klivenyi, P., Ferrante, R.J., Matthews, R.T., Bogdanov, M.B., Klein, A.M., Andreassen, O.A., Mueller, G., Wermer, M., Kaddurah-Daouk, R., and Beal, M.F., 1999, Neuroprotective effects of creatine in a transgenic animal model of amyotrophic lateral sclerosis. *Nat. Med.* **5:** 347–350.

Klivenyi, P., Gardian, G., Calingasan, N.Y., Yang, L., and Beal, M.F., 2003, Additive neuroprotective effects of creatine and a cyclooxygenase 2 inhibitor against dopamine depletion in the 1-methyl-4-phenyl-1,2,3,6-tetrahydropyridine (MPTP) mouse model of Parkinson's disease. *J. Mol. Neurosci.* **21:** 191–198.

Klivenyi, P., Calingasan, N.Y., Starkov, A., Stavrovskaya, I.G., Kristal, B.S., Yang, L., Wieringa, B., and Beal, M.F., 2004, Neuroprotective mechanisms of creatine occur in the absence of mitochondrial creatine kinase. *Neurobiol. Dis.* **15:** 610–617.

Klivenyi, P., Kiaei, M., Gardian, G., Calingasan, N.Y., and Beal, M.F., 2004, Additive neuroprotective effects of creatine and cyclooxygenase 2 inhibitors in a transgenic mouse model of amyotrophic lateral sclerosis. *J. Neurochem.* **88:** 576–582.

Komura, K., Hobbiebrunken, E., Wilichowski, E.K., and Hanefeld, F.A., 2003, Effectiveness of creatine monohydrate in mitochondrial encephalomyopathies. *Pediatr. Neurol.* **28:** 53–58.

Komura, K., Nakano, K., Ishigaki, K., Tarashima, M., Nakayama, T., Sasaki, K., Saito, K., and Osawa, M., 2006, Creatine monohydrate therapy in a Leigh syndrome patient with A8344G mutation. *Pediatr. Int.* **48:** 409–412.

Kong, J., and Xu, Z., 1998, Massive mitochondrial degeneration in motor neurons triggers the onset of amyotrophic lateral sclerosis in mice expressing a mutant SOD1. *J. Neurosci.* **18:** 3241–3250.

Koroshetz, W.J., Jenkins, B.G., Rosen, B.R., and Beal, M.F., 1997, Energy metabolism defects in Huntington's disease and possible therapy with coenzyme Q10. *Ann. Neurol.* **41:** 160–165.

Kremer, B., Squitieri, F., Telenius, H., Andrew, S.E., Theilmann, J., Spence, N., Goldberg, Y.P., and Hayden, M.R., 1993, Molecular analysis of late onset Huntington's disease. *J. Med. Genet.* **30:** 991–995.

Kremer, B.P., Goldberg, S., Andrew, J., Theilmann, J., Telenius, H., and Zeisler, J., Squitieri, F., Lin, B., Bassett, A., Almqvist, E., Bird, T., and Hayden, M.R., 1994, A worldwide study of the Huntington's disease mutation—the sensitivity and specificity of measuring CAG repeats. *New Engl. J. Med.* **330:** 1401–1406.

Kuhl, D.E., Markham, C.H., Metter, E.J., Riege, W.H., Phelps, M.E., and Mazziotta, J.C., 1985, Local cerebral glucose utilization in symptomatic and presymptomatic Huntington's disease. *Res. Publ. Assoc. Res. Nerv. Ment. Dis.* **63:** 199–209.

Kuhl, D.E., Phelps, M.E., Markham, C.H., Metter, E.J., Riege, W.H., and Winter, J., 1982, Cerebral metabolism and atrophy in Huntington's disease determined by 18FDG and computed tomographic scan. *Ann. Neurol.* **12:** 425–434.

Kuwert, T., Lange, H.W., Langer, K.J., Herzog, H., Aulich, A., and Feinendegen, L.E., 1990, Cortical and subcortical glucose consumption measured by PET in patients with Huntington's disease. *Brain* **113:** 1405–1423.

Lawler, J.M., Barnes, W.S., Wu, G., Song, W., and Demaree, S., 2002, Direct antioxidant properties of creatine. *Biochem. Biophys. Res. Comm.* **290:** 47–52.

Liang, L.P., Ho, Y.S., and Patel, M., 2000, Mitochondrial superoxide production in kainate-induced hippocampal damage. *Neuroscience* **101:** 563–570.

Liang, L.P., and Patel, M., 2004, Mitochondrial oxidative stress and increased seizure susceptibility in *Sod2*$^{-/+}$ mice. *Free Radic. Biol. Med.* **36:** 542–554.

Lofberg, M., Lindholm, H., Naveri, H., Majander, A., Suomalainen, A., Paetau, A., Sovijarvi, A., Harkonen, M., and Somer, H., 2001, ATP, phosphocreatine and lactate in exercising muscle in mitochondrial disease and McArdle's disease. *Neuromusc. Disord.* **11:** 370–375.

Loike, J.D., Zalutsky, D.L., Kaback, E., Miranda, A.F., and Silverstein, S.C., 1988, Extracellular creatine regulates creatine transport in rat and human muscle cells. *Proc. Natl. Acad. Sci. USA* **85:** 807–811.

Loiseau, J., Loiseau, P., Guyot, M., Duche, B., Dartigues, J.F., and Aublet, B., 1990, Survey of seizure disorders in the French southwest. I. Incidence of epileptic syndromes. *Epilepsia* **31:** 391–396.

Ludolph, A.C., He, F., Spencer, P.S., Hammerstad, J., and Sabri, M., 1991, 3-Nitropropionic acid—exogenous animal neurotoxin and possible human striatal toxin. *Can. J. Neurol. Sci.* **18:** 492–498.

Ludolph, A.C., Seelig, M., Ludolph, A.G., Sabri, M.I., and Spencer, P.S., 1992, ATP deficits and neuronal degeneration induced by 3-nitropropionic acid. *Ann. N.Y. Acad. Sci.* **648:** 300–302.

Malcon, C., Kaddurah-Daouk, R., and Beal, M.F., 2000, Neuroprotective effects of creatine administration against NMDA and malonate toxicity. *Brain Res.* **860:** 195–198.

Mancini, G.M., Catsman-Berrevoets, C.E., and deCoco, I.F., 2004, Two novel mutations in SLC6A8 deficiency in X-linked mental retardation. *Am.J.Hum.Genet.* **75:** 97–105.

Mann, V.M., Cooper, J.M., Javoy-Agid, F., Agid, F., Jenner, P., and Schapira, A.H.V., 1990, Mitochondrial function and parental sex effect in Huntington's disease. *Lancet* **336:** 749.

Mark, R.J., Pang, Z., Geddes, J.W., Uchida, K., and Mattson, M.P., 1997, Amyloid peptide impairs glucose transport in hippocampal and cortical neurons: Involvement of membrane lipid peroxidation. *J. Neurosci.* **17:** 1046–1054.

Mastrogiacomo, F., LaMarche, J., Dozic, S., Lindsay, G., Bettendorff, L., Robitaille, Y., Schut, L., and Kish, S.J., 1996, Immunoreactive levels of alpha-ketoglutarate dehydrogenase subunits in Friedreich's ataxia and spinocerebellar ataxia type 1. *Neurodegeneration* **5:** 27–33.

Masui, Y., Mozai, T., and Kakehi, K., 1985, Functional and morphometric study of the liver in motor neuron disease. *J. Neurol.* **232:** 15–19.

Masumizu, T., Noda, Y., Mori, A., and Packer, L., 2005, Electron spin resonance assay of ascorbyl radical generation in mouse hippocampal slices during and after kainate-induced seizures. *Brain Res. Brain Res. Protoc.* **16:** 65–69.

Matsuishi, T., Sakai, T., Naito, E., Nagamitsu, S., Kuroda, Y., Iwashita, H., and Kato, H., 1996, Elevated cerebrospinal fluid lactate/pyruvate ratio in Machado-Joseph disease. *Acta Neurol. Scand.* **93:** 72–75.

Matthews, R.T., Yang, L., Jenkins, B.G., Ferrante, R.J., Rosen, B.R., Kaddurah-Daouk, R., and Beal, M.F., 1998, Neuroprotective effects of creatine and cyclocreatine in animal models of Huntington's disease. *J. Neurosci.* **18:** 156–163.

Matthews, R.T., Ferrante, R.J., Klivenyi, P., Yang, L., Klein, A.M., Mueller, G., Kaddurah-Daouk, R., and Beal, M.F., 1999, Creatine and cyclocreatine attenuate MPTP neurotoxicity. *Exp. Neurol.* **157:** 142–149.

Mattiazzi, M., D'Aurelio, M., Gajewski, C.D., Martushova, K., Kiaei, M., Beal, M.F., and Manfredi, G., 2002, Mutated human SOD1 causes dysfunction of oxidative phosphorylation in mitochondria of transgenic mice. *J. Biol. Chem.* **277:** 29626–29633.

Mazziotta, J.C., Phelps, M.E., Pahl, J.J., Huang, S.C., Baxter, L.R., and Riege, W.H., Hoffman, J.M., Juhl, D.E., Lanto, A.B., Wapenski, J.A., et al., 1987, Reduced cerebral glucose metabolism in asymptomatic patients at risk for Huntington's disease. *New Engl. J. Med.* **316:** 357–362.

McGeer, E.G., and McGeer, P.L., 2005, Pharmacologic approaches to the treatment of amyotrophic lateral sclerosis. *BioDrugs 19:* 31.

Mecocci, P., MacGarvey, U., and Beal, M.F., 1994, Oxidative damage to mitochondrial DNA is increased in Alzheimer's disease. *Ann. Neurol.* **36:** 747–751.

Mercimek-Mahmutoglu, S., Stoeckler-Ipsiroglu, S., Adami, A., Appleton, R., Araujo, H.C., Duran, M., Ensenauer, R., Fernandez-Alvarez, E., Garcia, P., Grolik, C., Item, C.B., Leuzzi, V., Marquardt, I., Muhl, A., Saelke-Kellermann, R.A., Salomons, G.S., Schulze, A., Surtees, R., van der Knaap, M.S., Vasconcelos, R., Verhoeven, N.M., Vilarinho, L., Wilichowski, E., and Jakobs, C., 2006, GAMT deficiency: Features, treatment, and outcome in an inborn error of creatine synthesis. *Neurology* **67:** 480–484.

Mesulam, M., 2000, A plasticity-based theory of the pathogenesis of Alzheimer's disease. *Ann. N.Y. Acad. Sci.* **924:** 42–52.

Mettler, F.A., 1972, Neuropathological effects of sodium azide administration in primates. *Fed. Proc.* **31**: 1504–1507.

Meyer, R.A., Sweeney, H.L., and Kushmerick, M.J., 1984, A simple analysis of the "phosphocreatine shuttle". *Am. J. Physiol.* **246:** C365–C377.

Mikati, M.A., Kurdit, R.M., Rahmeh, A.A., Farhat, F., Abu Rialy, S., Lteif, L., Francis, E., Geha, G., and Maraashli, W., 2004, Effects of creatine and cyclocreatine supplementation on kainate induced injury in pre-pubescent rats. *Brain Inj.* **18:** 1229–1241.

Miller, R.G., Mitchell, J.D., Lyon, M., and Moore, D.H., 2003, Riluzole for amyotrophic lateral sclerosis (ALS)/motor neuron disease (MND). *Amyotroph. Lateral Scler. Other Motor Neuron Disord.* **4:** 191.

Mueller, S.G., Laxer, K.D., Barbakos, J.A., Cashdollar, N., Flenniken, D.L., Vermathen, P., Matson, G.B., and Weiner, M.W., 2005, Metabolic characteristics of cortical malformations causing epilepsy. *J. Neurol.* **252:** 1082–1092.

Mutisya, E.M., Bowling, A.C., and Beal, M.F., 1994, Cortical cytochrome oxidase activity is reduced in Alzheimer's disease. *J. Neurochem.* **63:** 2179–2184.

Myers, R.H., MacDonald, M.E., Koroshetz, W.J., Duyao, M.P., Ambrose, C.M., Taylor, S.A., Barnes, G., Srinidhi, J., Lin, C.S., Whaley, W.L., Lazzarini, A.M., Schwarz, M., Wolff, G., Bird, E.D., Vonsattel, J.-P.G., and Gusella, J.F., 1993, De novo expansion of a $(CAG)_n$ repeat in sporadic Huntington's disease. *Nat. Genet.* **5:** 168–173.

Nakano, K., Hirayama, K., and Terai, K., 1987, Hepatic ultrastructural changes and liver dysfunction in amyotrophic lateral sclerosis. *Arch. Neurol.* **44:** 103–106.

Newbery, H.J., and Abbott, C.M., 2002, Of mice, men and motor neurons. *Trends Mol. Med.* **8:** 88.

Nicklas, W.J., Vyas, I., and Heikkila, R.E., 1985, Inhibition of NADH-linked oxidation in brain mitochondria by 1-methyl-4-phenyl-pyridine, a metabolite of the neurotoxin, 1-methyl-4-phenyl-1,2,5,6-tetrahydropyridine. *Life Sci.* **36:** 2503–2508.

NINDS NET-PD Investigators, 2006, A randomized, double-blind, futility clinical trial of creatine and minocycline in early Parkinson disease. *Neurology* **66:** 664–671.

Novelli, A., Reilly, J.A., Lysko, P.G., and Henneberry, R.C., 1988, Glutamate becomes neurotoxic via the N-methyl-D-aspartate receptor when intracellular energy levels are reduced. *Brain Res.* **451:** 205–212.

Nunes, I., Tovmasian, L.T., Silva, R.M., Burke, R.E., and Goff, S.P., 2003, Pitx3 is required for development of substantia nigra dopaminergic neurons. *Proc. Natl. Acad. Sci. USA* **100:** 4245–4250.

O'Gorman, E., Beutner, G., Wallimann, T., and Brdiczka, D., 1996, Differential effects of creatine depletion on the regulation of enzyme activities and on creatine-stimulated mitochondrial respiration in skeletal muscle, heart, and brain. *Biochim. Biophys. Acta* **1276:** 161–170.

O'Gorman, E., Beutner, G., Dolder, M., Koretsky, A.P., Brdiczka, D., and Wallimann, T., 1997, The role of creatine kinase in inhibition of mitochondrial permeability transition. *FEBS Lett.* **414:** 253–257.

Okamoto, K., Hirai, S., Shoji, M., Senoh, Y., and Yamazaki, T., 1990, Axonal swellings in the corticospinal tracts in amyotrophic lateral sclerosis. *Acta Neuropathol.* **80:** 222–226.

Olney, J.W., and de Gubareff, T., 1978a, Glutamate neurotoxicity and Huntington's chorea. *Nature* **271:** 557–559.

Olney, J.W., and de Gubareff, T., 1978b, The fate of synaptic receptors in the kainite-lesioned striatum. *Brain Res.* **140:** 340–343.

Palfi, S., Ferrante, R.J., Brouillet, E., Beal, M.F., Dolan, R., Guyot, M.C., Peschanski, M., and Hantraye, P., 1996, Chronic 3-nitropropionic acid treatment in baboons replicates the cognitive and motor deficits of Huntington's disease. *Neuroscience* **16:** 3019–3025.

Pan, J.W., Kim, J.H., Cohen-Gadol, A., Pan, C., Spencer, D.D., and Hetherington, H.P., 2005, Regional energetic dysfunction in hippocampal epilepsy. *Acta Neurol. Scand.* **111:** 218–224.

Panov, A.V., Gutekunst, C.A., Leavitt, B.R., Hayden, M.R., Burke, J.R., Strittmatter, W.J., and Greenamyre, J.T., 2002, Early mitochondrial calcium defects in Huntington's disease are a direct effect of polyglutamines. *Nat. Neurosci.* **5:** 731–736.

Pappolla, M.A., Chyan, Y.J., Omar, R.A., Hsiao, K., Perry, G., Smith, M.A., and Bozner, P., 1998, Evidence of oxidative stress and in vivo neurotoxicity of beta-amyloid in a transgenic mouse model of Alzheimer's disease: a chronic oxidative paradigm for testing antioxidant therapies in vivo. *Am. J. Pathol.* **152:** 871–877.

Parker, W.D. Jr., and Parks, J.K., 1995, Cytochrome c oxidase in Alzheimer's disease brain: purification and characterization. *Neurology* **45:** 482–486.

Parker, W.D. Jr., Boyson, S.J., and Parks, J.K., 1989, Abnormalities of the electron transport chain in idiopathic Parkinson's disease. *Ann. Neurol.* **26:** 719–723.

Parker, W.D. Jr., Boyson, S.J., Luder, A.S., and Parks, J.K., 1990, Evidence for a defect in NADH: ubiquinone oxidoreductase (complex I) in Huntington's disease. *Neurology* **40:** 1231–1234.

Pasinelli, P., Belford, B., Lennon, N., Bacskai, B., Hyman, B., Trotti, D., and Brown, R., 2004, Amyotrophic lateral sclerosis-associated SOD1 mutant proteins bind and aggregate with Bcl-2 in spinal cord mitochondria. *Neuron* **43:** 19–30.

Patel, M., 2004, Mitochondrial dysfunction and oxidative stress: cause and consequence of epileptic seizures. *Free Radic. Biol. Med.* **37:** 1951–1962.

Paulsen, J.S., Zhao, H., Stout, J.C., Brinkman, R.R., Guttman, M., Ross, C.A., Como, P., Manning, C., Hayden, M.R., Shoulson, I., and Huntington Study Group, 2001, Clinical markers of early disease in persons near onset of Huntington's disease. *Neurology* **57:** 658–662.

Persky, A.M., and Brazeau, G.A., 2001, Clinical pharmacology of the dietary supplement creatine monohydrate. *Pharmacol. Rev.* **53:** 161–176.

Pettegrew, J.W., Panchalingam, K., Klunk, W.E., McClure, R.J., and Muenz, L.R., 1994, Alterations of cerebral metabolism in probable Alzheimer's disease: a preliminary study. *Neurobiol. Aging* **15:** 117–132.

Prass, K., Royl, G., Lindauer, U., Freyer, D., Megow, D., Dirnagl, U., Stockler-Ipsiroglu, G., Wallimann, T., and Priller, J., 2007, Improved reperfusion and neuroprotection by creatine in a mouse model of stroke. *J. Cereb. Blood Flow Metab.* **27:** 452–459.

Pratico, D., Uryu, K., Leight, S., Trojanoswki, J.Q., and Lee, V.M., 2001, Increased lipid peroxidation precedes amyloid plaque formation in an animal model of Alzheimer amyloidosis. *J. Neurosci.* **21:** 4183–4187.

Rabchevsky, A.G., Sullivan, P.G., Fugaccia, I., and Scheff, S.W., 2003, Creatine diet supplement for spinal cord injury: influences on functional recovery and tissue sparing in rats. *J. Neurotrauma* **20:** 659–669.

Robitaille, Y., Lopes-Cendes, I., Becher, M., Rouleau, G., and Clark, A.W., 1997, The neuropathology of CAG repeat diseases: review and update of genetic and molecular features. *Brain Pathol.* **7:** 901–926.

Rodrigo, J., Fernandez-Vizarra, P., Castro-Blanco, S., Bentura, M.L., Nieto, M., Gomez-Isla, T., Martinez-Murillo, R., MartInez, A., Serrano, J., and Fernandez, A.P., 2004, Nitric oxide in the cerebral cortex of amyloid-precursor protein (SW) Tg2576 transgenic mice. *Neuroscience* **128:** 73–89.

Rosen, D.R., Sapp, P., O'Regan, J., McKenna-Yasek, D., Schlumpf, K.S., Haines, J.L., Gusella, J.F., Horvitz, H.R., and Brown, R.H. Jr., 1993, Mutations in Cu/Zn superoxide dismutase gene are associated with familial amyotrophic lateral sclerosis. *Nature* **362:** 59.

Ross, C.A., 1997, Intranuclear neuronal inclusions: a common pathogenic mechanism for glutamine-repeat neurodegenerative diseases? *Neuron* **19:** 1147–1150.

Ross, C.A., 2004, Huntington's disease: new paths to pathogenesis. *Cell* **118:** 4–7.

Rothstein, J.D., 2003, Of mice and men: reconciling preclinical ALS mouse studies and human clinical trials. *Ann. Neurol.* **53:** 423–426.

Rothstein, J.D., Martin, L.J., and Kuncl, R.W., 1992, Decreased glutamate transport by the brain and spinal cord in amyotrophic lateral sclerosis. *New Engl. J. Med.* **326:** 1464–1468.

Rothstein, J., Van Kammen, M., Levey, A., Martin, L., and Kuncl, R., 1995, Selective loss of glial glutamate transporter GLT-1 in amyotrophic lateral sclerosis. *Ann. Neurol.* **38:** 73–84.

Rowland, L., and Shneider, N., 2001, Amyotrophic lateral sclerosis. *New Engl. J. Med.* **344:** 1688.

Roy, S., and Nicholson, D.W., 2000, Cross-talk in cell death signaling. *J. Exp. Med.* **192:** 21–26.

Royes, L.F., Fighera, M.R., Furian, A.F., Oliveira, M.S., da Silva, L.G., Malfatti, C.R., Schneider, P.H., Braga, A.L., Wajner, M., and Mello, C.F., 2003, Creatine protects against the convulsive behavior and lactate production elicited by the intrastriatal injection of methylmalonate. *Neuroscience* **118:** 1079–1090.

Royes, L.F., Fighera, M.R., Furian, A.F., Oliveira, M.S., Myskiw, J de C, Fiorenza, N.G., Petry, J.C., Coelho, R.C., and Mello, C.F., 2006, Effectiveness of creatine monohydrate on seizures and oxidative damage induced by methylmalonate. *Pharmacol. Biochem. Behav.* **83:** 136–144.

Ryu, H., and Ferrante, R.J., 2005, Emerging chemotherapeutic strategies for Huntington's disease. *Expert Opin. Emerg. Drugs* **10:** 345–363.

Ryu, H., Rosas, H.D., Hersch, S.M., and Ferrante, R.J., 2005, The therapeutic role of creatine in Huntington's disease. *Pharmacol. Ther.* **108:** 193–207.

Sacco, R.L., Benjamin, E.J., Broderick, J.P., Dyken, M., Easton, J.D., Feinberg, W.M., Goldstein, L.B., Gorelick, P.B., Howard, G., Kittner, S.J., Manolio, T.A., Whisnant, J.P., and Wolf, P.A., 1997, American Heart Association Prevention Conference. IV. Prevention and rehabilitation of stroke risk factors. *Stroke* **28:** 1507–1517.

Sasaki, S., and Iwata, M., 1996, Impairment of fast axonal transport in the proximal axons of anterior horn neurons in amyotrophic lateral sclerosis. *Neurology* **47:** 535–540.

Scaglia, F., and Northrop, J.L., 2006, The mitochondrial myopathy encephalopathy, lactic acidosis with stroke-like episodes (MELAS) syndrome: a review of treatment options. *CNS Drugs* **20:** 443–464.

Scheff, S.W., and Dhillon, H.S., 2004, Creatine-enhanced diet alters levels of lactate and free fatty acids after experimental brain injury. *Neurochem. Res.* **29:** 469–479.

Schloss, P., Mayser, W., and Betz, H., 1994, The putative rat choline transporter CHOT1 transports creatine and is highly expressed in neural and muscle-rich tissues. *Biochem. Biophys. Res. Commun.* **198:** 637–645.

Schulz, J.B., Matthews, R.T., Jenkins, B.G., Ferrante, R.J., Siwek, D., Henshaw, D.R., Cipolloni, P.B., Mecocci, P., Kowall, N.W., and Rosen, B.R., 1995, Blockade of neuronal nitric oxide synthase protects against excitotoxicity in vivo. *J. Neurosci.* **15:** 8419–8429.

Shear, D.A., Haik, K.L., and Dunbar, G.L., 2000, Creatine reduces 3-nitropropionic-acid-induced cognitive and motor abnormalities in rats. *NeuroReport* **11:** 1833–1837.

Shefner, J.M., Cudkowicz, M.E., Schoenfeld, D., Conrad, T., Taft, J., Chilton, M., Urbinelli, L., Qureshi, M., Zhang, H., Pestronk, A., Caress, J., Donofrio, P., Sorenson, E., Bradley, W., Lomen-Hoerth, C., Pioro, E., Rezania, K., Ross, M., Pascuzzi, R., Heiman-Patterson, T., Tandan, R., Mitsumoto, H., Rothstein, J., Smith-Palmer, T., MacDonald, D., Burke, D.; NEALS Consortium, 2004, A clinical trial of creatine in ALS. *Neurology* **63:** 1656–1661.

Shin, E.J., Suh, S.K., Lim, Y.K., Jhoo, W.K., Hjelle, O.P., Ottersen, O.P., Shin, C.Y., Ko, K.H., Kim, W.K., Kim, D.S., Chun, W., Ali, S., and Kim, H.C., 2005, Ascorbate attenuates trimethyltin-induced oxidative burden and neuronal degeneration in the rat hippocampi by maintaining glutathione homeostasis. *Neuroscience* **133:** 715–727.

Shinkai, T., Nakashima, M., Ohmori, O., Terao, T., Nakamura, J., Hiramatsu, N., Hashiguchi, H., and Tsuji, S., 2000, Coenzyme Q10 improves psychiatric symptoms in adult-onset mitochondrial myopathy, encephalopathy, lactic acidosis and stroke-like episodes: a case report. *Aust. NZ J. Psychiatry* **34:** 1034–1035.

Shoffner, J.M., Lott, M.T., Lezza, A.M.S., Seibel, P., Ballinger, S.W., and Wallace, D.C., 1990, Myoclonic epilepsy and red-ragged fiber disease (MERRF) is associated with a mitochondrial DNA tRNA(Lys) mutation. *Cell* **61:** 931–937.

Shoffner, J., 1997, Oxidative phosphorylation defects and Alzheimer's disease. *Neurogenetics* **1:** 13–19.

Simon, H.H., Saueressig, H., Wurst, W., Goulding, M.D., and O'Leary, D.D., 2001, Fate of midbrain dopaminergic neurons controlled by the engrailed genes. *J. Neurosci.* **21:** 3126–3134.

Smith, G.E., Bohac, D.L., Waring, S.C., Kokmen, E., Tangalos, E.G., Ivnik, R.J., and Petersen, R.C., 1998, Apolipoprotein E genotype influences cognitive 'phenotype' in patients with Alzheimer's disease but not in healthy control subjects. *Neurology* **50:** 355–362.

Smith, M.A., Hirai, K., Hsiao, K., Pappolla, M.A., Harris, P.L., Siedlak, S.L., Tabaton, M., and Perry, G., 1998, Amyloid-beta deposition in Alzheimer transgenic mice is associated with oxidative stress. *J. Neurochem.***70:** 2212–2215.

Smith, K.M., Matson, S., Matson, W.R., Cormier, K., Del Signore, S.J., Hagerty, S.W., Stack, E.C., Ryu, H., and Ferrante, R.J., 2006, Dose ranging and efficacy study of high-dose coenzyme Q10 formulations in Huntington's disease mice. *Biochim. Biophys. Acta* **1762:** 616–626.

Snow, R.J., and Murphy, R.M., 2001, Creatine and the creatine transporter: a review. *Mol. Cell. Biochem.* **224:** 169–181.

Stöckler, S., Hanefeld, F., and Frahm, J., 1996a, Creatine replacement therapy in guanidinoacetate methyltransferase deficiency, a novel inborn error of metabolism. *Lancet* **348:** 789–790.

Stöckler, S., Isbrandt, D., Hanefeld, F., Schmidt, B., and von Figura, K., 1996b, Guanidinoacetate methyltransferase deficiency: the first inborn error of creatine metabolism in man. *Am. J. Hum. Genet.* **58:** 914–922.

Stockler, S., Schutz, P.W., and Salomons, G.S., 2007, Cerebral creatine deficiency syndromes: clinical aspects, treatment and pathophysiology. *Subcell. Biochem.* **46:** 149–166.

Stromberger, C., Bodamer, O.A., and Stöckler-Ipsiroglu, S., 2003, Clinical characteristics and diagnostic clues in inborn errors of creatine metabolism. *J. Inher. Metab. Dis.* **26:** 299–308.

Sullivan, P.G., Keller, J.N., Mattson, M.P., and Scheff, S.W., 1998, Traumatic brain injury alters synaptic homeostasis: implications for impaired mitochondrial and transport function. *J. Neurotrauma* **15:** 789–798.

Sullivan, P.G., Bruce-Keller, A.J., Rabchevsky, A.G., Christakos, S., Clair, D.K., Mattson, M.P., and Scheff, S.W., 1999, Exacerbation of damage and altered NF-kappaB activation in mice lacking tumor necrosis factor receptors after traumatic brain injury. *J. Neurosci.* **19:** 6248–6256.

Sullivan, P.G., Geiger, J.D., Mattson, M.P., and Scheff, S.W., 2000, Dietary supplement creatine protects against traumatic brain injury. *Ann. Neurol.* **48:** 723–729.

Sullivan, P.G., Rippy, N.A., Dorenbus, K., Concepcion, R.C., Agarwal, A.K., and Rho, J.M., 2004, The ketogenic diet increases mitochondrial uncoupling protein levels and activity. *Ann. Neurol.* **55:** 576–580.

Sugars, K.L., and Rubinsztein, D.C., 2003, Transcriptional abnormalities in Huntington disease. *Trends Genet.* **19:** 233–238.

Swerdlow, R.H., Parks, J.K., Cassarino, D.S., Maguire, D.J., Maguire, R.S., Bennett, J.P. Jr, Davis, R.E., and Parker, W.D. Jr., 1997, Cybrids in Alzheimer's disease: a cellular model of the disease? *Neurology* **49:** 918–925.

Swerdlow, R.H., Parks, J.K., Cassarino, D.S., Trimmer, P.A., Miller, S.W., Maguire, D.J., Sheehan, J.P., Maguire, R.S., Pattee, G., Juel, V.C., Phillips, L.H., Tuttle, J.B., Bennett, J.P., Jr., Davis, R.E., and Parker, W.D. Jr., 1998, Mitochondria in sporadic amyotrophic lateral sclerosis. *Exp. Neurol.* **153:** 135–142.

Sykut-Cegielska, J., Gradowska, W., Mercimek-Mahmutoglu, S., and Stöckler-Ipsiroglu, S., 2004, Biochemical and clinical characteristics of creatine deficiency syndromes. *Acta Biochim. Pol.* **51:** 875–882.

Szebenyi, G., Morfini, G.A., Babcock, A., Gould, M., Selkoe, K., Stenoien, D.L., Young, M., Faber, P.W., MacDonald, M.E., McPhaul, M.J., and Brady, S.T., 2003, Neuropathogenic forms of huntingtin and androgen receptor inhibit fast axonal transport. *Neuron* **40:** 41–52.

Tabrizi, S.J., Blamire, A.M., Manners, D.N., Rajagopalan, B., Styles, P., Schapira, A.H., and Warner, T.T., 2003, Creatine therapy for Huntington's disease: clinical and MRS findings in a 1-year pilot study. *Neurology* **61:** 141–142.

Tan, D., Manchester, L.C., Reiter, R.J., Qi, W., Kim, S.J., and El-Sokkary, G.H., 1998, Melatonin protects hippocampal neurons in vivo against kainic acid-induced damage in mice. *J. Neurosci. Res.* **54:** 382–389.

Taoka, Y., and Okajima, K., 1998, Spinal cord injury in the rat. *Prog. Neurobiol.* **56:** 341–358.

Tarnopolsky, M.A., and Beal, M.F., 2001, Potential for creatine and other therapies targeting cellular energy dysfunction in neurological disorders. *Ann. Neurol.* **49:** 561–574.

Tarnopolsky, M., and Martin, J., 1999, Creatine monohydrate increases strength in patients with neuromuscular disease. *Neurology* **52:** 854–857.

Tarnopolsky, M.A., Atkinson, S.A., MacDougall, J.D., Chesley, A., Phillips, S., and Schwarcz, H.P., 1992, Evaluation of protein requirements for trained strength athletes. *J. Appl. Physiol.* **73:** 1986–1995.

Tarnopolsky, M.A., Roy, B.D., and MacDonald, J.R., 1997, A randomized, controlled trial of creatine monohydrate in patients with mitochondrial cytopathies. *Muscle Nerve* **20:** 1502–1509.

Tarnopolsky, M.A., Maguire, J., Myint, T., Applegarth, D., and Robinson, B.H., 1998, Clinical, physiological, and histological features in a kindred with the T3271C melas mutation. *Muscle Nerve* **21:** 25–33.

Tatsumi, T., Shiraishi, J., Keira, N., Akashi, K., Mano, A., Yamanaka, S., Matoba, S., Fushiki, S., Fliss, H., and Nakagawa, M., 2003, Intracellular ATP is required for mitochondrial apoptotic pathways in isolated hypoxic rat cardiac myocytes. *Cardiovasc. Res.* **59:** 428–440.

Tellez-Nagel, I., Johnson, A.B., and Terry, R.D., 1974, Studies on brain biopsies of patients with Huntington's chorea. *J. Neuropathol. Exp. Neurol.***33:** 308–332.

Theodore, W.H., 1999, Cerebral blood flow and glucose metabolism in human epilepsy. *Adv. Neurol.* **79:** 873–881.

Tombes, R.M., and Shapiro, B.M., 1985, Metabolite channeling: a phosphorylcreatine shuttle to mediate high energy phosphate transport between sperm mitochondrion and tail. *Cell* **41:** 325–334.

Trushina, E., Dyer, R.B., Badger, 2nd J.D., Ure, D., Eide, L., Tran, D.D., Vrieze, B.T., Legendre-Guillemin, V., McPherson, P.S., Mandavilli, B.S., Van Houten, B., Zeitlin, S., McNiven, M., Aebersold, R., Hayden, M., Parisi, J.E., Seeberg, E., Dragatsis, I., Doyle, K., Bender, A., Chacko, C., and McMurray, C.T., 2004, Mutant huntingtin impairs axonal trafficking in mammalian neurons in vivo and in vitro. *Mol. Cell. Biol.* **24:** 8195–8209.

Ueda, Y., Doi, T., Tokumaru, J., Makajima, A., and Nagatomo, K., 2005, In vivo evaluation of the effect of zonisamide on the hippocampal redox state during kainate acid-induced seizure status in rats. *Neurochem. Res.* **30:** 1117–1121.

Van Brussel, E., Yang, J.J., and Seraydarian, M.W., 1983, Isozymes of creatine kinase in mammalian cell cultures. *J. Cell. Physiol.* **116:** 221–226.

van den Munckhof, P., Luk, K.C., Ste-Marie, L., Montgomery, J., Blanchet, P.J., Sadikot, A.F., and Drouin, J., 2003, Pitx3 is required for motor activity and for survival of a subset of midbrain dopaminergic neurons. *Development* **130:** 2535–2542.

Verbessem, P., Lemiere, J., Eijnde, B.O., Swinnen, S., Vanhees, L., Van Leemputte, M., Hespel, P., and Dom, R., 2003, Creatine supplementation in Huntington's disease: a placebo-controlled pilot trial. *Neurology* **61:** 925–930.

Verhoeven, N.M., Salomons, G.S., and Jakobs, C., 2005, Laboratory diagnosis of creatine synthesis and biotransport. *Clin. Chim. Acta* **361:** 1–9.

Walker, J.B., 1979, Creatine: biosynthesis, regulation, and function. *Adv. Enzymol. Relat. Areas Mol. Biol.* **50:** 177–242.

Walter, M.C., Reilich, P., Lochmuller, H., Kohnen, R., Schlotter, B., Hautmann, H., Dunkl, E., Pongratz, D., and Muller-Felber, W., 2002, Creatine monohydrate in myotonic dystrophy: a double-blind, placebo-controlled clinical study. *J. Neurol.* **249:** 1717–1722.

Webster, M.T., Pearce, B.R., Bowen, D.M., and Francis, P.T., 1998, The effects of perturbed energy metabolism on the processing of amyloid precursor protein in PC12 cells. *J. Neural Transm.* **105:** 839–853.

Weisskopf, M.G., O'Reilly, E.J., McCullough, M.L., Calle, E.E., Thun, M.J., Cudkowicz, M., and Ascherio, A., 2005, Prospective study of military service and mortality from ALS. *Neurology* **64:** 32.

Wiedemann, F.R., Manfredi, G., Mawrin, C., Beal, M.F., and Schon, E.A., 2002, Mitochondrial DNA and respiratory chain function in spinal cords of ALS patients. *J. Neurochem.* **80:** 616–625.

Wilken, B., Ramirez, J.M., Probst, I., Richter, D.W., and Hanefeld, F., 2000, Anoxic ATP depletion in neonatal mice brainstem is prevented by creatine supplementation. *Arch. Dis. Child Fetal Neonatal Ed.* **82:** F224–F227.

Williamson, A., Patrylo, P.R., Pan, J., Spencer, D.D., and Hetherington, H., 2005, Correlations between granule cell physiology and bioenergects in human temporal lobe. *Brain* **128:** 1199–1208.

Willott, C.A., Young, M.E., Leighton, B., Kemp, G.J., Boehm, E.A., Radda, G.K., and Clarke, K., 1999, Creatine uptake in isolated soleus muscle: kinetics and dependence on sodium, but not on insulin. *Acta Physiol. Scand.* **166:** 99–104.

Wong, P., Pardo, C., Borchelt, D., Lee, M., Copeland, N., Jenkins, N., Sisodia, S., Cleveland, S., and Price, D., 1995, An adverse property of a familial ALS-linked SOD1 mutation causes motor neuron disease characterized by vacuolar degeneration of mitochondria. *Neuron* **14:** 1105–1116.

Wyss, M., and Kaddurah-Daouk, R., 2000, Creatine and creatinine metabolism. *Physiol. Rev.* **80:** 1107–1213.

Wyss, M., and Schulze, A., 2002, Health implications of creatine: can oral creatine supplementation protect against neurological and atherosclerotic disease? *Neuroscience* **112:** 243–260.

Xiong, Y., Peterson, P.L., Muizelaar, J.P., and Lee, C.P., 1997, Amelioration of mitochondrial function by a novel antioxidant U-101033E following traumatic brain injury in rats. *J. Neurotrauma* **14:** 907–917.

Xiong, Y., Peterson, P.L., Verweij, B.H., Vinas, F.C., Muizelaar, J.P., and Lee, C.P., 1998, Mitochondrial dysfunction after experimental traumatic brain injury: combined efficacy of SNX-111 and U-101033E. *J. Neurotrauma* **15:** 531–544.

Xu, C.J., Klunk, W.E., Kanfer, J.N., Xiong, Q., Miller, G., and Pettegrew, J., 1996, Phosphocreatine-dependent glutamate uptake by synaptic vesicles. A compaison with ATP-dependent glutamate uptake. *J. Biol. Chem.* **271:** 13435–13440.

Zeevalk, G.D., and Nicklas, W.J., 1991, Mechanisms underlying initioation of excitotoxicity associated with metabolic inhibition. *J. Pharmacol. Exp. Ther.* **257:** 870–878.

Zhang, W., Narayanan, M., and Friedlander, R.M., 2003, Additive neuroprotective effects of minocycline with creatine in a mouse model of ALS. *Ann. Neurol.* **53:** 267–270.

Zhu, S., Stavrovskaya, I.G., Drozda, M., Kim, B.Y.S., Ona, V., Li, M., Sarang, S., Liu, A.S., Hartley, D.M., Wuk, D.C., Gullans, S., Ferrante, R.J., Przedborskikq, S., Kristal, B.S., and Friedlander, R.M., 2002, Minocycline inhibits cytochrome c release and delays progression of amyotrophic lateral sclerosis in mice. *Nature* **417:** 74–78.

Zhu, S., Li, M., Figueroa, B.E., Liu, A., Stavrovskaya, I.G., Pasinelli, P., Beal, M.F., Brown, R.H. Jr., Kristal, B.S., Ferrante, R.J., and Friedlander, R.M., 2004, Prophylactic creatine administration mediates neuroprotection in cerebral ischemia in mice. *J. Neurosci.* **24:** 5909–5912.

CHAPTER 12

ERGOGENIC EFFECTS OF CREATINE IN SPORTS AND REHABILITATION

PETER HESPEL[1] AND WIM DERAVE[2]

[1] *Research Center for Exercise and Health, Faculty of Kinesiology and Rehabilitation Sciences, K.U. Leuven, Leuven, Belgium*

[2] *Department of Movement and Sport Sciences, Faculty of Medicine and Health Sciences, Ghent University, Ghent, Belgium*

Abstract: The daily oral ingestion of supplementary creatine monohydrate can substantially elevate the creatine content of human skeletal muscle. This chapter aims to summarize the current knowledge regarding the impact muscle creatine loading can have on exercise performance and rehabilitation. The major part of the elevation of muscle creatine content is already obtained after one week of supplementation, and the response can be further enhanced by a concomitant exercise or insulin stimulus. The elevated muscle creatine content moderately improves contractile performance in sports with repeated high-intensity exercise bouts. More chronic ergogenic effects of creatine are to be expected when combined with several weeks of training. A more pronounced muscle hypertrophy and a faster recovery from atrophy have been demonstrated in humans involved in resistance training. The mechanism behind this anabolic effect of creatine may relate to satellite cell proliferation, myogenic transcription factors and insulin-like growth factor-1 signalling. An additional effect of creatine supplementation, mostly when combined with training, is enhanced muscle glycogen accumulation and glucose transporter (GLUT4) expression. Thus, creatine may also be beneficial in sport competition and training characterized by daily glycogen depletion, as well as provide therapeutic value in the insulin-resistant state

1. INTRODUCTION

Creatine is a natural guanidino compound which is found in meat and fish in concentrations ranging between 3 and 7 gram per kg (Walker, 1979). In 1992, Harris and his co-workers showed for the first time that supplementary oral creatine intake can increase muscle creatine content and enhance muscle power (Greenhaff *et al.*, 1993; Harris *et al.*, 1992). Since 1992, extensive research (for reviews see

G.S. Salomons and M. Wyss (eds.), Creatine and Creatine Kinase in Health and Disease, 245–259.

Derave *et al.* (2003a); Hespel *et al.* (2001a); Nissen and Sharp (2003); Terjung *et al.* (2000)) has proven that creatine, in contrast with many other 'pseudo-ergogenic' supplements, is a true ergogenic aid. However, one should also recognize that creatine supplementation as a rule has little effect. Therefore, this supplementation is probably most relevant to adult athletes seeking a minor improvement in performance.

A classical creatine loading regimen in the context of either sports or rehabilitation consists of an initial 'loading phase' (15–20 g per day for 4–7 days) followed by a maintenance dose (2–5 g per day). However, available findings indicate that the effects of creatine supplementation may decrease during prolonged supplementation (> 2–3 months) (Derave *et al.*, 2003a; Eijnde *et al.*, 2003; Olsen *et al.*, 2006). Therefore, it might be an option to interrupt 2–3 month periods of daily creatine intake with washout episodes of about 4–5 weeks to restore the responsiveness of muscle to creatine supplementation (Vandenberghe *et al.*, 1997). Contrary to many commercial claims, there is no evidence that so-called 'special' creatine formulations like creatine-citrate or creatine-pyruvate, administered either in solid, liquid or gel form (see chapter 15; Pischel and Gastner, 2007), yield any better results than ingestion of simple creatine monohydrate powder. Available literature implies that creatine intake in healthy adults, when used according to the aforementioned guidelines, is generally safe (Poortmans and Francaux, 2000). Initial concerns were raised with regards to the effects of high-dose creatine intake at the level of the kidneys. However, based on extensive observations in a number of studies it can be concluded that creatine intake, used at the above recommended dosages, does not impair renal function. Still, one should also be aware that sound long-term safety data on creatine supplementation are lacking and thus caution is warranted. For instance, in some rare cases, creatine loading may cause compartment syndrome due to an increased intramuscular pressure resulting from creatine-induced muscle water influx (Schroeder *et al.*, 2001). In addition, since athletes often tend to use supplement dosages far above the scientifically recommended range under the assumption that "if a little is good, more is probably better", it must be emphasized that the potential side-effects of excessive creatine intake are still unknown. In addition, the interaction of creatine intake with either legal as well as illegal ergogenic substances has remained virtually unexplored.

In this chapter, we summarize the currently available literature regarding ergogenic effects of creatine supplementation. First, we discuss the short-term effects, namely the effects that can already be observed after 5–7 days of supplementation and that result from the higher total creatine content in the muscle cells. Secondly, we describe the long-term effects of creatine supplementation over a period of several weeks to months, usually in combination with training, on muscle growth. Furthermore, we highlight the emerging role of creatine in muscle glycogen metabolism in relationship to the effects of insulin as well as other factors involved in muscle creatine loading.

2. SHORT-TERM EFFECTS OF CREATINE SUPPLEMENTATION ON SKELETAL MUSCLE CONTRACTILITY

The benefits of short-term creatine supplementation are probably more relevant in the context of sports than in rehabilitation, because rehabilitation programs usually aim for long-term improvement rather than a short and transient increase in muscular performance. Creatine has probably become one of the most popular ergogenic supplements for athletes involved in sports requiring very high power outputs. During sprint-type exercises lasting for several seconds, phosphocreatine breakdown through the creatine kinase reaction is the primary pathway of ATP provision in contracting muscles. The relative contribution of phosphocreatine to energy turnover is negatively related to exercise duration (Gaitanos *et al.*, 1993; Hultman *et al.*, 1996; Spriet *et al.*, 1995). It has been shown that increased muscle creatine content due to short-term creatine intake can prevent net ATP degradation during high-intensity muscle contractions, which conceivably results from increased flux through the creatine kinase reaction (Casey *et al.*, 1996). Nonetheless, the effects of short-term creatine ingestion on single bouts of short maximal exercise are equivocal, with most studies showing no beneficial effect (Juhn and Tarnopolsky, 1998; Terjung *et al.*, 2000). However, creatine ingestion has been consistently found to enhance power output during intermittent exercise modes, where maximal exercise bouts are interspersed with short recovery intervals (Greenhaff *et al.*, 1993; Vandebuerie *et al.*, 1998; Vandenberghe *et al.*, 1996, 1997). For instance, we have previously demonstrated in healthy volunteers that creatine intake (0.5 g per kg b.w. per day for 6 days) markedly increased torque production during repeated series of maximal knee-extensions with short recovery intervals in between (Figure 1). This performance effect can probably be explained – at least in part – by enhanced phosphocreatine resynthesis during recovery (Greenhaff *et al.*, 1994). In addition, as shown in Figure 2, we have also demonstrated that muscle relaxation time is decreased after creatine intake (Hespel *et al.*, 2002; Van Leemputte *et al.*, 1999). This may indicate that sarcoplasmic reticulum Ca^{2+}-ATPase pump activity is enhanced by muscle creatine loading. A likely explanation for this effect is improved maintenance of free energy of ATP hydrolysis due to increased creatine kinase activity at the level of the sarcoplasmic reticulum. Muscle relaxation time is important for power output in high-intensity exercise involving repeated high-velocity muscle contractions. Interestingly, in this regard, we have also shown that caffeine intake negates the beneficial effect of short-term creatine supplementation on both muscle power output and on muscle relaxation time (Hespel *et al.*, 2002; Vandenberghe *et al.*, 1996).

It is also clear from the literature that short-term creatine intake is not effective in enhancing force output during single maximal muscle contractions or in improving endurance exercise capacity. Data indicate that creatine intake in conjunction with a high rate of carbohydrate intake can increase muscle glycogen content (see below). However, there is no evidence for a beneficial effect of such intervention on endurance capacity or endurance performance. Still, we have shown that creatine can enhance

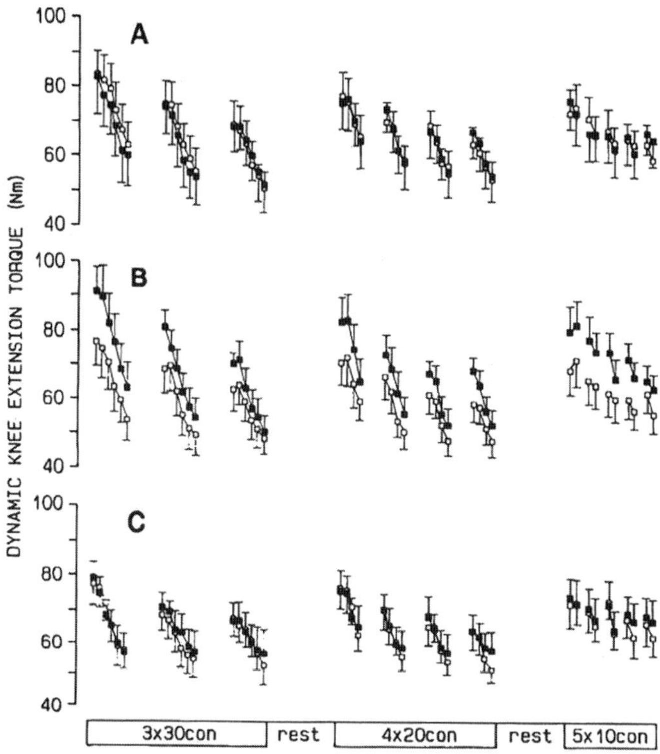

Figure 1. Effect of short-term creatine supplementation, alone or in combination with caffeine intake, on muscle force output during intermittent maximal muscle contractions. Young healthy subjects performed 3 series of maximal dynamic knee extensions on an isokinetic dynamometer with 2-min rest pauses in between. Series 1 consisted of 3 bouts of 30 contractions, series 2 of 4 bouts of 20 contractions, and series 3 of 5 bouts of 10 contractions, interrupted by rest intervals of 60, 40 and 20 sec, respectively. The data-points represent means of 5 consecutive contractions (con) before (□) and after (■) 6 days of either placebo (panel A), creatine (0.5 g per kg b.w. per day; panel B), or creatine + caffeine (5 mg per kg b.w. per day; panel C). Values are means ± SEM of 9 observations. Figure reproduced from Vandenberghe *et al.* (1996) with kind permission from the American Physiological Society.

intermittent sprint performance at the end of an endurance exercise bout (Vandebuerie *et al.*, 1998). Finally, it is also important to note that no benefit should be expected from a single dose of creatine intake before exercise (Preen *et al.*, 2002; Vandebuerie *et al.*, 1998). Due to the low rate of muscle creatine uptake, it is unlikely that this can cause a significant increase in muscle creatine content.

3. LONG-TERM EFFECTS OF CREATINE ON SKELETAL MUSCLE ANABOLISM

Several studies have specifically addressed the long-term effects of creatine supplementation on adaptations to resistance training. Early studies clearly showed that

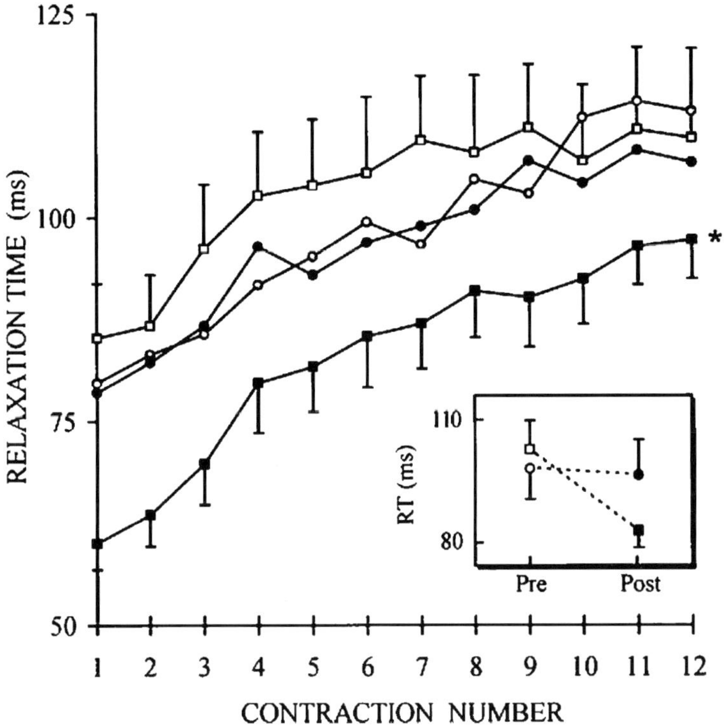

Figure 2. Effect of short-term creatine supplementation on muscle relaxation time. Young healthy subjects performed 12 maximal isometric contractions with the elbow flexors, interspersed by 10-sec rest intervals. Torque output was continuously measured (1000 Hz) and the relaxation time from 75% to 25% of peak tension was registered for each contraction. Values are means ± SEM of 8 observations either before (□) and after (■) creatine intake, or before (○) and after (●) placebo intake. Figure reproduced from Van Leemputte *et al.* (1999) with kind permission from the American Physiological Society.

creatine intake can stimulate fat-free mass, which also resulted in increased muscle force and power production (Kreider *et al.*, 1998; Vandenberghe *et al.*, 1997). Most probably, increased training performance contributes to this effect (Kreider *et al.*, 1998). Moreover, also our studies (Hespel *et al.*, 2001b; Vandenberghe *et al.*, 1997) showed that supplementary creatine intake can stimulate muscular adaptations upon a given training load. Since then, research findings have been equivocal, yet recent meta-analyses of published studies concluded that creatine supplementation in conjunction with resistance training indeed yields a small 'anabolic' effect (Nissen and Sharp, 2003). Support for this conclusion also comes from *in vivo* and *in vitro* studies, indicating that creatine supplementation can influence cellular pathways involved in muscle growth. First, creatine intake combined with resistance training in young healthy volunteers has been found to stimulate proliferation of satellite cells (Dangott *et al.*, 2000; Olsen *et al.*, 2006). Furthermore, several studies using various protocols indicate that creatine can stimulate the IGF-1 signaling pathway and affect myogenic regulatory factors (Deldicque *et al.*, 2005; Hespel *et al.*, 2001b; Louis *et al.*, 2004; Willoughby and Rosene, 2003). Nevertheless, direct evidence that creatine can stimulate protein synthesis in human muscle is missing (Louis *et al.*, 2003a,b; Parise *et al.*, 2001). Furthermore, one study showed that creatine intake can enhance exercise-induced increases in myosin heavy-chain mRNA content (Willoughby and Rosene, 2001). However, solid data to prove that this translates into increased myosin heavy-chain protein expression or increased muscle fiber size are lacking. It is also important to note that at least part of the increase in fat-free mass, which typically occurs after creatine loading, does not reflect protein accretion but comes from accumulation of intracellular water (Ziegenfuss *et al.*, 1998), which may be detrimental to performance in some sports where a low body weight is critical for optimal performance. Furthermore, one should also acknowledge that creatine intake alone, if not combined with heavy resistance training, does not induce muscle growth.

The potential of creatine to stimulate training-induced muscle hypertrophy is – for obvious reasons – not only relevant in the context of elite sports performance, but even more so in the context of prevention or rehabilitation of muscle atrophy. Studies performed in our laboratories have demonstrated that oral creatine supplementation at a rate of 5 g per day stimulates recovery of both leg muscle mass (Figure 3) and

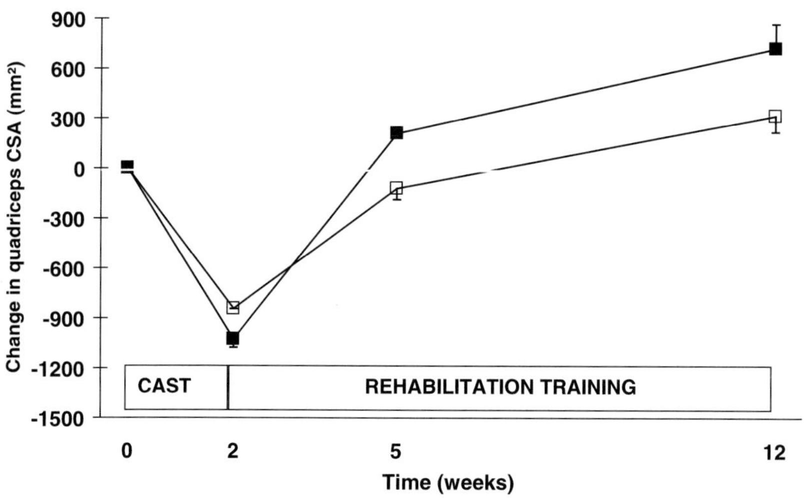

Figure 3. Effect of creatine supplementation during leg immobilisation and subsequent rehabilitation training on cross-sectional surface area of the quadriceps muscle. The right leg of 20 healthy subjects was immobilised with a cast from groin to ankle during 2 weeks. Thereafter the subjects were enrolled in a well-controlled rehabilitative weight training program for the knee-extensor muscles. Throughout the intervention period subjects received either placebo (□, n = 9) or creatine (■, n = 9) supplements. Cross-sectional area (CSA) of the m. quadriceps was measured by NMR-imaging. Creatine dosage was 4 × 5 g per day during the immobilisation period, and was decreased to 5 g per day during rehabilitation training. Values are means ± SEM. See Hespel *et al.* (2001b) for more details.

power output after 2 weeks of leg immobilization-induced muscle atrophy (Hespel *et al.*, 2001b). The recovery of muscle mass was due to hypertrophy in both type I and type II muscle fibers. It was also shown that the creatine-induced increase in muscle fiber area was closely correlated with an increase in MRF4 expression. This indicates that regulation of myogenic transcription factors, particularly MRF4, may be implicated in the action of creatine to stimulate muscle hypertrophy. This study indicates that creatine supplementation may serve as a useful adjuvant therapy to exercise training in the prevention and rehabilitation of muscle atrophy in various clinical conditions.

4. EFFECTS OF CREATINE ON MUSCLE INSULIN SENSITIVITY AND GLYCOGEN ACCUMULATION

As a crucial substrate for energy metabolism in active muscles, glycogen content determines the time-to-fatigue and performance in exercise of >1 h duration. Moreover, between subsequent days with daily glycogen-depleting exercise, complete glycogen resynthesis is difficult to attain. As a matter of fact, performance in multi-stage competitions, such as in cycling tours, seems to be directly correlated with the capacity for glycogen resynthesis. Therefore, a compound that enhances and/or accelerates glycogen accumulation and insulin-induced glycogen synthesis is certainly ergogenic. It has been shown that combined creatine and carbohydrate supplementation following exercise results in a greater muscle glycogen accumulation than carbohydrate alone (Robinson *et al.*, 1999). Several studies have confirmed the enhancing effect of creatine on glycogen synthesis and accumulation in human and rat skeletal muscle, especially when combined with exercise training (Derave *et al.*, 2003b; Nelson *et al.*, 2001; Op 't Eijnde *et al.*, 2001a,b). Despite intensive investigation, a conclusive mechanistic explanation for this effect is not yet available.

Creatine supplementation does not directly influence basal or insulin-stimulated glucose uptake in rodent muscle (Op 't Eijnde *et al.*, 2001a; Young and Young, 2002) nor in L6 myoblasts (Ceddia and Sweeney, 2004). On the other hand, creatine supplementation in humans enhances the training-induced increase and attenuates the disuse-induced decrease in expression of GLUT4 (Derave *et al.*, 2003b; Op 't Eijnde *et al.*, 2001b), the predominant insulin-sensitive glucose transporter in skeletal muscle. A solid explanation for enhanced GLUT4 expression is not available at present, although the involvement of 5'-AMP-activated protein kinase (AMPK) is hypothesized (Zheng *et al.*, 2001). *In vitro* studies on L6 myoblasts have suggested that creatine supplementation increases glucose oxidation and AMPK phosphorylation, but in human muscle such an enhancing effect of creatine on AMPK phosphorylation and expression was not observed (Eijnde *et al.*, 2005). Despite increasing GLUT4 expression, oral creatine supplementation during 4–8 weeks in healthy subjects does not enhance oral glucose tolerance (Derave *et al.*, 2003b; Newman *et al.*, 2003). Interestingly, however, combined creatine and protein supplementation together with resistance training results in improved glucose tolerance compared with resistance training alone (Derave *et al.*, 2003b).

In the insulin-resistant state, it is not clear whether creatine supplementation per se can improve glucose tolerance and insulin sensitivity. Initial studies on hyperglycemic transgenic Huntington's disease mice first suggested a glucose-lowering and insulin-sensitizing effect of creatine supplementation (Ferrante *et al.*, 2000). Recently, oral creatine supplementation in an animal model of type-2 diabetes was found to improve the insulinogenic index during an intravenous glucose tolerance test (Op 't Eijnde *et al.*, 2006). These observations emphasize the potential therapeutic value of creatine in insulin-resistant patients, which currently remains to be explored. As for healthy subjects and athletes, it can be generally concluded that creatine supplementation can enhance the training-induced increase in muscle glycogen content and GLUT4 expression, but probably not glucose tolerance. In this way, creatine is an ergogenic aid during training or competition periods that include frequent muscle glycogen depletion.

5. DETERMINANTS OF RESPONSIVENESS TO CREATINE SUPPLEMENTATION

The above-mentioned data clearly indicate that increased muscle creatine content can have a beneficial impact on a number of physiological processes that are related to exercise performance. Extensive experience from the lab and the field has shown that it is not always easy to obtain a significant increase in muscle creatine content by use of oral creatine supplementation. Two problems generally occur: a) the inability of supplementation to raise muscle creatine content in all subjects, i.e. the occurrence of the 'non-responders' (Greenhaff *et al.*, 1994), and b) the inability to keep muscle creatine content elevated for longer periods than a few weeks (Derave *et al.*, 2003a). Adequate tackling of these problems will further optimize the use of creatine in sports and rehabilitation.

5.1. Exercise

Since it was shown that oral creatine supplementation can increase muscular creatine content, researchers have searched for possible factors that can facilitate this process. Harris *et al.* (1992) have compared the effects of creatine supplementation with or without a daily 60-min submaximal cycling exercise bout. The mean muscle total creatine (TCr) concentration increased by 37% when creatine was combined with exercise, compared to 26% with creatine alone. Thus, submaximal exercise promotes muscle creatine accumulation, an effect that is restricted to the exercising muscles, as later shown by Robinson *et al.* (1999). In a more recent study by Preen *et al.* (2003), however, a daily 60-min repeated sprint exercise during 5-day creatine loading (20 g/day) did not elevate human vastus lateralis TCr content above creatine supplementation without exercise.

One can hypothesize about the underlying mechanism for the more pronounced accumulation of muscle creatine in exercised vs. non-exercised muscles, as shown in the above-mentioned supplementation studies (Harris *et al.* 1992; Robinson *et al.*, 1999). Most likely, the phenomenon is either due to enhanced perfusion of exercised muscles and the concomitantly higher availability of extracellular creatine, or alternatively to a direct stimulatory effect of contractions on the transsarcolemmal creatine transport process. As for the latter, such mechanisms of enhanced metabolite uptake evoked by isolated muscle contractions have long been established for glucose (Holloszy and Narahara, 1967) and amino acids (Narahara and Holloszy, 1974). In order to investigate whether a similar mechanism exists for creatine transport, we recently exposed isolated incubated mouse muscles to ^{14}C-labelled creatine and electrically-stimulated contractions. We found that electrical stimulation caused a 2- to 4-fold increase in creatine transport rate over basal levels (Derave *et al.*, 2006). Surprisingly, however, the effect was not caused by muscle contractions *per se*, but was conceivably initiated by electrolysis-induced generation of reactive oxygen species (ROS), inherent to electrical discharge between two electrodes immersed in a well-oxygenated physiological buffer (Derave *et al.*, 2006). Interestingly, the electrolysis-induced increase in creatine uptake was paralleled by an increased expression of the creatine transporter on the cell surface, suggesting that creatine transport activity in muscle is enhanced by a translocation process. Further studies will have to determine whether the enhanced creatine accumulation in exercised muscles *in vivo* is caused or accompanied by an increased cell surface expression of the creatine transporter.

5.2. Carbohydrates and Insulin

Apart from exercise, also carbohydrate ingestion and insulin have been suggested as promoting factors for creatine accumulation in human muscle. Green *et al.* (1996b) observed a much lower urinary creatine excretion when simple carbohydrates were co-ingested with creatine, compared with creatine alone. Subsequent studies by these and other researchers confirmed and strengthened the finding that carbohydrate ingestion augments muscle creatine uptake by virtue of its insulin-releasing effects (Green *et al.*, 1996a; Preen *et al.*, 2003). As both protein ingestion- and carbohydrate ingestion-induced insulin release stimulate creatine accumulation (Steenge *et al.*, 2000), it is proposed that insulin itself, rather than carbohydrates, is involved in this facilitative process. As in the case of exercise, also for the effect of insulin it can be questioned whether the effect is caused by enhanced muscle perfusion due to the vasodilatory action of insulin (Baron, 1994) or rather by a direct effect of insulin on the transsarcolemmal creatine transport process.

The literature is strikingly equivocal with respect to the effects of insulin on muscle creatine uptake. Two older studies indicate that insulin stimulates the uptake of radioactively labeled creatine in rat skeletal muscle *in vivo* (Koszalka and Andrew, 1972) and *in vitro* (Haughland and Chang, 1975). Two more recent studies, however, failed to show a stimulatory effect of insulin on creatine uptake in isolated mouse muscles (Willott *et al.*, 1999) and perfused rat hindlimb muscles (Brault and Terjung, 2003). In addition, the ability of creatine supplementation to elevate muscle creatine stores is maintained in animal models of diabetes (Haughland and Chang, 1975; Op 't Eijnde *et al.*, 2006), which makes an important role of insulin in muscular creatine uptake unlikely. Thus, the presently available literature on rodents is inconsistent and does not allow to unravel the underlying mechanism for the enhancement of muscle creatine accumulation by carbohydrates and insulin, as observed in human muscle (Green *et al.*, 1996a).

5.3. Muscle Fiber-type Distribution

Fast-twitch muscle fibers rely more on the creatine kinase system for energy delivery than slow-twitch fibers, as illustrated by higher CK activity, higher concentrations of TCr, and a higher rate of phosphocreatine degradation during high-intensity exercise (Casey *et al.*, 1996). The baseline phosphocreatine and total creatine content is 20% higher in human fast-twitch than slow-twitch fibers (Casey *et al.*, 1996), and the difference is even more pronounced in rodents (Op 't Eijnde *et al.*, 2001a). Two opposite expectations can be formulated regarding the responsiveness of different fiber-types to creatine supplementation. First, one could expect creatine content to increase more in fast-twitch than slow-twitch fibers, because the former is more dependent on this energy system and therefore may benefit most. Second, one could expect creatine content to increase more in slow-twitch fibers, because baseline creatine content is lower and the responsiveness to creatine supplementation is inversely related to the initial creatine content (Harris *et al.*, 1992).

Currently, only one study has investigated the effect of creatine supplementation on creatine content in different fiber types of humans, and concluded that both fast-twitch and slow-twitch fibers respond equally well to supplementation (Casey *et al.*, 1996). Likewise, studies in rats and mice generally report a 5–25% increase in creatine content in both slow-twitch soleus and fast-twitch plantaris, extensor digitorum longus (EDL) and gastrocnemius muscles. However, some authors reported the increases to be significant in only soleus but not in fast-twitch muscles (Eijnde *et al.*, 2004; Op 't Eijnde *et al.*, 2001a); to be significant in both muscle types (Brannon *et al.*, 1997); or to be significant only in fast-twitch muscles (McMillen *et al.*, 2001). Thus, it can be concluded that both fiber-types are roughly equally responsive to creatine supplementation. However, recent evidence from our laboratory suggests that the rate of creatine loading is higher in slow-twitch compared with fast-twitch muscles. As illustrated in Figure 4A, one week of creatine supplementation (repletion) following two weeks of creatine depletion (with β-GPA) made total creatine content return to near-baseline levels in soleus, whereas total creatine content in EDL still remained significantly below baseline (Eijnde *et al.*, 2004). This could be related to and explained by the higher rate of creatine transport in soleus vs. EDL muscles, when evaluated by uptake of radiolabeled creatine in incubated muscles (Derave *et al.*, 2006; Willott *et al.*, 1999), as illustrated in Figure 4B. Some studies (Brault and Terjung, 2003; Murphy *et al.*, 2001)

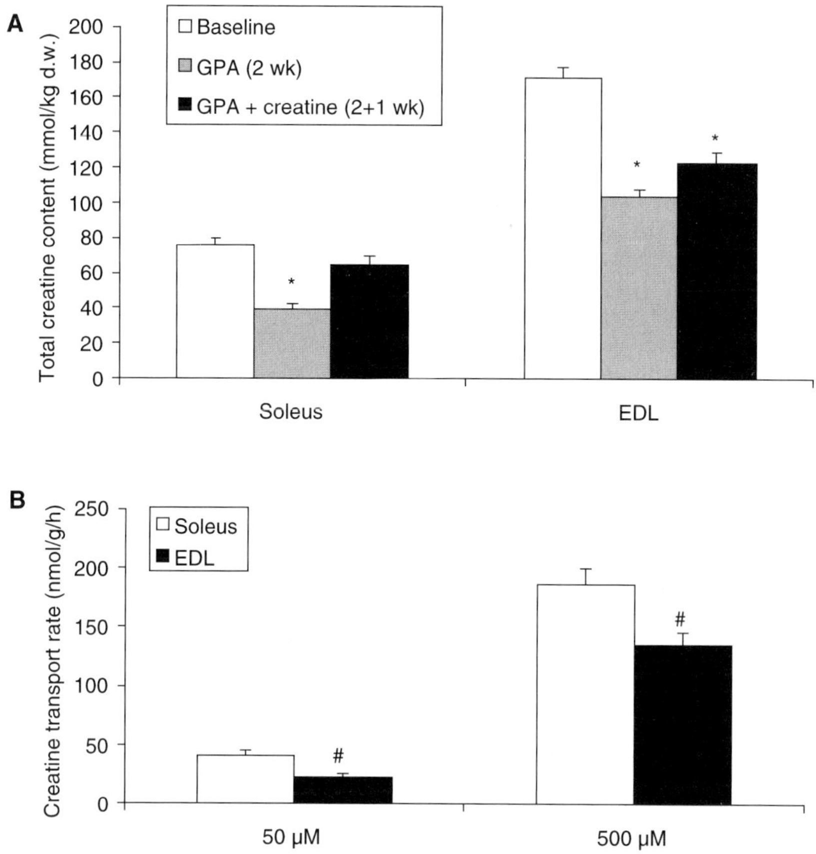

Figure 4. Creatine content and uptake rate in mouse soleus and EDL muscles. Panel A: Total creatine content in soleus and EDL muscles of NMRI mice at baseline, when fed for 2 weeks with β-guanidinoproprionic acid (GPA; 1% w/w of food intake) and when fed for 2 weeks with GPA and subsequently for 1 week with creatine monohydrate (5% w/w of food intake). Data are from Eijnde *et al.* (2004). * indicates significant difference from baseline ($P < 0.05$). Panel B: Creatine transport rate in incubated soleus and EDL muscles of NMRI mice with 50 µM and 500 µM creatine in the incubation medium. Data are from Derave *et al.* (2006). # indicates significant difference from soleus ($P < 0.05$). All values are means ± SEM.

have suggested that the higher V_{max} of creatine transport in slow-twitch vs. fast-twitch muscles parallels a higher protein expression of the creatine transporter. However, the latter notion needs to be questioned because creatine transporter mRNA expression is not different between fiber types (Brault and Terjung, 2003; Murphy *et al.*, 2001), and the currently available creatine transporter antibodies seem to cross-react with mitochondrial proteins such as the E2 components of pyruvate dehydrogenase (Speer *et al.*, 2004). In conclusion, it seems that creatine supplementation can elevate the creatine content in both slow-twitch and fast-twitch muscle fibers by a similar order of magnitude (~20%), but that the elevated level can be reached more rapidly in slow-twitch fibers.

For a more thorough review of (i) the literature regarding these and other factors influencing creatine loading in human muscle (e.g. vegetarianism, gender, age) and of (ii) the functional insights into the creatine transporter, we refer to two relevant reviews (Hespel *et al.*, 2001a; Snow and Murphy, 2003) and to chapter 6 of this book (Christie, 2007), respectively.

6. GENERAL CONCLUSIONS

Although its effects are usually limited, creatine supplementation can be beneficial for certain types and aspects of exercise performance. The increase in total creatine content of muscle, already obtained after one week of oral creatine monohydrate supplementation, results in improved maximal muscle torque in repeated contraction series. Furthermore, creatine facilitates muscle glycogen accumulation. Finally, longer periods of creatine supplementation in conjunction with resistance training can induce muscle hypertrophy, an aspect that benefits both strength performance and rehabilitation.

REFERENCES

Baron, A.D., 1994, Hemodynamic actions of insulin. *Am. J. Physiol.* **267:** E187–E202.

Brannon, T.A., Adams, G.R., Conniff, C.L., and Baldwin, K.M., 1997, Effects of creatine loading and training on running performance and biochemical properties of rat skeletal muscle. *Med. Sci. Sports Exerc.* **29:** 489–495.

Brault, J.J., and Terjung, R.L., 2003, Creatine uptake and creatine transporter expression among rat skeletal muscle fiber types. *Am. J. Physiol. Cell Physiol.* **284:** C1481–C1489.

Casey, A., Constantin-Teodosiu, D., Howell, S., Hultman, E., and Greenhaff, P.L., 1996, Creatine ingestion favorably affects performance and muscle metabolism during maximal exercise in humans. *Am. J. Physiol. Endocrinol. Metab.* **271:** E31–E37.

Ceddia, R.B., and Sweeney, G., 2004, Creatine supplementation increases glucose oxidation and AMPK phosphorylation and reduces lactate production in L6 rat skeletal muscle cells. *J. Physiol.* **555:** 409–421.

Christie, D.L., 2007, Functional insights into the creatine transporter. *Subcell. Biochem.* **46:** 99–118.

Dangott, B., Schultz, E., and Mozdziak, P.E., 2000, Dietary creatine monohydrate supplementation increases satellite cell mitotic activity during compensatory hypertrophy. *Int. J. Sports Med.* **21:** 13–16.

Deldicque, L., Louis, M., Theisen, D., Nielens, H., Dehoux, M., Thissen, J.P., Rennie, M.J., and Francaux, M., 2005, Increased IGF mRNA in human skeletal muscle after creatine supplementation. *Med. Sci. Sports Exerc.* **37:** 731–736.

Derave, W., Eijnde, B.O., and Hespel, P., 2003a, Creatine supplementation in health and disease: what is the evidence for long-term efficacy? *Mol. Cell. Biochem.* **244:** 49–55.

Derave, W., Eijnde, B.O., Verbessem, P., Ramaekers, M., Van Leemputte, M., Richter, E.A., and Hespel, P., 2003b, Combined creatine and protein supplementation in conjunction with resistance training promotes muscle GLUT-4 content and glucose tolerance in humans. *J. Appl. Physiol.* **94:** 1910–1916.

Derave, W., Straumann, N., Olek, R.A., and Hespel, P., 2006, Electrolysis stimulates creatine transport and transporter cell surface expression in incubated mouse skeletal muscle: potential role of ROS. *Am. J. Physiol. Endocrinol. Metab.* **291:** E1250–E1257.

Eijnde, B.O., Derave, W., Wojtaszewski, J.F., Richter, E.A., and Hespel, P., 2005, AMP kinase expression and activity in human skeletal muscle: effects of immobilization, retraining, and creatine supplementation. *J. Appl. Physiol.* **98:** 1228–1233.

Eijnde, B.O., Lebacq, J., Ramaekers, M., and Hespel, P., 2004, Effect of muscle creatine content manipulation on contractile properties in mouse muscles. *Muscle Nerve* **29:** 428–435.

Eijnde, B.O., Van Leemputte, M., Goris, M., Labarque, V., Taes, Y., Verbessem, P., Vanhees, L., Ramaekers, M., Vanden Eynde, B., Van Schuylenbergh, R., Dom, R., Richter, E.A., and Hespel, P., 2003, Effects of creatine supplementation and exercise training on fitness in males 55 to 75 years old. *J. Appl. Physiol.* **95:** 818–828.

Ferrante, R.J., Andreassen, O.A., Jenkins, B.G., Dedeoglu, A., Kuemmerle, S., Kubilus, J.K., Kaddurah-Daouk, R., Hersch, S.M., and Beal, M.F., 2000, Neuroprotective effects of creatine in a transgenic mouse model of Huntington's disease. *J. Neurosci.* **20:** 4389–4397.

Gaitanos, G.C., Williams, C., Boobis, L.H., and Brooks, S., 1993, Human muscle metabolism during intermittent maximal exercise. *J. Appl. Physiol.* **75:** 712–719.

Green, A.L., Hultman, E., MacDonald, I.A., Sewell, D.A., and Greenhaff, P.L., 1996a, Carbohydrate ingestion augments skeletal muscle creatine accumulation during creatine supplementation in humans. *Am. J. Physiol. Endocrinol. Metab.* **271:** E821–E826.

Green, A.L., Simpson, E.J., Littlewood, J.J., MacDonald, I.A., and Greenhaff, P.L., 1996b, Carbohydrate ingestion augments creatine retention during creatine feeding in humans. *Acta Physiol. Scand.* **158:** 195–202.

Greenhaff, P.L., Bodin, K., Soderlund, K., and Hultman, E., 1994, Effect of oral creatine supplementation on skeletal muscle phosphocreatine resynthesis. *Am. J. Physiol.* **266:** E725–E730.

Greenhaff, P.L., Casey, A., Short, A.H., Harris, R., and Söderlund, K., 1993, Influence of oral creatine supplementation on muscle torque during repeated bouts of maximal voluntary exercise in man. *Clin. Sci.* **84:** 565–571.

Harris, R.C., Söderlund, K., and Hultman, E., 1992, Elevation of creatine in resting and exercised muscle of normal subjects by creatine supplementation. *Clin. Sci.* **83:** 367–374.

Haugland, R.B., and Chang, D.T., 1975, Insulin effect on creatine transport in skeletal muscle. *Proc. Soc. Exp. Biol. Med.* **148:** 1–4.

Hespel, P., Op 't Eijnde, B., Derave, W., and Richter, E.A., 2001a, Creatine supplementation: exploring the role of the creatine kinase/phosphocreatine system in human muscle. *Can. J. Appl. Physiol.* **26 (Suppl.):** S79–S102.

Hespel, P., Op 't Eijnde, B., and Van Leemputte, M., 2002, Opposite actions of caffeine and creatine on muscle relaxation time in humans. *J. Appl. Physiol.* **92:** 513–518.

Hespel, P., Op 't Eijnde, B., Van Leemputte, M., Urso, B., Greenhaff, P.L., Labarque, V., Dymarkowski, S., Van Hecke, P., and Richter, E.A., 2001b, Oral creatine supplementation facilitates the rehabilitation of disuse atrophy and alters expression of muscle myogenic factors in humans. *J. Physiol. (London)* **536:** 625–633.

Holloszy, J., and Narahara, H., 1967, Enhanced permeability to sugar associated with muscle contraction. *J. Gen. Physiol.* **50:** 551–562.

Hultman, E., Soderlund, K., Timmons, J.A., Cederblad, G., and Greenhaff, P.L., 1996, Breakdown and resynthesis of phosphorylcreatine and adenosine triphosphate in connection with muscular work in man. *Scand. J. Clin. Lab. Invest.* **19:** 56–66.

Juhn, M.S., and Tarnopolsky, M., 1998, Oral creatine supplementation and athletic performance: a critical review. *Clin. J. Sport Med.* **8:** 286–297.

Koszalka, T.R., and Andrew, C.L., 1972, Effect of insulin on the uptake of creatine-1-14C by skeletal muscle in normal and X-irradiated rats. *Proc. Soc. Exp. Biol. Med.* **139:** 1265–1271.

Kreider, R.B., Ferreira, M., Wilson, M., Grindstaff, P., Plisk, S., Reinardy, J., Cantler, E., and Almada, A.L., 1998, Effects of creatine supplementation on body composition, strength, and sprint performance. *Med. Sci. Sports Exerc.* **30:** 73–82.

Louis, M., Poortmans, J.R., Francaux, M., Berre, J., Boisseau, N., Brassine, E., Cuthbertson, D.J., Smith, K., Babraj, J.A., Waddell, T., and Rennie, M.J., 2003a, No effect of creatine supplementation on human myofibrillar and sarcoplasmic protein synthesis after resistance exercise. *Am. J. Physiol. Endocrinol. Metab.* **285:** E1089–E1094.

Louis, M., Poortmans, J.R., Francaux, M., Hultman, E., Berre, J., Boisseau, N., Young, V.R., Smith, K., Meier-Augenstein, W., Babraj, J.A., Waddell, T., and Rennie, M.J., 2003b, Creatine supplementation has no effect on human muscle protein turnover at rest in the postabsorptive or fed states. *Am. J. Physiol. Endocrinol. Metab.* **284:** E764–E770.

Louis, M., Van Beneden, R., Dehoux, M., Thissen, J.P., and Francaux, M., 2004, Creatine increases IGF-I and myogenic regulatory factor mRNA in C_2C_{12} cells. *FEBS Lett.* **557:** 243–247.

McMillen, J., Donovan, C.M., Messer, J.I., and Willis, W.T., 2001, Energetic driving forces are maintained in resting rat skeletal muscle after dietary creatine supplementation. *J. Appl. Physiol.* **90:** 62–66.

Murphy, R., McConell, G., Cameron-Smith, D., Watt, K., Ackland, L., Walzel, B., Wallimann, T., and Snow, R., 2001, Creatine transporter protein content, localization, and gene expression in rat skeletal muscle. *Am. J. Physiol. Cell Physiol.* **280:** C415–C422.

Narahara, H.T., and Holloszy, J.O., 1974, The actions of insulin, trypsin, and electrical stimulation on amino acid transport in muscle. *J. Biol. Chem.* **249:** 5435–5443.

Nelson, A.G., Arnall, D.A., Kokkonen, J., Day, A.R., and Evans, J., 2001, Muscle glycogen supercompensation is enhanced by prior creatine supplementation. *Med. Sci. Sports Exerc.* **33:** 1096–1100.

Newman, J.E., Hargreaves, M., Garnham, A., and Snow, R.J., 2003, Effect of creatine ingestion on glucose tolerance and insulin sensitivity in men. *Med. Sci. Sports Exerc.* **35:** 69–74.

Nissen, S.L., and Sharp, R.L., 2003, Effect of dietary supplements on lean mass and strength gains with resistance exercise: a meta-analysis. *J. Appl. Physiol.* **94:** 651–659.

Olsen, S., Aagaard, P., Kadi, F., Tufekovic, G., Verney, J., Olesen, J.L., Suetta, C., and Kjaer, M., 2006, Creatine supplementation augments the increase in satellite cell and myonuclei number in human skeletal muscle induced by strength training. *J. Physiol.* **573:** 525–534.

Op 't Eijnde, B., Jijakli, H., Hespel, P., and Malaisse, W.J., 2006, Creatine supplementation increases soleus muscle creatine content and lowers the insulinogenic index in an animal model of inherited type 2 diabetes. *Int. J. Mol. Med.* **17:** 1077–1084.

Op 't Eijnde, B., Richter, E.A., Henquin, J.C., Kiens, B., and Hespel, P., 2001a, Effect of creatine supplementation on creatine and glycogen content in rat skeletal muscle. *Acta Physiol. Scand.* **171:** 169–176.

Op 't Eijnde, B., Urso, B., Richter, E.A., Greenhaff, P.L., and Hespel, P., 2001b, Effect of oral creatine supplementation on human muscle GLUT4 protein content after immobilization. *Diabetes* **50:** 18–23.

Parise, G., Mihic, S., MacLennan, D., Yarasheski, K.E., and Tarnopolsky, M.A., 2001, Effects of acute creatine monohydrate supplementation on leucine kinetics and mixed-muscle protein synthesis. *J. Appl. Physiol.* **91:** 1041–1047.

Pischel, I., and Gastner, T., 2007, Creatine – its chemical synthesis, chemistry, and legal status. *Subcell. Biochem.* **46:** 291–307.

Poortmans, J.R., and Francaux, M., 2000, Adverse effects of creatine supplementation: fact or fiction? *Sports Med.* **30:** 155–170.

Preen, D., Dawson, B., Goodman, C., Beilby, J., and Ching, S., 2003, Creatine supplementation: a comparison of loading and maintenance protocols on creatine uptake by human skeletal muscle. *Int. J. Sport Nutr. Exerc. Metab.* **13:** 97–111.

Preen, D., Dawson, B., Goodman, C., Lawrence, S., Beilby, J., and Ching, S., 2002, Pre-exercise oral creatine ingestion does not improve prolonged intermittent sprint exercise in humans. *J. Sports Med. Phys. Fitness* **42:** 320–329.

Robinson, T.M., Sewell, D.A., Hultman, E., and Greenhaff, P.L., 1999, Role of submaximal exercise in promoting creatine and glycogen accumulation in human skeletal muscle. *J. Appl. Physiol.* **87:** 598–604.

Schroeder, C., Potteiger, J., Randall, J., Jacobsen, D., Magee, L., Benedict, S., and Hulver, M., 2001, The effects of creatine dietary supplementation on anterior compartment pressure in the lower leg during rest and following exercise. *Clin. J. Sport Med.* **11:** 87–95.

Snow, R.J., and Murphy, R.M., 2003, Factors influencing creatine loading into human skeletal muscle. *Exerc. Sport Sci. Rev.* **31:** 154–158.

Speer, O., Neukomm, L.J., Murphy, R.M., Zanolla, E., Schlattner, U., Henry, H., Snow, R.J., and Wallimann, T., 2004, Creatine transporters: a reappraisal, *Mol. Cell. Biochem.* **256–257:** 407–424.

Spriet, L.L., 1995, Anaerobic metabolism during high-intensity exercise. In: Hargreaves, M., ed., Exercise Metabolism, Human Kinetics, Champaign, IL, USA, pp. 1–40.

Steenge, G.R., Simpson, E.J., and Greenhaff, P.L., 2000, Protein- and carbohydrate-induced augmentation of whole body creatine retention in humans. *J. Appl. Physiol.* **89:** 1165–1171.

Terjung, R.L., Clarkson, P., Eichner, E.R., Greenhaff, P.L., Hespel, P.J., Israel, R.G., Kraemer, W.J., Meyer, R.A., Spriet, L.L., Tarnopolsky, M.A., Wagenmakers, A.J., and Williams, M.H., 2000, American College of Sports Medicine roundtable. The physiological and health effects of oral creatine supplementation. *Med. Sci. Sports Exerc.* **32:** 706–717.

Van Leemputte, M., Vandenberghe, K., and Hespel, P., 1999, Shortening of muscle relaxation time after creatine loading. *J. Appl. Physiol.* **86:** 840–844.

Vandebuerie, F., Vanden Eynde, B., Vandenberghe, K., and Hespel, P., 1998, Effect of creatine loading on endurance capacity and sprint power in cyclists. *Int. J. Sports Med.* **19:** 490–495.

Vandenberghe, K., Gillis, N., Van Leemputte, M., Van Hecke, P., Vanstapel, F., and Hespel, P., 1996, Caffeine counteracts the ergogenic action of muscle creatine loading. *J. Appl. Physiol.* **80:** 452–457.

Vandenberghe, K., Goris, M., Van Hecke, P., Van Leemputte, M., Vangerven, L., and Hespel, P., 1997, Long-term creatine intake is beneficial to muscle performance during resistance training. *J. Appl. Physiol.* **83:** 2055–2063.

Walker, J.B., 1979, Creatine: Biosynthesis, regulation and function. *Adv. Enzymol. Relat. Areas Mol. Biol.* **50:** 177–242.

Willott, C.A., Young, M.E., Leighton, B., Kemp, G.J., Boehm, E.A., Radda, G.K., and Clarke, K., 1999, Creatine uptake in isolated soleus muscle: kinetics and dependence on sodium, but not on insulin. *Acta Physiol. Scand.* **166:** 99–104.

Willoughby, D.S., and Rosene, J., 2001, Effects of oral creatine and resistance training on myosin heavy chain expression. *Med. Sci. Sports Exerc.* **33:** 1674–1681.

Willoughby, D.S., and Rosene, J.M., 2003, Effects of oral creatine and resistance training on myogenic regulatory factor expression. *Med. Sci. Sports Exerc.* **35:** 923–929.

Young, J.C., and Young, R.E., 2002, The effect of creatine supplementation on glucose uptake in rat skeletal muscle. *Life Sci.* **71:** 1731–1737.

Zheng, D., Maclean, P.S., Pohnert, S.C., Knight, J.B., Olson, A.L., Winder, W.W., and Dohm, G.L., 2001, Regulation of muscle GLUT-4 transcription by AMP-activated protein kinase. *J. Appl. Physiol.* **91:** 1073–1083.

Ziegenfuss, T.M., Lowery, L.M., and Lemon, P.W., 1998, Acute fluid volume changes in men during three days of creatine supplementation. *J. Exerc. Physiol.* **1:** 1–9.

CHAPTER 13

PHARMACOKINETICS OF CREATINE

WESLEY MCCALL AND ADAM M. PERSKY
Division of pharmacotherapy and Experimental Therapeutics, School of Pharmacy, University of North Carolina at Chapel Hill, Chapel Hill, NC 27599-7360, USA

Abstract: Research has demonstrated that creatine supplementation has some therapeutic benefit with respect to muscle function and more recently neurological function. Despite the growing body of literature on the pharmacologic effect of creatine, very little is known about the disposition of creatine after supraphysiologic doses. The movement of creatine throughout the body is governed by transport processes which impact the absorption of creatine from the intestine, clearance of creatine from the kidney, and access of creatine to target tissues. With repeated doses of creatine, it appears that the clearance of creatine decreases mainly due to the saturation of skeletal muscle stores. Insulin and insulin-stimulating foods appear to enhance muscle uptake of creatine but at the same time, high carbohydrate meals may slow the absorption of creatine from the intestine. Little is known about creatine disposition in special populations including the elderly and patients with neuromuscular disease. Knowledge of creatine disposition in these clinically relevant populations can help remove some of the guess work of dose selection during clinical trials

1. INTRODUCTION

Pharmacokinetics is the study of the behavior of a substance (e.g., drug) in an organism through quantifying the processes of absorption, distribution, metabolism and excretion. The fate of substances after their administration is important for the sole reason that it is the concentration of a substance at the site of action that drives physiologic, pharmacologic or toxicologic effects. Beyond basic knowledge about the time course of concentrations, pharmacokinetics can provide useful information for health-care professionals related to dose adjustments. Understanding pharmacokinetics allows clinicians to design dosing regimens to obtain desired drug concentrations, thus maximizing drug effectiveness while at the same time decreasing drug toxicity.

Though the positive physiologic effects of creatine supplementation have been well-studied, there is still much to learn regarding the pharmacokinetics of exogenous, supraphysiologic doses of creatine. Specifically, relatively little is

G.S. Salomons and M. Wyss (eds.), Creatine and Creatine Kinase in Health and Disease, 261–273.

known about the relationship between blood and tissue concentrations of creatine in intact organisms. Since creatine is known to have beneficial effects on tissues with high energy demands (e.g., skeletal muscle, nervous tissue, cardiac tissue), it is important to elucidate the pharmacokinetic profile of creatine at its sites of action. This information will give researchers new insights into how to optimize the use of creatine to treat specific diseases or unveil potential barriers of creatine as a therapeutic agent. In addition, further understanding of creatine pharmacokinetics will enable the design of optimal dosing regimens for varied patient populations.

2. ABSORPTION

Creatine can be obtained through an omnivorous diet. Because creatine is a nutrient, it is not surprising that the literature suggests creatine is absorbed from the gastrointestinal tract via a process similar to other nutrients (e.g., amino acids, glucose, vitamins). Transporters mediating creatine flux through the intestinal wall have been identified in rodents in the ileum (Peral *et al.*, 2005) and jejunum (Tosco *et al.*, 2004) and on the apical (Peral *et al.*, 2002; Tosco *et al.*, 2004) and basolateral membranes of enterocytes (Orsenigo *et al.*, 2005). Orsengio *et al.* (2005) also suggested possible paracellular movement of creatine by solvent drag as a mechanism of creatine absorption. However using the Caco-2 monolayer as a model of intestinal absorption, creatine showed poor apical to basolateral movement (Dash *et al.*, 2001), suggesting that the contribution of the paracellular route to the transcelluar route is minimal and that the Caco-2 cell may not express creatine transporters. One study modeled concentration-time data after oral doses of creatine and found that the model supported saturable absorption kinetics consistent with transport processes in the intestine (Persky *et al.*, 2003c). This study also seems to indicate potential two peaks in the concentration-time profile for a single dose of creatine (Figure 1); this

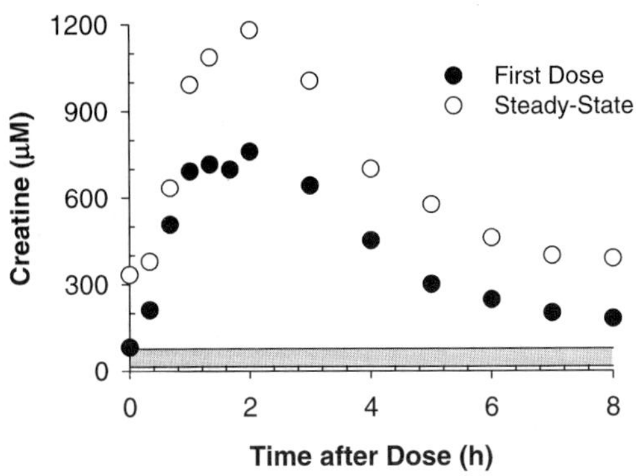

Figure 1. Concentration-time profile of creatine in the blood after administration of 5 g of creatine monohydrate to healthy volunteers. Closed circles represent single-dose concentrations. Open circles represent steady-state concentrations (5 g four times a day for 6 d). Grey box indicates the K_m of sarcolemmal creatine transporters (Snow and Murphy, 2001). Data taken from Persky *et al.* (2003c).

phenomenon has been be noted in another study (Green *et al.*, 1996b). Although the reason for the small double peak is unclear, it may suggest more than one absorptive process. Clearly, more research needs to be performed before the exact means of creatine absorption is determined, especially in humans as species differences may play a role in creatine disposition (Green *et al.*, 1996a; Kreider, 2003; Tarnopolsky *et al.*, 2003b).

Although it can be difficult to quantify absorption rate, the time to maximal concentration (t_{max}) and the maximal concentration, C_{max}, can be used as surrogate markers of absorption kinetics (Table 1). Although both t_{max} and C_{max} give some assessment of absorption processes, they are relatively inaccurate measures of absorption rate because they are secondary parameters dependent on other pharmacokinetic parameters (e.g., elimination rate, dose). Regardless of inaccuracies, these parameters can be used for relative comparisons as typically seen in bioequivalence studies.

2.1. Bioavailability

The absolute bioavailability (i.e., the fraction of dose absorbed from a non-intravenous route compared to an intravenous route) of supraphysiologic doses of creatine is unknown. There are four potential reasons why creatine bioavailability could be less than 100%. The first reason could be that creatine is susceptible to degradation in an acid environment with the highest rate of degradation at pH 3 (Cannan and Shore, 1928). Regardless, the combination of stomach pH (pH 1) and the relatively short time creatine might spend in the stomach means that very little of the oral dose of creatine should be lost. The second reason for less than 100% bioavailability could be that creatine is degraded by intestinal bacteria (Twort, and Mellanby, 1912). It has been noted that intestinal bacteria can degrade creatine and that creatine and its metabolite, creatinine, are lost in the feces (Wixom *et al.*, 1979).

Table 1. Pharmacokinetics after a single oral dose of creatine .

Dose (g)	t_{max}(hr)	C_{max}(μM)	Reference
2	0.5 to 2	180 to 390	Harris *et al.*, 2002
2.5	1	400	Schedel *et al.*, 1999
5	0.75 to 1.6	620 to 1300	Green *et al.*, 1996b; Harris *et al.*, 1992; Persky *et al.*, 2003c; Rawson *et al.*, 2002; Schedel *et al.*, 1999; Steenge *et al.*, 2000
10	2.25	1000	Schedel *et al.*, 1999
15	3	2100	Schedel *et al.*, 1999
20	3 to 4	2200	Schedel *et al.*, 1999

t_{max}: time to maximal concentration; C_{max}: maximal concentration.

The third reason could be that creatine may not be absorbed due to the kinetics of transport across the intestinal epithelia. If creatine absorption is limited to distinct anatomical locations along the intestine and absorption is a time consuming process, it is possible that flow down the intestinal tract pushes creatine past the transporters in the gastrointestinal tract. The efficiency of creatine absorption within the intestine then becomes a function of transit time through the intestine and the uptake clearance into enterocytes. The fourth reason for decreased bioavailability could be incomplete dissolution of creatine solid dosage forms (e.g., powder, tablet); only dissolved creatine will be absorbed, and thus if creatine does not enter solution it will not be absorbed.

Although there is no data to support the absolute bioavailability of creatine, it is most likely that lower doses of creatine (< 5 g) will have a larger bioavailability than higher doses (> 10 g) based on kinetics of a saturable process (i.e., Michaelis-Menten). Most studies that discuss the bioavailability of creatine base their discussion on urinary output of creatine. Although this is a fairly convenient measure of bioavailability, it is assuming 1) complete recovery of the creatine (and its degradation products) in the urine; 2) that there is no other loss of creatine within the system or all other loss is accounted for (i.e., mass balance) and 3) that endogenous creatine production and creatinine formation do not change. This mass balance approach to estimate bioavailability is at best a rough estimate because one or more of the three assumptions may be violated.

2.2. Impact of Dosage form on Absorption

Creatine may be consumed in several different forms beyond meat and fish. Creatine supplements sold over the counter are typically sold as solutions, powders, suspensions, capsules, or chewable tablets. One study compared the relative bioavailability of several of these dosage forms (Harris *et al.*, 2002). The authors found that ingesting 2 g of creatine, in different dosage forms, resulted in plasma peak concentration (C_{max}) where solution > suspension = lozenge > meat. There was little difference in time to maximal concentration (t_{max}) across dosage forms but the shape of the concentration-time profile suggested possible differences in absorption kinetics. It is not surprising that the solution yielded the shortest t_{max} and highest C_{max} as a molecule is required to be in solution prior to absorption; the solid dosage forms (e.g., lozenge, meat) would require disintegration and dissolution while suspensions would require dissolution of the suspended particles. Since solutions eliminate these other steps (i.e., degradation and dissolution), they have a more rapid rate of absorption. The authors also compared the area under the concentration-time curve (AUC), which can be helpful in determining bioequivalence and systemic exposure. They concluded that creatine in solution and meat had similar AUC to one another, but that the AUC of creatine solution was significantly higher than a lozenge formulation or suspension. This finding suggests that the solution delivers more creatine than either the lozenge or suspension. Though this study suggests not all formulations of creatine are equivalent, AUC was calculated only for measured

Table 2. Potential barriers to creatine absorption from the intestine .

Barrier	Potential Impact on Blood Concentrations
Low pH of the stomach	✓
Dissolution of solid dosage forms	✓
Movement across apical and basolateral membranes of epithelial cells	✓✓✓✓
Intracellular degradation in epithelial cells	✓
Degradation by intestinal bacteria	✓

✓ = minimal to ✓✓✓✓✓ substantial

time points; in order to fully establish bioequivalence, the AUC would have to be extrapolated through infinity in order to fully define systemic exposure. Therefore it is difficult to say which formulation of creatine delivers the largest fraction of the administered dose.

2.3. Impact of Food on Absorption

Research suggests that ingesting carbohydrates, as simple sugars or as a carbohydrate/protein combination, enhances muscle uptake of creatine. These findings will be discussed later in the clearance section, but ingestion of sugars and protein can impact the absorption of creatine. Although relatively low concentrations of sugar show no negative effect on absorption (< 4.5%), high sugar concentrations can slow absorption (Vist and Maughan, 1995). Studies examining the concomitant intake of sugars or sugar and protein with creatine have found slight increases in t_{max} by up to 40 min (Green *et al.*, 1996b; Steenge *et al.*, 2000). The use of carbohydrates might also impact possible gastrointestinal side effects if those carbohydrates draw water into the lumen or delay gastric emptying (e.g., high fructose fruit juices). It is still not fully clear if food impacts the oral bioavailability (i.e., fraction of the dose absorbed) or absorption kinetics of creatine and whether those changes significantly impact systemic concentrations to ultimately affect target tissue concentrations. Table 2 summarizes potential barriers of creatine absorption.

3. DISTRIBUTION

Creatine is distributed throughout the body in tissues such as the brain, eyes, cardiac muscle, testes, and kidneys (Walker, 1979; Wyss and Schulze, 2002). The primary site of creatine distribution is skeletal muscle, where greater than 95% of creatine is located (Walker, 1979). The volume of distribution of creatine approaches that of total body water (approximately 45 L) (Persky *et al.*, 2003a). Distribution of molecules throughout the body is, in part, determined by the binding affinity of plasma proteins. The binding of creatine to plasma protein in humans is insignificant, at less than 10% (Persky *et al.*, 2003b). Thus a large unbound fraction of creatine is free to distribute into tissues and is available as substrate for the creatine transporter that allows creatine to access tissues. Table 3 summarizes potential barriers to creatine entry into the biophase of pharmacologic tissue targets.

Skeletal muscle is the most studied tissue with respect to creatine disposition. Interestingly, the ability of skeletal muscle to accumulate creatine is finite; that is, it is possible to saturate muscle stores of creatine (Harris *et al.*, 1992). One reason that skeletal muscle may become saturated with creatine is possible down-regulation of either creatine transporter number or function. To date there is little evidence to support either conclusion (Tarnopolsky *et al.*, 2003a). Further discussion of the creatine transporter and its regulation can be found in chapter 6 (Christie, 2007).

Brain has become an increasingly popular tissue with respect to creatine disposition in the investigation of neurological disease. To date there is little evidence that brain creatine accumulation is saturable like skeletal muscle accumulation. Several studies have examined brain concentrations in Huntington's patients. One study found that patients receiving creatine (8 g/d for 16 weeks) had a 7.5 to 13% increase in brain creatine depending on brain region (Hersch *et al.*, 2006). A second study found an ~8% increase in brain creatine after 6 months of 10 g/d creatine (Tabrizi *et al.*, 2003). However, a third study in this patient population found no increase in brain creatine when patients received 20 g/d for 5 d followed by 6 g/d for 8 to 10 weeks (Bender *et al.*, 2005). In athletes receiving 20 g/d for 5 days, creatine supplementation did not appear to change brain concentrations of creatine (Wilkinson *et al.*, 2006) compared to individuals given 20 g/d for 5–7 days followed by 2 g/d for 7 d. This latter supplementation protocol (i.e., 20 g/d for 5–7 d followed by 2 g/d for 7 d) resulted in a 8-9% increase in brain creatine and an ~4% increase in brain phosphocreatine (Lyoo *et al.*, 2003). Although creatine distributes into the brain, similar doses that raise muscle creatine >20% tend to only increase brain creatine by <10%. This difference between brain and muscle uptake of creatine may be a reflection of a lower intrinsic clearance (i.e., V_{max}/K_m) for creatine transport into the brain via the blood-brain barrier compared to muscle.

Table 3. Potential barriers to creatine uptake into the biophase (i.e., pharmacologic effect compartment) and pharmacologic effect .

Barrier	**Potential Impact on Biophase Concentrations and Pharmacologic Effect**
Transport across the plasma membrane	✓✓✓✓✓
Transport into the mitochondria	✓✓✓✓
Conversion to phosphocreatine	✓✓✓
Intracellular degradation	✓✓
Phosphate pool	✓✓

✓ = minimal to ✓✓✓✓✓ substantial

4. CLEARANCE

Clearance is defined as the volume of a reference fluid (usually blood) irreversibly removed of a compound (e.g., drug) per unit time. In healthy individuals, creatine is irreversibly removed from the blood by both skeletal muscle and kidney. Muscle is typically not thought of as a clearing organ for drugs but because creatine is trapped and utilized in skeletal muscle of healthy individuals, it can be thought of as a clearance process; this may not be the case in patients with muscle disease. Early studies have shown that patients with muscle disease have ineffective trapping of creatine, leading to creatine efflux back into circulation (Fitch *et al.*, 1968). Creatine is irreversibly trapped in muscle because its polar nature prevents passive efflux back into circulation. Within the muscle, creatine and phosphocreatine spontaneously degrade into creatinine with an approximate half-life of 40 days, at physiologic pH. It has been previously hypothesized that muscle clearance is around 17 L/h using the well-stirred model of organ clearance (Persky *et al.*, 2003a,c); similar values can be estimated using the parallel-tube model derived by Pang and Rowland (1977). This latter model might be more appropriate for skeletal muscle because of the anatomical structure of muscle relative to blood flow. Creatine clearance by muscle is probably affected by the same factors that may affect the creatine transporter, which include insulin, IGF-1, catecholamines, exercise, and intracellular creatine levels (see chapter 6; Christie, 2007). The effects of insulin on creatine uptake by skeletal muscle have been studied in humans both by direct administration of insulin (Steenge *et al.*, 1998) and by insulin-stimulating foods such as carbohydrates and certain proteins/amino acids (Green *et al.*, 1996a,b; Steenge *et al.*, 2000). It should be noted that fructose tends to have a lower insulin response than glucose (Tappy *et al.*, 1986; Truswell, 1992) and thus fruit juices, in theory, are not as recommended as glucose containing products for enhancing creatine uptake through insulin stimulation. In addition, the amount of muscle mass may affect creatine clearance (Persky *et al.*, 2003a). Larger muscle mass likely correlates to a greater number of transporters and more storage area for creatine, suggesting that when dosing creatine, it may be more appropriate to scale the dose to bodyweight, ideal bodyweight, or lean body mass.

Creatine is also eliminated from the body by the kidney. At first it was thought that creatine underwent renal clearance at rates equivalent to glomerular filtration rate (GFR) (Pitts, 1934). However, later evidence revealed that creatine is reabsorbed in the kidney. In addition, the presence of creatine transporter in the kidney further supports creatine reabsorption from urine (Wyss and Schulze, 2002), assuming that the creatine transporter is effluxing creatine from the renal tubule back into systemic circulation. Studies looking at renal clearance of creatine have shown varied values. Poortmans and co-workers (Poortmans *et al.*, 1997, 2005; Poortmans and Francaux, 1999) reported renal clearance of creatine under unsupplemented conditions of 0.3–0.8 L/h, much less than GFR ($\sim$ 7.0 L/h) which again supports the reabsorption of creatine in the kidney. Under conditions of supplementation, renal clearance of creatine increased to 9–22 L/h; values higher than GFR would imply secretion of creatine from the blood into the renal tubules. Vandenberghe *et al.* (1997) reported creatine excretion rate under unsupplemented conditions of 0.038 g/day, compared to 3.6 g/day after supplementing with 5 g of creatine per day for 10 weeks. Unfortunately, steady-state blood levels were not assessed to calculate renal clearance in this study. Most studies estimate the renal clearance of creatine from 24 h urine collection but from a data analysis standpoint, smaller windows of urine collection and blood sampling would be needed for a more accurate estimate of renal clearance. Recommended collection windows should be less than a half-life of the compound. In any case, the evidence suggests that creatine is reabsorbed in the kidney which causes dose-dependent renal clearance, and it is possible that there is a secretory mechanism for creatine as well. As blood concentrations increase and more creatine is filtered, less reabsorption occurs and a greater percentage of creatine will be lost in the urine.

Few studies have estimated systemic clearance of creatine. After a single dose of creatine in healthy volunteers, the apparent oral clearance was 14 L/h which is close to the predicted contribution of skeletal muscle (Persky *et al.*, 2003c). At steady-state, however, clearance was estimated to decrease to around 7 L/h, which is close to the prediction of renal clearance (Persky *et al.*, 2003c). The non-stationary behavior, i.e., parameters governing creatine pharmacokinetics change over time, suggests that as skeletal muscle approaches its capacity to store creatine, the kidney and possibly other tissues are responsible for the removal of creatine from the blood. This decrease in clearance over time should translate to smaller doses over time. For example, during early doses (i.e., doses within the first one to three days) when clearance is high, doses of 10 to 15 g per day will give blood concentrations greater than the K_m for the creatine transporter. As the muscle becomes saturated and clearance decreases, it may be necessary to ingest 3 to 5 g of creatine a day to maintain similar blood concentrations. Figure 1 represents plasma concentrations of creatine after a single dose and after steady-state has been reached for a 5 g dose of creatine monohydrate in healthy volunteers.

As mentioned, both muscle and the kidney contribute to the systemic removal of creatine. Exactly how much the kidneys and muscle contribute to creatine clearance is probably dependent on dose and dose frequency (Persky *et al.*, 2003a). As a person continues to take creatine, muscle stores become saturated and clearance shifts from muscle uptake to primarily renal elimination.

5. INTRACELLULAR PHARMACOKINETICS

As stated earlier in the chapter, it is the concentration of drug at the site of action that drives pharmacologic response. For creatine, the main site of action is thought to be skeletal muscle and, more recently, the brain. Within the muscle cell, creatine may potentially exert its pharmacologic effects in the cytosol or mitochondria. Creatine's pharmacologic effects are mainly due to the formation of phosphocreatine. If not converted to phosphocreatine, it is unlikely that creatine supplementation would be of benefit. There are three rate-determining steps to intracellular kinetics: entry into the cytosol through the membrane-bound creatine transporter, entry into the mitochondria perhaps via a creatine transporter, and the creatine kinase reaction.

Previous chapters have focused on the creatine transporter but some discussion on the pharmacokinetic implications is warranted. In humans, the K_m for plasma membrane transport is on the order of 20 to 100 μM. Under normal conditions plasma concentrations are 25 to 50 μM, which is in the range of the K_m of the transporter. The kinetics of the system keeps interstitial concentrations at approximately 40% of plasma concentrations (Persky *et al.*, 2003c). Once in the cell, creatine is taken up into mitochondria through a creatine transporter, although the evidence for this is limited (Speer *et al.*, 2004). Finally the creatine kinase reaction is at equilibrium under normal conditions with approximately two-thirds of cytosolic creatine being in the phosphorylated form (Meyer *et al.*, 1984). The kinetics of the intracellular system have been previously described [see (Meyer *et al.*, 1984; Wallimann *et al.*, 1992)] but these studies did not take into account possible saturable processes of creatine entry into the cell or into the mitochondria. Additionally, the model did not examine the impact of increased plasma creatine concentrations seen under supplementation conditions. The kinetics of the intracellular system may dictate which individuals or populations respond to creatine therapy.

6. CREATINE DISPOSITION IN SPECIAL POPULATIONS

Historically, creatine research has focused on improving exercise performance in young, healthy males and females. However, accruing evidence shows that creatine may have a therapeutic role for the elderly or in individuals with certain diseases – especially muscular or neurological-based disease. Despite the potential therapeutic effect, not much is known regarding creatine disposition in special populations.

Rawson *et al.* (2002) showed that after a 5 g dose of creatine in young and elderly men, there was no increase in intramuscular phosphocreatine levels in the geriatric group despite similar plasma exposure and urinary excretion. This may be due to lack of conversion of creatine to phosphocreatine in either the cytosol or mitochondria. It is also possible that there is large variability in creatine uptake in the elderly but this would potentially lead to differences both in blood concentrations (higher in elderly) and muscle concentrations (lower in elderly). On the other hand, Brose *et al.* (2003) found that phosphocreatine did increase in the elderly when resistance training was concomitant with creatine ingestion; however no plasma data was available. Exercise has been shown to increase mitochondrial and creatine kinase content (Menshikova *et al.*, 2006). Thus, in the absence of exercise, elderly muscle may not respond to creatine supplementation due to lower mitochondrial function or creatine kinase activity.

Beyond its proposed use in elderly individuals, creatine supplementation is also being investigated as a therapeutic option for certain diseases. The implications of disturbances in creatine metabolism have been previously reviewed (Wyss and Schulze, 2002) and are outlined elsewhere in this book. Focusing specifically on the pharmacokinetic implications of creatine in disease, it has been previously shown that hypercreatinemia and creatinuria are present in patients with muscular dystrophy; in addition, these patients show lower muscle levels of creatine and phosphocreatine (Fitch *et al.*, 1968). Similarly, patients with heart failure and myopathies show diminished expression of the creatine transporter, CreaT1, as well as lower muscle concentrations of creatine and phosphocreatine (Neubauer *et al.*, 1999; Tarnopolsky *et al.*, 2001). Fitch *et al.* (1968) previously proposed that the deficits in creatine and phosphocreatine levels were caused by either ineffective trapping of creatine in muscle or by lack of uptake. Despite recent studies evaluating creatine supplementation in muscular dystrophy, ALS and other neurological diseases, no formal pharmacokinetic study has been conducted in these patient populations.

7. FUTURE DIRECTIONS

Pharmacokinetic research on creatine is important to elucidate the reasons and mechanisms for the differences between individuals and populations. As research focuses on clinical application it is important to understand the disposition of creatine so predictions can be made 1) with respect to optimizing dosing regimens to obtain therapeutic tissue concentrations and 2) on the impact of disease on those tissue concentrations. While studies have looked at regulation of creatine turnover under normal conditions, it is necessary to investigate the systemic regulation of creatine turnover when supraphysiologic doses are administered. Studies using labeled creatine can be useful in this area to differentiate endogenous from exogenous creatine. Understanding the impact of tissue and blood concentrations on the uptake of creatine into target tissues (e.g., muscle, brain) *in vivo* in humans would lead to insight on how best to dose creatine to achieve optimal therapeutic effects.

8. SUMMARY

Creatine is absorbed from the gastrointestinal tract. Regardless of dosage form, creatine exhibits rapid absorption ($t_{max} < 2$ h); however, consuming very high doses of creatine results in a prolonged t_{max}. Ingestion of high amounts of carbohydrates appears to delay the time to peak concentration but, at the same time, decreases C_{max}. The absolute bioavailability is unknown for creatine but the fraction of creatine absorbed may be dictated by the creatine transporter kinetics. Although creatine is found in several metabolically active tissues, skeletal muscle is the major site of accumulation. Movement of creatine into tissue is governed by a transport process and regulation of this transport process is not fully elucidated to date. Removal of creatine from systemic circulation is governed by irreversible uptake into skeletal muscle and filtration by the kidney. Clearance of creatine appears to be nonstationary with clearance decreasing with repeated dosing. There is very little data differentiating the disposition of supraphysiologic doses of creatine between healthy adults and patient or special populations (e.g., the elderly). The disposition of creatine may differ in these clinically relevant populations due to changes in total body water, renal function and skeletal muscle function.

9. RECOMMENDATIONS

Based on clearance of creatine during the initial days of supplementation, ingestion of 5 g three times a day (15 g/d) will maintain concentrations well above the K_m for the plasma membrane transporter throughout the day. After two days of the loading phase, muscle stores will approach saturation, resulting in an increased fraction of the dose being lost in the urine. Thus, after a short loading phase, doses can be reduced to 3 to 5 g/d to compensate for the reduction in clearance of creatine and to compensate for the loss of creatine via creatinine formation which is on average 2 g/d. Loading doses of creatine are probably best taken with glucose-based foods to maximize muscle uptake; maintenance doses may not require the carbohydrate/insulin stimulating component. For the elderly, physical activity might be an important parameter in increasing creatine uptake into muscle as exercise will improve the mitochondrial function within muscle. For patients requiring increased brain concentrations, doses of 5 to 10 grams for extended periods of time (e.g, >10 weeks) after an initial loading phase (e.g., 5 g three times a day for 2 to 5 days) might be necessary to attain the increase in brain creatine; despite this extended period of dosing, brain concentrations may only increase by 10%. There is little data to evaluate whether brain uptake of creatine changes with repeated dosing nor if higher doses are more effective; however, higher doses for extended periods of time may increase the risk of adverse events.

REFERENCES

Bender, A., Auer, D.P., Merl, T., Reilmann, R., Saemann, P., Yassouridis, A., Bender, J., Weindl, A., Dose, M., Gasser, T., and Klopstock, T., 2005, Creatine supplementation lowers brain glutamate levels in Huntington's disease. *J. Neurol.* **252:** 36–41.

Brose, A., Parise, G., and Tarnopolsky, M.A., 2003, Creatine supplementation enhances isometric strength and body composition improvements following strength exercise training in older adults. *J. Gerontol. A Biol. Sci. Med. Sci.* **58:** 11–19.

Cannan, R.K., and Shore, A., 1928, The creatine-creatinine equilibrium. The apparent dissociation constants of creatine and creatinine. *Biochem. J.* **22:** 920–929.

Christie, D.L., 2007, Functional insights into the creatine transporter. *Subcell. Biochem.* **46:** 99–118.

Dash, A.K., Miller, D.W., Huai-Yan, H., Carnazzo, J., and Stout, J.R., 2001, Evaluation of creatine transport using Caco-2 monolayers as an in vitro model for intestinal absorption. *J. Pharm. Sci.* **90:** 1593–1598.

Fitch, C.D., Lucy, D.D., Bornhofen, J.H., and Dalrymple, G.V., 1968, Creatine metabolism in skeletal muscle. II. Creatine kinetics in man. *Neurology* **18:** 32–42.

Green, A.L., Hultman, E., Macdonald, I.A., Sewell, D.A., and Greenhaff, P.L., 1996a, Carbohydrate ingestion augments skeletal muscle creatine accumulation during creatine supplementation in humans. *Am. J. Physiol.* **271:** E821–E826.

Green, A.L., Simpson, E.J., Littlewood, J.J., Macdonald, I.A., and Greenhaff, P.L., 1996b, Carbohydrate ingestion augments creatine retention during creatine feeding in humans. *Acta Physiol. Scand.* **158:** 195–202.

Harris, R.C., Nevill, M., Harris, D.B., Fallowfield, J.L., Bogdanis, G.C., and Wise, J.A., 2002, Absorption of creatine supplied as a drink, in meat or in solid form. *J. Sports Sci.* **20:** 147–151.

Harris, R.C., Soderlund, K., and Hultman, E., 1992, Elevation of creatine in resting and exercised muscle of normal subjects by creatine supplementation. *Clin. Sci. (London)* **83:** 367–374.

Hersch, S.M., Gevorkian, S., Marder, K., Moskowitz, C., Feigin, A., Cox, M., Como, P., Zimmerman, C., Lin, M., Zhang, L., Ulug, A.M., Beal, M.F., Matson, W., Bogdanov, M., Ebbel, E., Zaleta, A., Kaneko, Y., Jenkins, B., Hevelone, N., Zhang, H., Yu, H., Schoenfeld, D., Ferrante, R., and Rosas, H.D., 2006, Creatine in Huntington disease is safe, tolerable, bioavailable in brain and reduces serum 8OH2'dG. *Neurology* **66:** 250–252.

Kreider, R.B., 2003, Species-specific responses to creatine supplementation. *Am. J. Physiol. Regul. Integr. Comp. Physiol.* **285:** R725–R726.

Lyoo, I.K., Kong, S.W., Sung, S.M., Hirashima, F., Parow, A., Hennen, J., Cohen, B.M., and Renshaw, P.F., 2003, Multinuclear magnetic resonance spectroscopy of high-energy phosphate metabolites in human brain following oral supplementation of creatine-monohydrate. *Psychiatry Res.* **123:** 87–100.

Menshikova, E.V., Ritov, V.B., Fairfull, L., Ferrell, R.E., Kelley, D.E., and Goodpaster, B.H., 2006, Effects of exercise on mitochondrial content and function in aging human skeletal muscle. *J. Gerontol. A Biol. Sci. Med. Sci.* **61:** 534–540.

Meyer, R.A., Sweeney, H.L., and Kushmerick, M.J., 1984, A simple analysis of the "phosphocreatine shuttle". *Am. J. Physiol.* **246:** C365–377.

Neubauer, S., Remkes, H., Spindler, M., Horn, M., Wiesmann, F., Prestle, J., Walzel, B., Ertl, G., Hasenfuss, G., and Wallimann, T., 1999, Downregulation of the Na^+-creatine cotransporter in failing human myocardium and in experimental heart failure. *Circulation* **100:** 1847–1850.

Orsenigo, M.N., Faelli, A., De Biasi, S., Sironi, C., Laforenza, U., Paulmichl, M., and Tosco, M., 2005, Jejunal creatine absorption: what is the role of the basolateral membrane? *J. Membr. Biol.* **207:** 183–195.

Pang, K.S., and Rowland, M., 1977, Hepatic clearance of drugs. I. Theoretical considerations of a "well-stirred" model and a "parallel tube" model. Influence of hepatic blood flow, plasma and blood cell binding, and the hepatocellular enzymatic activity on hepatic drug clearance. *J. Pharmacokinet. Biopharm.* **5:** 625–653.

Peral, M.J., Galvez, M., Soria, M.L., and Ilundain, A.A., 2005, Developmental decrease in rat small intestinal creatine uptake. *Mech. Ageing Dev.* **126:** 523–530.

Peral, M.J., Gar cia-Delgado, M., Calonge, M.L., Duran, J.M., De La Horra, M.C., Wallimann, T., Speer, O., and Ilundain, A., 2002, Human, rat and chicken small intestinal Na^+-Cl^--creatine transporter: functional, molecular characterization and localization. *J. Physiol.* **545:** 133–144.

Persky, A.M., Brazeau, G.A., and Hochhaus, G., 2003a, Pharmacokinetics of the dietary supplement creatine. *Clin. Pharmacokinet.* **42:** 557–574.

Persky, A.M., Hochhaus, G., and Brazeau, G.A., 2003b, Validation of a simple liquid chromatography assay for creatine suitable for pharmacokinetic applications, determination of plasma protein binding and verification of percent labeled claim of various creatine products. *J. Chromatogr. B Analyt. Technol. Biomed. Life Sci.* **794:** 157–165.

Persky, A.M., Muller, M., Derendorf, H., Grant, M., Brazeau, G.A., and Hochhaus, G., 2003c, Single- and multiple-dose pharmacokinetics of oral creatine. *J. Clin. Pharmacol.* **43:** 29–37.

Pitts, R., 1934, The clearance of creatine in dog and man. *Am. J. Physiol.* **109:** 532–541.

Poortmans, J.R., Auquier, H., Renaut, V., Durussel, A., Saugy, M., and Brisson, G.R., 1997, Effect of short-term creatine supplementation on renal responses in men. *Eur. J. Appl. Physiol. Occup. Physiol.* **76:** 566–567.

Poortmans, J.R., and Francaux, M., 1999, Long-term oral creatine supplementation does not impair renal function in healthy athletes. *Med. Sci. Sports Exerc.* **31:** 1108–1110.

Poortmans, J.R., Kumps, A., Duez, P., Fofonka, A., Carpentier, A., and Francaux, M., 2005, Effect of oral creatine supplementation on urinary methylamine, formaldehyde, and formate. *Med. Sci. Sports Exerc.* **37:** 1717–1720.

Rawson, E.S., Clarkson, P.M., Price, T.B., and Miles, M.P., 2002, Differential response of muscle phosphocreatine to creatine supplementation in young and old subjects. *Acta Physiol. Scand.* **174:** 57–65.

Schedel, J.M., Tanaka, H., Kiyonaga, A., Shindo, M., and Schutz, Y., 1999, Acute creatine ingestion in human: consequences on serum creatine and creatinine concentrations. *Life Sci.* **65:** 2463–2470.

Snow, R.J., and Murphy, R.M., 2001, Creatine and the creatine transporter: a review. *Mol. Cell. Biochem.* **224:** 169–181.

Speer, O., Neukomm, L.J., Murphy, R.M., Zanolla, E., Schlattner, U., Henry, H., Snow, R.J., and Wallimann, T., 2004, Creatine transporters: a reappraisal. *Mol. Cell. Biochem.* **256–257:** 407–424.

Steenge, G.R., Lambourne, J., Casey, A., Macdonald, I.A., and Greenhaff, P.L., 1998, Stimulatory effect of insulin on creatine accumulation in human skeletal muscle. *Am. J. Physiol.* **275:** E974–979.

Steenge, G.R., Simpson, E.J., and Greenhaff, P.L., 2000, Protein- and carbohydrate-induced augmentation of whole body creatine retention in humans. *J. Appl. Physiol.* **89:** 1165–1171.

Tabrizi, S.J., Blamire, A.M., Manners, D.N., Rajagopalan, B., Styles, P., Schapira, A.H., and Warner, T.T., 2003, Creatine therapy for Huntington's disease: clinical and MRS findings in a 1-year pilot study. *Neurology* **61:** 141–142.

Tappy, L., Randin, J.P., Felber, J.P., Chiolero, R., Simonson, D.C., Jequier, E., and DeFronzo, R.A., 1986, Comparison of thermogenic effect of fructose and glucose in normal humans. *Am. J. Physiol.* **250:** E718–724.

Tarnopolsky, M., Parise, G., Fu, M.H., Brose, A., Parshad, A., Speer, O., and Wallimann, T., 2003a, Acute and moderate-term creatine monohydrate supplementation does not affect creatine transporter mRNA or protein content in either young or elderly humans. *Mol. Cell. Biochem.* **244:** 159–166.

Tarnopolsky, M.A., Bourgeois, J.M., Snow, R., Keys, S., Roy, B.D., Kwiecien, J.M., and Turnbull, J., 2003b, Histological assessment of intermediate- and long-term creatine monohydrate supplementation in mice and rats. *Am. J. Physiol. Regul. Integr. Comp. Physiol.* **285:** R762–769.

Tarnopolsky, M.A., Parshad, A., Walzel, B., Schlattner, U., and Wallimann, T., 2001, Creatine transporter and mitochondrial creatine kinase protein content in myopathies. *Muscle Nerve* **24:** 682–688.

Tosco, M., Faelli, A., Sironi, C., Gastaldi, G., and Orsenigo, M.N., 2004, A creatine transporter is operative at the brush border level of the rat jejunal enterocyte. *J. Membr. Biol.* **202:** 85–95.

Truswell, A.S., 1992, Glycaemic index of foods. *Eur. J. Clin. Nutr.* **46 Suppl 2:** S91–S101.

Twort, F.W., and Mellanby, E., 1912, On creatin-destroying Bacilli in the intestine and their isolation. *J. Physiol.* **44:** 43–49.

Vandenberghe, K., Goris, M., Van Hecke, P., Van Leemputte, M., Vangerven, L., and Hespel, P., 1997, Long-term creatine intake is beneficial to muscle performance during resistance training. *J. Appl. Physiol.* **83:** 2055–2063.

Vist, G.E., and Maughan, R.J., 1995, The effect of osmolality and carbohydrate content on the rate of gastric emptying of liquids in man. *J. Physiol.* **486:** 523–531.

Walker, J.B., 1979, Creatine: biosynthesis, regulation, and function. *Adv. Enzymol. Relat. Areas Mol. Biol.* **50:** 177–242.

Wallimann, T., Wyss, M., Brdiczka, D., Nicolay, K., and Eppenberger, H.M., 1992, Intracellular compartmentation, structure and function of creatine kinase isoenzymes in tissues with high and fluctuating energy demands: the 'phosphocreatine circuit' for cellular energy homeostasis. *Biochem. J.* **281:** 21–40.

Wilkinson, I.D., Mitchel, N., Breivik, S., Greenwood, P., Griffiths, P.D., Winter, E.M., and Van Beek, E.J., 2006, Effects of creatine supplementation on cerebral white matter in competitive sportsmen. *Clin. J. Sport Med.* **16:** 63–67.

Wixom, R.L., Davis, G.E., Flynn, M.A., Tsutakawa, R.T., and Hentges, D.J., 1979, Excretion of creatine and creatinine in feces of man. *Proc. Soc. Exp. Biol. Med.* **161:** 452–457.

Wyss, M., and Schulze, A., 2002, Health implications of creatine: can oral creatine supplementation protect against neurological and atherosclerotic disease? *Neuroscience* **112:** 243–260.

CHAPTER 14

SAFETY OF CREATINE SUPPLEMENTATION

ADAM M. PERSKY[1] AND ERIC S. RAWSON[2]

[1]*Division of Pharmacotherapy and Experimental Therapeutics, University of North Carolina at Chapel Hill, Chapel Hill, NC 27599-7360, USA*

[2]*Department of Exercise Science and Athletics, Bloomsburg University, Bloomsburg, PA 17815, USA*

Abstract: The literature on creatine supplementation supporting its efficacy has grown rapidly and has included studies in both healthy volunteers and patient populations. However, the first rule in the development of therapeutic agents is safety. Creatine is well-tolerated in most individuals in short-term studies. However, isolated reports suggest creatine may be associated with various side effects affecting several organ systems including skeletal muscle, the kidney and the gastrointestinal tract. The majority of clinical studies fail to find an increased incidence of side effects with creatine supplementation. To date, studies have not found clinically significant deviations from normal values in renal, hepatic, cardiac or muscle function. Few data are available on the long-term consequences of creatine supplementation

1. INTRODUCTION

Creatine is an endogenous molecule whose primary role is to act as an 'energy buffer'. Supraphysiologic doses of creatine have been shown to increase muscle stores of creatine and phosphocreatine, enhance muscle strength, reduce fatigue during exercise and improve exercise performance [see Branch (2003) for a meta-analysis]. Furthermore, there is a growing body of literature on pre-clinical and clinical evidence that creatine has a positive therapeutic outcome in various neurological or muscular diseases.

While creatine is well tolerated by most individuals in short-term studies, anecdotal reports and a small number of case-reports suggest that creatine may be associated with various side effects ranging from muscle cramping and gastrointestinal discomfort to renal dysfunction. Safety, whether it is a drug, food or supplement, does not imply the product is harmless, but that the therapeutic (or nutritional) benefits outweigh the risks of side effects for the intended population. As an example, the cholesterol-lowering agent Baycol (cerivastatin) was as effective

G.S. Salomons and M. Wyss (eds.), Creatine and Creatine Kinase in Health and Disease, 275–289.

as other statin medications in lowering cholesterol; however, cerivastatin caused severe muscle damage prompting its removal from the market in August 2001 (Charatan, 2001). Adverse events such as severe muscle damage are unacceptable for drug classes for non-life threatening diseases. This risk to benefit ratio helps classify compounds into categories of high therapeutic index (e.g., most over-the-counter medications) or low therapeutic index (e.g., medications that are routinely monitored such as warfarin or lithium). Creatine would be clinically useful if the desired therapeutic effects (e.g., increased muscle strength, reduced fatigue) outweighed the undesired pharmacologic/toxicologic effects (e.g., muscle cramping).

The purpose of this chapter is to summarize the evidence related to the safety of creatine supplements. Although there are safety data available from studies of creatine supplementation in animals, experimental data from animal models may not be directly applicable to the effects of creatine in humans. For instance, Green *et al.* (1996b) reported no increase in muscle creatine content in rats ingesting creatine in amounts equivalent to the dosages used in human studies. Additionally, creatine uptake is significantly enhanced in the presence of insulin in human skeletal muscle (Green *et al.*, 1996a,b; Preen *et al.*, 2003; Robinson *et al.*, 2000), but to a much lesser extent in rat skeletal muscle (Koszalka *et al.*, 1972). Most recently, species differences (rats vs. mice) in the response to creatine supplementation have been reported (Kreider, 2003; Tarnopolsky *et al.*, 2003). With that in mind, this chapter focuses on data regarding the safety of creatine supplementation from clinical trials and case studies.

2. SAFETY FINDINGS

2.1. Muscle Dysfunction

Creatine supplementation has been associated with increased muscle dysfunction (i.e. cramps, muscle strains, etc.) in the popular media. Increased muscle creatine content subsequent to creatine supplementation is associated with increased total body water (Powers *et al.*, 2003) and increased compartment pressure (Hile *et al.*, 2006). This resulted in the speculation that creatine supplementation could cause muscle dysfunction. In theory, increased muscle phosphocreatine levels resulting from creatine supplementation may reduce muscle dysfunction, as it is known that exogenous phosphocreatine reduces muscle damage in cardiac tissue as evidenced by decreased efflux of cardiac muscle proteins into the blood (Saks *et al.*, 1996; Saks and Strumia, 1993). In fact, phosphocreatine is used as a cardio-protective agent during heart surgery and to reduce infarct size after myocardial infarction (Saks *et al.*, 1996; Saks and Strumia, 1993). The effects of creatine supplementation on mild and severe indices of skeletal muscle dysfunction have been studied in cross-sectional studies, clinical trials, and case studies, as outlined below.

2.1.1. *Cramping*

Muscle cramping in creatine users has been reported by some groups. However, placebo or non-creatine user groups were not included for comparison, so that the relationship between creatine and muscle cramps cannot be determined from these studies (Greenwood *et al.*, 2000; Juhn *et al.*, 1999). A series of open label studies by Greenwood *et al.* (2003b,c) and one retrospective study (Schilling *et al.*, 2001) reported either similar instances of muscle dysfunction (i.e. cramping, muscle tightness, strains, injuries, etc.) between creatine and non-creatine users or fewer instances of muscle dysfunction in creatine users (Greenwood *et al.*, 2003a). Recently, Watson *et al.* (2006) reported no increase in cramping in creatine-supplemented subjects following dehydration and an 80 minute exercise heat tolerance test (33.5°C, 41% relative humidity). In this study, plasma sodium and potassium, and the dehydration levels following exercise, were unaffected by creatine ingestion, so it is not surprising that there was no increase in cramping (Watson *et al.*, 2006). Currently, there appears to be no empirical evidence linking creatine to muscle cramps.

2.1.2. *Muscle damage*

Reportedly, creatine ingestion has no effect on indices of muscle damage (e.g. blood creatine kinase or lactate dehydrogenase under resting conditions) (Kreider *et al.*, 2003; Mihic *et al.*, 2000; Robinson *et al.*, 2000; Schilling *et al.*, 2001). Three clinical studies have examined the interaction between creatine supplementation, extreme exercise, and muscle damage (Rawson *et al.*, 2001, 2007; Santos *et al.*, 2004). Rawson *et al.* (2001, 2007) found no effect of creatine on markers of muscle damage (decreased strength, decreased range of motion, increased muscle soreness, and increased serum creatine kinase and lactate dehydrogenase activity) following 50 high-force eccentric contractions of the elbow flexors or a high-repetition squat exercise challenge. These data were supported in a study by Warren *et al.* (2000) using an animal model. Santos *et al.*, (2004) reported that creatine supplementation attenuated the increase in plasma creatine kinase (by 19%), prostaglandin E_2 (by 61%), and tumor necrosis factor-α (by 34%) and eliminated the increase in plasma lactate dehydrogenase following a 30 km run, indicating less muscle damage. Collectively, clinical studies of the interactions between creatine supplementation and extreme exercise stress indicate that creatine supplementation does not exacerbate muscle damage (Rawson *et al.*, 2001, 2007) and might protect muscle from damage during certain types of stressful exercise (Santos *et al.*, 2004). It is important to note that serious adverse events associated with severe muscle damage (i.e. rhabdomyolysis), which may occur infrequently (1 in 10,000 exposures), are difficult to detect in the small clinical trials typically conducted on creatine supplementation ($n < 50$). The effects of creatine supplementation on indirect markers of muscle damage from double-blind placebo-controlled trials are described in Table 1.

Table 1. Effects of creatine supplementation on indices of muscle damage in double-blind placebo-controlled trials .

Outcome Variable	Exercise Challenge	Reference
Ø in resting plasma CK	-	Mihic *et al.*, 2000
Ø in resting serum CK	-	Robinson *et al.*, 2000
Ø in post-exercise serum CK and LDH, ROM, strength, DOMS	50 maximal eccentric contractions of the elbow flexors	Rawson *et al.*, 2001
Ø in post-exercise serum CK and LDH, ROM, strength, DOMS	5 sets of 15-20 squats with 50% of 1-RM	Rawson *et al.*, 2007
↓ in post-exercise plasma CK (19%), PGE_2 (61%), TNF-α (34%) and LDH (100%)	30 km run	Santos *et al.*, 2004

Note: Ø indicates no effect of creatine; ↓ indicates a decrease following creatine ingestion; ↑ indicates an increase following creatine ingestion. CK = creatine kinase, LDH = lactate dehydrogenase, ROM = range of motion, DOMS = delayed onset muscle soreness, PGE_2 = prostaglandin E_2, TNF-α = tumor necrosis factor α.

2.1.3. *Rhabdomyolysis*

Despite the intense media scrutiny on creatine supplementation, few cases of severe rhabdomyolysis in creatine users have been reported in the literature (Kuklo *et al.*, 2000; Robinson, 2000; Sandhu *et al.*, 2002; Sheth *et al.*, 2006). Robinson (2000) reported compartment syndrome and severe rhabdomyolysis in a patient who had been taking five times the recommended dosage of creatine (25 g/day) for one year and had performed three hours of lower extremity exercise the day before. Thus, it is unclear if the rhabdomyolysis was precipitated by the high-dose creatine supplementation, was a result of stressful unaccustomed resistance exercise, or a combination of the two. Similarly, rhabdomyolysis and compartment syndrome have been reported following stressful exercise in patients who had been ingesting creatine in combination with other supplements (e.g. ephedrine, natural diuretics) (Kuklo *et al.*, 2000; Sandhu *et al.*, 2002). Sheth *et al.* (2006) described a case of rhabdomyolysis in a creatine user in the days following arthroscopic knee surgery. It is uncommon, but post-operative rhabdomyolysis has been reported. Overall, it is unclear what role creatine supplementation played in these cases of severe rhabdomyolysis. It may be that in those who are predisposed to exercise-induced rhabdomyolysis, the combination of creatine supplementation and stressful unaccustomed exercise worsens symptoms compared to exercise alone, although this is speculative.

2.2. Dehydration

Creatine is often incorrectly associated with dehydration. In fact, creatine supplementation increases total body water (Powers *et al.*, 2003) and is more correctly referred to as a hyper-hydrating agent. Powers *et al.* (2003) used deuterium oxide and sodium bromide dilution analyses to demonstrate a 1.4 liter and 2.0 liter increase in total body water following 7 and 21 days of creatine supplementation, respectively. It has been proposed that creatine supplementation may bind water inside the muscle cell, making it unavailable for heat loss through the evaporation of sweat. However, Powers *et al.* (2003) reported that fluid distribution was unaffected by creatine supplementation (i.e. the intra- and extracellular water ratios were unaltered).

Several researchers have investigated thermoregulatory responses and exercise performance (Kern *et al.*, 2001; Kilduff *et al.*, 2004; Mendel *et al.*, 2005; Vogel *et al.*, 2000; Volek *et al.*, 2001; Watson *et al.*, 2006) in creatine-supplemented individuals following heat stress and/or hypohydration (Table 2). Collectively, these studies demonstrate that creatine-induced hyper-hydration does not impair thermoregulatory or metabolic responses to prolonged exercise in the heat. In fact, creatine may attenuate thermoregulatory responses and prevent heat related injuries and performance decrements (Kilduff *et al.*, 2004). For instance, Kilduff *et al.* (2004) reported decreased heart rate, perceived leg fatigue, rectal and body temperature, and sweat rate in creatine-supplemented individuals cycling to exhaustion in the heat (see Table 2). In the most comprehensive study to date, Watson *et al.* (2006) demonstrated that creatine supplementation does not adversely affect a number of variables spanning thermoregulatory, metabolic, and perceived responses to exercise in the heat in hypohydrated individuals (see Table 2). Although it has been theorized that increased total body water associated with creatine ingestion may cause thermoregulatory disturbances when exercise is combined with heat stress, the data do not support this.

2.3. Renal Dysfunction

Although the kidney has many physiologic functions it is most noted for its excretory function. The kidneys are responsible for predominantly removing small, hydrophilic molecules from the blood. Creatine and its major metabolite, creatinine, are both cleared from the blood by the kidney. Creatinine is an important clinical marker for renal function; more specifically, creatinine clearance is an indicator of glomerular filtration rate (GFR). Creatinine clearance is most frequently calculated from a single serum creatinine sample and the subsequent use of the Cockcroft-Gault equation, although other equations are available. The use of serum creatinine as a marker for renal function requires several assumptions, including: 1) the daily anabolic production of creatine is constant and 2) the conversion of creatine to creatinine is constant, and non-constant sources do not exist. During creatine supplementation, the first assumption is violated because supraphysiologic doses consumed far exceed daily endogenous production (daily liver production ~1 g/d, typical dosing 3–20 g/day). This in turn reduces endogenous creatine synthesis (Walker and Hannan, 1976). The second assumption of constant conversion is not violated however, as creatine stores increase with creatine dosing, so does serum creatinine. This is expected assuming the rate of degradation into creatinine is

Table 2. Effects of creatine supplementation on thermoregulatory responses and exercise performance in heat-stressed and hypohydrated individuals .

Effect of Creatine on Outcome Variables	Exercise/Environmental stress	Reference
Ø resting & post-exercise HR; Ø resting & post-exercise BP; Ø exercise T_{rect}; Ø exercise sweat rate; Ø post-exercise body mass; Ø post-exercise hemoglobin, hematocrit, & plasma volume; Ø post-exercise cortisol, aldosterone, renin, vasopressin, angiotensin I & II, & atrial peptide; Ø post-exercise urine volume, sodium, potassium, creatinine, and specific gravity; Ø reported muscle dysfunction; Ø RPE; Ø exercise performance	30 min cycling at 60 to 70% $VO_{2,peak}$ followed by three 10 s maximal sprints in an environmental chamber set at 37°C and 80% relative humidity	Volek *et al.*, 2001
Ø post-exercise body mass; Ø plasma volume; Ø exercise performance; Ø reported muscle dysfunction	Exercise and rest conducted in an environmental chamber set at 32°C and 50% relative humidity; 20 min rest in environmental chamber, five 5 s maximal cycling sprints, 75 min intermittent cycling (to reduce body mass 3 to 5%), 20 min rest in environmental chamber, five 5 s maximal cycling sprints, 75 min intermittent cycling, 20 min in environmental chamber, five 5 s maximal cycling sprints	Vogel *et al.*, 2000
Ø resting hematocrit; Ø resting & exercise HR; ↓ exercise T_{rect}	60 min cycling at 60% $VO_{2,max}$ in an environmental chamber set at 37°C and 25% relative humidity	Kern *et al.*, 2001
Ø time to exhaustion; ↓ exercise HR; ↓ RPE; ↓ exercise T_{rect} & T_{body}; Ø T_{skin}; ↓ exercise sweat rate; Ø sweat loss; Ø exercise metabolic rate, VO_2, VCO_2, V_E, RER; Ø exercise plasma volume	Cycling to exhaustion at 63% $VO_{2,max}$ in an environmental chamber set at 30°C and 70% relative humidity	Kilduff *et al.*, 2004
Ø exercise T_{rect}, T_{skin}, T_{body}; Ø reported thermal sensation	40 min cycling at 55% $VO_{2,max}$ in an environmental chamber set at 39°C and 26% relative humidity	Mendel *et al.*, 2005
Ø post-dehydration and post-exercise body mass; Ø exercise sweat loss; Ø exercise T_{rect}, T_{skin}; Ø exercise VO_2, HR, BP, MAP, lactate; Ø perceived environmental symptoms; Ø exercise plasma osmolality, volume, lactate, protein, sodium, potassium; ↑ post-dehydration plasma osmolality; ↑ exercise plasma glucose; ↓ post-exercise urine osmolality; ↑ resting, pre-exercise, and post-exercise urine specific gravity; Ø resting urine osmolality, specific gravity, color, volume	120 min cycling/treadmill walking in an environmental chamber set to 33.5°C and 41% relative humidity (to reduce body mass 2%), 80 min of alternatively running, walking, and standing	Watson *et al.*, 2006

Note: Ø indicates no effect of creatine; ↓ indicates a decrease following creatine ingestion; ↑ indicates an increase following creatine ingestion. HR = heart rate, BP = blood pressure, T_{rect} = rectal temperature, T_{body} = body temperature, T_{skin} = skin temperature, RPE = rating of perceived exertion, VO_2 = oxygen consumption, VCO_2 = carbon dioxide production, V_E= ventilation, RER = respiratory exchange ratio, MAP = mean arterial pressure.

concentration-dependent (i.e., a first-order process). This increase in stores can make the body appear to have a larger muscle mass than accounted for in the ideal body weight parameter of the Cockroft-Gault equation.

Clinical studies examining changes in serum creatinine with supplementation have found serum creatinine either does not change or increases but remains in the normal range (~ 0.5 to 1.5 mg/dL for adults) (Table 3). Some concern has been raised by this increase because it is assumed a rise in serum creatinine indicates reduced kidney function. However, studies using both serum and urine creatinine to estimate renal function in healthy individuals (Kreider *et al.*, 2003; Mihic *et al.*, 2000; Poortmans and Francaux, 1999) and patients (Groeneveld *et al.*, 2005; Louis *et al.*, 2003; Tarnopolsky *et al.*, 2004; Tarnopolsky and Raha, 2005) have found no change in kidney function. Recently, it was reported that 16 months of creatine supplementation had no effect on plasma urea and micro-albuminuria (indirect markers of renal dysfunction) in 175 patients with amyotrophic lateral sclerosis (ALS) (Groeneveld *et al.*, 2005). No changes in renal function have been noted in dystrophic patients as well (Louis *et al.*, 2003).

Since Harris *et al.* (1992) first demonstrated that muscle creatine levels could be increased with oral creatine supplementation, there have been over 200 clinical studies (through 2006) examining the impact of creatine supplementation. Of the hundreds of clinical studies examining the effects of creatine supplementation, and the thousands of exposures to creatine through these studies and through use by the general population, we are aware of three case studies where individuals developed renal dysfunction during creatine ingestion (Koshy *et al.*, 1999; Pritchard and Kalra, 1998; Revai *et al.*, 2003).

In the first case study, a 20-year old male with nausea, vomiting and bilateral flank pain was consuming 20 g/d creatine (5 g four times a day) for four weeks prior to hospital admittance (Koshy *et al.*, 1999). His serum creatinine was 1.4 mg/dL and urine analysis was positive for protein and red blood cells. Renal biopsy revealed acute focal interstitial nephritis and focal tubular injury. The patient did recover during his hospital admission. Most cases of interstitial nephritis are hypersensitivity reactions to medications such as non-steroidal anti-inflammatory drugs or antibiotics; in addition, obstruction of the tubules can cause this pathology as well. There was no evidence of inflammation hypersensitivity to creatine or renal obstruction as possible causes of the nephritis in this patient. It is possible that the dysfunction was caused by changes in osmotic gradient as seen with compounds such as mannitol.

The second case study involved a 25-year old male with focal segmental glomerulosclerosis with relapsing steroid responsive nephrotic syndrome (Pritchard and Kalra, 1998); he was taking cyclosporine for the previous 5 years to minimize nephrotic episodes and drug concentrations were within the therapeutic range. The patient had a history of normal renal function but the patient's creatinine clearance started to decline over time. The patient admitted to taking 15 g/d creatine (5 g three times a day) for 1 week followed by 2 g/d maintenance therapy. One month after stopping the creatine supplement, creatinine clearance returned to normal. A later

Table 3. Clinical studies examining serum creatinine (or creatinine clearance) responses during creatine supplementation. Serum creatinine is used as a clinical marker of renal function (normal limits: 0.5 to 1.5 mg/dL). The table summarizes data from 21 studies with creatine supplementation regimes ranging from 2 to 30 g/d for 1 d to 5.6 yrs. "load" = loading dose; "maint" = maintenance dose .

Study Finding	Number of Studies	Dose Amount	Duration Range	References
Studies finding no change in serum creatinine	12	15.75 g/d (load) 5 g/d (maint)	5 d up to 21 months	Kreider *et al.*, 2003
		20 g/d	5 d	Robinson *et al.*, 2000 #
		20 g/d	5 d	Poortmans *et al.*, 1997
		3 to 30 g/d	10 mo to 5 yr	Poortmans and Francaux, 1999
		10 g/d	Up to 310 d	Groeneveld *et al.*, 2005
		2 g	1 d	Harris *et al.*, 1992
		21 g/d	10 d	Poortmans *et al.*, 2005
		2.5 g	1 d	Harris *et al.*, 2004
		20 g/d (load) 5 g/d (maint)	5 d 3 yr	Schroder *et al.*, 2005
		20 g/d	2 d	Mendes *et al.*, 2004
		3 g/d	3 mo	Louis *et al.*, 2003
		20 g/d (load) 5 g/d (maint)	5 d 3-4 d	Parise *et al.*, 2001
Studies finding an increase in serum creatinine but within normal limits	8	20 g/d (load) 3 g/d (maint)	5 d 8 wk	Robinson *et al.*, 2000 #
		20 g	1 d	Schedel *et al.*, 1999
		20 g/d	5 d	Kamber *et al.*, 1999
		20 g/d	5 d	Mihic *et al.*, 2000
		5 g/d	16 wk	Tarnopolsky *et al.*, 2004
		0.3 g/kg/d	7 d	Volek *et al.*, 2001
		10 g/d	10 wk	Tarnopolsky and Raha, 2005
		13.7 ± 10.0 g/d (load) 9.7 ± 5.7 g/d (maint)	0.8 to 4 yr	Schilling *et al.*, 2001
Studies finding an increase in serum creatinine above normal limits	2	5 to 20 g/d	0.25 to 5.6 yr	Mayhew *et al.*, 2002*
		20 g/d	5 d	Skare *et al.*, 2001

*However, serum creatinine was not different from control group.
There was no difference in serum creatinine after the initial 20 g/d loading phase, but serum creatinine did increase after the 8-week maintenance phase.

study found that cyclosporine does impact the kinetics of creatine transport, and the interaction of cyclosporine with creatine might explain the renal dysfunction (Tran *et al.*, 2000).

In the third case study (full publication is available in Hungarian), the patient was "continuously" taking a "large quantity" of the anabolic-androgenic steroid methandion and 200 grams of creatine per day and subsequently developed diffuse membranoproliferative glomerulonephritis type I (Revai *et al.*, 2003). In all three case studies, the patients either had previous renal disease (i.e., glomerulosclerosis with relapsing nephrotic syndrome) or ingested 4 to 100 times the recommended daily amount of creatine for extended periods of time with or without other anabolic agents. Conversely, several human clinical trials have been published demonstrating that creatine has no adverse effects on renal health (see Table 3) in individuals ingesting creatine supplements for up to five years.

2.4. Other

Although the effects of creatine supplementation on muscle function, thermoregulation, and renal function comprise the bulk of the available safety data, several other areas have been studied. These include the effects of creatine on gastrointestinal, hepatic, and cardiovascular health, production of undesirable metabolites subsequent to creatine supplementation, and product impurities.

2.4.1. Gastrointestinal, hepatic, and cardiovascular health

Anecdotal reports of gastrointestinal distress and diarrhea have been associated with creatine ingestion. Potentially, this could result from the ability of creatine to draw water into the intestine in a similar manner to how creatine draws water into the muscle. This could be prevented by ingestion of smaller quantities of creatine per serving, ingesting creatine in a liquid dosage form compared to a solid dosage form, and avoiding fruit juice consumption which may further increase the discomfort because of the osmotic potential of fructose.

To date, there are no reports of hepatic or cardiovascular dysfunction from clinical trials of creatine supplementation. Several studies have shown that creatine does not impact blood-based liver function tests (Kamber *et al.*, 1999; Kreider, 2003; Mayhew *et al.*, 2002; Robinson *et al.*, 2000). Further, systolic and diastolic blood pressure appear to be unaffected by creatine supplementation in young (Mihic *et al.*, 2000) and older subjects (Rawson *et al.*, unpublished observations). Earnest *et al.* (1996) reported decreased total cholesterol (6%), triglycerides (26%), and very low density lipoprotein in hypercholesterolemic men and women (32 to 70 yrs) supplemented with creatine for 56 days. However, Volek *et al.* (2000) found no additional effect of creatine on blood lipids when combined with resistance exercise training.

There is one case study related to creatine ingestion and cardiac dysfunction. A 30 year-old vegetarian male developed diarrhea and cramps after one month of creatine supplementation, and subsequently changed to a different creatine supplement and developed palpitations and dyspnea (Kammer, 2005). The patient underwent chemical conversion and cardiac catheterization to correct the arrhythmia. It is unknown what role creatine played in this case.

2.4.2. *Metabolites*

There is some concern that creatine may form formaldehyde through a minor metabolic pathway, and in this regard, it was recently hypothesized that creatine supplementation could be cytotoxic (Yu and Deng, 2000). Creatine can be converted to formaldehyde and hydrogen peroxide, and formaldehyde has the potential to cross-link proteins and DNA leading to cytotoxicity. Yu and Deng (2000) did find an increase in urine formaldehyde after creatine administration; however, they did not measure markers of protein or DNA cross-linking or measures of oxidative stress. Another study examined urinary methylamine, formaldehyde and formate with creatine supplementation (Poortmans *et al.*, 2005). These investigators found increases in both methylamine and formaldehyde when subjects ingested 21 g/d for 14 d. The increase in methylamine was still below the upper-limit of normal but for formaldehyde, no safety range has been established.

2.4.3. *Impurities*

In the United States, creatine is not regulated for its processing and impurities, so as with other dietary supplements, there is some concern that contaminants may lead to adverse effects with creatine use. Benzi (2000) theorized that because creatine is produced from the reaction of sarcosine and cyanamide, several possible contaminants such as creatinine, dicyandiamide, dihydrotriazines, and ions such as arsenic could be produced. Although many creatine manufacturers provide a certificate of analysis with their products that addresses the issue of impurities, these findings have not been confirmed in many independent analyses. At least two studies examined creatine product quality with respect to percent labeled claim (Dash and Sawhney, 2002; Persky *et al.*, 2003). Both studies found powdered creatine products contained >99% creatine and did not note any unidentified peaks upon liquid chromatography.

3. FUTURE DIRECTIONS

The majority of information regarding the safety of creatine supplementation is available from relatively small clinical studies that lasted short periods of time, and examined healthy volunteers. Studies involving patients, larger numbers of subjects, long periods of time, and over a dose range are lacking. Investigators should assess and report on adverse events even by simple questionnaire. If possible, conclusions should be drawn whether adverse events were likely or unlikely related to the treatment. In addition, blood or muscle creatine levels should be evaluated to help relate adverse events to a systemic concentration of creatine, which could help assess whether side effects are dose-related. Finally, placebo controls allow judgments to be made whether the use of creatine is more likely to cause an adverse event than in the un-supplemented condition. For example, Table 4 summarizes adverse events from a study in patients with amyotrophic lateral sclerosis treated with creatine (Groeneveld *et al.*, 2005). The use of a placebo group allows conclusions to be drawn based on a potentially increased risk of a certain side effect. Muscle cramps

Table 4. Percentage of ALS patients who had an adverse event. Time average ratio indicates the relative prevalence of adverse events during the study. 1.0 = equal prevalence in the creatine (Cr) and placebo groups (Pl); $<$ 1.0: less events in the creatine group; $>$ 1.0: more events in the creatine group. Data adapted from Groeneveld *et al.* (2005) .

System	Event	Treatment	1 month	2 months	4 months	8 months	12 months	At any time	Time Average Ratio
Muscle	Cramps	Pl	61	70	79	69	73	91	
		Cr	62	73	78	73	67	95	1.0
	Cramps on	Pl	37	33	28	38	45	62	
	Exertion	Cr	24	36	38	42	38	70	1.1
	Limb Edema	Pl	18	9	18	29	45	46	
		Cr	21	26	35	42	42	54	1.4
Gastrointestinal	Nausea	Pl	13	7	7	7	14	24	
		Cr	6	6	9	9	8	23	0.99
	Vomiting	Pl	2	4	0	4	9	10	
		Cr	3	3	3	0	8	10	0.76
	Diarrhea	Pl	11	8	8	9	9	24	
		Cr	14	3	9	13	17	35	1.3
	Constipation	Pl	11	9	16	22	23	35	
		Cr	15	4	9	22	42	38	1.1
	General	Pl	11	7	8	4	5	18	
	Discomfort	Cr	3	6	6	9	13	19	1.4
	Reflux	Pl	7	12	15	16	9	27	
		Cr	12	9	7	13	17	28	0.85
Skin	Rash	Pl	6	7	7	20	23	24	
		Cr	5	8	7	13	13	19	0.71
	Pruritus	Pl	7	8	25	27	23	35	
		Cr	13	13	15	18	33	24	0.85

appear to be equally likely in this patient population whether on creatine or placebo whereas general gastrointestinal discomfort appears more prevalent in the creatine group.

4. CONCLUSION

The available data suggest that there are few adverse effects associated with creatine supplementation when ingested at recommended doses. Anecdotally, muscle dysfunction appears to be commonly associated with creatine supplementation, but data do not support this. Additionally, anecdotal reports of an association between creatine supplementation and impaired thermoregulation or dehydration are not supported by data. Although case reports suggest possible renal related side effects, most clinical studies show no indication of renal dysfunction with creatine use. As creatine supplementation becomes increasingly used as a potential intervention in clinical populations, larger scale studies should provide useful information into potential side effects, their severity and their incidence rate.

REFERENCES

Benzi, G., 2000, Is there a rationale for the use of creatine either as nutritional supplementation or drug administration in humans participating in a sport? *Pharmacol. Res.* **41:** 255–264.

Branch, J.D., 2003, Effect of creatine supplementation on body composition and performance: a meta-analysis. *Int. J. Sport Nutr. Exerc. Metab.* **13:** 198–226.

Charatan, F., 2001, Bayer decides to withdraw cholesterol lowering drug. *BMJ* **323:** 359.

Dash, A.K., and Sawhney, A., 2002, A simple LC method with UV detection for the analysis of creatine and creatinine and its application to several creatine formulations. *J. Pharm. Biomed. Anal.* **29:** 939–945.

Earnest, C.P., Almada, A.L., and Mitchell, T.L., 1996, High-performance capillary electrophoresis-pure creatine monohydrate reduces blood lipids in men and women. *Clin. Sci. (London)* **91:** 113–118.

Green, A.L., Hultman, E., Macdonald, I.A., Sewell, D.A., and Greenhaff, P.L., 1996a, Carbohydrate ingestion augments skeletal muscle creatine accumulation during creatine supplementation in humans. *Am. J. Physiol.* **271:** E821–E826.

Green, A.L., Simpson, E.J., Littlewood, J.J., Macdonald, I.A., and Greenhaff, P.L., 1996b, Carbohydrate ingestion augments creatine retention during creatine feeding in humans. *Acta Physiol. Scand.* **158:** 195–202.

Greenwood, M., Farris, J., Kreider, R., Greenwood, L., and Byars, A., 2000, Creatine supplementation patterns and perceived effects in select division I collegiate athletes. *Clin. J. Sport Med.* **10:** 191–194.

Greenwood, M., Kreider, R.B., Greenwood, L., and Byars, A., 2003a, Cramping and injury incidence in collegiate football players are reduced by creatine supplementation. *J. Athl. Train.* **38:** 216–219.

Greenwood, M., Kreider, R.B., Greenwood, L., Willoughby, D.S., and Byars, A., 2003b, The effects of creatine supplementation on cramping and injury occurrence during college baseball training and competition. *J. Ex. Physiol. Online* **6:** 16–23.

Greenwood, M., Kreider, R.B., Melton, C., Rasmussen, C., Lancaster, S., Cantler, E., Milnor, P., and Almada, A., 2003c, Creatine supplementation during college football training does not increase the incidence of cramping or injury. *Mol. Cell. Biochem.* **244:** 83–88.

Groeneveld, G.J., Beijer, C., Veldink, J.H., Kalmijn, S., Wokke, J.H., and van den Berg, L.H., 2005, Few adverse effects of long-term creatine supplementation in a placebo-controlled trial. *Int. J. Sports Med.* **26:** 307–313.

Harris, R.C., Almada, A.L., Harris, D.B., Dunnett, M., and Hespel, P., 2004, The creatine content of Creatine Serum and the change in the plasma concentration with ingestion of a single dose. *J. Sports Sci.* **22:** 851–857.

Harris, R.C., Soderlund, K., and Hultman, E., 1992, Elevation of creatine in resting and exercised muscle of normal subjects by creatine supplementation. *Clin. Sci. (London)* **83:** 367–374.

Hile, A.M., Anderson, J.M., Fiala, K.A., Stevenson, J.H., Casa, D.J., and Maresh, C.M., 2006, Creatine supplementation and anterior compartment pressure during exercise in the heat in dehydrated men. *J. Athl. Train.* **41:** 30–35.

Juhn, M.S., O'Kane, J.W., and Vinci, D.M., 1999, Oral creatine supplementation in male collegiate athletes: a survey of dosing habits and side effects. *J. Am. Diet. Assoc.* **99:** 593–595.

Kamber, M., Koster, M., Kreis, R., Walker, G., Boesch, C., and Hoppeler, H., 1999, Creatine supplementation–part I: performance, clinical chemistry, and muscle volume. *Med. Sci. Sports Exerc.* **31:** 1763–1769.

Kammer, R.T., 2005, Lone atrial fibrillation associated with creatine monohydrate supplementation. *Pharmacotherapy* **25:** 762–764.

Kern, M., Podewils, L.J., Vukovich, M., and Buono, M.J., 2001, Physiological responses to exercise in the heat following creatine supplementation. *J. Ex. Physiol. Online* **4:** 18–27.

Kilduff, L.P., Georgiades, E., James, N., Minnion, R.H., Mitchell, M., Kingsmore, D., Hadjicharlambous, M., and Pitsiladis, Y.P., 2004, The effects of creatine supplementation on cardiovascular, metabolic, and thermoregulatory responses during exercise in the heat in endurance-trained humans. *Int. J. Sport Nutr. Exerc. Metab.* **14:** 443–460.

Koshy, K.M., Griswold, E., and Schneeberger, E.E., 1999, Interstitial nephritis in a patient taking creatine. *N. Engl. J. Med.* **340:** 814–815.

Koszalka, T.R., Andrew, C.L., and Brent, R.L., 1972, Effect of insulin on the uptake of creatine-1-14C by skeletal muscle in normal and X-irradiated rats. *Proc. Soc. Exp. Biol. Med.* **139:** 1265–1271.

Kreider, R.B., 2003, Species-specific responses to creatine supplementation. *Am. J. Physiol. Regul. Integr. Comp. Physiol.* **285:** R725–R726.

Kreider, R.B., Melton, C., Rasmussen, C.J., Greenwood, M., Lancaster, S., Cantler, E.C., Milnor, P., and Almada, A.L., 2003, Long-term creatine supplementation does not significantly affect clinical markers of health in athletes. *Mol. Cell. Biochem.* **244:** 95–104.

Kuklo, T.R., Tis, J.E., Moores, L.K., and Schaefer, R.A., 2000, Fatal rhabdomyolysis with bilateral gluteal, thigh, and leg compartment syndrome after the Army Physical Fitness Test. A case report. *Am. J. Sports Med.* **28:** 112–116.

Louis, M., Lebacq, J., Poortmans, J.R., Belpaire-Dethiou, M.C., Devogelaer, J.P., Van Hecke, P., Goubel, F., and Francaux, M., 2003, Beneficial effects of creatine supplementation in dystrophic patients. *Muscle Nerve* **27:** 604–610.

Mayhew, D.L., Mayhew, J.L., and Ware, J.S., 2002, Effects of long-term creatine supplementation on liver and kidney functions in American college football players. *Int. J. Sport Nutr. Exerc. Metab.* **12:** 453–460.

Mendel, R.W., Blegen, M., Cheatham, C., Antonio, J., and Ziegenfuss, T., 2005, Effects of creatine on thermoregulatory responses while exercising in the heat. *Nutrition* **21:** 301–307.

Mendes, R.R., Pires, I., Oliveira, A., and Tirapegui, J., 2004, Effects of creatine supplementation on the performance and body composition of competitive swimmers. *J. Nutr. Biochem.* **15:** 473–478.

Mihic, S., MacDonald, J.R., McKenzie, S., and Tarnopolsky, M.A., 2000, Acute creatine loading increases fat-free mass, but does not affect blood pressure, plasma creatinine, or CK activity in men and women. *Med. Sci. Sports Exerc.* **32:** 291–296.

Parise, G., Mihic, S., MacLennan, D., Yarasheski, K.E., and Tarnopolsky, M.A., 2001, Effects of acute creatine monohydrate supplementation on leucine kinetics and mixed-muscle protein synthesis. *J. Appl. Physiol.* **91:** 1041–1047.

Persky, A.M., Hochhaus, G., and Brazeau, G.A., 2003, Validation of a simple liquid chromatography assay for creatine suitable for pharmacokinetic applications, determination of plasma protein binding and verification of percent labeled claim of various creatine products. *J. Chromatogr. B Analyt. Technol. Biomed. Life Sci.* **794:** 157–165.

Poortmans, J.R., Auquier, H., Renaut, V., Durussel, A., Saugy, M., and Brisson, G.R., 1997, Effect of short-term creatine supplementation on renal responses in men. *Eur. J. Appl. Physiol. Occup. Physiol.* **76:** 566–567.

Poortmans, J.R., and Francaux, M., 1999, Long-term oral creatine supplementation does not impair renal function in healthy athletes. *Med. Sci. Sports Exerc.* **31:** 1108–1110.

Poortmans, J.R., Kumps, A., Duez, P., Fofonka, A., Carpentier, A., and Francaux, M., 2005, Effect of oral creatine supplementation on urinary methylamine, formaldehyde, and formate. *Med. Sci. Sports Exerc.* **37:** 1717–1720.

Powers, M.E., Arnold, B.L., Weltman, A.L., Perrin, D.H., Mistry, D., Kahler, D.M., Kraemer, W., and Volek, J., 2003, Creatine supplementation increases total body water without altering fluid distribution. *J. Athl. Train.* **38:** 44–50.

Preen, D., Dawson, B., Goodman, C., Beilby, J., and Ching, S., 2003, Creatine supplementation: a comparison of loading and maintenance protocols on creatine uptake by human skeletal muscle. *Int. J. Sport Nutr. Exerc. Metab.* **13:** 97–111.

Pritchard, N.R., and Kalra, P.A., 1998, Renal dysfunction accompanying oral creatine supplements. *Lancet* **351:** 1252–1253.

Rawson, E.S., Conti, M.P., and Miles, M.P., 2007, Creatine supplementation does not reduce muscle damage or enhance recovery from resistance exercise. *J. Strength Cond. Res.*, in press.

Rawson, E.S., Gunn, B., and Clarkson, P.M., 2001, The effects of creatine supplementation on exercise-induced muscle damage. *J. Strength Cond. Res.* **15:** 178–184.

Revai, T., Sapi, Z., Benedek, S., Kovacs, A., Kaszas, I., Viranyi, M., and Winkler, G., 2003, [Severe nephrotic syndrome in a young man taking anabolic steroid and creatine long term]. *Orv. Hetil.* **144:** 2425–2427.

Robinson, S.J., 2000, Acute quadriceps compartment syndrome and rhabdomyolysis in a weight lifter using high-dose creatine supplementation. *J. Am. Board Fam. Pract.* **13:** 134–137.

Robinson, T.M., Sewell, D.A., Casey, A., Steenge, G., and Greenhaff, P.L., 2000, Dietary creatine supplementation does not affect some haematological indices, or indices of muscle damage and hepatic and renal function. *Br. J. Sports Med.* **34:** 284–288.

Saks, V.A., Stepanov, V., Jaliashvili, I.V., Konorev, E.A., Kryzkanovsky, S.A., Strumia, E., and Clark, J.F.S.D., 1996, Molecular and cellular mechanisms of action for the cardioprotective and therapeutic role of creatine phosphate. In: Conway, M.A. and Clark, J.F. (eds.), Creatine and Creatine Phosphate: Scientific and Clinical Perspectives. Academic Press, San Diego.

Saks, V.A., and Strumia, E., 1993, Phosphocreatine: molecular and cellular aspects of the mechanism of cardioprotective action. *Curr. Ther. Res.* **53:** 565–598.

Sandhu, R.S., Como, J.J., Scalea, T.S., and Betts, J.M., 2002, Renal failure and exercise-induced rhabdomyolysis in patients taking performance-enhancing compounds. *J. Trauma* **53:** 761–764.

Santos, R.V., Bassit, R.A., Caperuto, E.C., and Costa Rosa, L.F., 2004, The effect of creatine supplementation upon inflammatory and muscle soreness markers after a 30km race. *Life Sci.* **75:** 1917–1924.

Schedel, J.M., Tanaka, H., Kiyonaga, A., Shindo, M., and Schutz, Y., 1999, Acute creatine ingestion in human: consequences on serum creatine and creatinine concentrations. *Life Sci.* **65:** 2463–2470.

Schilling, B.K., Stone, M.H., Utter, A., Kearney, J.T., Johnson, M., Coglianese, R., Smith, L., O'Bryant, H.S., Fry, A.C., Starks, M., Keith, R., and Stone, M.E., 2001, Creatine supplementation and health variables: a retrospective study. *Med. Sci. Sports Exerc.* **33:** 183–188.

Schroder, H., Terrados, N., and Tramullas, A., 2005, Risk assessment of the potential side effects of long-term creatine supplementation in team sport athletes. *Eur. J. Nutr.* **44:** 255–261.

Sheth, N.P., Sennett, B., and Berns, J.S., 2006, Rhabdomyolysis and acute renal failure following arthroscopic knee surgery in a college football player taking creatine supplements. *Clin. Nephrol.* **65:** 134–137.

Skare, O.C., Skadberg, and Wisnes, A.R., 2001, Creatine supplementation improves sprint performance in male sprinters. *Scand. J. Med. Sci. Sports* **11:** 96–102.

Tarnopolsky, M., Mahoney, D., Thompson, T., Naylor, H., and Doherty, T.J., 2004, Creatine monohydrate supplementation does not increase muscle strength, lean body mass, or muscle phosphocreatine in patients with myotonic dystrophy type 1. *Muscle Nerve* **29:** 51–58.

Tarnopolsky, M.A., Bourgeois, J.M., Snow, R., Keys, S., Roy, B.D., Kwiecien, J.M., and Turnbull, J., 2003, Histological assessment of intermediate- and long-term creatine monohydrate supplementation in mice and rats. *Am. J. Physiol. Regul. Integr. Comp. Physiol.* **285:** R762–R769.

Tarnopolsky, M.A., and Raha, S., 2005, Mitochondrial myopathies: diagnosis, exercise intolerance, and treatment options. *Med. Sci. Sports Exerc.* **37:** 2086–2093.

Tran, T.T., Dai, W., and Sarkar, H.K., 2000, Cyclosporin A inhibits creatine uptake by altering surface expression of the creatine transporter. *J. Biol. Chem.* **275:** 35708–35714.

Vogel, R.A., Webster, M.J., Erdmann, L.D., and Clark, R.D., 2000, Creatine supplementation: Effect on supramaximal exercise performance at two levels of acute hypohydration. *J. Strength Cond. Res.* **14:** 214–219.

Volek, J.S., Duncan, N.D., Mazzetti, S.A., Putukian, M., Gomez, A.L., and Kraemer, W.J., 2000, No effect of heavy resistance training and creatine supplementation on blood lipids. *Int. J. Sport Nutr. Exerc. Metab.* **10:** 144–156.

Volek, J.S., Mazzetti, S.A., Farquhar, W.B., Barnes, B.R., Gomez, A.L., and Kraemer, W.J., 2001, Physiological responses to short-term exercise in the heat after creatine loading. *Med. Sci. Sports Exerc.* **33:** 1101–1108.

Walker, J.B., and Hannan, J.K., 1976, Creatine biosynthesis during embryonic development. False feedback suppression of liver amidinotransferase by N-acetimidoylsarcosine and 1-carboxymethyl-2-iminoimidazolidine (cyclocreatine). *Biochemistry* **15:** 2519–2522.

Warren, G.L., Fennessy, J.M., and Millard-Stafford, M.L., 2000, Strength loss after eccentric contractions is unaffected by creatine supplementation. *J. Appl. Physiol.* **89:** 557–562.

Watson, G., Casa, D.J., Fiala, K.A., Hile, A., Roti, M.W., Healey, J.C., Armstrong, L.E., and Maresh, C.M., 2006, Creatine use and exercise heat tolerance in dehydrated men. *J. Athl. Train.* **41:** 18–29.

Yu, P.H., and Deng, Y., 2000, Potential cytotoxic effect of chronic administration of creatine, a nutrition supplement to augment athletic performance. *Med. Hypotheses* **54:** 726–728.

CHAPTER 15

CREATINE – ITS CHEMICAL SYNTHESIS, CHEMISTRY, AND LEGAL STATUS

IVO PISCHEL[1] AND THOMAS GASTNER[2]

[1]*Finzelberg GmbH & Co. KG, Koblenzer Straße 48-56, D-56626 Andernach, Germany*
[2]*Degussa AG, Dr. Albert-Frank-Straße 32, D-83308 Trostberg, Germany*

Abstract: Creatine, a small molecule present in muscular tissue of many vertebrates, evolves to one of the most widely used and successful dietary supplements of recent decades (Graham and Hatton, 1999). Importantly, in the industrial manufacturing process, a high quality standard must be maintained. Validated analytical methods capable of providing reliable and consistent analysis of the main substance, side products and potentially harmful impurities must be employed. The principles of those determinations and the nature of possible by-products will be elucidated in this chapter. In addition, the pure creatine produced may be unstable under certain conditions, e.g. within special formulations or galenical forms. Some hints how to deal with this fact and how to avoid instability will also be discussed. Thus, this chapter will serve as a survey of the paths of chemical synthesis of creatine, its chemistry, properties, stability, analytical determination methods and legal status

1. THE HISTORY OF CREATINE

In 1832, the French scientist Chevreul discovered a new ingredient of beef tea and muscular tissue, which he named creatine, according to the source from which it was extracted (κρέας, Greek for flesh) (Chevreul, 1834). The German scientist von Liebig confirmed that creatine is a regular constituent of flesh. Creatine levels in wild animals were ten times higher compared to captive animals suggesting that physical activity might have an influence on the amount of creatine present in flesh (Liebig, 1847). Animals were exclusively used as a source for creatine for about the next century. Thus, meat extracts (Liebigs Fleischextrakt) and urine – via isolation by means of creatinine-zinc chloride double salt, followed by alkaline conversion of creatinine to creatine (see below) – were employed as sources for creatine (Mulder and Mouthaan, 1869; Folin, 1914; Benedict, 1914). If particularly pure material was needed, e.g. for organic preparations and analytical standards, creatine was prepared from meat extract. Twenty-five kilograms of fat-free flesh contains 1 kilogram of this

G.S. Salomons and M. Wyss (eds.), Creatine and Creatine Kinase in Health and Disease, 291–307.

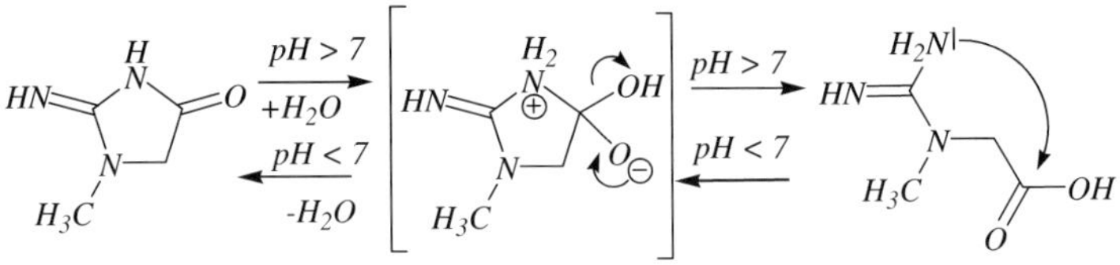

Figure 1. pH-dependent conversion between creatine and creatinine.

extract which, after extraction for 3 times with 2 litres of absolute ethanol each and re-crystallization from water, yields 25-30 grams of pure creatine (Steudel, 1921).

Dessaignes (1857) and Liebig (1858) found that creatine can be obtained from creatinine in basic solutions, e.g. with lime milk (Figure 1). The chemists Strecker and Volhard both described in 1868 the synthesis of the guanidino compound creatine by reacting sarcosine with cyanamide (Figure 2) over several hours or days, in slightly ammoniacal alkaline aqueous (Strecker, 1868) or in heated alcoholic solutions (Volhard, 1868). Those non-optimized reactions led to numerous by-products, so for the next decades, the more efficacious preparation was still the isolation from meat extract. In the first half of the 20th century, with an increase in knowledge about production of cyanamide and other guanylation reagents, technical manufacturing of creatine could be realized.

2. CHEMICAL SYNTHESES AND INDUSTRIAL PRODUCTION OF CREATINE

Nowadays, due to the growing interest in natural or synthetic guanidine compounds including creatine as potential pharmaceutical and bioactive leads, a large number of guanylation reagents are described in literature (Gastner, 2002; Yet, 1999). Medical chemistry and pharmaceutical research are the main driving forces in the development of gentle synthetic pathways for novel chemical structures including artificial peptide molecules and peptidomimetics possessing guanidino functionalities and having biological or even medicinal activity (Berlinck, 2002; Baker *et al.*, 2000). Most of the synthetic routes are of academic value and are used for lab scale preparation of new molecules.

Among the various synthetic paths described today, only a few could be implemented within an industrial setting for the production of large quantities of creatine monohydrate. All of the herein elucidated technical processes are based on guanylation, precisely called amidination, of the amino acid sarcosine or its salts (sarcosinates). The guanylation agents (Grambow *et al.*, 2003) are cyanamide (Volhard, 1868), O-methylisourea (Kapfhammer and Müller, 1934) or S-methylisothiourea salts (Schütte, 1943). Interestingly, mainly cyanamide is used in Western countries, e.g. Europe and the United States, whereas in the Eastern world, e.g. Japan and China, O-methylisourea or S-methylisothiourea salts are preferably used for commercial manufacturing of creatine.

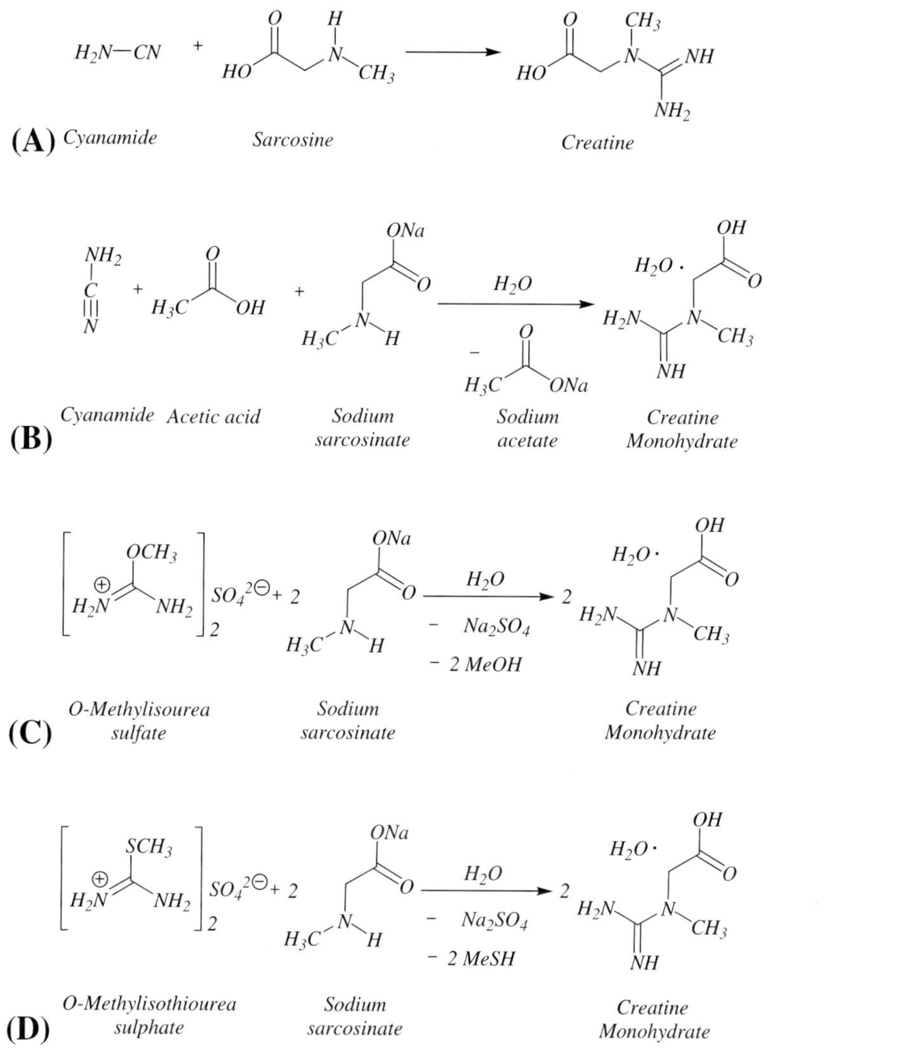

Figure 2. Chemical methods of creatine synthesis. (A) Early chemical synthesis of creatine described by Strecker (1868) and Volhard (1868). (B) Creatine synthesis from sarcosinate and cyanamide. (C) Creatine synthesis from sodium sarcosinate and O-methylisourea sulphate. (D) Creatine synthesis from sodium sarcosinate and O-methylisothiourea sulphate.

2.1. Creatine Synthesis from Sarcosinates and Cyanamide

Based on the early discovery by Strecker and Volhard, creatine can be obtained by reaction of cyanamide and sarcosinates using water as an inexpensive solvent. The raw materials of this process are readily available as basic chemical components. Sarcosinate sodium or potassium salts are produced on an industrial scale by reaction of methylamine, formaldehyde and sodium or potassium cyanide. On the other hand, cyanamide is derived from calcium cyanamide, which became industrially available by nitrogenation of calcium carbide, known as the Frank-Caro process established around 1900 (Hartmann and Zieke, 1954).

A state-of-the-art manufacturing process for creatine can be summarized as follows (Weiss and Krommer, 1995) (Figure 3): Technical aqueous sodium sarcosinate solution is adjusted with acetic acid under intensive cooling and vigorous stirring to a pH value of about 10. The reaction takes place in a stirring vessel at a temperature around 80 °C by careful addition of aqueous cyanamide solution under agitation followed by a time period of stirring to complete the reaction. After cooling, the crystalline creatine monohydrate is filtered or centrifuged off and washed several times in warm water to remove the reaction solution. Subsequently, the product is dried under vacuum at slightly elevated temperature. The yield of this process is about 75% of HPLC-pure creatine monohydrate.

A variety of manufacturing processes for creatine based on cyanamide are disclosed in the patent literature. Vassel and Garst (1953) describe generally the guanylation of amino acids, including sarcosine, using calcium cyanamide or free cyanamide. Kessel *et al.* (2004) disclose the use of carbon dioxide as acid to adjust the pH value of

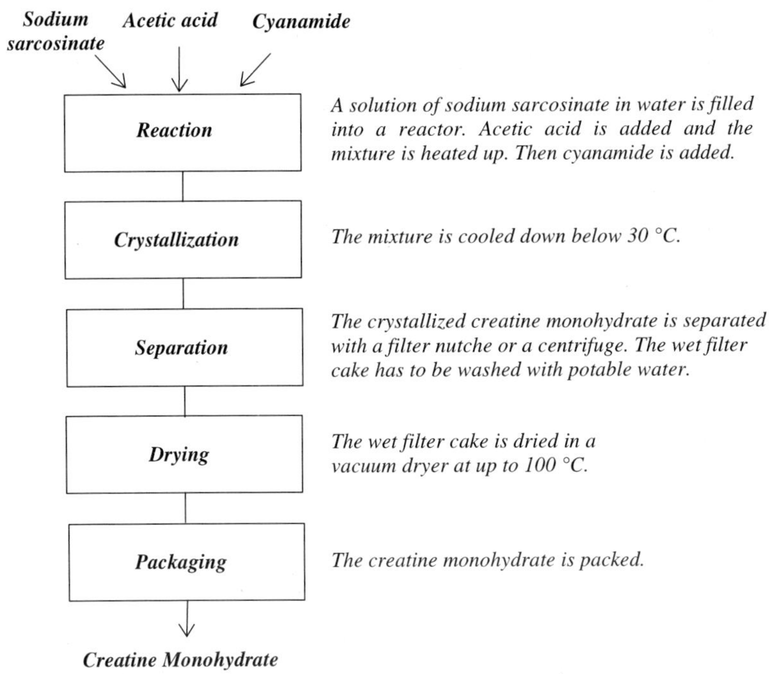

Figure 3. Industrial process flow chart for the synthesis of creatine monohydrate (Weiss and Krommer, 1995).

the reaction solution, whereas Kessel *et al.* (2003) describe a continuous process for creatine production.

2.2. Creatine Synthesis from Sarcosinates and O-alkylisourea

The use of O-alkylisourea as a guanylation agent for amino compounds was described for the first time by Kapfhammer and Müller (1934). Both a Japanese (Iwai and Tsunoda, 1980) and a German patent (Kessel and Kluge, 1998) describe creatine synthesis by continuous introduction of O-alkylisourea, especially O-methylisourea sulphate, into an aqueous sarcosine solution at a constantly maintained pH from 10 to 12 and at 5 to 25 °C. Creatine monohydrate crystals are obtained and purified via a sequence of washing steps using water and alcohols. A German patent discloses a similar process for preparing creatine by O-methylation of urea and reaction of the resultant O-methylisourea salt with sodium sarcosinate. The resultant creatine monohydrate crystals are washed with water (Greindl *et al.*, 1999).

2.3. Creatine Synthesis from Sarcosinates and S-alkylisothiourea

The first synthesis of guanidino acids from amino acids employing S-methylisothiourea was described by Wheeler and Merriam (1903). Although the guanylation agent is readily available and cost-efficient, the reaction has the disadvantage of the liberation of methyl mercaptan, a toxic and environmentally hazardous gas that needs to be annihilated before discharging the reaction vessel after the reaction is completed.

A Chinese patent discloses a process for the chemical production of creatine monohydrate (An *et al.*, 1999), which can be summarized as follows: To a diluted unprocessed aqueous solution of sodium N-methylglycinate, concentrated hydrochloric acid is added under stirring and cooling below 15 °C to adjust the pH to 9.5. At a temperature of less than 35 °C, S-methylisothiourea sulphate is slowly added into the solution under stirring. At about 30 °C, the reaction is completed in several hours. Then hot vapor is fed into the reaction solution. The crystalline mass is separated by centrifugation, followed by several washing steps with water, acids, and alcohol. After drying under vacuum at 40 °C, creatine monohydrate is obtained at a yield of 72.0% and a purity (by HPLC) of 99.3%.

All manufacturing processes described so far, especially those disclosed in the patent literature, can be performed under a rather wide range of process conditions, but generally lead to creatine monohydrate in yields of around 75%. Two additional patents on creatine production shall be mentioned here as well. Kessel *et al.* (2002) describe a purification procedure for creatine, whereas Greindl and Scherr (1999) disclose a process for creatine synthesis based on haloformamidium salts as guanylation agents.

3. PROPERTIES OF CREATINE

Creatine crystallizes from water as monoclinic prisms holding one molecule of water of crystallisation per molecule of creatine. The crystals easily loose this water of crystallisation at around 100 °C. The solubility of creatine in water increases with temperature and the correlation between solubility and temperature is almost linear. One litre of water dissolves 6 g of creatine at 4 °C, 14 g at 20 °C, 34 g at 50 °C, and 45 g at 60 °C. Creatine, as a small ampholytic substituted amino acid, is only very slightly soluble in pure ethanol (1 part in 9410 parts) and insoluble in ethyl ether. Table 1 shows the major features and key data of creatine.

One of the most important chemical reactions of creatine is related to its stability in aqueous solutions. Creatine is not stable in aqueous solution due to intra-molecular cyclisation to creatinine (see Figure 1). The velocity of creatine degradation is not dependent on its concentration, but on pH (the lower the pH, the faster) and temperature (the higher the temperature, the faster). This conversion is known as the creatine-creatinine-equilibrium and was previously thoroughly investigated by Edgar and Shiver (1925) and Cannan and Shore (1928). A more practical approach to this topic was undertaken by Howard and Harris (1999): their patent discloses formulations of creatine in acidic compositions. Figures 4 and 5 confirm, in correspondence with older experimental data, a decrease in creatine concentration in water under various conditions; for instance, they reveal degradation of creatine at 25 °C over a period of up to three days at different pH values (3.5–7.5) and at 4 °C over a period of up to thirty days at pH values between 3.5 and 7.0. The pH of the samples was adjusted to the desired value using acetic acid or potassium hydroxide, and the pH of the samples was also tested to ensure that pH did not change during the experiment. It was shown that creatine in aqueous solution is reasonably stable for up to 8 h at 25 °C and at a pH of 6.5 or 7.5. The breakdown observed after 3 days at pH 5.5, 4.5 and 3.5 was

Table 1. Characteristics of creatine. CAS = Chemical Abstract Service index name; http://www.cas.org .

Chemical structure	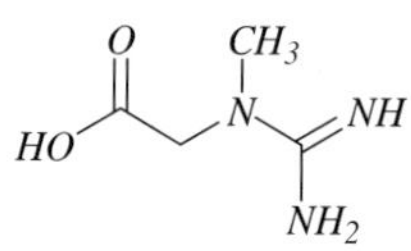
Chemical name (CAS, 2006)	N-(aminoiminomethyl)-N-methyl-glycine
Trival names	Engl. creatine; Ger. Kreatin; Fren. créatine
Synonyms	(α-methylguanido)acetic acid; methylguanidino-acetic acid; N-amidinosarcosine; α–methyl-guanidino acetic acid; N-methyl-N-guanylglycine; methylglycocyamine; N-amidino-sarcosine; N-(aminoiminomethyl)-N-methyl-glycine
Chemical formula	$C_4H_9N_3O_2$
Molecular mass	131.13 g/mol
CAS number	[57-00-1]
EINECS number	200-306-6
Melting point	decay at 303 °C

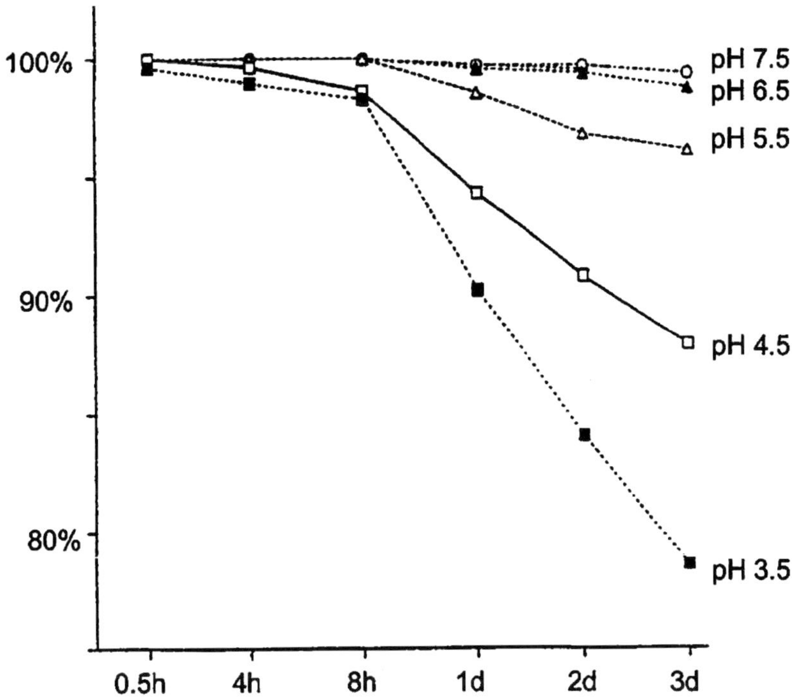

Figure 4. Degradation of creatine over time (0–3 days) at 25 °C and at different pH values (3.5–7.5) (Howard and Harris, 1999).

4%, 12% and 21%, respectively. At 4 °C, the reaction slows down so that even at pH 3.5 and after 30 days, the concentration of creatine in the solution is still above 90%.

Metal salt solutions and acids favour conversion of creatine to creatinine (Beilstein Crossfire Research, 2006, http://www.beilstein.com). Finally, the treatment of creatine with strong oxidative agents and bases cleaves the molecule to small nitrogenous compounds or oxidized molecules like carbonic acids.

Therefore, in respect to aqueous creatine formulations, their use should be as immediate as possible. Otherwise, after preparation, such formulations should be stored at a low temperature to slow down the conversion to creatinine. The solubility of creatine monohydrate at the chosen temperature should be considered, and is 14 g per litre at 25 °C and 8.5 g per litre at 4 °C.

Recently, an intensive investigation into the interesting characteristic of creatine as an anti-oxidant contributed to the understanding of its cell protective activity. Lawler *et al.* (2002) showed that creatine is able to scavenge a number of reactive oxygen species (ROS) having physiological significance in living matter.

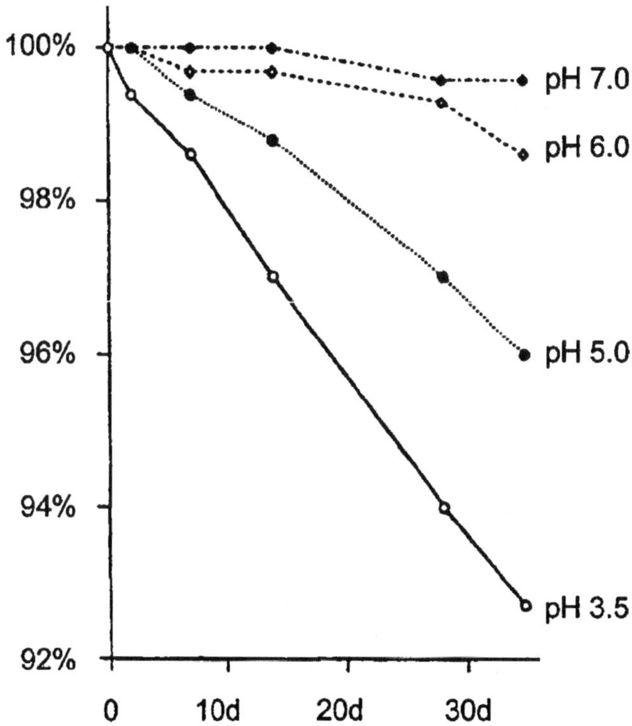

Figure 5. Degradation of creatine over time (0–30 days) at 4 °C and at different pH values (3.5–7.0) (Howard and Harris, 1999).

4. CHEMICAL ANALYSIS OF CREATINE

Various chemical analytical methods for creatine have been developed already. Soon after the discovery of creatine, gravimetric analysis was introduced which was based on formation of insoluble precipitates from creatine and a particular reagent, followed by separation of this precipitate and evaluation of its weight which correlated with the creatine content in the analyzed solution. Those methods became obsolete after the introduction of spectrometric procedures, like photometry, infrared, and magnetic resonance spectroscopy. These methods are still used on their own or in combination with highly efficacious separation methodologies, especially liquid chromatography. Some of the numerous methods are described here with respect to creatine analysis, especially those that have proven their utility in an industrial setting at Degussa.

4.1. Photometric Determination Methods for Creatine

Photometry is a method in which the concentration of an analyte, usually in solution, is rendered measurable by the addition of a reagent that forms a coloured product with the analyte under investigation. The intensity of the colour is dependent on the analyte's

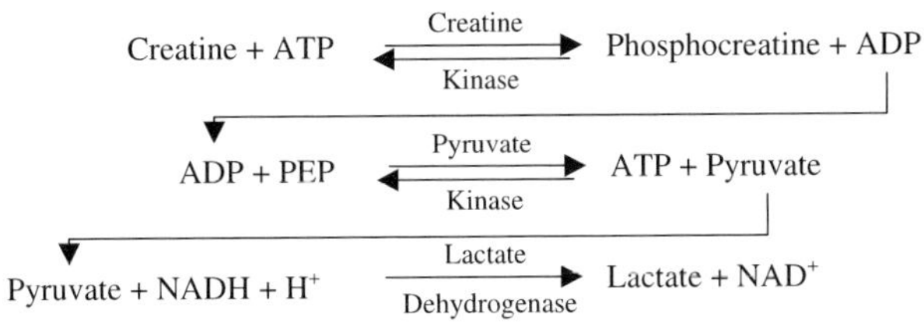

Figure 6. Cascade of enzymatic reactions for creatine determination. ATP: adenosine triphosphate; ADP: adenosine diphosphate; PEP: phosphoenolpyruvate; NADH: nicotine adenine dinucleotide (reduced form); NAD^+: nicotine adenine dinucleotide (oxidized form); H^+: proton.

concentration, and hence can be used to measure it. Photometric determination of creatine can be achieved according to the Folin method (Folin, 1904) which is based on the previously discovered Jaffé colour reaction of creatinine and picric acid in alkaline solution (Jaffé, 1886) which forms an orange-red complex that can be measured at about 500 nm (Merck, 1974, 1987). Creatine needs to be converted to creatinine by acid treatment in solution before measurement. This method is still commonly used to analyze creatine as well as creatinine in biological samples like blood serum, tissue and urine (Davidsohn and Henry, 1969). Due to the fact that this reaction is not highly specific and can be disturbed by other organic compounds found in body fluids, more specific methods for creatine determination were developed.

4.2. Enzymatic Analyses of Creatine

In general, enzymatic determination methods are more specific and accurate than photometric ones based on colour reactions. Enzymes are used to catalyse a specific biochemical reaction. The change in concentration of an involved constituent of the reaction or reaction cascade is measured commonly by a photometer.

Bernt *et al.* (1970) described a method based on Tanzer and Gilvarg (1959) where creatine is converted with ATP and creatine kinase into creatine phosphate, followed by a reaction cascade shown in Figure 6. The associated NADH decrease is photometrically measured at a wavelength of 340 nm, and is proportional to the creatine concentration. An enzymatic procedure developed by Kodak Ektachem uses the enzyme creatinine iminohydrolase, which catalyzes the hydrolysis of creatinine to N-methylhydantoin and ammonia. The ammonia reacts with bromophenol blue to form a blue dye that is measured by reflectance photometry (Blick and Liles, 1985). Several other enzymatic determination methods for creatine are described in literature, e.g. based on creatine kinase (Beyer, 1993).

4.3. Chromatographic Methods for Creatine Determination

Chromatography involves passing a sample – i.e. a mixture which contains the compound in question dissolved in the mobile phase, mostly a solvent – through the stationary phase that retards the passage of the components of the sample by adsorption phenomena. The components of the sample become separated in time because their passage through the system happens at different rates, so that they finally can be detected successively.

In earlier times, methods for detection of creatine by means of paper and thin layer chromatography were developed but they do no longer play a role nowadays. More advanced determination methods were developed and are described hereafter.

4.3.1. High performance liquid chromatography (HPLC)

Recently, Dash and Sawhney (2002) developed a simple liquid chromatographic method with UV detection for the analysis of creatine and creatinine and applied it to various commercially available creatine supplement formulations. In this method, the mobile phase consists of 0.045 M ammonium sulphate in water. A C-18 column is employed for the chromatographic separation at ambient temperature. The eluent is monitored at 205 nm. As an internal standard, 4-(2-aminoethyl) benzene sulfonamide is used. This method requires less than 7 min of chromatographic time. The method can be used to determine ~1–100 mg/ml of creatine or creatinine. The precision and accuracy for creatine were found to be sufficient for measurements of creatine in various marketed products as well as for analyses of the solubility of various creatine salts and of the stability of creatine in aqueous solution.

Reliable and accurate determination of sample components can be accomplished with high-performance liquid chromatography. HPLC techniques employ fine particles of highly porous materials as an absorbing stationary phase packed typically in steel columns. A solvent phase is forced at high pressure through the packed column with concurrent separation of the transported constituents of the sample. HPLC-determination of creatine can be achieved, e.g., by a method developed by Degussa AG (Manfred Wildenauer, personal communication) which is also suitable for the simultaneous determination of the potential impurities dicyandiamide, creatinine and dihydrotriazine. The mobile phase consists of 0.2 M ammonium dihydrogen phosphate in water which is adjusted to pH 4 with phosphoric acid. Chromatographic separation is achieved at ambient temperature on a nucleosil column 5 SA (ET 250/4, Macherey-Nagel). The flow rate is maintained at 1.0 ml/min and the effluent is monitored at 225 nm. The concentration of creatine is calculated through comparison with external standards. The retention time of creatine is approx. 3.5 min (dicyandiamide approx. 2.8 min, dihydrotriazine approx. 5.4 min, and creatinine approx. 8.6 min).

4.4. NMR and IR Spectroscopy of Creatine

4.4.1. ^{1}H-NMR-spectroscopy of creatine

Nuclear magnetic resonance, or NMR, is a physical phenomenon based upon the magnetic property of an atom's nucleus. It allows the determination of the number, type and relative positions of certain atoms in a molecule. NMR spectroscopy is a powerful instrumental technique used to obtain physical, chemical, electronic and

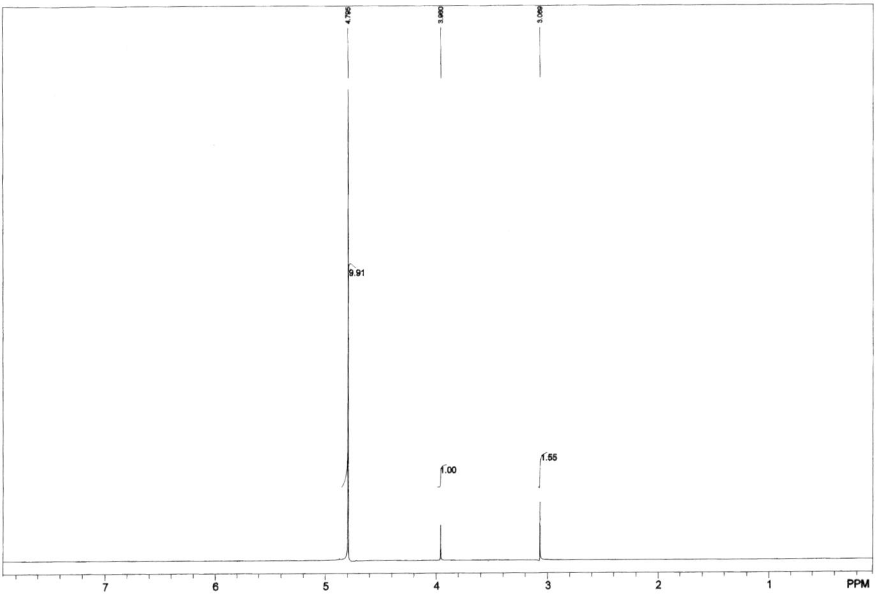

Figure 7. ^{1}H-NMR spectrum of creatine monohydrate in D_2O.

structural information about a molecule. NMR spectroscopy can be applied to atoms having nuclear spin; the most important for NMR analysis of living matter are the hydrogen (^{1}H), carbon (^{13}C) and phosphorus (^{31}P) atoms.

Figure 7 shows a ^{1}H-NMR spectrum of creatine monohydrate. The singlets at 3.069 ppm and 3.960 ppm are caused by the protons of the N-CH_3 group and the methylene protons of the acetate group, respectively. As expected, the integrals indicate a ratio of 3:2. The signal at 4.795 ppm represents the sum of the NH-protons, the protons of water of crystallization, and the protons of HDO contained in D_2O.

4.4.2. ^{13}C-NMR-spectroscopy of creatine

The ^{13}C-NMR spectrum (Figure 8) consists of 4 signals. The signals at 36.935 ppm and 53.875 ppm are caused by the carbon atoms of the methyl group and the methylene group, respectively. The chemical shifts of the guanidine carbon at 157.127 ppm and the carboxylate carbon at 174.604 ppm are in agreement with expectations.

4.4.3. IR-spectroscopy of creatine

Infrared (IR) spectroscopy measures the vibrations of molecules. Each functional group, or structural characteristic, of a molecule has a unique vibrational frequency that can be used for its identification in a sample. When the effects of all the different functional groups are taken together, the result is a unique molecular "fingerprint" that can be used to confirm the identity of a sample.

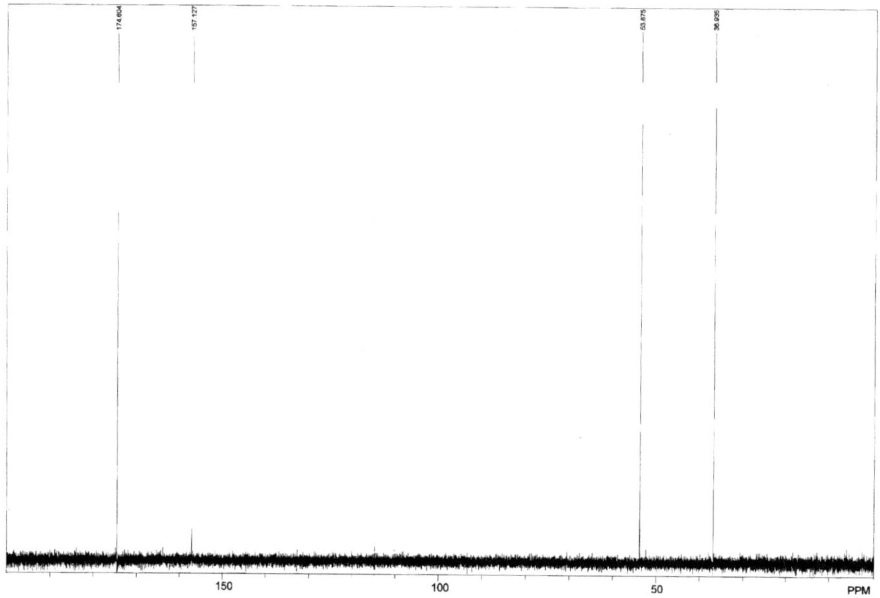

Figure 8. ^{13}C-NMR spectrum of creatine monohydrate in D_2O.

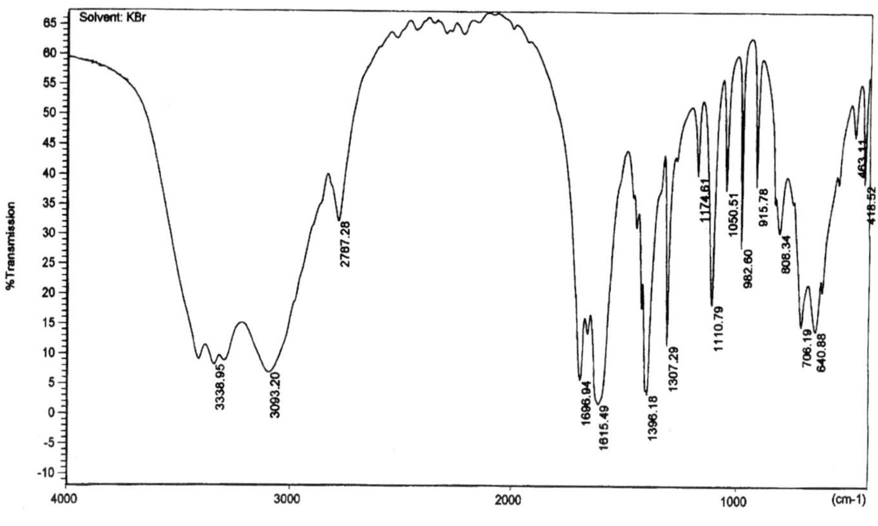

Figure 9. IR-spectrum of creatine monohydrate.

The infrared spectrum of creatine monohydrate in KBr is shown in Figure 9. The broad and intense absorption band between 3500 and 2800 cm^{-1} is due to the valence vibrations of the O-H and N-H bonds of crystal water and the guanidine group, respectively. The absorption band at 2787 cm^{-1} is caused by the valence vibrations of the C-H bonds of the methyl group. The intense bands between 1700 and 1600 cm^{-1} are caused by the valencc vibrations of the C–O bonds of the carboxyl group and the C–N bonds of the guanidine group.

5. POTENTIAL IMPURITIES

The analytical methodologies described above are tools for the investigation of synthesised products like creatine monohydrate and its derivatives. This includes both evaluation in respect of the product's quality, more specifically its purity, and the detection of possible impurities. Due to the rather high amount of creatine monohydrate that is consumed for its sports application or when used as supplementation in certain disease conditions, it is obvious that even small proportions of foreign matter, by-products or impurities may be harmful.

As described in section 3 of this chapter, several processes are currently in use to manufacture creatine monohydrate on an industrial scale. Therefore, creatine monohydrate may be contaminated with specific impurities depending on the used process. Analytical investigation of commercially available creatine formulations was undertaken using HPLC to elucidate the differences in quality among marketed products (Figure 10). Distressingly, among those marketed products, some were found to contain the following levels of impurities (for the chemical structures, see Figure 11): up to 54,000 ppm (5.4%) dicyandiamide, 860 ppm (0.09%) dihydrotriazine, and 13,000 ppm (1.3%) creatinine. In addition, elevated levels of heavy metals like mercury and

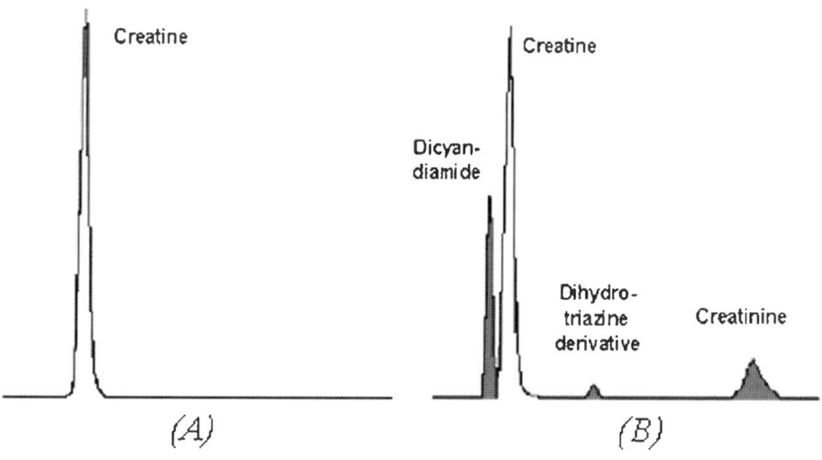

Figure 10. HPLC spectrum of two commercial creatine monohydrate products: (A) pure; (B) contaminated.

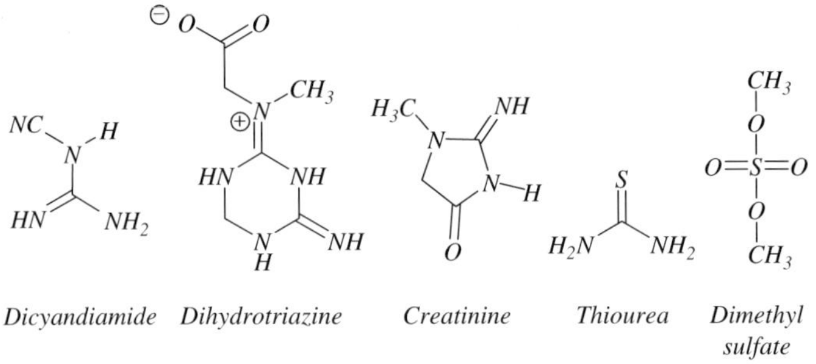

Figure 11. Potential impurities found in commercial creatine monohydrate preparations due to poorly controlled manufacturing processes.

lead have been detected (http://www.creapure.com, August 2006). The contaminations can be explained by poorly controlled synthesis or unsuccessful purification during production. The chemical mechanisms for the formation of by-products during manufacturing have been elucidated and experimentally confirmed at Degussa (Thomas Gastner, unpublished data).

If sodium sarcosinate and cyanamide or O-methylisourea are used as raw materials, the potential by-products are dihydrotriazine and creatinine. Creatinine is a conversion product of creatine, especially when kept in acidic conditions. Dihydrotriazine is formed from contaminations within the sarcosinate solution and has unknown pharmaceutical and toxicological properties. Dihydrotriazine might be one of the most dangerous impurities in creatine monohydrate, since structurally related compounds are known to be carcinogenic. Furthermore, if cyanamide is used as a raw material, the dimerization product dicyandiamide can be formed, especially when cyanamide is used in excess or introduced into the reaction medium at a too high pH value.

If sodium sarcosinate and O-methylisothiourea are used as raw materials, the potential by-products are dihydrotriazine, creatinine, dimethyl sulphate and thiourea. O-Methylisothiourea is made by reaction of thiourea with dimethyl sulphate which explains potential contamination with these two compounds. Thiourea and dimethyl sulphate are harmful if ingested and are known carcinogens. For that reason, it is an obvious and unavoidable postulate not only to control and guard the manufacturing process, but also to implement an appropriate quality assurance policy to maintain high quality standards.

6. RECENT DEVELOPMENTS

In the last decade, a number of innovations and developments were described to overcome technical limitations that creatine bears. One major obstacle of creatine as an ampholytic amino acid is its rather low solubility in water. A solution to this was the formulation of more soluble creatine salts and complexes based on mono-, di- or tricarboxylic acids to form for instance the citrate, maleate, fumarate, tartrate (Negrisoli and Del Corona, 1997), pyruvate (Pischel and Weiss, 1996), ascorbate

(Pischel *et al.*, 1999) or orotate salts (Abraham and Jiang, 2005) of creatine. A few of those novel chemical compounds have – aside from a higher solubility – other advantages, like additive or synergistic effects, e.g. a combination of the ergogenic effect of creatine with anti-oxidative properties of ascorbic acid. The creation of carnitine creatinate (Fang, 1999) was likely undertaken to combine the benefits of both constituents, creatine and carnitine. Similar advantages are claimed for creatine hydroxycitrate that may be used as a dietary supplement for the purposes of reducing adiposity and of appetite suppression, as well as for improvement of muscle and exercise performance (Gardiner *et al.*, 2006). Carnazzo (1999) describes in his patent the use of edible creatine citrate and its formulation for effervescent tablets. Solid and stable creatine/citric acid compositions with carbohydrate were developed by Purpura *et al.* (2005).

Recently, creatine products with an alkaline pH value came onto the market, which claim a higher stability and better bioavailability. However, no clinical studies have been performed and there is no scientific evidence supporting these claims. Due to their weak buffering capacity, these products are also adjusted to an acidic pH in the stomach and will therefore be degraded in the same way as pure creatine monohydrate. For creatine monohydrate, it has been shown that less than 0.1 g of a 5 g dose is lost within 1 hour at pH 3.5. The stability of creatine at pH 1 is even better. Therefore, conversion of creatine to creatinine in the gastrointestinal tract is minimal regardless of transit time (Persky *et al.*, 2003).

7. REGULATORY STATUS OF CREATINE MONOHYDRATE AND ITS SALTS

In the European Union, creatine and its salts fall under the definition of "food" according to Art. 2 of regulation 178/2002 (http://eur-lex.europa.eu/LexUriServ/site/de/oj/2002/l_031/l_03120020201de00010024.pdf) of the European parliament and of the council as they are a "substance or product intended to be, or reasonably expected to be ingested by humans." Taking Art. 1 of Council Directive 89/107/EEC (amendment 1994; http://www.fsai.ie/legislation/food/eu_docs/Food_additives/Dir89.107.pdf) concerning additives approved for use in foodstuffs intended for human consumption into consideration, it is clear that creatine and its salts are not food additives. Their most precise definition is found in the food supplements directive 2002/46/EC (http://europa.eu.int/eur-lex/pri/en/oj/dat/2002/l_183/l_18320020712en00510057.pdf) as they meet the criteria of Art. 2 (a): "foodstuffs… which are concentrated sources of nutrients or other substances with a nutritional or physiological effect …". In the United States, creatine monohydrate and its salts are classified as "dietary supplements" under the 1994 "Dietary Supplements Health and Education Act" (http://www.fda.gov/opacom/laws/dshea.html).

REFERENCES

Abraham, S., and Jiang, S., 2005, Process for preparing a creatine heterocyclic acid salt and method of use. US patent no. 6,838,562.

An, L., Zheng, Y., and Zhang, G., 1999, Process for producing creatine or creatine monohydrate. Chinese patent no. 99118985.

Baker, T.J., Rew, Y., and Goodman, M., 2000, Novel reagents and reactions for drug design. *Pure Appl. Chem.* **72:** 347–354.

Benedict, S.R., 1914, Studies in creatine and creatinine metabolism. I. The preparation of creatine and creatinine from urine. *J. Biol. Chem.* **18:** 183–190.

Berlinck, R.G.S., 2002, Natural guanidine derivatives. *Nat. Prod. Rep.* **19:** 617–649.

Bernt, E., Bergmeyer, H.U., and Möllering, H., 1970, Creatin. In: Bergmeyer, H.U., Methoden der Enzymatischen Analyse, 2. Aufl., Verlag Chemie, Weinheim, Germany, pp. 1724–1728.

Beyer, C., 1993, Creatine measurement in serum and urine with an automated enzymatic method. *Clin. Chem.* **39:** 1613–1619.

Blick, K.E., and Liles, S.M., 1985, Principles of Clinical Chemistry, John Wiley & Sons, New York, p. 598.

Cannan, R.K., and Shore, A., 1928, The creatine-creatinine equilibrium: The apparent dissociation constants of creatine and creatinine. *Biochemistry* **22:** 920–929.

Carnazzo, J.W., 1999, Method for enhancing delivery and uniformity of concentration of cellular Creatine. US patent no. 5,925,378.

Chevreul, M.E., 1834, Creatin, eine neue im Muskelfleisch enthaltene Substanz. *Liebigs Ann. Chem.* **4:** 293–294.

Dash, A.K., and Sawhney, A., 2002, A simple LC method with UV detection for the analysis of creatine and creatinine and its application to several creatine formulations. *J. Pharm. Biomed. Anal.* **29:** 939–945.

Davidsohn, I., and Henry, J.B., 1969, Clinical Diagnosis by Laboratory Methods, W.B. Saunders Company, Philadelphia, pp. 544–546.

Dessaignes, M., 1857, In: *Jber. Fortschr. Chem.*, p. 543.

Edgar, G., and Shiver, H.E., 1925, The equilibrium between creatine and creatinine, in aqueous solution: the effect of hydrogen ion. *J. Am. Chem. Soc.* **47:** 1179–1188.

Fang, S., 1999, Carnitine creatinate. US patent no. 5,994,581.

Folin, O., 1904, Beitrag zur Chemie des Kreatinins und Kreatins im Harne. *Z. physiol. Chem.* **41:** 223–242.

Folin, O., 1914, On the preparation of creatine, creatinine and standard creatinine solutions. *J. Biol. Chem.* **17:** 463–468.

Gardiner, P., Heuer, M.A., and Molino, M., 2006, Creatine hydroxycitric acids salts and methods for their production and use in individuals. US patent application no. 20060194877 A1.

Gastner, T., 2002, Amidination reagents for the synthesis of guanidines. Cyanamide Newsletter, Degussa AG, http://www.finechemicals.de/Service/PDF/cyanamide_newsletter_1_2002.pdf.

Graham, A.S., and Hatton, R.C., 1999, Creatine: a review of efficacy and safety. *J. Am. Pharm. Assoc.* **39:** 803–810.

Grambow, C., Weiss, S., Youngman, R., Antelmann, B., Mertschenk, B., Stengele, K.P., 2003, Guanidine and Derivatives. In: *Ullmann's Encyclopedia of Industrial Chemistry*, Wiley-VCH-Verlag, Weinheim, Germany, 6th ed., Vol. 16, pp. 73–86.

Greindl, T., Scherr, G., Schneider, R., and Mundinger, K., 1999, Preparation of substituted guanidine derivative. US patent no. 5,994,582.

Greindl, T., and Scherr, G., 1999, Verfahren zur Herstellung von substituierten Guanidinderivaten. German patent no. 19748696.

Hartmann, F., and Zieke, K., 1954, Calciumcyanamid. In: Foerst, W. (ed.), Ullmanns Encyklopädie der technischen Chemie, Vol. 5, pp. 43–68.

Howard, A.N., and Harris, R.C., 1999, Compositions containing creatine. US patent no. 5,968,544.

Iwai, T., and Tsunoda, T., 1980, Preparation of creatine. Japanese patent no. 55004349.

Jaffé, M., 1886, Über den Niederschlag, welchen Pikrinsäure in normalem Harn erzeugt und über eine neue Reaktion des Kreatinins. *Z. physiol. Chem.* **10:** 391–400.

Kapfhammer, J., and Müller, H., 1934, Guanidinosäuren und Guanidinopeptide. *Z. physiol. Chem.* **225:** 1–12.

Kessel, K., and Kluge, M., 1998, Synthesis of guanidines from O-alkylisoureas using a crystal seed. German patent no. 19860048.

Kessel, K., Kluge, M., Bogenstatter, T., and Scherr, G., 2003, Process for preparing guanidine derivatives. US patent no. 6,509,497.

Kessel, K., Scherr, G., Bogenstatter, T., Orsten, S., and Franke, D., 2002, Process for purifying creatine. US patent no. 6,399,819.

Kessel, K., Scherr, G., Kluge, M., Biedermann, N., Greindl, T., Bogenstatter, T., and Hahnlein, W., 2004, Process for the preparation of creatine or creatine monohydrate. US patent no. 6,759,553.

Lawler, J.M., Barnes, W.S., Wu, G., Song, W., and Demaree, S., 2002, Direct antioxidant properties of creatine. *Biochem. Biophys. Res. Commun.* **290:** 47–52.

Liebig, J.v., 1847, Ueber die Bestandtheile der Flüssigkeiten des Fleisches. *Liebigs Ann. Chem.* **62:** 257–369.

Liebig, J.v., 1858, Ueber Kreatin und Kynurensäure im Hundeharn, *Liebigs Ann. Chem.* **108:** 354–356.

Merck, E., 1974, Klinisches Labor. E. Merck, Darmstadt, pp. 125–127.

Merck, E., 1987, Arbeitsanleitung Diagnostika Merck, Darmstadt, pp. 46–47.

Mulder, E., and Mouthaan, N., 1869, Kreatin und Aldehyde. *Z. Chem.*, p. 341.

Negrisoli, G., and Del Corona, L., 1997, Hydrosoluble organic salts of creatine. Italian patent no. 1271687.

Persky, A.M., Brazeau, G.A., and Hochhaus, G., 2003, Pharmacokinetics of the dietary supplement creatine. *Clin. Pharmacokinet.* **42:** 557–574.

Pischel, I., and Weiss, S., 1996, New creatine pyruvate derivatives from crystallisation in polar solvents. German patent no. 19653225.

Pischel, I., Weiss, S., Gloxhuber, C., and Mertschenk, B., 1999, Creatine ascorbates and a method of producing them. US patent no. 5,863,939.

Purpura, M., Pischel, I., Jaeger, R., and Ortenburger, G., 2005, Solid and stable creatine/citric acid composition(s) and compositions carbohydrate(s) or hydrates thereof, method for the production and use thereof. US patent application no. 20050037069 A1.

Schütte, E., 1943, Preparation of guanido acids and guanido peptides. *Z. physiol. Chem.* **279:** 52–59.

Steudel, H., 1921, A simple method for the preparation of creatinine from meat extract. *Z. physiol. Chem.* **112:** 53–54.

Strecker, A., 1868, In: *Jber. Fortschr. Chem.*, p. 686.

Tanzer, M.L., and Gilvarg, C., 1959, Creatine and creatine kinase measurement. *J. Biol. Chem.* **234:** 3201–3204.

Vassel, B., and Garst, R., 1953, Method of preparation of guanidino fatty acids. US patent no. 2,654,779.

Volhard, J., 1868, In: *Jber. Fortschr. Chem.*, p. 685.

Weiss, S., and Krommer, H., 1995, Process for the preparation of creatine or creatine-monohydrate. European patent application no. 0754679.

Wheeler, H.I., and Merriam, H.F., 1903, On the condensation-products of the pseudothioureas: synthesis of uracil, thymine, and similar compounds. *Am. Chem. J.* **29:** 478–492.

Yet, L., 1999, Recent developments in guanylation chemistry. Albany Molecular Research, Inc., Technical Reports, Vol. 3, No. 6, http://www.albmolecular.com/chemlinks/vol3.html.

CHAPTER 16

CREATINE AND CREATINE KINASE IN HEALTH AND DISEASE – A BRIGHT FUTURE AHEAD?

MARKUS WYSS[1], OLIVIER BRAISSANT[2], IVO PISCHEL[3], GAJJA S. SALOMONS[4], ANDREAS SCHULZE[5], SYLVIA STOCKLER[6] AND THEO WALLIMANN[7]

[1] *DSM Nutritional Products Ltd., Biotechnology R&D, Bldg. 203/17B, P.O. Box 3255, CH-4002 Basel, Switzerland*

[2] *Clinical Chemistry Laboratory, Centre Hospitalier Universitaire Vaudois and University of Lausanne, CH-1011 Lausanne, Switzerland*

[3] *Finzelberg GmbH & Co. KG, Koblenzer Straße 48-56, D-56626 Andernach, Germany*

[4] *Department of Clinical Chemistry, Metabolic Unit, VU University Medical Center, De Boelelaan 1117, NL-1081 HV Amsterdam, The Netherlands*

[5] *University of Toronto, Department of Paediatrics, Division of Clinical and Metabolic Genetics, and Research Institute, The Hospital for Sick Children, 555 University Avenue, Toronto, ON. M5G 1X8, Canada*

[6] *Department of Pediatrics, University of British Columbia, Division of Biochemical Diseases, British Columbia Children's Hospital, Vancouver, B.C., V6H 3V4, Canada*

[7] *Institute of Cell Biology, ETH Zurich, Hönggerberg HPM-D24.1, Schafmattstrasse 18, CH-8093 Zurich, Switzerland*

Abstract: Many links are reported or suspected between the functioning of creatine, phosphocreatine, the creatine kinase isoenzymes or the creatine biosynthesis enzymes on one hand, and health or disease on the other hand. The aim of the present book was to outline our current understanding on many of these links. In this chapter, we summarize the main messages and conclusions presented in this book. In addition, we refer to a number of recent publications that highlight the pleiotropy in physiological functions of creatine and creatine kinase, and which suggest that numerous discoveries on new functions of this system are still ahead of us. Finally, we present our views on the most promising future avenues of research to deepen our knowledge on creatine and creatine kinase. In particular, we elaborate on how state-of-the-art high-throughput analytical ("omics") technologies and systems biology approaches may be used successfully to unravel the complex network of interdependent physiological functions related to creatine and creatine kinase

G.S. Salomons and M. Wyss (eds.), Creatine and Creatine Kinase in Health and Disease, 309–334.

1. SHORT SUMMARY OF KNOWLEDGE PRESENTED IN THIS BOOK

For this book, a number of eminent players in the creatine (Cr) and creatine kinase (CK) research arena joined forces to summarize the state-of-the-art in their respective fields of expertise, and to share their views on future scientific developments in these fields. The first six chapters of the book discuss our current knowledge on the origins and physiological functions of the main players of Cr metabolism, namely Cr, phosphocreatine (PCr), L-arginine:glycine amidinotransferase (AGAT), guanidinoacetate methyltransferase (GAMT), the Cr transporter (SLC6A8/CT1) and the different CK isoenzymes. In chapter 2, the evolutionary history of CK isoenzymes, of the Cr transporter and of the Cr biosynthesis enzymes is presented, suggesting that all these processes may have evolved already in the earliest (invertebrate) animals to facilitate high-energy phosphate transport in flagellated cells (Ellington and Suzuki, 2007). Thermodynamic and kinetic aspects relevant for the CK phosphotransfer network are discussed in chapter 3, as are attempts to devise informative mathematical models that appropriately describe the physiological functions of the CK system (Saks *et al.*, 2007). Besides reporting on the modeling achievements already made, this chapter also provides a glimpse at the challenges faced when attempting to find the appropriate balance between (i) a proper description of a highly compartmentalized enzyme system that is functionally coupled to a multitude of dynamic physiological processes and (ii) the level of abstraction required to make the modeling both feasible from a computational perspective and amenable to experimental validation. Chapters 4–7 provide insights into the tissue-specific expression, localization and function of AGAT, GAMT, the Cr transporter and the different CK isoenzymes (Braissant *et al.*, 2007; Christie, 2007; Heerschap *et al.*, 2007; Tachikawa *et al.*, 2007). They report, among others, on new structural insights into the mechanism of action of the Cr transporter based on a comparison with the recently published high-resolution structure of a prokaryotic homologue of the SLC6 transporters; on the distinct cell type-specific expression of AGAT, GAMT and the Cr transporter in the brain; on the functional relevance of transport processes for Cr between different cell types and across the blood-brain and blood-retinal barriers; and on the use of gene knock-out mouse models to elucidate the physiological functions of GAMT and of the different CK isoenzymes.

The second part of the book focuses on clinical aspects of Cr metabolism. As a logical continuation and complementation of chapter 7 on GAMT- and CK-knockout animals (Heerschap *et al.*, 2007), chapters 8 and 9 summarize our current knowledge on inherited disorders of Cr biosynthesis and transport in man – i.e., on (cerebral) creatine deficiency syndromes (CDSs) that are due to pathogenic mutations in the *AGAT*, *GAMT* or Cr transporter genes –, together with therapeutic approaches to treat Cr biosynthesis defects (Schulze and Battini, 2007; Stockler *et al.*, 2007). Key messages from these chapters are that CDSs may be more prevalent than typically presumed; that the severe neurological sequelae due to AGAT or GAMT deficiency may be prevented completely if appropriate therapeutic intervention is initiated early after birth; and that, thus, at least some CDSs may qualify for inclusion in neonatal screening programs. In chapters 10 and 11, the promise of Cr supplementation as a therapeutic strategy in neurological, neuromuscular and neurometabolic disorders is discussed (Klein and Ferrante, 2007; Tarnopolsky, 2007). While benefits of Cr supplementation were reported in some studies and for several disorders, other trials failed to provide a positive effect. Nevertheless, the authors conclude that Cr supplementation is a potentially powerful therapeutic approach, but that additional, larger-scale clinical trials on homogeneous patient groups are required to determine whether these disorders are responsive to Cr supplementation, and which supplementation regime is optimal for what kind of disorder. Since the benefits of Cr supplementation in these disorders are most likely rather modest, it would seem desirable to pursue combination therapies together with other neuroprotective agents that display different mechanisms of action than Cr. The evidence for ergogenic actions of Cr is presented in chapter 12 (Hespel and Derave, 2007). It is shown that – although Cr's effects are usually limited – consistent improvement in exercise performance can be obtained with Cr supplementation, mostly in explosive sports with repeated contraction series. In addition, Cr supplementation shows promise as an adjuvant therapy in rehabilitation from immobilization-induced muscle atrophy.

In the last part of the book, several aspects relevant for Cr supplementation in man are featured. Chapters 13 and 14 discuss the pharmacokinetics of Cr and the safety of oral Cr supplementation, respectively (McCall and Persky, 2007; Persky and Rawson, 2007). Surprisingly little is as yet known on the pharmacokinetics of Cr, especially for clinically relevant patient populations and for special populations such as the elderly. Although many 'clinical' studies on the safety of oral Cr supplementation have been published, most of them involved only a small number of subjects, thus allowing at best preliminary conclusions. So far, Cr supplementation at the recommended dosages has proven to be safe, although more extended studies on large cohorts are required to allow definitive conclusions on the long-term safety of chronic oral Cr supplementation especially at higher dosages. In chapter 15, an industrial perspective is provided on the commercial production and use of supplementary Cr (Pischel and Gastner, 2007). The different chemical production methods as well as analytical methods used for quality control are presented, as is the regulatory status for Cr as a dietary supplement in the USA and Europe. Emphasis is placed on implementing an appropriate quality assurance policy to maintain high quality standards for oral Cr supplements.

Overall, the various chapters of this book provide a very good overview on many of the key topics that are currently addressed in Cr and CK research, and bear testimony to the attractiveness of the field. The last ten years have shown that in man, Cr and CK may play their most crucial functions in the brain. Thus, the neuroprotective and neurostimulatory effects of Cr supplementation may be(come) more relevant for mankind than the ergogenic effects that are currently so popular.

2. FURTHER CONSIDERATIONS AND HIGHLIGHTS FROM THE RECENT LITERATURE

It is far beyond the scope of this book to comprehensively cover all aspects of Cr and CK research. This is understandably so when considering that Cr and CK were discovered about 175 and 70 years ago, respectively. For reviews on some other aspects of Cr and CK research, the reader is referred to Wyss and Kaddurah-Daouk (2000), Wyss and Schulze (2002), Schlattner *et al.* (2006), Vial (2006), Saks (2007), and Brosnan and Brosnan (2007). The purpose of the current section is to convey some additional thoughts and to pinpoint at some recently published studies on aspects not covered otherwise in this book that provide further insights into the pleiotropy of functions exerted by Cr and CK.

Given the recent interest in (cerebral) creatine deficiency syndromes (CDSs) and the realization that this group of inherited diseases may be rather frequent, the question arises of the prevalence and functional implications of CK deficiency. In principle, both Cr and CK deficiency might be expected to similarly compromise the function of the CK/PCr/Cr system. Nevertheless, only five CK-deficient patients have been reported so far (Feng *et al.*, 2007; Nagai, 2000; Oita *et al.*, 1988; Shibuya *et al.*, 1992; Yamamichi *et al.*, 2001). Four of the five patients were diagnosed as suffering from acute myocardial infarction, and three patients died of acute myocardial infarction at age 49–56 years. In two patients, a point mutation in the M-CK gene has been identified, causing the amino acid substitution D54G (p.Asp54Gly) (Nagai, 2000; Yamamichi *et al.*, 2001). Compared to wild-type M-CK, the D54G mutant revealed substantially lower enzymatic activity, substrate affinity and stability (Feng *et al.*, 2007).

The primary reason for the low number of CK-deficient patients recognized so far seems to be the multiplicity of CK isoenzymes in higher eukaryotes, comprising two cytosolic CK isoenzymes, M-CK and B-CK, and two mitochondrial CK isoenzymes, ubiquitous and sarcomeric MtCK (Wallimann *et al.*, 2007). In either muscle or brain, at least two CK isoenzymes are co-expressed: M-CK and sMtCK in muscle, B-CK and uMtCK in brain. In transgenic mice with knock-outs in a given CK isoenzyme, the remaining CK isoenzyme(s) can take over the function of the deleted CK isoenzyme, at least in part (de Groof *et al.*, 2001; in 't Zandt *et al.*, 2003; Streijger *et al.*, 2005). Therefore, the phenotype of CK-deficient patients may be milder than that of patients with deficiencies in either AGAT, GAMT or the Cr transporter. The likelihood of a simultaneous deficiency in two or even more CK isoenzymes can be expected to be extremely low. It is somewhat intriguing, though, that while the brain seems to be the most affected tissue in CDSs in man, only M-CK-deficient, but no B-CK- or uMtCK-deficient patients have been identified so far.

The brain is probably the tissue that is currently most intensively investigated for the physiological functions of Cr and CK. This is reflected in several chapters of this book – most prominently in chapters 4, 5, 7–9, and 11 (Braissant *et al.*, 2007; Heerschap *et al.*, 2007; Klein and Ferrante, 2007; Schulze and Battini, 2007; Stockler *et al.*, 2007; Tachikawa *et al.*, 2007) –, but also in a significant number of further recent publications. One report indicated that Cr concentrations in the brain may play a role in regulating appetite and weight (Galbraith *et al.*, 2006). According to an earlier suggestion, Cr may act as one of the most relevant osmolytes in brain (Bothwell *et al.*, 2001, 2002; Videen *et al.*, 1995). This may explain the stimulation of Cr transporter expression in conditionally immortalized mouse brain capillary endothelial cells exposed to ammonia, an *in vitro* model system for studying the impact of hyperammonemia (due, e.g., to acute liver failure) on the blood-brain barrier (Bélanger *et al.*, 2007). In tissue culture of human fetal spinal cord, chronic exposure to Cr resulted in significantly higher densities of GABA-immunoreactive neurons while total neuronal cell density was unchanged (Ducray *et al.*, 2007). This finding points to an important role of Cr in inducing differentiation towards the GABAergic phenotype, and for normal brain development and function in general. In addition, it suggests that Cr supplementation may become a promising strategy in cell replacement strategies for central nervous system (CNS) tissue repair.

Several studies have provided evidence for a link between Cr and/or high-energy phosphate metabolism and mental performance (Ferrier *et al.*, 2000; Hoyer *et al.*, 2004; McMorris *et al.*, 2006, 2007; Rae *et al.*, 2003; Valenzuela *et al.*, 2003; Watanabe *et al.*, 2002). Among others, oral Cr supplementation was shown to positively affect working memory and intelligence (Rae *et al.*, 2003). On the other hand, CK has been identified as a primary target for a number of potential neurotoxins: for instance, both mitochondrial and cytosolic CK from rat cerebral cortex were found to be inhibited by 3-hydroxykynurenine, a neurotoxin implicated in neurodegenerative disorders such as Huntington's disease and Parkinson's disease (Cornelio *et al.*, 2006). Similarly, human B-CK was inhibited by micromolar concentrations of 4-hydroxy-2-nonenal, a lipid peroxidation product that is known to be increased in the brain of, e.g., Parkinson's disease patients (Markus Wyss, unpublished data). Administration of arginine to rats significantly reduced the cytosolic and total CK activities in cerebellum, while not affecting mitochondrial CK activity (Delwing *et al.*, 2007). Thus, these results suggest that inhibition of CK by arginine may contribute to the neurotoxicity seen in hyperargininemic patients. Similarly, intrastriatal administration of guanidinoacetate, supposed to be responsible for at least part of the neurological symptoms in GAMT-deficient patients (chapter 9; Schulze and Battini, 2007), was shown to decrease mitochondrial and total CK activity in rat striatum, whereas cytosolic CK activity was not affected (Zugno *et al.*, 2006). Finally, 2-methyl-3-hydroxybutyrate, a substance that accumulates in patients with either mitochondrial β-ketothiolase or 2-methyl-3-hydroxybutyryl-CoA dehydrogenase deficiency (which are inherited neurometabolic disorders), was demonstrated to inhibit both total and mitochondrial CK activity from cerebral cortex of developing rats, while not affecting cytosolic CK activity (Rosa *et al.*, 2005).

Together with the neuroprotective effects of Cr seen under a variety of different conditions (see chapter 11; Klein and Ferrante, 2007; Andres *et al.*, 2005a,b; Bender *et al.*, 2007; Braissant *et al.*, 2002; Burklen *et al.*, 2006; Ducray *et al.*, 2006; Lensman *et al.*, 2006; Morton *et al.*, 2005; Pena-Altamira *et al.*, 2005; Sakellaris *et al.*, 2006; Vasques *et al.*, 2006), the above findings provide a flavour of the potentially tight dependency of optimal brain performance from proper functioning of the CK/Cr system. Although some of the listed studies need independent corroboration, the wealth of data on a critical role of the CK/Cr system in the brain may be sufficient rationale to more thoroughly consider oral Cr supplementation as a means to improve mental performance, mood, and/or general well-being.

Aging and neurodegeneration are thought to share pathophysiological pathways (Bender *et al.*, 2007). Therefore, testing the impact of Cr on the aging process suggested itself. In fact, in aged mice, oral Cr supplementation was found to increase the healthy life span by 9%, to improve neurobehavioral functions, and to reduce accumulation of the "aging pigment" lipofuscin (Bender *et al.*, 2007).

A surprising finding has recently been made on the expression of AGAT in human myocardium, suggesting that this organ is able to synthesize Cr locally (Cullen *et al.*, 2006). AGAT mRNA levels and enzyme activity were found to be increased in end-stage heart failure. Moreover, upon combination therapy consisting of mechanical unloading using a left ventricular assist device and pharmacological intervention, functional recovery of the heart was correlated with a decrease in AGAT expression to baseline values. These findings may suggest a shortage of Cr in the failing heart and are further proof for the relevance of the CK/Cr system in this organ, a conclusion that is supported by other recent findings (chapter 7; Heerschap *et al.*, 2007; Ingwall, 2006; Lygate *et al.*, 2007; Neubauer, 2007; Saks *et al.*, 2006a; Schlattner *et al.*, 2006).

Besides the tissues analyzed extensively for the functions of Cr and CK such as skeletal and cardiac muscle, brain, retina and spermatozoa (Wyss and Kaddurah-Daouk, 2000), new, relevant roles for the CK/Cr system have recently been identified in tissues not studied in detail before. In rat osteoblast-like cells, addition of Cr to the culture medium had a stimulatory effect on metabolic activity, differentiation and mineralization (Gerber *et al.*, 2005). Similarly, supplementation of rats for 8 weeks with Cr increased lumbar bone mineral density as well as the mechanical load at failure of the femur (Antolic *et al.*, 2007). In a study on older men, preliminary evidence was obtained for a positive effect of Cr supplementation on bone mineral content (Chilibeck *et al.*, 2005). Together with previous findings showing that in boys with Duchenne and Becker muscular dystrophy, Cr supplementation increased bone mineral density and reduced urinary excretion of N-telopeptide (Louis *et al.*, 2003; Tarnopolsky *et al.*, 2004), these studies suggest that Cr supplementation might be used as an adjuvant therapy for bone repair *in vivo*. A study on the impact of Cr supplementation, eventually in combination with vitamin D and calcium, on the mineral content and density loss of compact and spongeous bones of post-menopausal women might be most warranted, given their high incidence of osteoporosis.

Aging of the skin is characterized by a decline in cellular energy metabolism, which seems to be caused predominantly by free radicals that are generated by exogenous noxes such as UV radiation. Both the Cr transporter and CK isoenzymes are expressed in human skin (Lenz *et al.*, 2005; Schlattner *et al.*, 2002; see also

Zemtsov, 2007). In cultured human skin fibroblasts and keratinocytes, addition of Cr to the medium increased CK activity and mitochondrial function, and protected against free oxygen radical damage and, in particular, DNA mutations (Berneburg *et al.*, 2005; Lenz *et al.*, 2005). In studies with healthy old volunteers, topical application of Cr prevented the decline in mitochondrial membrane potential due to UV irradiation (Lenz *et al.*, 2005), increased the density of dermal papillae, and reduced the appearance and depth of fine lines and wrinkles (Blatt *et al.*, 2005). Although Cr is already used commercially in skin creams (http://www.beiersdorf.com), more in-depth studies are required to corroborate the relevance of the CK/Cr system for proper function and aging of the skin. It will be particularly interesting to evaluate the beneficial effects of Cr on wound healing, as well as on a variety of pathological skin conditions, such as toxic epidermal necrolysis, psoriasis, chronic ulcers, burns and neurodermitis.

A shotgun proteomics approach identified B-CK as the second most-abundant protein in the sensory hair cells of the inner ear (Shin *et al.*, 2007). Theoretical considerations and experiments in which B-CK was selectively inactivated demonstrated that high-energy phosphate transport in these elongated mechanoreceptor cells is crucially dependent on the CK/Cr system. In another study, detailed analysis of the gene expression profiles in the superior olivary complex, a structure of the mammalian auditory brainstem, as well as in striatum, hippocampus and extraocular muscle tissue revealed increased expression of several genes of Cr metabolism in the superior olivary complex (Nothwang *et al.*, 2006). Expression of the Cr transporter, of AGAT and of B-CK was higher in the superior olivary complex than in the other tissues analyzed, thus also suggesting an involvement of CK and Cr in auditory function. The findings of both studies likely explain the hearing loss and the vestibular dysfunctions observed in transgenic knock-out mice lacking either B-CK alone, or both B-CK and ubiquitous mitochondrial CK (uMtCK) (Shin *et al.*, 2007). In mice lacking both CK isoenzymes, the sensitivity of the auditory system to 8–32 kHz tone bursts was reduced 10- to 30-fold as compared to wild-type mice. Notably, supplementation of guinea pigs with high doses of Cr (3% w/w) has recently been shown to exert remarkable protection from noise-induced hearing loss and degeneration of hair cells (Minami *et al.*, 2007). Therefore, it might be attractive to evaluate Cr supplementation in patients with tinnitus or other degenerative hearing loss conditions, preferably at early time points of the disease. In addition, Cr supplementation might be tested as a preventive measure for people who are chronically exposed to high noise levels.

Investigations on the regulation of methyl balance in humans have shown that approximately 40% of methyl groups from *S*-adenosylmethionine (SAM) are used for Cr biosynthesis (see Brosnan *et al.*, 2007a,b). In addition, about 10% of dietary glycine and as much as 20% of dietary arginine are used for Cr synthesis. Since elevated plasma homocysteine levels have been implicated as a risk factor for a number of chronic diseases, Cr supplementation may prove useful in situations of compromised SAM homeostasis and /or increased demand for methyl groups from SAM.

Recent studies also provide further evidence for a critical role of the CK/Cr system in diseases other than those discussed elsewhere in this book. *S*-adenosylhomocysteine hydrolase deficiency is a rare disease with primarily neuromuscular symptomatology (among others, hypotonia, psychomotor delay, and delayed myelination). In the three patients described so far (Baric *et al.*, 2004, 2005; Buist *et al.*, 2006), plasma concentrations of SAM and *S*-adenosylhomocysteine (AdoHcy) were largely increased. A slight increase in guanidinoacetate and a slight reduction in phosphatidylcholine and free choline concentrations were also observed occasionally. Dietary methionine restriction and supplementation with Cr and phosphatidylcholine markedly reduced plasma SAM and AdoHcy concentrations, and improved myelination and psychomotor development (Baric *et al.*, 2004, 2005). In patients with argininosuccinate synthetase or argininosuccinate lyase deficiency treated with dietary arginine supplementation, cerebral guanidinoacetate concentrations were found to be increased (Sijens *et al.*, 2006; van Spronsen *et al.*, 2006). Considering the neurotoxicity of guanidinoacetate (see chapters 8 and 9; Schulze and Battini, 2007; Stockler *et al.*, 2007), these findings – if confirmed – may advocate a reduction in arginine supplementation as well as complementary Cr supplementation in these urea cycle defects (van Spronsen *et al.*, 2006). In a rat model of endotoxin-induced sepsis, MtCK content, MtCK activity and Cr-stimulated mitochondrial respiration were all largely reduced in diaphragm and heart (Callahan and Supinski, 2007). MtCK inactivation may be linked to the increase in mitochondrial free radical generation observed in sepsis, and may contribute to the functional abnormalities occurring in this syndrome. Again, it might be worthwhile to test whether Cr is able to protect against the sequelae of sepsis. Cystinosis is an autosomal recessive disorder associated with lysosomal cystine accumulation and involves, among others, visual deficits. In cytosolic and mitochondrial fractions isolated from pig retina, CK activity was inhibited in a dose- and time-dependent manner by cystine (Pereira Oliveira *et al.*, 2007). Cysteamine prevented and reversed the inactivation caused by cystine, suggesting oxidative modification of the reactive sulfhydryl groups of CK by cystine as the underlying mechanism for the observed effects. Given the likely relevance of the CK/Cr system for high-energy phosphate transport in photoreceptor cells of the retina (Hemmer *et al.*, 1993; Wegmann *et al.*, 1991), inhibition of CK by cystine may contribute to visual impairment in cystinosis patients. Cachexia is prevalent in cancer and in certain infectious diseases such as AIDS or tuberculosis. In patients with colorectal cancer undergoing milder chemotherapy, preliminary experiments suggested oral Cr supplementation to be a potentially promising adjuvant therapy for preventing cachexia (Norman *et al.*, 2006). Patients on aggressive chemotherapy, however, did not seem to benefit from Cr supplementation. Further experiments are required to evaluate the potential benefits of Cr for increasing body weight and muscle mass in cachexia patients and whether it would seem desirable to add Cr to high-caloric clinical nutrition. Finally, as an extension to previous studies, Ghosh *et al.* (2006) demonstrated that in tumor-bearing mice, a combination of methylglyoxal, ascorbic acid (both injected intraperitoneally) and Cr (intraperitoneally or supplied in the drinking water) completely inhibited proliferation of tumor cells and significantly increased the life span of the animals. Nearly 80% of the tumor-bearing mice were completely cured. Therefore, such a combination therapy might be considered as a low-cost alternative to the exquisite cancer drugs marketed by the leading pharmaceutical companies.

A number of recent publications also provide new insights into the mechanisms of action of Cr and CK. Epand *et al.* (2007) presented striking experimental evidence for octameric MtCK to mediate lipid transfer between two membranes. Through their particular symmetry and membrane-binding properties, MtCK octamers can simultaneously bind to two nearby membrane bilayers, e.g. in mitochondrial contact sites. In the *in vitro* experiments performed, the lipid transfer process was suggested not to involve vesicle fusion or loss of the internal contents of the liposomes used. These findings are remarkable when considering that interbilayer transfer of lipids between the inner and outer mitochondrial membranes may play an important role in the regulation and initiation of apoptosis (Epand *et al.*, 2007).

Exposure of C2C12 murine myoblasts or porcine endothelial cells to hypertonic stress induced an increase in the expression of the Cr transporter as well as in Cr transport activity (Alfieri *et al.*, 2006). When exposed to hypertonic conditions, survival of C2C12 cells was enhanced by Cr in a manner similar to that of betaine, taurine, or *myo*-inositol. Thus, in muscle as in brain (see above), Cr may act as a compatible osmolyte.

In rats selectively bred for high vs. low aerobic running capacity, muscle mitochondria displayed higher Cr-stimulated respiration despite similar mitochondrial density (Walsh *et al.*, 2006). This suggests increased functional coupling between MtCK and adenine nucleotide translocase of the inner mitochondrial membrane. Remarkably, rats bred for high aerobic capacity also revealed a lower susceptibility to ischemia-reperfusion-mediated ventricular tachyarrhythmias (Lujan *et al.*, 2006) which is reminiscent of the anti-arrhythmic effects of PCr reported previously (Cisowski *et al.*, 1996; Fagbemi *et al.*, 1982; Hearse *et al.*, 1986; Ruda *et al.*, 1988). In a further series of experiments on isolated rat brain mitochondria and cultured embryonic rat cortical neurons, MtCK was demonstrated to favour internal ADP cycling in mitochondria, resulting in tight coupling of the respiratory chain to ATP synthesis, and thereby to diminish production of reactive oxygen species, particularly under hyperglycemic conditions (Meyer *et al.*, 2006). The preventive anti-oxidant role depended on both the PCr/Cr ratio and MtCK activity, as evidenced by a lack of anti-oxidant action of Cr in rat liver mitochondria which are devoid of MtCK.

In cultured MIN-6 β-cells, addition of Cr in the presence of glucose elicited a significant increase in insulin secretion (Rocic *et al.*, 2007). On the other hand, Cr had no effect in the absence of glucose, suggesting that Cr acts as a potentiator rather than an initiator of insulin secretion. In addition, Cr increased cellular ATP levels, independently of the presence of glucose. These mechanisms of action are likely an underlying basis for the observed hypoglycemic effects of Cr.

Cr supplementation has been shown in rat skeletal muscle cells to increase the phosphorylation of AMP-activated protein kinase (AMPK) (Ceddia and Sweeney, 2004), a key sensor and fuel gauge of cellular energetics (Hardie *et al.*, 2006), as well as to stimulate expression of the glucose transporter GLUT4 in rat muscle (Ju *et al.*, 2005). These findings are in accord with the fact that AMPK is a well-known regulator of glucose transport (Fujii *et al.*, 2004). According to preliminary data, AMPK also seems to regulate the Cr transporter (Darrabie *et al.*, 2007). Elucidation of the detailed interplay between AMPK and regulation of the Cr transporter on one hand, and between Cr supplementation and AMPK signalling on the other hand (Neumann *et al.*, 2007) deserves further attention in the future.

Let us conclude this section on recent Cr- and CK-related reports with some rather unrelated and preliminary findings. Given the ergogenic effects of oral Cr supplementation (see chapter 12; Hespel and Derave, 2007) and the popular use of Cr as a strength- and muscle mass-enhancing aid in explosive sports and bodybuilding, it seemed logical to test the impact of dietary Cr on growth performance and meat quality in livestock production. Disappointingly, however, current evidence does not seem to provide sufficient rationale for regular, economical use of Cr in livestock feed (Lindahl *et al.*, 2006; Stahl *et al.*, 2007). One reason for the inconclusive data so far might be – as in man – the occurrence of Cr "responders" and "non-responders" among different breeds of, e.g., pigs (Young *et al.*, 2007).

An intriguing hypothesis was raised by Zhou *et al.* (2006) who linked a genetic polymorphism in the 3'-region of the M-CK gene to differences in the response of running economy to endurance training in man. Finally, in a study on a multiethnic population involving 1444 Dutch citizens, a significant correlation was seen between serum CK activity and both systolic and diastolic blood pressure (Brewster *et al.*, 2006). Higher serum CK activity may be a reflection of higher tissue CK content which, in turn, may trigger higher pressor responses in heart and vasculature. Clearly, many of the findings reported in this section and, in particular, the last-mentioned hypotheses require further experimental corroboration.

The collection of articles presented in this section is by no means meant to be exhaustive. Rather, it should provide a flavour and further proof for the pleiotropy of effects exerted by Cr and CK. Despite 175 years since the discovery of Cr, new disclosures are still being made on a regular basis on the localization and function of CK and Cr in many different organs, tissues and cell types of living organisms across large evolutionary distances.

3. FUTURE AVENUES OF RESEARCH

Considering the recent emphasis on (cerebral) Cr deficiency syndromes (CDSs) and the multitude of cellular processes with which the CK/Cr system interacts, what may be the future avenues of research that promise the most significant gains in our knowledge on the relevance of the CK/Cr system in both health and disease? What obstacles may be encountered, and how may they be circumvented? The following facets of the picture will be discussed in this section: (i) the relevance of suitable animal models; (ii) the need for better understanding of compartmentalization, transmembrane transport processes and cell-to-cell trafficking involved in Cr metabolism; (iii) the expected contributions of "omics" and "systems biology" to knowledge gain; (iv) neuroprotection and CDSs – what is next?; (v) suggestions for optimized Cr dosages and supplementation regimes; and (vi) why we cannot expect the large pharmaceutical companies to drive future developments in this field.

3.1. Relevance of Suitable Animal Models

Given the functional redundancy among different CK isoenzymes and among different high-energy phosphate transport systems, standardization of experimental conditions is a key prerequisite for the detection of subtle functional differences. There are limits to such standardization in human studies (different ethnicities, genotypes, living styles or diets), so that suitable animal models will remain a crucial instrument in furthering our knowledge on the functions of the CK/Cr system in health and disease. In knock-out studies done so far, certain CK isoenzymes or GAMT were deleted globally by disruption of the corresponding gene in the genome. Thus, the phenotypic expression of the genetic deficit in these studies may have been a composite effect of compromised cellular functions in many different cell types at once. This makes proper dissection of cause-and-effect relationships very difficult if not impossible. Studies on the liver of transgenic mice expressing either cytosolic or mitochondrial CK isoenzymes have been one example to address the functions of CK and Cr in one particular tissue only (see chapter 7; Heerschap *et al.*, 2007). Similarly, CK genes have been introduced into distant model organisms that do not normally express these enzymes, i.e. *Escherichia coli* and *Saccharomyces cerevisiae* (Canonaco *et al.*, 2002, 2003).

In the future, with further improvements in genetic tools, more targeted and better controlled changes in expression of genes involved in Cr metabolism should be attempted, for instance specific expression of such enzymes in cell types or tissues in which they are not normally expressed, or tissue-specific gene knock-outs. In addition, conditional knock-out and knock-in strategies are desirable to study the time-course of functional responses to such modifications at different developmental stages. Furthermore, rather than studying just all-or-none effects, more graded changes in gene expression should be targeted to establish dose-response relationships, and to provide better resolution for establishing the logical sequence of events in response to the given genetic derangement. RNA interference strategies may be a suitable instrument in many of these endeavors.

Studies on cell cultures, either with a single cell type or with, e.g., brain cell primary 3D cultures composed of mixed oligodendrocytes, astrocytes and neurons (chapter 4; Braissant *et al.*, 2007), may be a means of attaining even better experimental standardization. However, extrapolation of results obtained in cell culture to the *in vivo* situation may not be trivial. Finally, we still await the first (transgenic) animal models with targeted alterations in expression of AGAT or the Cr transporter.

3.2. Relevance of Transport Processes in Cr Metabolism

It is broadly believed that in the human body, sites of Cr biosynthesis are typically separated from tissues and cell types that most rely on the CK/Cr system for high-energy phosphate recycling and transport such as striated muscle. The reason for this may be the need for independent regulation of Cr biosynthesis and CK function. The direct implication is that transport processes between different tissues and cell types and across cellular membranes play a crucial role in overall Cr metabolism. Possibly due to the fact that membrane proteins and processes are more difficult to study than soluble, cytosolic enzymes, our knowledge on transport processes important for Cr metabolism is still limited. For instance, it is still largely unknown how guanidinoacetate enters hepatocytes in the liver and how Cr is exported from the liver into the blood to be directed to the target organs.

Tissues most relying on the CK/Cr system take up Cr from the bloodstream against a large concentration gradient, a process that is mediated by a Na^+- and Cl^--dependent Cr transporter (chapter 6; Christie, 2007). Increasing the extracellular Cr concentration – either by addition of Cr to the culture medium *in vitro* or by oral Cr supplementation *in vivo* – is known to downregulate the Cr transporter in striated muscle cells (Guerrero-Ontiveros and Wallimann, 1998; Loike *et al.*, 1988). This is one of the reasons why in the sports and bodybuilding arena, the daily dose of oral Cr supplementation is reduced from 15–20 g/day in the initial loading phase lasting 3–5 days to 2–5 g/day in the maintenance phase. In addition, it is often recommended to regularly pause Cr supplementation for several weeks, so as to allow the body to restore baseline expression of the Cr transporter and, thus, to prepare the cells and tissues for a next cycle of Cr supplementation. However, the dynamics of downregulation and recovery of Cr transporter expression have been studied only poorly so far, and almost nothing is known in this respect about tissues other than striated muscle. Furthermore, Cr transporter expression *per se* as well as its regulation may differ between health and disease, an area that is also explored to only a very limited extent so far (chapter 6; Christie, 2007).

Recent evidence summarized in chapters 4 and 5 of this book (Braissant *et al.*, 2007; Tachikawa *et al.*, 2007) suggests that Cr can be synthesized locally in the brain, but that also there, Cr biosynthesis is physically separated from cells most relying on the CK/Cr system. How Cr biosynthesis is regulated in the brain, how the different cell types in the brain communicate to balance Cr biosynthesis to demand, and which role transport processes play in the overall scheme requires further in-depth studies. The Cr transporter is expressed in the brain and is very likely responsible for uptake of Cr into cells depending on the CK/Cr system. However, neither in the brain nor elsewhere in the body, it is known how guanidinoacetic acid and Cr are exported from cells and tissues involved in Cr biosynthesis.

It will be interesting to explore whether this is just an example of passive diffusion through non-specific membrane channels, whether a specific Cr exporter exists, and/or whether the Cr transporter also contributes to Cr export in these cells and tissues.

The Cr transporter is also supposed to be responsible for uptake of Cr across the blood-brain and blood-retina barriers (chapters 4 and 5; Braissant *et al.*, 2007; Tachikawa *et al.*, 2007). However, uptake of Cr from the bloodstream into the brain is very slow, requiring several weeks or even months for seeing significant benefits of oral Cr supplementation on the Cr content in brain. Considering (i) that the Cr transporter plays such a crucial role in delivering Cr to its sites of action, (ii) that increasing extracellular Cr concentrations reduce rather than promote cellular uptake of Cr, (iii) that Cr transport across the blood-brain barrier is severely restricted, and (iv) that, on the other hand, Cr supplementation offers a plethora of potential health-beneficial effects as well as clinical benefits in patients with CDSs (see this book and Wyss and Schulze, 2002), well-designed comprehensive studies addressing the regulation of Cr transport are clearly warranted. In addition, the large body of evidence for a neuroprotective role of Cr (chapter 11, Klein and Ferrante, 2007) offers great opportunities for Cr analogues with better blood-brain barrier transport properties. Either these analogues might be neuroprotective themselves, or they should be readily converted to Cr selectively in the brain. On a more hypothetical level, Cr analogues that are potent substrates of CK but cannot undergo cyclization and/or the phosphorylated forms of which have a higher phosphorylation potential than PCr, might perceivably offer even better health benefits than Cr itself. Finally, Cr analogues that would allow more rapid uptake into cells and tissues might allow applications not only in prophylaxis, but also in cure. Towards these goals, first Cr analogues have been designed (Lunardi *et al.*, 2006); however, their properties do not yet warrant testing in clinical studies.

3.3. The Promises of "Omics" and "Systems Biology"

Functional genomics technologies aim at the global quantification of all mRNA transcripts (transcriptomics), all proteins (proteomics), or all metabolites (metabolomics) in a given biological sample. The target of "systems biology" is to understand a living cell at the system level, based on mathematical models of whole-cellular metabolic and regulatory networks, and using functional genomics and other experimental data to constrain the solution space of these models. It is tempting to believe that if you just measure 'everything', you should also be able to fully comprehend the physiological processes in a living cell. The reality is quite different, though. Therefore, as with any other technology, asking the right questions and taking care of the most appropriate experimental design is key to leverage most value from functional genomics technologies and systems biology concepts. How, then, can those up-to-date scientific approaches be used for studying the functions of the CK/Cr system?

We know the reaction catalyzed by CK, but we have serious problems comprehending Cr's overall role in a living cell. This is not entirely surprising, since Cr probably has only few direct effects. On the other hand, through regeneration of ATP from ADP – two molecules acting as substrates, products or effectors of a multitude of cellular reactions –, the CK/Cr system most certainly has pleiotropic effects which, in addition, may influence each other in complex interplays. Therefore, an in-depth understanding of Cr's role will depend on taking all those pleiotropic effects into consideration, in 3D-space and time. This will require complex models, which can only be validated with comprehensive sets of high-throughput functional genomics data. This will not be an easy win. Modeling at this level of complexity is not yet feasible for even a simple prokaryotic cell. It must be even less feasible then for a complex higher eukaryotic cell, and very difficult to forecast when we may reach this stage. However, the prospects of such models are tempting; once these models are able to meaningfully describe a biological system under a wide range of environmental conditions, they may potentially be used not only to describe the system, but also to predict its behavior under situations not tested experimentally before. This may, in turn, guide further experimental design and/or reduce the need for experimentation altogether. And eventually, it may become an ideal aid for predicting optimal therapeutic interventions in CDSs. For the current status of mathematical modeling and systems biology related to the CK/Cr system, the reader is referred to chapter 3 (Saks *et al.*, 2007) as well as to some recent articles (Beard *et al.*, 2006; Saks *et al.*, 2006a,b; Weiss *et al.*, 2006). Fast progress in this field will depend on further improvements in the collaboration between experimentalists and theoreticists (bioinformaticians, mathematicians, statisticians, etc.).

If whole-cellular modeling is not yet feasible, should we fully restrain from further studying the functions of the CK/Cr system, until the technologies have sufficiently advanced? Clearly not; we simply need to limit our aspirations to what is already feasible now. For instance, increasingly powerful genome (re-) sequencing technologies will allow to identify further genetic variants in the *AGAT*, *GAMT* and Cr transporter genes and to study their impact on the expressed phenotype. In addition, they may enable identification of further CK variants with compromised function.

Use of functional genomics technologies in combination with good (new) animal models with targeted alterations in CK, AGAT, GAMT and/or Cr transporter expression (see above) also offers great promise. It may allow to understand specific features of each of the Cr-deficiency syndromes (particularly in CNS), such as analysis of cell-type-specific alterations or identification of specific groups of genes or metabolites that are altered (neurotransmitter systems, metabolic pathways, signal transduction pathways, etc.). Comparative studies (e.g., between a knock-out mouse model and its reference strain) should be done not only under resting conditions, but also under stimulated or stressed conditions, so as to allow identification of functional dependencies over a broader range of physiological and pathological states. Eventually, such studies may help to pinpoint new ways for treating Cr-deficient patients, or to uncover new roles and functions of Cr in a high-throughput manner. Studies of this type have already been reported. For instance, a proteomics approach revealed protein-protein interactions *in vitro* and *in vivo* between ubiquitous MtCK preprotein and amyloid precursor protein (APP) family proteins (Li *et al.*, 2006). Likewise, proteomics studies resulted in the identification of CK as one of the primary targets of oxidative modification in the aging brain, and that prevention of aging-related learning and memory deficits by behavioral enrichment and antioxidant-fortified food is correlated with an increased CK content in the brain (Opii *et al.*, 2006; Poon *et al.*, 2004, 2005). These findings may, thus, suggest an involvement of CK in learning and memory (see also above).

What are the limitations of such studies, and what may need to be considered in their design? None of the currently existing or perceivable experimental models (either through genetic changes, Cr analogues or chemical inhibitors) allows an abrupt and specific perturbation of the CK/Cr system. Establishment of the perturbed state requires time intervals that leave sufficient room for compensatory, confounding metabolic adaptation processes. Given the redundancy of high-energy phosphate transport systems in higher eukaryotes, this is a significant complication. On the other hand, the functional adaptations may provide a signature of molecular changes that enables extrapolation back to the start of the perturbation. Use of chemical inhibitors, however specific, always bears the risk of unrecognized side effects and is, thus, recommended only in selected cases.

Carefully designed time-series experiments in which samples for functional genomics studies are taken at multiple time-points after application of a perturbation may be instrumental in discriminating between primary and secondary effects, in dissecting the logical sequence of events, and in unravelling potential compensatory effects that might preclude development of a more severe phenotype in response to the perturbation applied. In terms of maturity of the different "omics" technologies, DNA microarray technology and metabolomics are most advanced and comprehensive, whereas successful use of proteomics in studying higher eukaryotes requires considerably higher technical sophistication due to the high complexity of the proteome (due to posttranslational modification) and because of limited throughput and proteome coverage of the currently available technologies.

3.4. Neuroprotection and Creatine Deficiency Syndromes – What is Next?

Despite a significant number of studies on CDSs (including the development of models such as the GAMT knock-out mouse) and on the expression of AGAT, GAMT and the Cr transporter, important questions remain unsolved on the specific roles of these genes in different tissues and cell types, in particular in CNS. In addition, Cr deficiency characteristics remain to be elucidated, particularly at the cellular level. Advances in our understanding of the links between the CK/Cr system and brain function are expected to come from improved diagnostics, inclusion of CDSs in neonatal screening programs, establishment of CDS disease registries and multi-centre studies, and innovative new concepts for understanding, preventing and curing neurological diseases that involve derangements in Cr metabolism or CK function.

Screening for (cerebral) Cr deficiency syndromes (CDSs) is currently performed either by proton magnetic resonance spectroscopy (MRS), metabolite screening and/or molecular investigations. Nowadays, in most institutes, urinary analysis is the first line of screening for CDSs. However, it is expected that several cases with AGAT deficiency as well as the majority of females – and likely some males – with SLC6A8 deficiency will be missed by urinary metabolite screening. Despite this limitation, the relatively high prevalence of CDSs and the availability of potential treatment options will likely result, in due time, in a situation where every mentally retarded patient in the Western world, male or female, is screened for CDSs. The question remains how this will be achieved. One option is proton MRS that will certainly become more widely available and has the advantage of disclosing additional diseases. A marked reduction or absence of the Cr signal in MRS is diagnostic for primary Cr deficiency. A disadvantage is the fact that, usually, the patients need to be sedated and that specific training is needed for the analytical laboratories involved. Currently, molecular analysis for the AGAT, GAMT and SLC6A8 genes is mainly performed by direct DNA sequencing and, as yet, may be included only in well-equipped DNA diagnostic laboratories as the primary diagnostic approach; however, combination of molecular analysis with metabolic testing seems to be more preferable for the time being. The development of (re-) sequencing microarrays by companies such as Affymetrix or NimbleGen offers a new perspective in the diagnosis and screening of CDSs, although their diagnostic sensitivity is currently not yet acceptable for routine clinical use.

So far, CDSs are not included in neonatal screening programs. However, tandem mass spectrometry has the potential for simultaneous multi-disease screening and has been applied recently in some neonatal screening programs. Provided that tandem-MS measurement of guanidinoacetic acid in dried blood spots proves to be specific and sensitive enough for detection or exclusion of GAMT deficiency, this disorder should be included in neonatal screening (Bodamer *et al.*, 2001; Carducci *et al.*, 2002; Schulze *et al.*, 2006). On the other hand, AGAT and SLC6A8 deficiency are not yet eligible for neonatal screening since Cr and creatinine do not seem to be informative in the neonatal period (Schulze *et al.*, 2006).

Cr deficiency syndromes are relatively rare diseases. As with many other rare diseases, progress in understanding of the natural history and phenotype of the disease as well as in the efficacy of treatment has been delayed or is still being hampered by the lack of multi-centre studies. Only after half a century of history of inborn errors of metabolism, we begin to understand that worldwide networks and orphan disease registries are needed to facilitate rapid progress in the development of strategies for treatment and prevention. The group of Cr deficiency syndromes should be investigated using these tools in the future.

In many instances, our understanding of pathobiochemistry and, thus, of possible therapeutic interventions is still superficial. The pathobiochemical actions and pathophysiological consequences of the accumulation of guanidinoacetate,

guanidinoacetate's interaction with brain function, and pharmacological inhibition of guanidinoacetate's action in the brain need to be understood fundamentally to allow development of more effective treatment strategies for GAMT deficiency. New concepts are also needed for understanding the pathogenesis of brain dysfunction in Cr transporter deficiency. What is the pathogenic impact of Cr deficiency, and are there other functions of the Cr transporter which we do not yet know but contribute to pathogenesis? Deeper knowledge about the regulation of Cr transporter activity and its interaction with other genes might provide promising targets for alternative treatment strategies such as pharmacological gene therapy.

Treatment of SLC6A8 deficiency is one of the big challenges. Clinical improvement has been observed in patients with Cr biosynthesis defects (i.e., AGAT and GAMT deficiency) upon treatment comprising Cr supplementation, with almost complete restoration of Cr in brain. This proves that restoration of cerebral Cr levels is essential. Once a vehicle for Cr uptake into brain will be found (see above), treatment should also be successful in SLC6A8 deficiency. Further elucidation of Cr's biosynthesis and function in the brain may also increase the success rate of treatment. For example, "omics" studies may reveal increased levels of compensating genes and their products, which may lead to clever design of specific drugs for the restoration of Cr or its function(s).

AGAT knock-out mice would represent an ideal model system of Cr depletion and might be fundamental for understanding Cr's effects beyond its high-energy phosphate buffering function. This animal model may also allow to investigate the possible neuroprotective role of Cr which still has not been demonstrated convincingly in humans. The role of Cr as a neuroprotective substance has to be critically reconsidered in non-Cr-deficient conditions. It has been shown in numerous studies that under physiological conditions, Cr is present in high concentrations in brain and muscle, the sites where we most expect a neuroprotective or ergogenic effect, respectively. It has also been shown that in these organs with high baseline Cr concentrations, oral Cr supplementation only leads to a minor further increase in cellular Cr content. Therefore, the hypotheses that a further minor quantitative increase in cellular Cr content can be achieved and that it truly provides health benefits have to be revisited.

3.5. Suggestions for Optimized Cr Dosages and Supplementation Regimes

A considerable number of health-beneficial effects have been ascribed to oral Cr supplementation (chapters 9–12; Hespel and Derave, 2007; Klein and Ferrante, 2007; Schulze and Battini, 2007; Tarnopolsky, 2007; see also Wyss and Schulze, 2002). So far, the most clear-cut effects have been observed in animal models of disease, mostly in mice or rats. On the other hand, studies in humans often resulted in marginal to weak benefits only, e.g., in amyotrophic lateral sclerosis (Groeneveld *et al.*, 2003) or Parkinson's disease (Bender *et al.*, 2006). A reason for this discrepancy may be the largely different dosages of Cr used in experimental animals vs. man. In animal studies, Cr is typically administered at 2–3% (w/w) in the feed. Considering that a 40 g adult mouse consumes approximately 4.0 g feed per day, this corresponds to an estimated Cr intake of 2.0–3.0 g·(kg body weight)$^{-1}$·d^{-1}. On the other hand, daily dosages of 4–10 g Cr have mostly been employed in clinical studies which, for a 70-kg person, translates into an estimated intake of only 0.06–0.14 g·(kg body weight)$^{-1}$·d^{-1}, i.e. 14- to 50-fold less than in animal studies! Remarkably, in one of the few successful – but still preliminary – studies in humans (Sakellaris *et al.*, 2006), protection from complications of traumatic brain injury in children and adolescents was achieved by oral supplementation of Cr at 0.4 g·(kg body weight)$^{-1}$·d^{-1} for six months. Therefore, it would seem desirable to carefully consider significantly higher daily dosages of Cr for clinical studies. However, the potential additional benefits of such higher dosages would need to be balanced properly against the higher risk for potential side effects. As a matter of fact, no side effects of Cr supplementation have been noted in the study of Sakellaris *et al.* (2006), one important prerequisite being that enough liquid is consumed during Cr supplementation.

As pointed out in chapter 12 (Hespel and Derave, 2007), the favourable effects of Cr supplementation on athletic performance may decrease during prolonged supplementation, thus warranting alternating supplementation (2–3 months) and washout periods (4–5 weeks). It will be of great importance to test whether this also holds true for clinical studies involving patients with chronic diseases such as Parkinson's disease, amyotrophic lateral sclerosis or multiple sclerosis, where Cr supplementation would be indicated to last for years.

3.6. Whom can We Expect (or not) to Drive Knowledge Gain

Despite the multiple potential benefits reported for oral Cr supplementation, we cannot expect the large pharmaceutical or nutrition companies to make major investments in the further exploration of the true health-beneficial effects of Cr, the best supplementation regimes, and/or the most appropriate forms of delivery of this compound. The pleiotropic effects of Cr go against the general 'one chemical – one effect' policy of pharmaceutical enterprises. A successful drug should be highly specific at low dosage for a single, clearly defined target, with very limited or, preferably, no side effects. In addition, because Cr is known for about 175 years, product patents on the substance itself are no longer feasible. Last but not least, Cr can be produced easily and cheaply, is consumed orally, and is available over-the-counter worldwide. Therefore, even if application patents existed, an end-consumer would have easy means to circumvent a premium-tagged product with an attached health claim. An attractive business opportunity for pharmaceutical or nutrition companies would only emerge with new, patent-protectable Cr analogues with improved uptake into the brain or if, for distinct health benefits, sophisticated and protectable delivery strategies were required that would allow, for instance, selective uptake of Cr into one particular tissue. For these reasons, generous funding by public authorities will be essential for fully exploiting Cr's potential for human health and general well-being.

4. SOME FINAL WORDS

In conclusion, we hope that this volume of *Subcellular Biochemistry* and its individual chapters have managed to convey our admiration for nature's complexity and beauty as well as our firm conviction that the CK/Cr system plays an exquisite role in safeguarding proper functioning and an optimally balanced interplay between diverse physiological processes. We also hope that this volume manages to challenge some existing concepts and to stimulate new hypotheses. And last but not least, we are excited about the prospects of emerging scientific concepts and technologies for unraveling the true scope of functions exerted by the CK/Cr system, as well as for harnessing the full potential of dietary supplementation with Cr or its analogues for disease prevention or cure, and for improving general well-being. Although we have identified funding as a potential bottleneck, we are confident that a growing body of evidence for a tight link between the CK/Cr system and health will convince public authorities (and even private institutions?) to generously support the field and to let creative creatures create creatine-based nutritional strategies!

REFERENCES

Alfieri, R.R., Bonelli, M.A., Cavazzoni, A., Brigotti, M., Fumarola, C., Sestili, P., Mozzoni, P., De Palma, G., Mutti, A., Carnicelli, D., Vacondio, F., Silva, C., Borghetti, A.F., Wheeler, K.P., and Petronini, P.G., 2006, Creatine as a compatible osmolyte in muscle cells exposed to hypertonic stress. *J. Physiol.* **576:** 391–401.

Andres, R.H., Ducray, A.D., Huber, A.W., Perez-Bouza, A., Krebs, S.H., Schlattner, U., Seiler, R.W., Wallimann, T., and Widmer, H.R., 2005a, Effects of creatine treatment on survival and differentiation of GABA-ergic neurons in cultured striatal tissue. *J. Neurochem.* **95:** 33–45.

Andres, R.H., Huber, A.W., Schlattner, U., Perez-Bouza, A., Krebs, S.H., Seiler, R.W., Wallimann, T., and Widmer, H.R., 2005b, Effects of creatine treatment on the survival of dopaminergic neurons in cultured fetal ventral mesencephalic tissue. *Neuroscience* **133:** 701–713.

Antolic, A., Roy, B.D., Tarnopolsky, M.A., Zernicke, R.F., Wohl, G.R., Shaughnessy, S.G., and Bourgeois, J.M., 2007, Creatine monohydrate increases bone mineral density in young Sprague-Dawley rats. *Med. Sci. Sports Exerc.* **39:** 816–820.

Baric, I., Cuk, M., Fumic, K., Vugrek, O., Allen, R.H., Glenn, B., Maradin, M., Pazanin, L., Pogribny, I., Rados, M., Sarnavka, V., Schulze, A., Stabler, S., Wagner, C., Zeisel, S.H., and Mudd, S.H., 2005, *S*-Adenosylhomocysteine hydrolase deficiency: a second patient, the younger brother of the index patient, and outcomes during therapy. *J. Inherit. Metab. Dis.* **28:** 885–902.

Baric, I., Fumic, K., Glenn, B., Cuk, M., Schulze, A., Finkelstein, J.D., James, S.J., Mejaski-Bosnjak, V., Pazanin, L., Pogribny, I.P., Rados, M., Sarnavka, V., Scukanec-Spoljar, M., Allen, R.H., Stabler, S., Uzelac, L., Vugrek, O., Wagner, C., Zeisel, S., and Mudd, S.H., 2004, *S*-Adenosylhomocysteine hydrolase deficiency in a human: a genetic disorder of methionine metabolism. *Proc. Natl. Acad. Sci. USA* **101:** 4234–4239.

Beard, D.A., 2006, Modeling of oxygen transport and cellular energetics explains observations on *in vivo* cardiac energy metabolism. *PLoS Comput. Biol.* **2:** e107.

Bélanger, M., Asashima, T., Ohtsuki, S., Yamaguchi, H., Ito, S., and Terasaki, T., 2007, Hyperammonemia induces transport of taurine and creatine and suppresses claudin-12 gene expression in brain capillary endothelial cells *in vitro*. *Neurochem. Int.* **50:** 95–101.

Bender, A., Beckers, J., Schneider, I., Holter, S.M., Haack, T., Ruthsatz, T., Vogt-Weisenhorn, D.M., Becker, L., Genius, J., Rujescu, D., Irmler, M., Mijalski, T., Mader, M., Quintanilla-Martinez, L., Fuchs, H., Gailus-Durner, V., de Angelis, M.H., Wurst, W., Schmidt, J., and Klopstock, T., 2007, Creatine improves health and survival of mice. *Neurobiol. Aging.*, in press.

Bender, A., Koch, W., Elstner, M., Schombacher, Y., Bender, J., Moeschl, M., Gekeler, F., Muller-Myhsok, B., Gasser, T., Tatsch, K., and Klopstock, T., 2006, Creatine supplementation in Parkinson disease: a placebo-controlled randomized pilot trial. *Neurology* **67:** 1262–1264.

Berneburg, M., Gremmel, T., Kurten, V., Schroeder, P., Hertel, I., von Mikecz, A., Wild, S., Chen, M., Declercq, L., Matsui, M., Ruzicka, T., and Krutmann, J., 2005, Creatine supplementation normalizes mutagenesis of mitochondrial DNA as well as functional consequences. *J. Invest. Dermatol.* **125:** 213–220.

Blatt, T., Lenz, H., Koop, U., Jaspers, S., Weber, T., Mummert, C., Wittern, K.P., Stab, F., and Wenck, H., 2005, Stimulation of skin's energy metabolism provides multiple benefits for mature human skin. *Biofactors* **25:** 179–185.

Bodamer, O.A., Bloesch, S.M., Gregg, A.R., Stöckler-Ipsiroglu, S., and O'Brien, W.E., 2001, Analysis of guanidinoacetate and creatine by isotope dilution electrospray tandem mass spectrometry. *Clin. Chim. Acta* **308:** 173–178.

Bothwell, J.H., Rae, C., Dixon, R.M., Styles, P., and Bhakoo, K.K., 2001, Hypo-osmotic swelling-activated release of organic osmolytes in brain slices: implications for brain oedema *in vivo*. *J. Neurochem.* **77:** 1632–1640.

Bothwell, J.H., Styles, P., and Bhakoo, K.K., 2002, Swelling-activated taurine and creatine effluxes from rat cortical astrocytes are pharmacologically distinct. *J. Membr. Biol.* **185:** 157–164.

Braissant, O., Bachmann, C., and Henry, H., 2007, Expression and function of AGAT, GAMT and CT1 in the mammalian brain. *Subcell. Biochem.* **46:** 67–81.

Braissant, O., Henry, H., Villard, A.M., Zurich, M.G., Loup, M., Eilers, B., Parlascino, G., Matter, E., Boulat, O., Honegger, P., and Bachmann, C., 2002, Ammonium-induced impairment of axonal growth is prevented through glial creatine. *J. Neurosci.* **22:** 9810–9820.

Brewster, L.M., Mairuhu, G., Bindraban, N.R., Koopmans, R.P., Clark, J.F., and van Montfrans, G.A., 2006, Creatine kinase activity is associated with blood pressure. *Circulation* **114:** 2034–2039.

Brosnan, J.T., and Brosnan, M.E., 2007, Creatine: endogenous metabolite, dietary, and therapeutic supplement. *Annu. Rev. Nutr.* **27:** 241–261.

Brosnan, J.T., da Silva, R., and Brosnan, M.E., 2007a, Amino acids and the regulation of methyl balance in humans. *Curr. Opin. Clin. Nutr. Metab. Care* **10:** 52–57.

Brosnan, M.E., Edison, E.E., da Silva, R., and Brosnan, J.T., 2007b, New insights into creatine function and synthesis. *Adv. Enzyme Regul.*, in press.

Buist, N.R., Glenn, B., Vugrek, O., Wagner, C., Stabler, S., Allen, R.H., Pogribny, I., Schulze, A., Zeisel, S.H., Baric, I., and Mudd, S.H., 2006, *S*-Adenosylhomocysteine hydrolase deficiency in a 26-year-old man. *J. Inherit. Metab. Dis.* **29:** 538–545.

Burklen, T.S., Schlattner, U., Homayouni, R., Gough, K., Rak, M., Szeghalmi, A., and Wallimann, T., 2006, The creatine kinase/creatine connection to Alzheimer's disease: CK-inactivation, APP-CK complexes and focal creatine deposits. *J. Biomed. Biotechnol.* **2006:** 35936.

Callahan, L.A., and Supinski, G.S., 2007, Diaphragm and cardiac mitochondrial creatine kinases are impaired in sepsis. *J. Appl. Physiol.* **102:** 44–53.

Canonaco, F., Schlattner, U., Pruett, P.S., Wallimann, T., and Sauer, U., 2002, Functional expression of phosphagen kinase systems confers resistance to transient stresses in *Saccharomyces cerevisiae* by buffering the ATP pool. *J. Biol. Chem.* **277:** 31303–31309.

Canonaco, F., Schlattner, U., Wallimann, T., and Sauer, U., 2003, Functional expression of arginine kinase improves recovery from pH stress of *Escherichia coli*. *Biotechnol. Lett.* **25:** 1013–1017.

Carducci, C., Birarelli, M., Leuzzi, V., Battini, R., Cioni, G., and Antonozzi, I., 2002, Guanidinoacetate and creatine plus creatinine assessment in physiologic fluids: an effective diagnostic tool for the biochemical diagnosis of arginine:glycine amidinotransferase and guanidinoacetate methyltransferase deficiencies. *Clin. Chem.* **48:** 1772–1778.

Ceddia, R.B., and Sweeney, G., 2004, Creatine supplementation increases glucose oxidation and AMPK phosphorylation and reduces lactate production in L6 rat skeletal muscle cells. *J. Physiol.* **555:** 409–421.

Chilibeck, P.D., Chrusch, M.J., Chad, K.E., Shawn Davison, K., and Burke, D.G., 2005, Creatine monohydrate and resistance training increase bone mineral content and density in older men. *J. Nutr. Health Aging* **9:** 352–353.

Christie, D.L., 2007, Functional insights into the creatine transporter. *Subcell. Biochem.* **46:** 99–118.

Cisowski, M., Bochenek, A., Kucewicz, E., Wnuk-Wojnar, A.M., Morawski, W., Skalski, J., and Grzybek, H., 1996, The use of exogenous creatine phosphate for myocardial protection in patients undergoing coronary artery bypass surgery. *J. Cardiovasc. Surg. (Torino)* **37:** 75–80.

Cornelio, A.R., Rodrigues-Junior Vda, S., Rech, V.C., de Souza Wyse, A.T., Dutra-Filho, C.S., Wajner, M., and Wannmacher, C.M., 2006, Inhibition of creatine kinase activity from rat cerebral cortex by 3-hydroxykynurenine. *Brain Res.* **1124:** 188–196.

Cullen, M.E., Yuen, A.H., Felkin, L.E., Smolenski, R.T., Hall, J.L., Grindle, S., Miller, L.W., Birks, E.J., Yacoub, M.H., and Barton, P.J., 2006, Myocardial expression of the arginine:glycine amidinotransferase gene is elevated in heart failure and normalized after recovery: potential implications for local creatine synthesis. *Circulation* **114:** I16–I20.

Darrabie, M.D., Santacruz-Toloza, L., Goers, L., Toloza, E.M., and Jacobs, D.O., 2007, AMPK-activation increases activity of the creatine transporter expressed in an immortalized adult cardiac cell line. *Biophys. J.*, Suppl. 21a, Abstract 284-Pos.

de Groof, A.J., Oerlemans, F.T., Jost, C.R., and Wieringa, B., 2001, Changes in glycolytic network and mitochondrial design in creatine kinase-deficient muscles. *Muscle Nerve* **24:** 1188–1196.

Delwing, D., Cornelio, A.R., Wajner, M., Wannmacher, C.M., and Wyse, A.T., 2007, Arginine administration reduces creatine kinase activity in rat cerebellum. *Metab. Brain Dis.* **22:** 13–23.

Ducray, A., Kipfer, S., Huber, A.W., Andres, R.H., Seiler, R.W., Schlattner, U., Wallimann, T., and Widmer, H.R., 2006, Creatine and neurotrophin-4/5 promote survival of nitric oxide synthase-expressing interneurons in striatal cultures. *Neurosci. Lett.* **395:** 57–62.

Ducray, A.D., Qualls, R., Schlattner, U., Andres, R.H., Dreher, E., Seiler, R.W., Wallimann, T., and Widmer, H.R., 2007, Creatine promotes the GABAergic phenotype in human fetal spinal cord cultures. *Brain Res.* **1137:** 50–57.

Ellington, W.R., and Suzuki, T., 2007, Early evolution of the creatine kinase gene family and the capacity for creatine biosynthesis and membrane transport. *Subcell. Biochem.* **46:** 17–26.

Epand, R.F., Schlattner, U., Wallimann, T., Lacombe, M.L., and Epand, R.M., 2007, Novel lipid transfer property of two mitochondrial proteins that bridge the inner and outer membranes. *Biophys. J.* **92:** 126–137.

Fagbemi, O., Kane, K.A., and Parratt, J.R., 1982, Creatine phosphate suppresses ventricular arrhythmias resulting from coronary artery ligation. *J. Cardiovasc. Pharmacol.* **4:** 53–58.

Feng, S., Zhao, T.J., Zhou, H.M., and Yan, Y.B., 2007, Effects of the single point genetic mutation D54G on muscle creatine kinase activity, structure and stability. *Int. J. Biochem. Cell Biol.* **39:** 392–401.

Ferrier, C.H., Alarcon, G., Glover, A., Koutroumanidis, M., Morris, R.G., Simmons, A., Elwes, R.D., Cox, T., Binnie, C.D., and Polkey, C.E., 2000, N-Acetylaspartate and creatine levels measured by ^{1}H MRS relate to recognition memory. *Neurology* **55:** 1874–1883.

Fujii, N., Aschenbach, W.G., Musi, N., Hirshman, M.F., and Goodyear, L.J., 2004, Regulation of glucose transport by the AMP-activated protein kinase. *Proc. Nutr. Soc.* **63:** 205–210.

Galbraith, R.A., Furukawa, M., and Li, M., 2006, Possible role of creatine concentrations in the brain in regulating appetite and weight. *Brain Res.* **1101:** 85–91.

Gerber, I., ap Gwynn, I., Alini, M., and Wallimann, T., 2005, Stimulatory effects of creatine on metabolic activity, differentiation and mineralization of primary osteoblast-like cells in monolayer and micromass cell cultures. *Eur. Cell. Mater.* **10:** 8–22.

Ghosh, M., Talukdar, D., Ghosh, S., Bhattacharyya, N., Ray, M., and Ray, S., 2006, *In vivo* assessment of toxicity and pharmacokinetics of methylglyoxal. Augmentation of the curative effect of methylglyoxal on cancer-bearing mice by ascorbic acid and creatine. *Toxicol. Appl. Pharmacol.* **212:** 45–58.

Groeneveld, G.J., Veldink, J.H., van der Tweel, I., Kalmijn, S., Beijer, C., de Visser, M., Wokke, J.H., Franssen, H., and van den Berg, L.H., 2003, A randomized sequential trial of creatine in amyotrophic lateral sclerosis. *Ann. Neurol.* **53:** 437–445.

Guerrero-Ontiveros, M.L., and Wallimann, T., 1998, Creatine supplementation in health and disease. Effects of chronic creatine ingestion *in vivo*: down-regulation of the expression of creatine transporter isoforms in skeletal muscle. *Mol. Cell. Biochem.* **184:** 427–437.

Hardie, D.G., Hawley, S.A., and Scott, J.W., 2006, AMP-activated protein kinase - development of the energy sensor concept. *J. Physiol.* **574:** 7–15.

Hearse, D.J., Tanaka, K., Crome, R., and Manning, A.S., 1986, Creatine phosphate and protection against reperfusion-induced arrhythmias in the rat heart. *Eur. J. Pharmacol.* **131:** 21–30.

Heerschap, A., Kan, H.E., Nabuurs, C.I.H.C., Renema, W.K., Isbrandt, D., and Wieringa, B., 2007, *In vivo* magnetic resonance spectroscopy of transgenic mice with altered expression of guanidinoacetate methyltransferase and creatine kinase isoenzymes. *Subcell. Biochem.* **46:** 119–148.

Hemmer, W., Riesinger, I., Wallimann, T., Eppenberger, H.M., and Quest, A.F., 1993, Brain-type creatine kinase in photoreceptor cell outer segments: role of a phosphocreatine circuit in outer segment energy metabolism and phototransduction. *J. Cell Sci.* **106:** 671–683.

Hespel, P., and Derave, W., 2007, Ergogenic effects of creatine in sports and rehabilitation. *Subcell. Biochem.* **46:** 245–259.

Hoyer, S., Lannert, H., Latteier, E., and Meisel, T., 2004, Relationship between cerebral energy metabolism in parietotemporal cortex and hippocampus and mental activity during aging in rats. *J. Neural Transm.* **111:** 575–589.

in 't Zandt, H.J., de Groof, A.J., Renema, W.K., Oerlemans, F.T., Klomp, D.W., Wieringa, B., and Heerschap, A., 2003, Presence of (phospho)creatine in developing and adult skeletal muscle of mice without mitochondrial and cytosolic muscle creatine kinase isoforms. *J. Physiol.* **548:** 847–858.

Ingwall, J.S., 2006, On the hypothesis that the failing heart is energy starved: lessons learned from the metabolism of ATP and creatine. *Curr. Hypertens. Rep.* **8:** 457–464.

Ju, J.S., Smith, J.L., Oppelt, P.J., and Fisher, J.S., 2005, Creatine feeding increases GLUT4 expression in rat skeletal muscle. *Am. J. Physiol. Endocrinol. Metab.* **288:** E347–E352.

Klein, A.M., and Ferrante, R.J., 2007, The neuroprotective role of creatine. *Subcell. Biochem.* **46:** 205–243.

Lensman, M., Korzhevskii, D.E., Mourovets, V.O., Kostkin, V.B., Izvarina, N., Perasso, L., Gandolfo, C., Otellin, V.A., Polenov, S.A., and Balestrino, M., 2006, Intracerebroventricular administration of creatine protects against damage by global cerebral ischemia in rat. *Brain Res.* **1114:** 187–194.

Lenz, H., Schmidt, M., Welge, V., Schlattner, U., Wallimann, T., Elsasser, H.P., Wittern, K.P., Wenck, H., Stab, F., and Blatt, T., 2005, The creatine kinase system in human skin: protective effects of creatine against oxidative and UV damage *in vitro* and *in vivo*. *J. Invest. Dermatol.* **124:** 443–452.

Li, X., Burklen, T., Yuan, X., Schlattner, U., Desiderio, D.M., Wallimann, T., and Homayouni, R., 2006, Stabilization of ubiquitous mitochondrial creatine kinase preprotein by APP family proteins. *Mol. Cell. Neurosci.* **31:** 263–272.

Lindahl, G., Young, J.F., Oksbjerg, N., and Andersen, H.J., 2006, Influence of dietary creatine monohydrate and carcass cooling rate on colour characteristics of pork loin from different pure breeds. *Meat Sci.* **72:** 624–634.

Loike, J.D., Zalutsky, D.L., Kaback, E., Miranda, A.F., and Silverstein, S.C., 1988, Extracellular creatine regulates creatine transport in rat and human muscle cells. *Proc. Natl. Acad. Sci. USA* **85:** 807–811.

Louis, M., Lebacq, J., Poortmans, J.R., Belpaire-Dethiou, M.C., Devogelaer, J.P., Van Hecke, P., Goubel, F., and Francaux, M., 2003, Beneficial effects of creatine supplementation in dystrophic patients. *Muscle Nerve* **27:** 604–610.

Lujan, H.L., Britton, S.L., Koch, L.G., and DiCarlo, S.E., 2006, Reduced susceptibility to ventricular tachyarrhythmias in rats selectively bred for high aerobic capacity. *Am. J. Physiol. Heart Circ. Physiol.* **291:** H2933–H2941.

Lunardi, G., Parodi, A., Perasso, L., Pohvozcheva, A.V., Scarrone, S., Adriano, E., Florio, T., Gandolfo, C., Cupello, A., Burov, S.V., and Balestrino, M., 2006, The creatine transporter mediates the uptake of creatine by brain tissue, but not the uptake of two creatine-derived compounds. *Neuroscience* **142:** 991–997.

Lygate, C.A., Fischer, A., Sebag-Montefiore, L., Wallis, J., Ten Hove, M., and Neubauer, S., 2007, The creatine kinase energy transport system in the failing mouse heart. *J. Mol. Cell. Cardiol.*, **42:** 1129–1136.

McCall, W., and Persky, A.M., 2007, Pharmacokinetics of creatine. *Subcell. Biochem.* **46:** 261–273.

McMorris, T., Harris, R.C., Howard, A.N., Langridge, G., Hall, B., Corbett, J., Dicks, M., and Hodgson, C., 2007, Creatine supplementation, sleep deprivation, cortisol, melatonin and behavior. *Physiol. Behav.* **90:** 21–28.

McMorris, T., Harris, R.C., Swain, J., Corbett, J., Collard, K., Dyson, R.J., Dye, L., Hodgson, C., and Draper, N., 2006, Effect of creatine supplementation and sleep deprivation, with mild exercise, on cognitive and psychomotor performance, mood state, and plasma concentrations of catecholamines and cortisol. *Psychopharmacology (Berlin)* **185:** 93–103.

Meyer, L.E., Machado, L.B., Santiago, A.P., da-Silva, W.S., De Felice, F.G., Holub, O., Oliveira, M.F., and Galina, A., 2006, Mitochondrial creatine kinase activity prevents reactive oxygen species generation: antioxidant role of mitochondrial kinase-dependent ADP re-cycling activity. *J. Biol. Chem.* **281:** 37361–37371.

Minami, S.B., Yamashita, D., Ogawa, K., Schacht, J., and Miller, J.M., 2007, Creatine and tempol attenuate noise-induced hearing loss. *Brain Res.* **1148:** 83–89.

Morton, A.J., Hunt, M.J., Hodges, A.K., Lewis, P.D., Redfern, A.J., Dunnett, S.B., and Jones, L., 2005, A combination drug therapy improves cognition and reverses gene expression changes in a mouse model of Huntington's disease. *Eur. J. Neurosci.* **21:** 855–870.

Nagai, T., 2000, Acute myocardial infarction without raised creatine kinase activity. *J. R. Soc. Med.* **93:** 315–316.

Neubauer, S., 2007, The failing heart - an engine out of fuel. *N. Engl. J. Med.* **356:** 1140–1151.

Neumann, D., Wallimann, T., Rider, M.A., Tokarska-Schlattner, M., Hardie, G.D., and Schlattner, U., 2007, Signalling by AMP-activated protein kinase. In "Molecular Systems Bioenergetics – Energy for Life", Saks, V.A., ed., Wiley-VCH, Weinheim, Germany.

Norman, K., Stubler, D., Baier, P., Schutz, T., Ocran, K., Holm, E., Lochs, H., and Pirlich, M., 2006, Effects of creatine supplementation on nutritional status, muscle function and quality of life in patients with colorectal cancer - a double blind randomised controlled trial. *Clin. Nutr.* **25:** 596–605.

Nothwang, H.G., Koehl, A., and Friauf, E., 2006, Comparative gene expression analysis reveals a characteristic molecular profile of the superior olivary complex. *Anat. Rec. A Discov. Mol. Cell. Evol. Biol.* **288:** 409–423.

Oita, T., Imoto, S., Soma, M., Sakizono, K., Nakamura, K., Hosomi, K., Fukuda, K., Yamamichi, H., Shirane, H., Uchida, H., Kasakura, S., Koizumi, K., and Yoshikawa, J., 1988, Deficiency of creatine kinase MM fraction. *Jpn. J. Clin. Pathol.* **9:** 1045–1050.

Opii, W.O., Joshi, G., Head, E., Milgram, N.W., Muggenburg, B.A., Klein, J.B., Pierce, W.M., Cotman, C.W., and Butterfield, D.A., 2006, Proteomic identification of brain proteins in the canine model of human aging following a long-term treatment with antioxidants and a program of behavioral enrichment: Relevance to Alzheimer's disease. *Neurobiol. Aging*, in press.

Pena-Altamira, E., Crochemore, C., Virgili, M., and Contestabile, A., 2005, Neurochemical correlates of differential neuroprotection by long-term dietary creatine supplementation. *Brain Res.* **1058:** 183–188.

Pereira Oliveira, P.R., Rodrigues-Junior, V., Rech, V.C., and Duval Wannmacher, C.M., 2007, Cystine inhibits creatine kinase activity in pig retina. *Arch. Med. Res.* **38:** 164–169.

Persky, A.M., and Rawson, E.S., 2007, Safety of creatine supplementation. *Subcell. Biochem.* **46:** 275–289.

Pischel, I., and Gastner, T., 2007, Creatine – its chemical synthesis, chemistry, and legal status. *Subcell. Biochem.* **46:** 291–307.

Poon, H.F., Castegna, A., Farr, S.A., Thongboonkerd, V., Lynn, B.C., Banks, W.A., Morley, J.E., Klein, J.B., and Butterfield, D.A., 2004, Quantitative proteomics analysis of specific protein expression and oxidative modification in aged senescence-accelerated-prone 8 mice brain. *Neuroscience* **126:** 915–926.

Poon, H.F., Farr, S.A., Thongboonkerd, V., Lynn, B.C., Banks, W.A., Morley, J.E., Klein, J.B., and Butterfield, D.A., 2005, Proteomic analysis of specific brain proteins in aged SAMP8 mice treated with alpha-lipoic acid: implications for aging and age-related neurodegenerative disorders. *Neurochem. Int.* **46:** 159–168.

Rae, C., Digney, A.L., McEwan, S.R., and Bates, T.C., 2003, Oral creatine monohydrate supplementation improves brain performance: a double-blind, placebo-controlled, cross-over trial. *Proc. R. Soc. Lond. B Biol. Sci.* **270:** 2147–2150.

Rocic, B., Lovrencic, M.V., Poje, M., and Ashcroft, S.J., 2007, Effect of creatine on the pancreatic beta-cell. *Exp. Clin. Endocrinol. Diabetes* **115:** 29–32.

Rosa, R.B., Schuck, P.F., de Assis, D.R., Latini, A., Dalcin, K.B., Ribeiro, C.A., da, C.F.G., Maria, R.C., Leipnitz, G., Perry, M.L., Filho, C.S., Wyse, A.T., Wannmacher, C.M., and Wajner, M., 2005, Inhibition of energy metabolism by 2-methylacetoacetate and 2-methyl-3-hydroxybutyrate in cerebral cortex of developing rats. *J. Inherit. Metab. Dis.* **28:** 501–515.

Ruda, M., Samarenko, M.B., Afonskaya, N.I., and Saks, V.A., 1988, Reduction of ventricular arrhythmias by phosphocreatine (Neoton) in patients with acute myocardial infarction. *Am. Heart J.* **116:** 393–397.

Sakellaris, G., Kotsiou, M., Tamiolaki, M., Kalostos, G., Tsapaki, E., Spanaki, M., Spilioti, M., Charissis, G., and Evangeliou, A., 2006, Prevention of complications related to traumatic brain injury in children and adolescents with creatine administration: an open label randomized pilot study. *J. Trauma* **61:** 322–329.

Saks, V., Dzeja, P., Schlattner, U., Vendelin, M., Terzic, A., and Wallimann, T., 2006a, Cardiac system bioenergetics: metabolic basis of the Frank-Starling law. *J. Physiol.* **571:** 253–273.

Saks, V., Favier, R., Guzun, R., Schlattner, U., and Wallimann, T., 2006b, Molecular system bioenergetics: regulation of substrate supply in response to heart energy demands. *J. Physiol.* **577:** 769–777.

Saks, V.A., ed., 2007, Molecular System Bioenergetics – Energy for Life. Wiley-VCH, Weinheim, Germany.

Saks, V., Kaambre, T., Guzun, R., Anmann, T., Sikk, P., Schlattner, U., Wallimann, T., Aliev, M., and Vendelin, M., 2007, The creatine kinase phosphotransfer network: thermodynamic and kinetic considerations, the impact of the mitochondrial outer membrane and modelling approaches. *Subcell. Biochem.* **46:** 27–65.

Schlattner, U., Mockli, N., Speer, O., Werner, S., and Wallimann, T., 2002, Creatine kinase and creatine transporter in normal, wounded, and diseased skin. *J. Invest. Dermatol.* **118:** 416–423.

Schlattner, U., Tokarska-Schlattner, M., and Wallimann, T., 2006, Mitochondrial creatine kinase in human health and disease. *Biochim. Biophys. Acta* **1762:** 164–180.

Schulze, A., and Battini, R., 2007, Pre-symptomatic treatment of creatine biosynthesis defects. *Subcell. Biochem.* **46:** 167–181.

Schulze, A., Hoffmann, G.F., Bachert, P., Kirsch, S., Salomons, G.S., Verhoeven, N.M., and Mayatepek, E., 2006, Presymptomatic treatment of neonatal guanidinoacetate methyltransferase deficiency. *Neurology* **67:** 719–721.

Shibuya, J., Matsumoto, T., Takahashi, K., Sugisawa, K., Yasutomi, N., Kawashima, S., Naruse, H., Tateishi, J., Iwasaki, T., and Tozawa, T., 1992, The first report of a case with acute myocardial infarction showing familial deficiency of creatine kinase. *Intern. Med.* **31:** 611–616.

Shin, J.B., Streijger, F., Beynon, A., Peters, T., Gadzala, L., McMillen, D., Bystrom, C., Van der Zee, C.E., Wallimann, T., and Gillespie, P.G., 2007, Hair bundles are specialized for ATP delivery via creatine kinase. *Neuron* **53:** 371–386.

Sijens, P.E., Reijngoud, D.J., Soorani-Lunsing, R.J., Oudkerk, M., and van Spronsen, F.J., 2006, Cerebral ^{1}H MR spectroscopy showing elevation of brain guanidinoacetate in argininosuccinate lyase deficiency. *Mol. Genet. Metab.* **88:** 100–102.

Stahl, C.A., Carlson-Shannon, M.S., Wiegand, B.R., Meyer, D.L., Schmidt, T.B., and Berg, E.P., 2007, The influence of creatine and a high glycemic carbohydrate on the growth performance and meat quality of market hogs fed ractopamine hydrochloride. *Meat Sci.* **75:** 143–149.

Stockler, S., Schutz, P.W., and Salomons, G.S., 2007, Cerebral creatine deficiency syndromes: clinical aspects, treatment and pathophysiology. *Subcell. Biochem.* **46:** 149–166.

Streijger, F., Oerlemans, F., Ellenbroek, B.A., Jost, C.R., Wieringa, B., and Van der Zee, C.E., 2005, Structural and behavioural consequences of double deficiency for creatine kinases BCK and UbCKmit. *Behav. Brain Res.* **157:** 219–234.

Tachikawa, M., Hosoya, K.-i., Ohtsuki, S., and Terasaki, T., 2007, A novel relationship between creatine transport at the blood-brain and blood-retinal barriers, creatine biosynthesis, and its use for brain and retinal energy homeostasis. *Subcell. Biochem.* **46:** 83–98.

Tarnopolsky, M.A., 2007, Clinical use of creatine in neuromuscular and neurometabolic disorders. *Subcell. Biochem.* **46:** 183–204.

Tarnopolsky, M.A., Mahoney, D.J., Vajsar, J., Rodriguez, C., Doherty, T.J., Roy, B.D., and Biggar, D., 2004, Creatine monohydrate enhances strength and body composition in Duchenne muscular dystrophy. *Neurology* **62:** 1771–1777.

Valenzuela, M.J., Jones, M., Wen, W., Rae, C., Graham, S., Shnier, R., and Sachdev, P., 2003, Memory training alters hippocampal neurochemistry in healthy elderly. *Neuroreport* **14:** 1333–1337.

van Spronsen, F.J., Reijngoud, D.J., Verhoeven, N.M., Soorani-Lunsing, R.J., Jakobs, C., and Sijens, P.E., 2006, High cerebral guanidinoacetate and variable creatine concentrations in argininosuccinate synthetase and lyase deficiency: implications for treatment? *Mol. Genet. Metab.* **89:** 274–276.

Vasques, V., Brinco, F., Viegas, C.M., and Wajner, M., 2006, Creatine prevents behavioral alterations caused by methylmalonic acid administration into the hippocampus of rats in the open field task. *J. Neurol. Sci.* **244:** 23–29.

Vial, C., ed., 2006, Creatine Kinase. NovaScience Publishers, New York, USA.

Videen, J.S., Michaelis, T., Pinto, P., and Ross, B.D., 1995, Human cerebral osmolytes during chronic hyponatremia. A proton magnetic resonance spectroscopy study. *J. Clin. Invest.* **95:** 788–793.

Wallimann, T., Tokarska-Schlattner, M., Neumann, D., Epand, R.M., Epand, R.F., Andres, R.H., Widmer, H.R., Hornemann, T., Saks, V.A., Agarkova, I., and Schlattner, U., 2007, The phosphocreatine circuit: molecular and cellular physiology of creatine kinases, sensitivity to free radicals and enhancement by creatine supplementation. In "Molecular Systems Bioenergetics – Energy for Life", Saks, V.A., ed., Wiley-VCH, Weinheim, Germany.

Walsh, B., Hooks, R.B., Hornyak, J.E., Koch, L.G., Britton, S.L., and Hogan, M.C., 2006, Enhanced mitochondrial sensitivity to creatine in rats bred for high aerobic capacity. *J. Appl. Physiol.* **100:** 1765–1769.

Watanabe, A., Kato, N., and Kato, T., 2002, Effects of creatine on mental fatigue and cerebral hemoglobin oxygenation. *Neurosci. Res.* **42:** 279–285.

Wegmann, G., Huber, R., Zanolla, E., Eppenberger, H.M., and Wallimann, T., 1991, Differential expression and localization of brain-type and mitochondrial creatine kinase isoenzymes during development of the chicken retina: Mi-CK as a marker for differentiation of photoreceptor cells. *Differentiation* **46:** 77–87.

Weiss, J.N., Yang, L., and Qu, Z., 2006, Systems biology approaches to metabolic and cardiovascular disorders: network perspectives of cardiovascular metabolism. *J. Lipid Res.* **47:** 2355–2366.

Wyss, M., and Kaddurah-Daouk, R., 2000, Creatine and creatinine metabolism. *Physiol. Rev.* **80:** 1107–1213.

Wyss, M., and Schulze, A., 2002, Health implications of creatine: can oral creatine supplementation protect against neurological and atherosclerotic disease? *Neuroscience* **112:** 243–260.

Yamamichi, H., Kasakura, S., Yamamori, S., Iwasaki, R., Jikimoto, T., Kanagawa, S., Ohkawa, J., Kumagai, S., and Koshiba, M., 2001, Creatine kinase gene mutation in a patient with muscle creatine kinase deficiency. *Clin. Chem.* **47:** 1967–1973.

Young, J.F., Bertram, H.C., Theil, P.K., Petersen, A.-G.D., Poulsen, K.A., Rasmussen, M., Malmendal, A., Nielsen, N.C., Vestergaard, M., and Oksbjerg, N., 2007, *In vitro* and *in vivo* studies of creatine monohydrate supplementation to Duroc and Landrace pigs. *Meat Sci.* **76:** 342–351.

Zemtsov, A., 2007, Skin phosphocreatine. *Skin Res. Technol.* **13:** 115–118.

Zhou, D.Q., Hu, Y., Liu, G., Gong, L., Xi, Y., and Wen, L., 2006, Muscle-specific creatine kinase gene polymorphism and running economy responses to an 18-week 5000-m training programme. *Br. J. Sports Med.* **40:** 988–991.

Zugno, A.I., Scherer, E.B., Schuck, P.F., Oliveira, D.L., Wofchuk, S., Wannmacher, C.M., Wajner, M., and Wyse, A.T., 2006, Intrastriatal administration of guanidinoacetate inhibits Na^+, K^+-ATPase and creatine kinase activities in rat striatum. *Metab. Brain Dis.* **21:** 41–50.

INDEX